Name	Symbol	Atomic number	Atomic weight	Name	Symbol	Atomic number	Atomic weight
Molybdenum	Mo	42	95.94	Samarium	Sm	62	150.35
Neodymium	Nd	60	144.24	Scandium	Sc	21	44.956
Neon	Ne	10	20.183	Selenium	Se	34	78.96
Neptunium	Np	93	[237]	Silicon	Si	14	28.086
Nickel	Ni	28	58.71	Silver	Ag	47	107.868
Niobium	Nb	41	92.906	Sodium	Na	11	22.9898
Nitrogen	N	7	14.0067	Strontium	Sr	38	87.62
Nobelium	No	102	[253]	Sulfur	S	16	32.064
Osmium	Os	76	190.2	Tantalum	Ta	73	180.948
Oxygen	O	8	15.9994	Technetium	Tc	43	[99]
Palladium	Pd	46	106.4	Tellurium	Te	52	127.60
Phosphorus	P	15	30.9738	Terbium	Tb	65	158.924
Platinum	Pt	78	195.09	Thallium	Tl	81	204.37
Plutonium	Pu	94	[242]	Thorium	Th	90	232.038
Polonium	Po	84	210	Thulium	Tm	69	168.934
Potassium	K	19	39.102	Tin	Sn	50	118.69
Praseodymium	Pr	59	140.907	Titanium	Ti	22	47.90
Promethium	Pm	61	[145]	Tungsten	W	74	183.85
Protactinium	Pa	91	231	Uranium	U	92	238.03
Radium	Ra	88	226.05	Vanadium	V	23	50.942
Radon	Rn	86	222	Xenon	Xe	54	131.30
Rhenium	Re	75	186.2	Ytterbium	Yb	70	173.04
Rhodium	Rh	45	102.905	Yttrium	Y	39	88.905
Rubidium	Rb	37	85.47	Zinc	Zn	30	65.37
Ruthenium	Ru	44	101.07	Zirconium	Zr	40	91.22

*A value given in brackets denotes the mass number of the longest-lived or best known isotope.

Form and Function

Natural Science in a Technological Society

Form and Function

Natural Science in a Technological Society

WITHDRAWN

Avrom A. Blumberg

Associate Professor of Chemistry
and Head of the Division of Natural Science
and Mathematics, DePaul University

Jon Stanley

Research Scientist
United States Department of Interior

McGraw-Hill Book Company

New York
St. Louis
San Francisco
Düsseldorf
Johannesburg
Kuala Lumpur
London
Mexico
Montreal
New Delhi
Panama
Rio de Janeiro
Singapore
Sydney
Toronto

Form and Function: Natural Science
In A Technological Society

Copyright © 1972 by McGraw-Hill, Inc. All rights reserved.
Printed in the United States of America. No part of this
publication may be reproduced, stored in a retrieval system,
or transmitted, in any form or by any means, electronic,
mechanical, photocopying, recording, or otherwise, without
the prior written permission of the publisher.

Library of Congress Catalog Card Number 71-174609

07-006186-6

1234567890   MURM   798765432

This book was set in Baskerville by Progressive
Typographers, printed by the Murray Printing
Company, and bound by Rand McNally & Company. The
designer was Marsha Cohen; the drawings were done by
Vantage Art, Inc. The editors were William P. Orr, Nancy
L. Marcus, and Andrea Stryker-Rodda. Peter Guilmette
supervised production.

to Judy, David, and Susan;
Carol, Cynthia, and Suzanne

Contents

Contents
viii

Contents

x

Contents
xii

A Proem for the Student

It is natural for a student to question studying any subject outside his own major discipline and outside his own interests. Why should science be a part of one's liberal education? What do we hope to accomplish in this course?

In popular use the word *science* has several meanings: a system of knowledge obtained and organized by something called the scientific method; special techniques and expertise; and a human activity to better understand natural phenomena. It is in this last sense that science is properly part of a liberal education.

Natural science comprises the traditional broad disciplines of astronomy, biology, chemistry, geology, and physics, as well as areas of cooperation, such as geochemistry, and narrower specializations, such as medicinal chemistry. There is a common purpose in all these disciplines: to discover physical reality, i.e., to describe our universe — how it is constructed and how it operates. In searching for the form and function in our world, natural science has much in common with the fine arts. W. H. Auden expressed it this way:

> Both science and art are spiritual activities, not practical, whatever practical applications may be derived from their results. Disorder, lack of meaning are spiritual not physical discomforts: order and sense are spiritual and not physical satisfactions. The subject and the methods of the scientists and the artist differ, but their impulse is the same. . . .making a memorable structure out of what was a structureless flux of novelty.

If natural science were no more than this — a great intellectual and esthetic activity of the human mind — its study should be necessary for a more complete, balanced education. But there is a further compelling reason to study natural science. There is an intimate relationship between science (how we describe our universe) and technology (how we control it). The combined effects of science and technology constitute the single most important influence on our lives.

An obvious example of the cooperation between science and technology is the computer. A complex assemblage of many simple components, a computer can perform a very large number of arithmetic operations in a very short time. Through scientific investigation, components were designed to make these rapid calculations possible. The fastest computer today can perform about 20 million times more quickly than a human being using pencil and paper. The amount of work requirng 135 men, each working 8 hours a day for 1 year, using only pencil and paper, can be done in 1 minute by a single man and a computer. Today, computer calculations are commonplace in science, technology, and industry. The enormous power available to an individual through the computer cannot be overstated. This technological revolution has changed the nature of science, technology, and of many other areas of human endeavor, sufficiently for our time in history to be called the Second Industrial Revolution.

At the beginning of this century a scientist could identify his work by a simple word, for example, chemistry. By World War II he more properly should have said he is a *biochemist* or a *nuclear physicist*. Today he may call himself an *enzyme kineticist*, a *membrane morphologist*, or a *fatty acid chemist*. There has been phenomenal specialization in science. Ironically, the more we know collectively about our universe, the less of this whole each of us can know individually.

It is also evident that a highly distinctive jargon has been developed to serve this specialization. Science is too often completely unintelligible to the nonspecialist. What is not always realized is that scientists, too, have trouble understanding each other if their work is not closely related.

Nevertheless, even the most casual observer today cannot avoid the scientific and technological character of our times. We take comfort in our high standard of living, but are concerned over pollution of air and water. Utilizing nuclear power, we must be anxious about nuclear weapons. Taking pride in man's walking on the moon, we worry about offensive and defensive missile systems. Conquering infectious diseases, we are horrified about chemical and biological warfare agents. Rejoicing in lower infant mortalities, we fear overpopulation and famine. Using pesticides to control harmful insects, we find these poisons accumulating in fish, birds, and even in man. It is evident, for good and ill, that ours is a technical age.

The suspicion has long existed that perhaps we are more clever than wise in the way we manipulate nature. A century ago, the critics Edmond and Jules de Goncourt remarked:

> We do believe that at that particular stage of scientific development, the good Lord, with a flowing white beard, will arrive on Earth with his chain of keys and will say to humanity, just as they do at the Art Gallery at five o'clock, "Gentlemen, it's closing time."

> —*Journals, April 7, 1869*

As a nation becomes technologically and scientifically more advanced, even the character of its government is affected. Implicit in the phase "deriving their just powers from the consent of the governed" is the idea of an approximately equal distribution of power between the governed and the governors. Things may be considerably different today:

> For the most important decisions the government may render today, in contrast to the past, are far removed both from the life experiences and the understanding of the man in the street. Thus the great national decisions of life and death are rendered by technological elites, and both the Congress and the people at large retain little more than the illusion of making the decision which the theory of democracy supposes them to make.
>
> —*Hans J. Morgenthau*
> *"What Ails America,"*
> *The New Republic,*
> *October 28, 1967*

There is a paradox in this: the more we collectively control our universe, the less individual control each of us has over our immediate circumstances.

Yet there is no escaping the twentieth century. Unless we destroy ourselves, we shall continue to enjoy its benefits and pay the price of this progress. It is to our advantage to make that price as small as possible; and, of course, to ensure our survival.

The nonspecialist can ignore and condemn the entire scientific-technological enterprise,

> (While you and i have lips and voices which
> are for kissing and to sing with
> who cares if some oneeyed son of a bitch
> invents an instrument to measure Spring with?
>
> —*E. E. Cummings,*
> *Is 5, 33 (1926)*

But the management of our affairs ought not be left solely to scientists and technologists. There are social, economic, political, esthetic, and moral aspects which transcend the special competence of scientists and engineers. Indeed, the purely technical parts to these solutions are frequently already known. Further, professional knowledge and skills are no guarantee of wisdom.

In summary, there are several reasons why natural science belongs in a general education curriculum. Science is among the great intellectual and esthetic activities of the human mind. The implications and consequences of this creative activity are enormous. Our best hope of guiding the purpose of this activity lies in understanding how it functions.

Designing a course in natural science especially for the nonscience major, we begin with two premises: it is both very important and not very easy to know something of science and technology. The jargon developed by specialists for their convenience also serves as a barrier to understanding by the nonspecialist who may be a scientist or engineer trained outside of that specialty.

A physical chemist, for example, who studies the heat released during combustion of petroleum can be as ignorant as a grocer or lawyer of exactly what it is a biologist does when he studies the life cycle of a microorganism. But he understands that both he and the biologist do their individual work according to the same general procedures. There seems to be an important difference between the scientist and nonscientist in their feeling toward the overall enterprise of science-and-technology. The feeling of alienation from this enterprise, and even of *anomie* in our society, appears to be less acute today among those technically trained. This is not to say that scientists and engineers cannot feel dismay and despair over our circumstances; they can, and often with more clearly defined objections in mind.

There may be a clue in this to what we hope to accomplish in such a course.

What we *do not* hope to have you do is accumulate a lot of facts. Not all the individual bits of information used in this book will be true forever. Unfortunately, we don't know which are going to remain true for some time and which will be superseded soon by better facts. It is important, however, for you to see *how* scientists obtain these facts. Laboratory techniques and several of the important experiments are described in detail. No great purpose is served in your memorizing these techniques and experiments. But if you can follow and understand what happens and what is learned, you will have a clearer idea of the significance of a scientific fact.

The major accomplishments in science are the creations of *laws* and *theories*. A law is a statement summarizing experimentally observed behavior. A theory is a description of how a law or a set of phenomena works. Thus facts are woven together by laws and theories. There is nothing sacred about either laws or theories. When something better is discovered either is modified or dropped entirely. It is not important to retain any law or theory permanently, but rather to see how they are discovered. It is not a difficult matter for anyone to recognize patterns of regularity in natural phenomena, just as a scientist does. Therefore, collections of experimental facts have been arranged to permit this recognition.

In this book, we have used a scientific vocabulary because there is often no other way of stating things simply and unambiguously. In addition, mastering a small scientific vocabulary for a short time will help make you feel more comfortable in a technological society. If you were abruptly and magically transported to a remote Greek village your experience there should be less strange if you knew the Greek alphabet and could, even imperfectly, read signs. Analogously by mastering a small scientific vocabulary for a short time your feeling about the strangeness of the jargon will abate.

Whenever we ask our students what they liked least about our courses, we got the answer "mathematics." Probing deeper we discover they mean simply using numbers. Scientists use numbers because their observations of nature require something more exact than big, bigger, very big, very very big, small, quite small, rather small. Scientists use mathematics because the concepts or facts cannot be handled readily with ordinary words. Mathematics is a very elegant language. In this text we use numbers quite a bit. Looking at a table of numbers is no more difficult than looking up a telephone number, or a price in a catalog. Numerical examples are carefully worked out with optional problems at the end of some chapters. Nowhere have we used mathematics more advanced than first year algebra. As with the vocabulary, simple experiences in using elementary mathematics will do much to show you how science operates.

It is not possible in a single textbook to teach you all about science. A science major doesn't learn all about science in four years. Only a small part of the entire enterprise of science can be handled. The very specialization and diversification which marks all science means that there can be no linear path, no one-topic-followed-by-another-single-topic sequence, which can cover a significant part of science. Any path which forms the course content of a natural science syllabus is ramified: it is complicated by nontrivial ingressions and digressions at almost every point. Rather than being linear, the organization of science is intricately reticulate.

Because the authors of this text are a physical chemist and a physiologist, we chose topics of interest to both of us frequently from a chemist's viewpoint. There are good reasons for this.

There are as many chemists as there are biologists and physicists combined in this country. Chemistry overlaps all the other natural sciences. We can illustrate some of the organization of science by looking at examples of how chemists use physics, and of how chemistry is used to probe into biological systems. In addition, since over half the chemists work in industry, there is an excellent opportunity to see how science and technology interact. We have digressed frequently in the text to show how science is applied to problems.

We use detail; our purpose can be clarified by looking briefly at Chapter 7 and the last third of Chapter 8. Chapter 7 is heavily illustrated with chemical formulas and molecular structures. Chemists believe all matter is composed of molecules. Further, they believe each substance has its own particular molecule with its own special structure. We don't suggest that you memorize all or any of these. This would be a formidable task even for a scientist. Rather, upon finishing Chapter 7, we hope you will have a better idea of why chemists can talk about molecules. You will have a better understanding of molecules as a part of our reality deduced, not by our immediate senses, but from experimental evidence and intuitive guesses.

Nuclear warfare is an important consideration in international politics. We can simply say: nuclear warfare is a horrible possibility, or it is unthinkable, or that we can destroy almost all of mankind. But we believe the implications of nuclear warfare can better be felt by a careful discussion of the detailed effects of nuclear weapons. This is why Chapter 8 ends as it does, providing a vicarious experience to reinforce the simple statement of horror.

We wrote this book especially for the nonscientist, not to proselytize, but to give him certain *experiences* in science. Our purpose is to make the enterprise of science and technology seem less strange. We are more concerned about the student's feelings toward than his knowledge about science and technology.

Acknowledgments

An author is a mirror, whose book is the reflection of a host of influences—several teachers (who have inspired by their examples), even more students (who teach by their questioning), and, especially, colleagues (who have been generous with their time and helpful with their advice), is a mirror. It is a pleasure to thank a few of our colleagues by name: Jurgis Anysas, Anthony Behof, Fred Breitbeil III, Thomas Murphy, William Pasterczyk, Franklin Prout, and James Vasa.

We are grateful to Hans J. Morgenthau, for letting us quote his writing; Sidney Pollack, for the x-ray diffraction pictures, and Harvey E. White, for letting us use his diffraction pattern.

Charlene Popek typed almost the entire manuscript, giving us a nonscientist's opinion. We are indebted to her.

And, because the first draft was done with her encouragement during her last days, to Eleanor Leah Simon Blumberg (1927–1967), whose delight in the form and function of things was a joy to her husband.

Finally, we extend our appreciation to our editor, Nancy L. Marcus, who made a formidable job seem less overwhelming.

David sang of "the children of one's youth; happy is the man who has his quiver full of them." His son, who was said to be wise in such matters, valued "a good wife far above rubies." Our families are this and more. In countless small and big ways they have sustained us. This book is dedicated to them.

Avrom A. Blumberg
Jon Stanley

Form
and
Function

Natural
Science
in a
Techno-
logical
Society

The Method
and Techniques
of Science

As an introduction to science we shall first see what scientists do. For illustration we shall look at a few examples and return to them in later chapters for a more detailed treatment. Next, we shall present some of the special vocabulary and devices scientists use to make the study of natural phenomena more intelligible.

Although natural science comprises such broad areas of study as astronomy, biology, chemistry, geology, and physics; an increasing number of interdisciplinary areas such as geophysics, physical biochemistry, and astrophysics; and subjects of high specialization such as enzyme chemistry, polymer physics, and membrane morphology; nevertheless there is a pattern of activity common to all of these called "the scientific method."

THE SCIENTIFIC METHOD

Briefly, the scientific method includes these more or less distinct steps.

1. The scientist isolates some small part of the universe for study.
2. Realizing that the universe is quite complicated, he invents a model or hypothesis which resembles in form or function the subject under study. The model is always a compromise. It is neither so detailed or so exact a replica of the universe as to defy comprehension, nor is it so simple as to have no relation to the real, observed universe.
3. From the model certain predictions about structure and behavior are made.
4. These predictions are compared to the structure and behavior observed experimentally in the universe.

5. Depending on how closely prediction and observation agree, a judgment is made about the correctness of the model.

Models are continually modified and, if necessary, eliminated. The entire collection of models from natural science is the best objective picture we have of the physical universe. Natural science is the means we have of knowing what is physical reality.

We can look briefly at three models which have been used quite successfully in describing natural phenomena. Each model will be discussed in greater detail later.

MODEL 1. Robert Boyle (1627–1691), using apparatus similar to that shown in Fig. 1.1, confined a sample of air in the short closed end of a J-shaped tube by means of a column of mercury. He increased the pressure on the air sample by adding more mercury to the longer, open end, and observed that the air was compressed to a smaller volume. When the pressure was decreased by removing mercury, the air volume expanded. By measuring the pressures due to the mercury column and the corresponding air volumes, all at the same temperature, he noticed a simple relationship: the product obtained from multiplying the imposed pressure p times the corresponding air volume V was very nearly a constant number, or

$$pV = \text{constant}$$

This is called "Boyle's law." A *law* is usually a summary of related observable facts.

In 1738 Daniel Bernoulli proposed the model of a gas as made up of a large number of very small particles moving rapidly and randomly, hitting each other and the walls of the container. He was able to show that the pressure exerted on the container walls by the colliding particles is related to the volume of the container in exactly the same way that Boyle's law relates these quantities.

Because the pressure-volume relationship obtained using the small-particle model for a gas agrees with the experimental pressure-volume relationship of Boyle, we say that Bernoulli's small-particle theory is valid. We believe, as part of our ideas of the physical universe, that a gas is made up of many small particles even though these are too small to be seen.

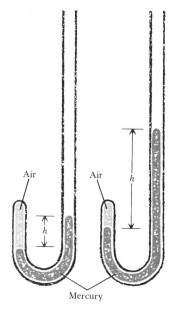

Figure 1.1. Boyle's experiment. Air is compressed in the sealed short end of the J tube by a column of mercury. The longer the column of mercury h, the more the air is compressed.

MODEL 2. At the end of the eighteenth century Benjamin Thompson (Count Rumford, 1753–1814) observed a relationship between the mechanical work expended in boring a cannon and the heat produced (as measured by the temperature increase in the water used to prevent overheating of the boring equipment). James Joule (1818–1889) made very careful measurements to see if there was a simple relationship between heat and work.

By passing an electric current through a wire immersed in water and by observing the increase in water temperature and the electric energy consumed, Joule observed an equivalence between the electric energy and the heat which caused the temperature increase. If the electric energy expenditure is doubled, twice as much heat is produced.

Joule then built a brass paddle wheel fitted in a can of water. The paddle wheel was set in motion by a system of weights and pulleys. These were wound up by raising the weights, which were then allowed to fall, turning the paddle wheel and agitating the water. Joule calculated exactly how much work was involved in winding up the weights and pulleys, which was the work used to turn the paddle wheel. He observed a water temperature increase and from this knew how much heat was needed to produce this increase. Joule found an equivalence between the mechanical work and the produced heat; the more work done, the more heat produced.

Julius Mayer and Joule observed that when work or energy is converted into heat there is an exact relationship between the work or energy and the heat. Heat is one form of energy; when heat is added to a substance its energy increases by exactly the amount of heat added. When work is done on a body, for example, by lifting, compressing or stretching against a force, the energy of that body is increased by exactly the amount of work done. More succinctly, energy cannot be created or destroyed; the energy of the universe is constant. This statement is called the "law of conservation of energy," or the "first law of thermodynamics." No exception to this statement has ever been observed. Therefore, we accept this law as true for all natural phenomena.

MODEL 3. The Austrian monk Abbot Gregor Mendel (1822–1884) observed in the breeding of garden peas that certain traits had two possible forms—the seed color was either yellow or green, and the flower color was either red or white. He saw that these traits in any pea plant were related to the form of the traits in the two parents, so that among all the offspring of a given pair of parent pea plants, the traits appeared according to some regular statistical pattern.

Mendel proposed that each parent contributed what he called an "elemente" and what was later called a *gene*, and that the manifest form of the trait depended upon the nature of the pair of genes found in each pea plant. Mendel was able to predict from the two parents' manifest trait form, the frequency of occurrence of each trait. His calculated frequency of occurrence for seven different kinds of traits agreed very well with the observed frequency of occurrence of these traits.

Mendel's rules for inherited traits were announced in 1865, but they did not attract much attention until they were rediscovered by three independent workers at the beginning of this century. Since the model of genes successfully describes how traits are passed from one generation to the next, the existence of genes was accepted as a real part of living systems. Only very recently have gene structure and function been elucidated.

SCIENCE AND TECHNOLOGY

A distinction can be made between science and technology. Science is the means we have of describing, and technology the means of controlling the universe. However, there is no sharp boundary between description and control. There is a mutual and strong influence between science and technology. The vigorous growth of both is a consequence of this cooperation.

NUMBERS

Whenever possible, natural scientists construct quantitative descriptions of the physical world. Therefore, whenever possible the natural scientist uses numbers. Several aspects of numbers will be mentioned: precision, accuracy, significant figures, and a standard scientific notation for numbers.

PRECISION

We can measure the height of someone by having him stand against a wall; placing a pencil level on his head; then marking a horizontal pencil line on the wall; finally measuring the floor-to-line distance with a yardstick or tape measure.

Table 1.1. Measurement of Height

Measured height, in.	Deviation from average, in.
70.4	0.0
70.2	−0.2
70.3	−0.1
70.5	+0.1
70.4	0.0
70.6	+0.2
Sum 422.4	0.6
Average 70.4	0.1

Suppose we have a special tape measure marked in inches and tenths of an inch. If the tape-measure distance is 70.4 in., we say the person is that tall. We mean, according to this measuring technique, his height is closer to that number than to either 70.3 or 70.5 in.

This person can be measured several times using exactly the same technique and the same tape measure. Six such measurements are listed in the first column of Table 1.1. The average or *mean* of these numbers is calculated by adding up the values and dividing by the number of measurements. In the second column are the deviations of each measurement from the average, 70.4 in. The measured values less than the averaged value have negative deviations; those greater have positive deviations. The deviations are averaged without considering the positive and negative signs. We can say that the average height of the person is 70.4 ± 0.1 in. The ± 0.1 inches is the measure of the fineness or *precision* of our measurement. It tells us with what confidence we can say the height is 70.4 in. If the average deviation were greater the average height would be less precisely known; the precision of the measurement would be less. Similarly, if the average deviation were less than ± 0.1 in. the precision of the height would be greater.

The fact that the six measured values are not all identical can be attributed to several factors: the roughness of the wall, failure to hold the pencil exactly horizontally, variations in the height of the person during the day, stretching of the tape measure, the thickness of the pencil mark, and so on.

ACCURACY

The *accuracy* is a measure of the correctness of the averaged value. For example, if someone cut off the first 2 in. of the tape and did not adjust for it, the measured heights would be inaccurate by 2 in. as read on the tape. Another inaccuracy is introduced because the pencil mark is aligned with the center of the pencil cross section while the outside of the pencil rests on the person's head. It is possible for a set of measurements to be very precise, i.e., reproducible, but at the same time to be inaccurate. The precision of a measurement refers to a collection of measurements done exactly the same way and indicates the consistency of the measured numbers.

SIGNIFICANT FIGURES

Usually there is more than one way to make measurements. Implicit in the statement that the height is 70.4 in. is the fact that the measuring instrument, the tape measure, was not able to be read to 0.01 in. We say that 70.4 has three *significant figures;* they are 7, 0, and 4. Suppose a second measuring tape is marked only in inches. The height read by this tape is 70, which means the height is closer to 70 than to either 69 or 71. This measurement has two significant figures, which corresponds to the accuracy of the second tape.

Still a third measuring tape, for example, might give the number 70.38 in. This number has four significant figures which corresponds to a measuring tape read to the nearest 0.01 in. The number means that this particular measurement is closer to 70.38 than to either 70.37 or 70.39.

However, it may not be realistic to measure a person's height to the nearest one-hundredth of an inch. A set of measurements using this last tape measure may give the set of numbers listed in Table 1.2. Then the average height by this technique is

Table 1.2. Measurement of Height

Measured height, in.	Deviation from average, in.
70.38	−0.03
70.18	−0.23
70.34	−0.07
70.54	+0.13
70.61	+0.20
70.41	0.00
Sum 422.46	0.66
Average 70.41	0.11

70.41 ± 0.11 in., which is not significantly different from the previously calculated average, 70.4 ± 0.1 in. The larger average deviation here, 0.11, indicates that measuring a person's height to the nearest 0.01 in. is an unnecessary refinement. The precision of this measuring technique becomes evident in averaging the set of measurements obtained by that technique.

For convenience we can *round off* 70.41 ± 0.11 to 70.4 ± 0.1 in., that is, drop the last digit in each average and the average deviation. *Rounding off* is the method we use when we want to reduce the number of significant figures. The rules are simple:

1. If the last digit, the one to be dropped, is 0, it is omitted without changing the value of the number (19.40 and 19.4 have the same value).
2. If the last digit is 1, 2, 3, or 4, it is dropped (19.82 becomes 19.8).
3. If the last digit is 6, 7, 8, or 9, the penultimate digit is increased by one (46.27 becomes 46.3).
4. If the last digit is 5, then the penultimate digit, if even, is unchanged, and if odd, is increased to the next even digit (82.45 becomes 82.4 and 61.75 becomes 61.8).

There is a more complicated way of stating the deviation in a set of measurements. First the individual deviations are squared; then the average of the squared deviations is obtained; finally, the square root of this is calculated. This is called the "standard deviation." It involves more work but it has a special significance. From the data in Table 1.2 the standard deviation is 0.14 and the measured height is 70.41 ± 0.14 in. This means we are 68 percent confident the height is within one standard deviation of the mean, between 70.27 and 70.55 in., provided we have eliminated all discoverable inaccuracies and are dealing with truly random errors. We are 95 percent confident the height is within two standard deviations, that is, between 70.13 and 70.69 in. And we are better than 99.73 percent sure the true height is within three standard deviations, that is, 69.99 to 70.83 in. We are never *absolutely confident* about any measured number. Even in simple counting we can make a mistake. One place where we are confident is in definitions: there are 3 ft in 1 yd, and so on. Here we can use as many significant numbers as we wish. But in practice we simply understand this when we say there are 3 ft in 1 yd.

We have already seen that 70.4 means that the measured

number is closer to 70.4 than to either 70.3 or 70.5; that 70.38 is closer to that than either to 70.37 or 70.39. An ambiguity exists, however, in numbers such as 7,200. Does this mean the measured number is closer to 7,200 than to either 7,199 or 7,201; or does it mean it is closer to 7,200 than to either 7,100 or 7,300? Is the fineness of measurement a unit, that is, ±1, or a hundred, that is, ±100? Are there four or two significant figures?

THE STANDARD NOTATION

Scientists and engineers commonly use a standard notation for numbers to resolve this ambiguity of significant figures and to make easier the handling of very small and very large numbers. Because 100 is 10×10 it can for convenience be written $100 = 10^2$; $1,000 = 10 \times 10 \times 10 = 10^3$; and so on. The superscripts 2 and 3 are called "exponents" and are the number of times 10 is used as a factor in the multiplication. Table 1.3 lists three ways of writing some numbers which are perfect multiples of 10. When any number which is a perfect multiple or power of 10 is multiplied by 10, the exponent increases by one; when divided by 10, the exponent is reduced by one. The notation should be obvious for positive exponents. If we start with $1,000 = 10^3$ and successively divide it by 10, we get $100, 10, 1, \frac{1}{10}, \frac{1}{100}, \frac{1}{1,000}$, and so on, and the exponent 3 is reduced to $2, 1, 0, -1, -2, -3$, and so on. Observe the curious but sound use of exponents in $10^0 = 1$. That 10 multiplied by itself zero times (i.e., not used as a factor, even once) is unity does not have an immediately obvious significance but is both a correct and convenient notation.

Table 1.3. Exponential Notation for Numbers

Decimal	Fraction	Exponential	Exponent
	Forms		
1,000,000	1,000,000	10^6	6
1,000	1,000	10^3	3
100	100	10^2	2
10	10	10^1 or 10	1
1	1	10^0	0
0.1	$\frac{1}{10}$	10^{-1}	-1
0.01	$\frac{1}{100}$	10^{-2}	-2
0.001	$\frac{1}{1000}$	10^{-3}	-3

Any number can be rewritten in such a way that the number is between 1 and 10 and then multiplied by 10 raised to some power.

Example. The speed of light is 186,272 mi/s and can be written as

$$
\begin{aligned}
186{,}272 &= 18{,}627.2 \times 10 \\
&= 1{,}862.72 \times 10^2 \\
&= 186.272 \times 10^3 \\
&= 18.6272 \times 10^4 \\
&= 1.86272 \times 10^5
\end{aligned}
$$

Moving the decimal point to the left is equivalent to dividing the number by 10; to keep the value constant it is necessary to compensate for the division by 10 by multiplying by 10, that is, by increasing the exponent by one.

Example. The radius (in inches) of a hydrogen atom is

$$
\begin{aligned}
0.00000000208 &= 0.0000000208 \times 10^{-1} \\
&= 0.000000208 \times 10^{-2} \\
&= 0.00000208 \times 10^{-3} \\
&= 0.0000208 \times 10^{-4} \\
&= 0.000208 \times 10^{-5} \\
&= 0.00208 \times 10^{-6} \\
&= 0.0208 \times 10^{-7} \\
&= 0.208 \times 10^{-8} \\
&= 2.08 \times 10^{-9}
\end{aligned}
$$

Moving the decimal point to the right multiplies the number by 10; to keep the number at a constant value it is necessary to divide the resultant number by 10, that is, to decrease the exponent of 10 by one.

Example. The number 7,200, when it has four significant figures, is written 7.200×10^3.

Example. The number 7,200, when it has two significant figures, is written 7.2×10^3.

Comparing numbers written in this standard notation is easier, and the ambiguity of significant figures is removed.

UNITS OF MEASUREMENT

If we are interested in how much something weighs and if we live in the United States or in the British Commonwealth, there is more than one system of weights in which to express that amount. Table 1.4 lists two such systems, the *avoirdupois* system and the *troy*, or *apothecary*, system. The former is used in ordinary commerce and the latter for precious metals and medicines.

The ounce has been a much-used unit. (Both ounce and inch are derived from the Latin *unus*, one.) In addition to meaning two different weights, it is also used as a volume or capacity measure. There are 128 fluid oz in a U.S. gal but 160 fluid oz in 1 British Imperial gal; and the United States fluid ounce is a little larger than the Imperial ounce. In the United States there are two systems of volume measure: dry and liquid; but a liquid quart is somewhat smaller than a dry quart which, in turn, is smaller than a British Imperial quart.

It is evident that these numbers are not the most convenient to use in our base 10 number system and that there is more than one kind of ounce, gallon, pound, and so on.

Happily, the *metric system* has been designed to make the entire system of weights and measures as uniform and simple as possible. Two important reasons for having such an international system are: (1) it is essential for exchange and comparison of precise experimental results that a common, accurate, and precise system of measures is used throughout the world; and (2) the decimal system of numbers is the most convenient one for calculations. All the nations of the world, except the United States, Nigeria, and

Table 1.4. Two Systems of Weights

Avoirdupois weights	Troy or apothecary weights
16 drams = 1 oz	20 grains = 1 scruple
16 oz = 1 lb	3 scruples = 1 dram
14 lb = 1 stone	8 drams = 1 oz
100 lb = 1 short hundredweight	20 pennyweights = 1 oz
112 lb = 1 long hundredweight	12 oz = 1 lb
2,000 lb = 1 short ton	
2,204.62 lb = 1 metric ton	
2,240 lb = 1 long ton	

1 oz avoirdupois = 0.9114562 oz troy
1 oz troy = 1.09714 oz avoirdupois

a half-dozen smaller countries, are either using or now converting to the metric system.

Measurements made by scientists and engineers almost always have dimensions; i.e., they consist of a number and some units, the exception being measurements which count the number of objects in a set. Examples of dimensioned numbers are the following.

1. *Speed:* 10 mph, a distance (in miles) divided by a time (in hours)
2. *Pressure:* 14.7 lb/in.², a force (in pounds) divided by an area (in square inches)
3. *Energy value:* 190 Cal/oz of butter, an energy (in Calories) divided by a weight (in ounces)
4. *Cooling capacity* (of an air-conditioner): 10,000 Btu/hr, an energy (in British thermal units) divided by a time (in hours)

All the dimensions used can be expressed as some combination of four basic units: length, time, temperature, and mass. Therefore, it is important that the same four basic units be defined as accurately and precisely as needed wherever accurate and precise measurements are made. For example, the speed of light in a vacuum, according to the best measurements, is $(2.997925 \pm 0.000001) \times 10^8$ m/s, where the meter m (about 39.37 in.) is the unit of length and the second s is the unit of time. It is evident that since this speed is known to seven significant figures, the length and time units must be defined with at least that precision. It is meaningless to write the speed, a length divided by a time, more precisely than either the length or the time is known. To understand precisely what is meant by 2.997925×10^8 m/s one must know precisely what is meant by a meter and by a second.

THE STANDARD OF LENGTH

The unit of length is the *meter.* Originally it was defined as *one ten-millionth of the distance from the equator to the North Pole.* Since it is not especially easy to measure distances along the earth with precision, it was decided, in 1889, to define the meter as *the distance between two engraved lines on a platinum-iridium bar kept at the freezing temperature of water at Sèvres outside of Paris.* Copies of this prototype bar were given to participant nations, who agreed to use the meter as the standard of length. The copies were compared with the Sèvres prototype and, in turn, were used in the participant nations to define their units of length. Even though the use of inches, feet, yards, miles, and so on, is common in the United States and the British Commonwealth, the inch itself is defined to be *exactly* 0.0254 m.

At the Eleventh General Conference on Weights and Measures (1960, Paris) it was decided to define the meter in terms of the wave properties of light emitted by the rare gas krypton at a well-defined temperature. This is a convenient experimental technique available to anyone with the necessary and not uncommon laboratory equipment. An exact definition of the meter is possible without comparison with a prototype bar or copy.

TIME

The unit of time is the *second*, originally 1/86,400 of a day. Because all days do not have a common duration in time, a better definition is 1/31,556,925.9747 of the tropical year 1900. A still better definition is in terms of the vibrating wave properties of light emitted by the metal cesium. Such an *atomic clock*, or metronome, is far superior to the erratic rotation of the earth. It has a precision of 1 trillionth s; in 6,000 yr the atomic clock will have an error of 1 s. The atomic clock can be built in any reasonably well-equipped laboratory.

TEMPERATURE

Temperature is the measure of how hot or cold something is. Since most substances expand when warmed and contract when cooled, the volume of a substance can be used as a *thermometer*, i.e., as an instrument to measure temperature. Gabriel Fahrenheit (1686–1736) made a thermometer by filling a glass bulb with liquid mercury and connecting it to a glass tube with a very fine bore, or capillary. When heated, the liquid in the bulb expands into the capillary bore. The warmer the thermometer liquid, the higher into the capillary bore the liquid rises. This thermometer can be placed in an ice and water mixture at the temperature of freezing water and melting ice, called the "ice point." Then, observing how high into the bore the mercury goes and marking that height, the ice point (the temperature at which water freezes) is identified. With any mercury and glass thermometer the ice point is reproducible:

the mercury rises to the same height each time, Fahrenheit called this temperature "32 degrees," or 32°.

By placing the thermometer in boiling water and marking the mercury height, the boiling point of water, or the "steam point," can be identified. On the Fahrenheit scale this is 212 degrees Fahrenheit, or 212°F. The steam point, too, is reproducible. The distance along these two marks is divided into 180 equally spaced distances, each of which is 1 degree Fahrenheit, or 1°F. This is the temperature scale used in the United States (but not in science) and in very few other countries.

All scientists and almost all of the rest of the world use another scale designed by Anders Celsius (1701–1744), who marked the freezing and boiling points of water as 0 and 100 and divided this distance along the capillary bore into 100 equally spaced distances, each one of which is called "1 degree centigrade," or, more recently, "1 degree Celsius." Water freezes at 0°C and boils at 100°C. (Figure 1.2 compares the Fahrenheit and centigrade thermometers.)

To describe a temperature scale which can be reproduced anywhere, it is necessary to describe two temperatures, e.g., the ice and the steam points, and the material of the thermometer, e.g.,

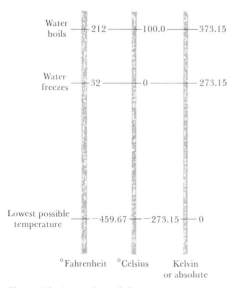

Figure 1.2. Comparison of three temperature scales.

mercury or colored alcohol. With this information the same temperature scale can be used by all scientists. However, there are several limitations to the reproducibility of the mercury–ice-point–steam-point system described. The normal freezing point of water is determined by an ice and water mixture open to the air and is reproducible to 0.001°C; we say the normal freezing point of water is 0.000 ± 0.001°C. The uncertainty is due to varying amounts of contaminants present in the air, which dissolve in the pure water and lower the freezing point slightly. The normal boiling point of pure water is not reproducible to more than 0.01°C.

In the next chapter we shall see that there is a minimum temperature below which a substance cannot be cooled: the *absolute zero* (−459.67°F, or −273.15°C). We shall also see that a much better thermometer can be made using a gas rather than a liquid.

MASS

The advantages of these standards for length, time, and temperature lie in their precise and accurate reproducibility. Unfortunately no one has devised a simple, precise, reproducible standard for mass.

Although the words *mass* and *weight* often are used interchangeably, there is a basic difference. Mass is a measure of how much matter there is in a body, and weight is a gravitational force which exists between two masses. This gravitational force depends upon the multiplication product of the two masses and the distance between the masses. The weight of an object on earth is the gravitational force of attraction between the object and the earth. The same object with exactly the same mass has less weight on the moon since the mass of the moon is less than the mass of the earth. Mass is a property of the object alone; weight depends on what other objects are around.

The unit of mass is the *kilogram*. The prototype is a platinum-iridium cylinder kept at the International Bureau of Weights and Measures at Sèvres. The *pound* has been used ambiguously both for mass and for weight or force; the kilogram is a unit of mass only. One instrument used for determining the mass of an object is the *analytical balance* (Fig. 1.3). The earlier models have two pans and a pointer, or examen, each suspended from a horizontal cross beam which is balanced on a knife-point fulcrum. The newer balances operate on the same principal but generally

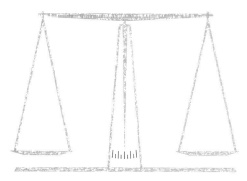

Figure 1.3. The analytical balance.

have only one visible pan. The object whose mass is to be measured is placed on one pan and units of known mass (compared with copies of the kilogram prototype) are placed on the other pan until the examen points exactly vertically down: the two pans are balanced. When this happens, we know that the force between the earth and the object equals the force between the earth and the added units of known mass. Then, the mass of the object is equal to the combined masses of the added units.

If the object, analytical balance, and set of known mass units are carried to the moon, the balancing will be accomplished with exactly the same units of mass as on the earth. At balancing, the force between the moon and the object equals that between the moon and the units of mass. The process of determining mass, called "weighing," is independent of the mass of the planet on which the weighing is done.

The analytical balance depends on the presence of a large mass to attract both the object to be weighed and the standard unit masses. It cannot be used, therefore, in free space far away from large masses such as planets. In addition to gravity, inertia can be used to determine mass. The force required to change the motion of a body depends on its mass.

The weight of a substance in a gravitational field (a region in which a gravitational force is exerted on a substance) can be calculated if we know the mass of the substance and the gravitational force acting on a unit mass in that gravitational field. Alternatively, since a force is required to stretch a spring and the greater the force the more the stretching, the weight can be determined from the spring extension. The common bathroom scale depends on this spring behavior.

THE INTERNATIONAL
SYSTEM OF UNITS

This system of basic units is called the "Système International d'Unites" (SI). Since the meter, kilogram, and second are used, it is sometimes called the "mks system." Table 1.5 lists the four standard units and their abbreviations.

Table 1.5. International System of Units, or mks System

Physical quantity	Name	Symbol
Length	Meter	m
Mass	Kilogram	kg
Time	Second	s
Temperature	Degree Kelvin, or Kelvin	K, or Kel

An advantage of using the metric system can be seen from looking at Table 1.6, which lists lengths and masses. There is only one number to remember—10—and there is a common set of prefixes. Table 1.7 lists the complete set of fractions and multiples, prefixes, and symbols.

Table 1.6. The Metric System

Metric lengths	Metric masses
10 milligrams = 1 centigram	10 millimeters = 1 centimeter
10 centigrams = 1 decigram	10 centimeters = 1 decimeter
10 decigrams = 1 gram	10 decimeters = 1 meter
10 grams = 1 dekagram	10 meters = 1 dekameter
10 dekagrams = 1 hectogram	10 dekameters = 1 hectometer
10 hectograms = 1 kilogram	10 hectometers = 1 kilometer

Table 1.7. Prefixes Indicating Magnitudes

Fraction of a unit	Prefix	Symbol	Multiple	Prefix	Symbol
10^{-18}	atto	a	10^{12}	tera	T
10^{-15}	femto	f	10^{9}	giga	G
10^{-12}	pico	p	10^{6}	mega	M
10^{-9}	nano	n	10^{3}	kilo	k
10^{-6}	micro	μ	10^{2}	hecto†	h
10^{-3}	milli	m	10	deka†	da
10^{-2}	centi†	c			
10^{-1}	deci†	d			

† The use of these is restricted to convenience; for example, the centimeter is a unit of length in some biological measurements, and the cubic centimeter is used in chemistry. The system using the centimeter-gram-second units is called the "cgs system."

Example. One-millionth of a second is 10^{-6} s or 1 μs.

Example. Six thousand meters is 6×10^3 m or 6 km.

CONVERSION OF UNITS

Because systems other than SI are still used, there are many oc-casions when conversions must be made from one system to another. Within SI only the number 10 is used, and conversions are made by varying the exponent to the base 10. The conversion from one system of units to another can be illustrated by two examples.

Example. Convert the distance 11.5 yd to meters, given the infor-mation

1 yd = 3 ft
1 ft = 12 in.
1 in.= 0.0254 m

Since 1 yd = 3 ft, we can write

$$1 = \frac{3 \text{ ft}}{1 \text{ yd}} \qquad \text{and also} \qquad \frac{1 \text{ yd}}{3 \text{ ft}} = 1$$

so that when 11.5 yd is multiplied by 3 ft/1 yd, the multiplication is formally equivalent to multiplying by unity and dimensionally equivalent to converting yards to feet by the formal cancellation of the dimension yards

$$11.5 \text{ yds} \times \frac{3 \text{ ft}}{1 \text{ yd}} = 34.5 \text{ ft}$$

The other factors—12 in./1 ft and 0.0254 m/1 in.—are introduced to convert feet to inches and then inches to meters:

$$11.5 \text{ yd} \times \frac{3 \text{ ft}}{1 \text{ yd}} \times \frac{12 \text{ in.}}{1 \text{ ft}} \times \frac{0.0254 \text{ m}}{1.000 \text{ in.}}$$

$$= \frac{11.5 \times 3 \times 12 \times 0.0254}{1 \times 1 \times 1} \text{ m}$$

$$- 1.15 \times 10 \times 3 \times 1.2 \times 10 \times 2.54 \times 10^{-2} \text{ m}$$
$$= 1.15 \times 3 \times 1.2 \times 2.54 \times 10^0 \text{ m}$$
$$= 10.51560 \text{ m}$$
$$= 10.5 \text{ m}$$

The numbers 3, 12, and 0.0254 are exact, being defining numbers; however, the number 11.5 yd has three significant figures, and so the equivalent 10.5 m can have no more than that.

Example. Convert the speed 60.0 mph to kilometers per second given:

$$60 \text{ s} = 1 \text{ min}$$
$$60 \text{ min} = 1 \text{ hr}$$
$$1 \text{ mile} = 5{,}280 \text{ ft}$$
$$1 \text{ ft} = 12 \text{ in.}$$
$$1 \text{ in.} = 0.0254 \text{ m}$$
$$1{,}000 \text{ m} = 1 \text{ km}$$

$$\frac{60.0 \text{ mile}}{1 \text{ hr}} \times \frac{1 \text{ hr}}{60 \text{ min}} \times \frac{1 \text{ min}}{60 \text{ s}}$$

$$\times \frac{5{,}280 \text{ ft}}{1 \text{ mile}} \times \frac{12 \text{ in.}}{1 \text{ ft}} \times \frac{0.0254 \text{ m}}{1 \text{ in.}} \times \frac{1 \text{ km}}{1000 \text{ m}}$$

$$= \frac{60.0 \times 5{,}280 \times 12 \times 0.0254 \text{ km}}{60 \times 60 \times 1{,}000 \text{ s}}$$

$$= \frac{6.00 \times 10 \times 5.280 \times 10^3 \times 1.2 \times 10 \times 2.54 \times 10^{-2} \text{ km}}{6 \times 6 \times 10^5 \text{ s}}$$

$$= \frac{6.00 \times 5.280 \times 1.2 \times 2.54 \times 10^3 \text{ km}}{6 \times 6 \times 10^5 \text{ s}}$$

$$= \frac{95.56 \times 10^{-2} \text{ km}}{36 \text{ s}}$$

$$= 2.682 \times 10^{-2} \text{ km/s}$$
$$= 0.0268 \text{ km/s}$$

Since the original speed 60.0 mph has three significant figures, so must the converted speed 0.0268 km/s.

THE MEASUREMENT OF PRESSURE

In his investigations of gas behavior, Robert Boyle measured the pressure of the confined gas by the height of a mercury column. In Fig. 1.1 the height h was the measure of that pressure; the greater h, the more the pressure. In this section we shall look at the methods and dimensions of pressure measurement in more detail.

Pumping air out of the top of a pipe, the bottom of which rested in water, Galileo Galilei (1564–1642) observed that it was not possible to raise water to a height greater than about 33 ft. He called this phenomenon to the attention of his pupil and assistant Evangelista Torricelli (1608–1647). Torricelli asserted that the atmosphere itself has a weight exactly equal to the weight of that 33-ft water column. The two weights balance each other perfectly: if the atmosphere weighed more, the height of the water column should be greater; if the atmosphere weighed less, the water column should be less.

To check this claim, Torricelli reasoned that since mercury is 13 times more dense than water, a column of mercury (which balances the atmosphere and which, therefore, weighs as much as the 33-ft column of water) ought to be only $\frac{1}{13}$ as high as 33 ft, or about 30 in. high. He filled a tube, 4 ft long and closed at one end, with mercury (Fig. 1.4(*b*)). He then put his finger on the open end and inverted the entire tube, placing the lower and open end beneath the surface of a dish of mercury. The mercury in the tube dropped to a height of about 30 in., or 76 cm, above the mercury surface in the dish. This demonstration is called the "Torricellian experiment," and the empty space above the mercury in the closed tube the "Torricellian vacuum."

Both René Descartes (1596–1650) and Blaise Pascal (1623–1662) predicted that if the Torricellian experiment were performed first at the base and then at the top of a mountain, the mercury column height would be greater at the base than at the top. At the higher elevation there is less atmosphere pressing down to be balanced by a shorter mercury column. Florin Perier (1605–1672) confirmed this prediction. Torricelli's equipment, called a "cistern barometer," can be used to measure elevation above sea level.

In addition to varying with altitude, the mercury column changes with weather conditions. The *barometer* is a principal instrument in the forecasting of weather. Precipitation of rain and snow are associated with short mercury columns, i.e., with lower barometric readings, and fair weather with higher barometric readings. Although the *cistern* barometer is widely used in laboratories (especially where there is no need to move the barometer), the *siphon* barometer is another and sometimes more convenient modification of Torricelli's original apparatus (Fig. 1.5).

We can see what feature of the atmosphere is measured by barometers by first imagining a symmetric U-tube, open at both ends, into one end of which we pour some mercury. Momentarily the two mercury levels are unequal, but there is an immediate flow and adjustment until the levels are equal. A *force*, when acting upon

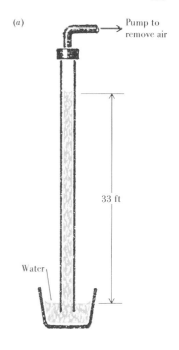

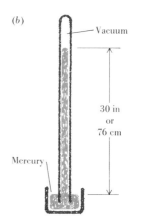

Figure 1.4. Torricellian experiment. (a) Atmospheric pressure using water; (b) atmospheric pressure using mercury

a body (such as mercury in the tube), may cause that body to move. The weight of a body is the force caused by the gravitational attraction between the masses of the body and of the earth. In this case, initially there is more mercury in one arm, and so there is more weight and a greater force directed down in one arm than in the other. This imbalance causes mercury to flow until the levels are equal (Fig. 1.6). When the levels are equal, the gravitational forces in the arms are also equal and balanced. There is no further flow. The weight of the atmosphere pressing on each mercury surface in the symmetric U tube is the same for the two arms, and this also contributes balanced forces.

If the same experiment is done with an asymmetric U tube, the same flow takes place until the levels are equal (Fig. 1.7). Then it is evident that the weight of both the mercury in each arm and the atmosphere pressing on each surface is not the important factor. The arm with the greater cross section has more mercury and also supports a wider column of atmosphere than the more narrow arm. The weight of mercury and the atmosphere is greater in and over the wider arm. The forces are not balanced, and yet there is no flow when the levels are equal.

What is equal is the force per cross-sectional area. If one arm in an asymmetric U tube has twice the cross-sectional area of the other, when the mercury levels are equal that arm has twice the weight of the atmosphere pressing down in and on it as the other arm. But the weight or force per area is the same. *Pressure* is the term for force per area, and that is what must be balanced when the mercury levels are equal.

A *siphon* barometer is made from a U tube by adding a tube, closed at its upper end, to one arm (Fig. 1.5). The pressure on the mercury in the open arm is the weight of the mercury in the open arm down to where the two arms join plus the weight of the atmosphere pressing on the mercury surface divided by the cross-sectional area of that arm. The pressure for the closed arm comes from the mercury weight in that arm divided by that arm's cross section. As in the Torricellian experiment, the space above the mercury in the closed end is empty and has no weight. The difference between the two pressures is that the open arm has the atmospheric pressure and the closed arm has the additional height h of mercury. The atmospheric pressure then equals the pressure of the column of mercury with this height. The siphon barometer can be connected to any vessel in place of the open atmosphere.

Figure 1.5. Siphon barometer. The open end is attached to the container whose pressure is to be measured. In this case, it is the atmosphere.

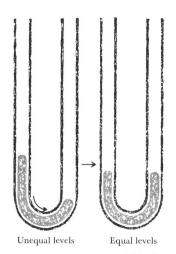

Unequal levels   Equal levels

Figure 1.6. Symmetric U tube.

The pressure of the contents of that vessel is measured by the indicated height h. In this use, the apparatus is called a "manometer" (Fig. 1.8).

There is a third and very common type of barometer, the *aneroid* barometer, made from a compressible and evacuated metal box which is deformed to varying extents by the varying atmospheric pressure squeezing it. The deformation of this aneroid barometer can be measured by the motion of an attached needle or pen on paper.

At sea level and at a latitude of 45° above the equator the average air pressure is about 76 cm of mercury. For this reason, the pressure of exactly 76 cm of mercury is called "one atmosphere" of pressure (1 atm). Mercury expands with an increase in temperature; for example, a column 76.00 cm high at 0°C will be 76.34 cm at 25°C. One atm is defined as 76 cm of mercury at 0°C. At whatever temperature the mercury height in the manometer or barometer is measured, this height is converted to what it should be at 0°C from volume-temperature data. For very precise work, variations in the gravitational force with geographic location must also be considered.

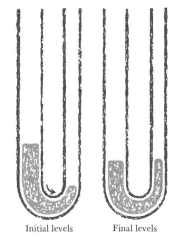

Initial levels   Final levels

Figure 1.7. Asymmetric U tube.

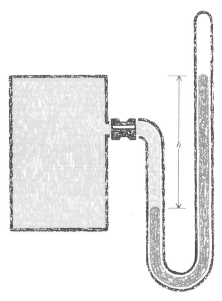

Figure 1.8. The manometer measures the pressure of the gas sample in the square container.

In addition to the units atmosphere and centimeters of mercury (at 0°C), other units are commonly employed. The mass of a column of mercury 76 cm high at 0°C and 1 in.² in cross section is 14.696 lb. The pressure 1 atm is equivalent to 14.696 lb/in.², or 14.696 psi. In honor of Torricelli, a pressure of 1 mm, or 0.1 cm, Hg at 0°C is defined as *1 torr;* 1 atm = 760 torr. Meteorologists report atmospheric pressure in units of bars and millibars (mb). The normal atmosphere, 1 atm, equals 1,013.25 mb.

In instances when it is not convenient to have a mercury manometer, a *Bourdon-type pressure gauge* is often used (Fig. 1.9). The open end of a curved metal tube is exposed to a pressure which tends to straighten the tube, moving a gear and pointer at the closed end. The pressure can be indicated by the pointer and a scale. By connecting the gauge to gas-filled vessels, the pressure of which is measured by a standard mercury manometer, and observing the position of the pointer on the scale, a mark can be made on the scale indicating the gas pressure producing this particular needle position. Repeating this with several known gas pressures, the needle-and-scale gauge is said to be *calibrated* in centimeters of mercury, atmospheres, or pounds per square inch. This gauge can be used to measure gas pressures.

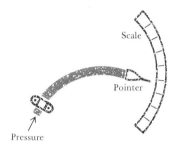

Figure 1.9. Bourdon-type pressure gauge. As the pressure is increased, the curved tube straightens and the pointer is moved along the scale.

> **Example.** A mercury manometer has a reading of 32.06 cm at 0°C. Calculate the pressure in torrs, atmospheres, and pounds per square inch.

$$32.06 \text{ cm Hg} \times \frac{10 \text{ mm Hg}}{1 \text{ cm Hg}} \times \frac{1 \text{ torr}}{1 \text{ mm Hg}} = 320.6 \text{ torr}$$

$$32.06 \text{ cm Hg} \times \frac{1 \text{ atm}}{76.00 \text{ cm Hg}} = 0.4218 \text{ atm}$$

$$32.06 \text{ cm Hg} \times \frac{14.696 \text{ psi}}{76.00 \text{ cm Hg}} = 6.199 \text{ psi}$$

CALCULATIONS

Although almost all the calculations in natural science can be done by ordinary longhand methods, there are a number of mathematical techniques available to reduce considerably the work and time in performing involved calculations. Before we consider two of these labor-saving methods, we shall briefly review some algebraic conventions.

EXPONENTS IN ALGEBRA. If we have an algebraic term b and we multiply it by itself, we can write $b \times b = b^2$. The superscript, or exponent, 2 indicates that the factor b was used twice in the multiplication. Similarly, $b \times b \times b = b^3$. The term b^m indicates that the factor b was used m times in multiplication; b^n indicates that b was used n times. Then $b^m \times b^n$ indicates that the term b was used a total of $m + n$ times: $b^m \times b^n = b^{m+n}$.

Consider the series of terms: b^3, b^2, b, b^0. Each member of the series is obtained from the preceding one by division by b; that is $b^3/b = b^2$; $b^2/b = b$, and $b/b = 1 = b^0$. When we divide b^m by b, the quotient is b^{m-1}. The exponent m is reduced by one because the division reduces the number of times b is used as a multiplication factor by one. Formally, then, we can define $b^0 = 1$.

A negative exponent, for example, b^{-m}, has this significance

$$b^m \times b^{-m} = b^{m-m} = b^0 = 1$$

Therefore

$$b^{-m} = \frac{1}{b^m}$$

and

$$b^m \times b^{-m} = b^m \times \frac{1}{b^m} = \frac{b^m}{b^m} = 1$$

If we multiply b^m by itself c times we have

$$b^m \times b^m \times b^m \times \cdots \times b^m = b^{m+m+m+\cdots+m} = b^{cm}$$

Then

$$(b^m)^c = b^{cm}$$

Fractional exponents have this significance: $b^{\frac{1}{2}} \times b^{\frac{1}{2}} = b^{\frac{1}{2}+\frac{1}{2}} = b^1 = b$. Therefore, $b^{\frac{1}{2}}$ multiplied by itself is equal to b; $b^{\frac{1}{2}}$ is the square root of b. Similarly, $b^{\frac{1}{4}} \times b^{\frac{1}{4}} \times b^{\frac{1}{4}} \times b^{\frac{1}{4}} = b^{\frac{1}{4}+\frac{1}{4}+\frac{1}{4}+\frac{1}{4}} = b^1 = b$. The term $b^{\frac{1}{4}}$, when multiplied four times, gives the product b; $b^{\frac{1}{4}}$ is the fourth root of b.

In summary, the terms of the form b^m, where b is called the "base" and m the "exponent," have certain properties of interest to us. When a set of numbers is written in the form of exponents to a common base, then multiplying these numbers involves adding the exponents, dividing involves subtracting exponents, raising a number to a power c involves multiplying the exponent by c, and extracting the cth root involves dividing the exponent by c.

LOGARITHMS. It is possible to express any positive number in the form of the base 10 raised to some power. We have seen, for example, that 1,000 can be written as 10^3 and that 0.0001 is 10^{-4}. The square root of 10 is 3.1623 to five significant figures, which means that $3.1623 \times 3.1623 = 10.000$. But also $10^{0.5000} \times 10^{0.5000} = 10^{1.0000} = 10$. It follows, then, that $3.1623 = 10^{0.5000}$.

By computation, it can be shown that $2 = 10^{0.3010}$. The number 10 raised to the power 0.3010 is equal to 2. We say that 0.3010 is the logarithm of 2; 0.5000 is the logarithm of 3.1623; 3 is the logarithm of 1,000; -4 is the logarithm of 0.0001. For convenience we write

$$\log 2 = 0.3010$$
$$\log 3.1623 = 0.5000$$
$$\log 1,000 = 3$$
$$\log 0.0001 = -4$$

In multiplying $10^m \times 10^n = 10^{m+n}$, if we let $M = 10^m$ and $N = 10^n$, then $MN = 10^{m+n}$. From our definition of logarithms, m is the logarithm of M, or $\log M = m$; $\log N = n$; $\log MN = m + n = \log M + \log N$. Therefore, *the multiplication of numbers involves adding the logarithms of these numbers.* It is easier to add than to multiply, under most circumstances, and it follows that the use of logarithms reduces the work of multiplication.

If we wish to know the logarithm of 200, we can first write $200 = 2.00 \times 10^2$. Then, $\log 200 = \log 2.00 + \log 10^2$. Since we know that $\log 2.00 = 0.3010$ and, from the definition of logarithm, that $\log 10^2 = 2$, it follows that $\log 200 = 0.3010 + 2 = 2.3010$.

We can write any number as the product of some number between 1 and 10 multiplied by some power of 10. Then, the logarithm of any number is the sum of two terms: the logarithm of a number between 1 and 10 and the logarithm of 10 to some whole-number power. The latter logarithm is always an integer. We require, therefore, a tabulation of the logarithms of numbers which are between 1 and 10. The logarithm of 1 is 0, and the logarithm of 10 is 1; the logarithms tabulated will lie between 0 and 1.

The log table in Appendix A lists two-digit numbers in the left column from 10 to 99. The third digit is read from the columns headed from 0 to 9. Thus, the table lists numbers in the range from 100 to 999. Since it is understood that the numbers are between 1.00 and 9.99, the decimal point is omitted. The logarithms of these numbers range from 0.0000 to 0.9996; again, the decimal point is omitted. The following examples illustrate the use of the logarithm table.

Example.

$\log 20.6 = \log 2.06 + \log 10 = 0.3139 + 1 = 1.3139$

Example.

$\log 760 = \log 7.60 + \log 10^2 = 0.8808 + 2 = 2.8808$

Example.

$\log 0.0013 = 1.3 + \log 10^{-3} = 0.1139 - 3 = -2.8861$

(Either of the last two forms is used.)

Example. Multiply $20.6 \times 760 \times 0.0013$ using logarithms. (The logarithm of the product is the sum of the logarithms of the three factors.)

$$\begin{aligned}
\log 20.6 &= 1.3139 \\
\log 760 &= 2.8808 \\
\log 0.0013 &= 0.1139 - 3 \\
\log \text{product} &= 4.3086 - 3 \\
\log \text{product} &= 0.3086 + 1 \\
\text{product} &= 2.035 \times 10^1 = 20.35 = 20
\end{aligned}$$

After obtaining the logarithm product, which is the logarithm of the product we wish to calculate, we have to find the number whose logarithm is 0.3086; that is, we want to find the antilogarithm of 0.3086. From the table:

$\log 2.03 = 0.3075$
$\log 2.04 = 0.3096$

and since 0.3086 lies midway between 0.3075 and 0.3096, we can estimate that the number lies midway between 2.03 and 2.04; that is, the number is 2.035. This estimation of a number which lies somewhere between those listed in the table is called "interpolation." Finally, 20.35 is rounded off to 20.

For division

$$\frac{M}{N} = 10^{m-n}$$

$$\log \frac{M}{N} = m - n = \log M - \log N$$

Therefore, *division is effected by the subtraction of logarithms.*

> **Example.** Divide 456 by 17.2, using logarithms.
>
> $$\begin{aligned} \log 456 = \quad & 2.6590 \\ -\log 17.2 = & -1.2355 \\ \log \text{quotient} = \quad & 1.4235 = 0.4235 + 1 \\ \text{quotient} = \quad & 26.5 \end{aligned}$$

If we multiply $M \times M \times M = M^3 = 10^m \times 10^m \times 10^m = 10^{3m}$, then $\log M^3 = 3m = 3 \log M$. In general, $\log M^p = p \log M$. Raising a number to a power p involves multiplying the logarithm of that number by p.

> **Example.** Calculate $(9.62)^3$ using logarithms.
>
> $$\begin{aligned} \log (9.62)^3 &= 3 \log 9.62 \\ &= 3 \times 0.9832 \\ &= 2.9496 \\ (9.62)^3 &= 890 \end{aligned}$$

From the definition of fourth root,

$$M^{\frac{1}{4}} \times M^{\frac{1}{4}} \times M^{\frac{1}{4}} \times M^{\frac{1}{4}} = M$$
$$4 \log M^{\frac{1}{4}} = \log M$$
$$\log M^{\frac{1}{4}} = \tfrac{1}{4} \log M$$

In general, $\log M^{1/c} = 1/c \log M$. The cth root of a number can be calculated by dividing the logarithm of that number by c.

> **Example.** On a conventional piano keyboard the note A_3 has 220 vibrations, or cycles per second (cps), and the note A_4 has 440 cps.

On an equally tempered scale with the notes A_3, $A\sharp$, B, C, $C\sharp$, D, $D\sharp$, E, F, $F\sharp$, G, $G\sharp$, and A_4, a simple relation exists among the cycles per second, or frequencies, associated with each note. We can let each note letter represent both the note and the frequency number. A constant factor exists so that

$$A\sharp = A_3 p$$
$$B = A\sharp p = A_3 p^2$$
$$C = B p = A_3 p^3$$
$$C\sharp = C p = A_3 p^4$$
$$\cdots$$
$$A_4 = G\sharp p = A_3 p^{12}$$

The frequency of any note is equal to the frequency of the preceding note multiplied by a factor p. This factor can be calculated using logarithms:

$$A_4 = A_3 p^{12}$$
$$440 = 220 p^{12}$$
$$\frac{440}{220} = p^{12}$$
$$2 = p^{12}$$
$$\log 2 = 12 \log p$$
$$\frac{\log 2}{12} = \log p$$
$$\frac{0.3010}{12} = \log p$$
$$0.02508 = \log p$$
$$p = 1.0594$$

Therefore

$$A\sharp = 1.0594 \ A_3 = (1.0594)(220) = 233.1 \text{ cps}$$
$$B = 1.0594 \ A\sharp = (1.0594)(233.1) = 246.9 \text{ cps}$$
$$C = 1.0594 \ B = (1.0594)(246.9) = 261.6 \text{ cps} \qquad \text{and so on}$$

Using logarithms as a time-saving technique because there are fewer numerical operations involved gives us another advantage: there are fewer opportunities for making mistakes. A series of multiplications and divisions can be performed together without the necessity of writing down intermediate answers.

Example. Calculate by logarithms

$$p = \frac{(0.00100)(0.0821)(298)}{4.78}$$

$$= \frac{(1.00 \times 10^{-3})(8.21 \times 10^{-2})(2.98 \times 10^{2})}{4.78}$$

$$= \frac{(1.00)(8.21)(2.98)}{4.78} \times 10^{-3}$$

$$\log p = \log 1.00 + \log 8.21 + \log 2.98 - \log 4.78 - 3$$
$$= 0 + 0.9143 + 0.4742 - 0.6794 - 3$$
$$= 0.7091 - 3$$

Therefore,

$$p = 5.12 \times 10^{-3}$$

THE SLIDE RULE. A very simple device for adding and subtracting numbers can be made from a pair of ordinary rulers, or rules (Fig. 1.10). Suppose we wish to add $3 + 2$. By placing the two rules as indicated, that is, by placing the beginning of the second rule (the upper rule) at the 3 mark of the first, or lower, rule and then by reading the number 5 on the first rule which is directly below the number 2 of the second rule, we have combined the lengths 3 and 2 to obtain a length of 5. Subtraction can be done by subtracting one length from another (Fig. 1.11). For example, $8 - 5$ can be performed as indicated by subtracting a length 5 marked on the upper rule from the length 8 marked on the lower rule, leaving a length 3 on the lower rule.

Because multiplication can be performed by adding logarithms and division by subtracting logarithms, we can construct rules with logarithmic scales to multiply and divide numbers. The common slide rules are 10 in. long. Since $\log 1 = 0$, the number 1 is

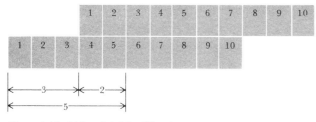

Figure 1.10. Adding $3 + 2$ by slide rule.

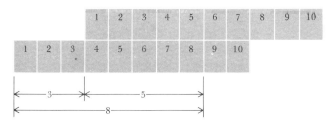

Figure 1.11. Subtracting 5 from 8 by slide rule.

placed at the beginning of the rule, i.e., at the position corresponding to 0 in.; since log 10 = 1, the number 10 is placed at the other end of the rule; since log 2 = 0.3010, the number 2 is placed exactly 3.010 in. from the beginning of the rule. In this manner the entire 10-in. rule is marked, or graduated, as illustrated in Fig. 1.12. Multiplication of 2×3 can be done by adding lengths corresponding to log 2 and log 3 to obtain a length corresponding to log 6 (Fig. 1.13). Division of 9 by 4 can be done by subtracting a length corresponding to log 4 from a length corresponding to log 9 to obtain a length corresponding to log 2.25 (Fig. 1.14).

As with logarithms on paper, it is possible to perform a series of multiplications and divisions without ever writing down the intermediate answers. The slide rule saves time and reduces error, but the common slide rule is limited to three-significant-figure precision.

Figure 1.12. Logarithmic scale

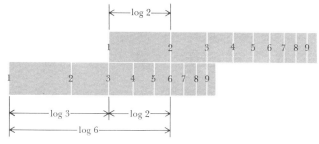

Figure 1.13. Multiplication of 3×2 as addition of log 3 plus log 2 by slide rule.

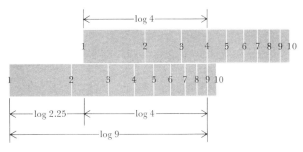

Figure 1.14. Division of 9 by 4 as the subtraction of log 4 from log 9 by slide rule.

PRESENTATION OF DATA IN TABLES

There are several common ways in which scientific data can be presented. To illustrate these we can imagine a simple experiment which will give us a collection of numbers. A round flask with a volume of 10,000 cm³, containing 1.308 g of the gas helium and connected to a gauge which can measure the pressure of the gas within the flask, is placed in a large cylindrical tub. The tub is filled with oil and fitted with a heater to warm the oil, a thermometer to measure the oil and flask temperature, and a stirrer to ensure well-mixed oil

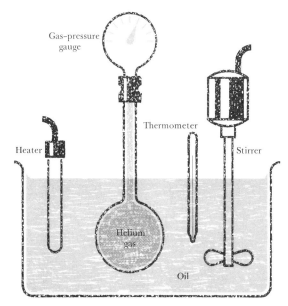

Figure 1.15. Apparatus to study the effect of temperature changes on the pressure of a gas.

Table 1.8. Dependence of Pressure on Temperature of 1.308 g of Helium Confined to a Volume of 10,000 cm.

Temperature t, °C	Pressure p, psi
−25.0	9.78
0.0	10.76
25.0	11.74
50.0	12.73
75.0	13.72
100.0	14.70
125.0	15.68
150.0	16.67

and uniform temperature throughout the tub-and-flask system (Fig. 1.15). As the temperature changes, so does the gas pressure. Table 1.8 lists the temperatures and corresponding gauge pressures.

DATA IN GRAPHS

In addition to a table, a graph may be used to present the data. Figure 1.16 shows two axes: a horizontal axis (along which temperature is marked) and a vertical axis (for pressure). The first point is plotted by drawing a dashed line up from the point on the t axis

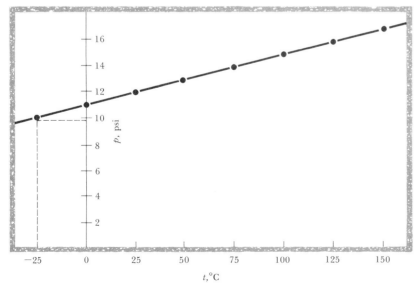

Figure 1.16. Graphic presentation of the pressure dependence of a gas on temperature.

corresponding to −25.0°C and by drawing a dashed line over from the point on the p axis corresponding to 9.78 psi. At the intersection of these two dashed lines the point corresponding to $t = -25.0$°C, $p = 9.78$ psi is marked (for simplicity called the "−25.0°C, 9.78 psi point"). All eight points, in turn, are plotted, and then a line is drawn through these points. By examination this particular graph appears to have points which all lie on a perfectly straight line.

Not all experimental data will form a straight line when plotted on graph paper. The straight-line arrangement of the data is the simplest type of relation between pairs of experimental values, and we say that here the pressure and temperature have a *linear* relationship. If the plotted line is curved, the relationship is said to be *nonlinear*.

DATA AS AN EQUATION

The pressure-temperature relationship can be expressed in a third way by the equation $p = 0.0394t + 10.76$, where t is expressed in degrees Celsius and p in pounds per square inch. This means that if we select a value for the temperature, multiply it by 0.0394, and add 10.76, the calculated result is the pressure.

Example. Calculate the pressure at $t = 75.0$°C.

$$0.0394 \times 75.0 = \quad 2.96$$
$$+10.76$$
$$\text{Pressure} = \quad 13.72$$

Both the graphical and equation form of data presentation have an advantage over the original tabular presentation in that pressures can be read from the graph or calculated from the equation for temperatures at which there was no direct pressure measurement. For example, if we wish to know the pressure at $t = 37.2$°C, either the graph or the equation will give us this information. We are able to obtain nontabulated values by interpolation.

Where the numbers 0.0394 and 10.76 come from can be seen by looking at the general equation for a straight line, using the variables x and y:

$$y = mx + b$$

which has exactly the same form as

$$p = 0.0394t + 10.76$$

in that there are two variables in each—(x,y) and (t,p)—which are related in the same manner. One variable (x in the former equation and t in the latter) is multiplied by a number (m and 0.0394, respectively), and then to this product is added a constant number (b and 10.76, respectively).

Consider the set of equations,

$$y = \tfrac{1}{2}x$$
$$y = x$$
$$y = 2x$$

that is, the set of linear equations where $b = 0$ and $m = \tfrac{1}{2}$, 1, and 2. These are plotted in Fig. 1.17. The three lines go through the origin, i.e., through the point (0, 0). The greater the value of m, the steeper the line. The number m, which is the constant multiplying x, is said to be the *slope* of the line $y = mx$.

Next, consider the set of equations:

$$y = x - 2$$
$$y = x$$
$$y = x + 2$$

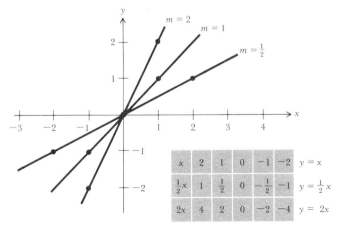

x	2	1	0	-1	-2	$y = x$
$\tfrac{1}{2}x$	1	$\tfrac{1}{2}$	0	$-\tfrac{1}{2}$	-1	$y = \tfrac{1}{2}x$
$2x$	4	2	0	-2	-4	$y = 2x$

Figure 1.17. Plot of $y = mx$. A set or family of straight lines all going through the origin but with different slopes m.

that is, where $m = 1$ and $b = -2$, 0, and $+2$. These lines can be plotted as shown in Fig. 1.18. These three lines are parallel and differ only where they cross the y axis. At the y axis, $x = 0$; and so in $y = mx + b$, $y = m \times 0 + b = b$. Therefore, b is the y-axis *intercept* of the equation $y = mx + b$.

Looking at the tabulated data in Fig. 1.16, we see that at $t = 0.0°C$, $p = 10.76$ psi, and so 10.76 is the p-axis intercept. The slope is calculated this way. It must have the units pounds per square inch per degree Celsius (psi/°C) since when it is multiplied by t in degrees Celsius, the product will have the units pound per square inch to be added to the p-axis intercept 10.76 psi to give the sum, also with the units pounds per square inch. The slope is a measure of how fast the pressure increases with an increase in temperature and has the units pounds per square inch per degree Celsius. When the temperature is increased from 0.0 to 100.0°C, the measured pressure is increased from 10.76 to 14.70 psi. There is a pressure increase of 3.94 psi accompanying a temperature increase of 100.0°C. This ratio, the slope

$$m = 3.94 \text{ psi}/100.0°C$$
$$= 0.0394 \text{ psi}/°C$$

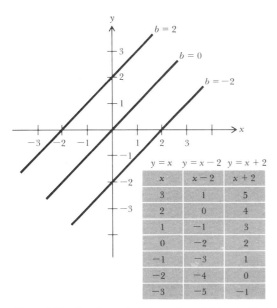

Figure 1.18. Plot of $y = x + b$. A set of parallel lines (with the same slope) but intercepting the axes at different points.

Since we know the intercept and the slope, we can write

$p = mt + b$
$p = 0.0394t + 10.76$

PROBLEMS

1.1. Round off these numbers to one *less* significant figure

    18.75   15327   0.00365
     9.004   74.9

1.2. Round off these numbers to one *less* significant figure

    14.92   467.95   38.301
     9.006   0.00734

1.3. Write each number of Prob. 1.1 in the standard scientific notation.

1.4. Write each number of Prob. 1.2 in the standard scientific notation.

1.5. What are the logarithms of the numbers in Prob. 1.1?

1.6. What are the logarithms of the numbers in Prob. 1.2?

1.7. A man weighs himself four times in one day on a scale that can be read to a tenth of a pound. The four weights he records are 152.4, 152.0, 151.8, and 152.6 lb. Using another scale of the same type, he finds that during the same day he weighs 151.6, 152.9, 153.1, and 152.0 lb. For each scale calculate the average mass and the average deviation. Which average is more precise?

1.8. A race horse runs around a 1-mile oval track. Three judges time this event using stopwatches. Their measured times are 1 min, 40.5 s; 1 min, 39.9 s; and 1 min, 40.2 s. On the next day the event is repeated, and the recorded times are 1 min, 40.2 s; 1 min, 39.7 s; and 1 min, 40.7 s. Calculate the average time for the mile run each day and the average deviations, and compare the precisions of each average time.

1.9. Using the table of troy weights, how many tablets each containing 5 grains of aspirin can be made from 4 troy of aspirin?

1.10. The importance of the analytical balance in investigation is hard to exaggerate. Consult a dictionary to learn the etymology of the word "examine."

1.11. There is about 1 grain of quinine in each bottle of quinine tonic. How many bottles can be prepared with 5 troy oz of quinine?

1.12. You are asked to build a cube-shaped box with a volume of 100.0 m^3. The volume of a cube is $V = s \times s \times s = s^3$, where s is the length of an edge. How long should each edge of the box be?

1.13. To construct a cube-shaped box containing 10.0 m^3 of space, how long should each box edge be?

1.14. An investment of 1 dollar at 4 percent per annum will be worth $1.04 after 1 yr; $(1.04)^2$, or $1.0816, after 2 yr; $(1.04)^3$, or $1.1249, at the end of 3 yr; and $(1.04)^n$ at the end of n years. After how many years will the investment be worth 2 dollars?

1.15. How long will it take an investment to double in value at 6 percent annual interest?

1.16. Calculate the number of seconds in 4 billion yr to three significant figures.

1.17. If light travels at 186,000 mi/s, how far can light travel in 4 billion yr?

1.18. If a typical grain of sand occupies a volume of $5 \times 10^{-4} \text{ cm}^3$ and if a beach is 1.6 km long, 0.4 km wide, and 12 cm deep, estimate how many grains of sand there are in the beach.

1.19. A competent human being, using pencil and paper, can correctly add 2 one-digit numbers (for example, $2 + 5 = 7$) at the rate of 100 additions/min. A modern high-speed computer can perform the same additions at the rate of 20 million/s. Almost all numerical problems are a collection of such simple additions. If a computer takes 1 min to do a particular problem, how long a time is required for one person to do the same problem using pencil and paper, working 8 hr a day and five days a week? If the problem must be done in one week, how many men must be employed if each works 8 hr a day?

The Concept of Atoms

We are now ready to look at one of the most successful notions in all of science—the model of matter as being composed of very small particles.

Democritus [460–362(?) B.C.] imagined that if we cut a coin into two pieces and cut either of these two pieces into two pieces, and so on, we will eventually come to a piece so small that it can no longer be further divided. He called this ultimate particle an "atom." Aristotle (384–322 B.C.), on the other hand, imagined that the cutting and dividing of particles could continue endlessly. We shall examine the experimental evidence which leads us to believe there are ultimate units called "atoms" that exist in our universe.

PROPERTIES OF SUBSTANCES

We can distinguish between one substance and another because each substance has its own unique set of characteristics, or properties. Sugar is sweet, and table salt is salty. At 0°C, one can dissolve about 180 g of sugar in 100 g of water but only about 36 g of salt in the same amount of water. Salt melts at 801°C into a clear colorless liquid; sugar melts at 186°C but turns into a dark brown substance, caramel.

A very important distinguishing property of a substance is its *density*, the mass contained in a unit volume of that substance. Density commonly has the units of grams per cubic centimeter (g/cm^3). A cubic centimeter is the volume of a cube 1 cm along each edge. Sugar has a density of 1.59 g/cm^3, and salt a density of 2.16 g/cm^3. In comparison, air has a density of $1.2 \times 10^{-3}\ g/cm^3$, and osmium (a metal and the most dense substance found naturally on earth) has a density of 22.48 g/cm^3. Water at 3.98°C has a density of

Table 2.1. Identifying Properties

Substance	Taste	Density, g/cm³	Melting point, °C	Solubility in 100 g of water at 0°C
Sugar	Sweet	1.59	186 (decomposes)	180
Table salt	Salty	2.16	801	36
Water	None	1.000	0	—

1.000 g/cm³. This is not surprising since the gram was originally defined as the mass of that volume of water.

Table 2.1 lists some common substances and a few of their characteristic properties. None of these properties depends on how much of the substance there is. One tiny grain or one very large piece of salt tastes salty; each has exactly the same density and melting point.

CHANGES IN SUBSTANCES

When table salt is melted and then cooled or frozen, the new solid has exactly the same set of properties as did the starting material: it is table salt. Melting and freezing does not alter the salt. However, when sugar melts, it turns into a brown liquid and remains brown when refrozen. This brown solid does not have the properties of sugar. It is not sugar; it is caramel. If the heating process is continued, the molten caramel will be transformed into a black solid substance, carbon. Carbon melts at a very high temperature (about 3650°C) and also burns in the presence of air until it is completely consumed.

The changes undergone by sugar are more drastic than those undergone by table salt. We refer to the changes illustrated by the conversion of sugar to caramel, then to carbon, and then to gaseous combustion products as "chemical changes"; changes such as simple melting and freezing are called "physical changes." However, the distinction between these changes is not always clear.

MIXTURES, COMPOUNDS, AND ELEMENTS

A *mixture* is a combination of substances which can be separated into its constituent substances by some physical method. Seawater is a *solution*, or clear mixture, of many substances and water. If seawater is cooled below its freezing point, a solid—pure water—initially separates out and can be removed from the mixture. A mixture of iron and sand can be separated by a magnet which will pick up iron and not sand.

Substances which cannot be further separated into constituents by physical means but which can be broken up into two or more simpler substances by some chemical means are said to be *compounds.* Water cannot be separated into constituents by ordinary physical means, but if two electrodes are placed in water in which a little salt is dissolved and an electric current is passed through the solution, gas is evolved at each electrode (Fig. 2.1); one gas is hydrogen, and the other is oxygen. If table salt is melted and an electric current is passed through it in the same way, a shiny liquid collects at one electrode and a green gas is evolved at the other; these are the metal sodium and the gas chlorine, respectively. Both water and table salt are compounds.

Elements cannot be further broken up by chemical means, nor can they be transformed into other elements by chemical means. Examples of elements are hydrogen, oxygen, carbon, sodium, and chlorine. Over one hundred elements are known; most of these are found in nature, although some elements are produced artificially. Table 2.2 lists the 12 most common elements, the chemical symbol for each, and the percent by weight of each in the earth's crust, in the oceans, and in the combined crust, oceans, and atmosphere. Although some of the symbols have a clear association with the English name of the element, for example, H for hy-

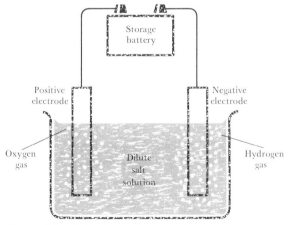

Figure 2.1. Separation of water into gaseous components—hydrogen and oxygen.

Table 2.2. Abundance of Elements

Element	Symbol	Weight percent		
		Crust†	Ocean	Crust, ocean, and atmosphere
Oxygen	O	46.71	85.79	49.52
Silicon	Si	27.69		25.75
Aluminum	Al	8.07		7.51
Iron	Fe	5.05		4.70
Calcium	Ca	3.65	0.05	3.39
Sodium	Na	2.75	1.14	2.64
Potassium	K	2.58	0.04	2.40
Magnesium	Mg	2.08	0.14	1.94
Hydrogen	H	0.14	10.67	0.88
Titanium	Ti	0.62		0.58
Chlorine	Cl	0.045	2.07	0.188
Phosphorus	P	0.13		0.12
Totals		99.52	99.90	99.62

† Percent of the element to a depth of approximately 10 miles.

drogen, in other cases, especially for elements which have been known for a long time, the symbol is related to the Latin name, for example, Fe from *ferrum* (iron), Na from *natrium* (sodium), and K from *kalium* (potassium).

MASS AND CHEMICAL PROCESSES

As long ago as about A.D. 800 Jabir ibn Hayyan of Baghdad was reported to have measured an increase in mass when a metal, heated in air, turns into a calx (a metal oxide, that is, a chemical combination of oxygen and metal atoms). Jean Rey (1630) observed, too, that the calx is heavier than the original metal and concluded that this additional mass came from the air. He found that air is not taken up endlessly by the metal but, rather, there is a limit to how much calx is formed from a given amount of original metal.

    Antoine Lavoisier (1734–1794), using a very sensitive analytical balance, was able to investigate with high precision the masses of substances involved in reactions. For example, he sealed both tin metal and air in a closed vessel and determined the mass of the vessel and its contents. Upon heating, the metal turned into a calx. Then the vessel and contents were reweighed. Lavoisier could detect no difference between the two masses. What was gained by the metal was lost by the air.

Reactions involving many substances can be investigated using a Landolt tube, a Y-shaped closed vessel. Two substances which are known to react chemically are placed in the Landolt tube, one in each leg, and the mass of the two substances and the tube is determined (Fig. 2.2). Then the Landolt tube is inverted, and the two substances are allowed to react (a chemical reaction is usually accompanied by a temperature change); next the mass of the Landolt tube and its contents is again determined.

All the evidence points to the fact that there are no measurable differences in the mass of the tube and contents before and after the chemical reaction. The total mass of the *reactants* (the substances which participate in and are used up in a reaction) is equal to the total mass of the *products* (the substances formed). In a chemical reaction, reactants are converted to products but there is no change in the total mass. The same kind of sealed-vessel experiment can be done with physical changes such as freezing, melting, and evaporating. Again, there is no measurable change in mass attending a change from solid to liquid to gas.

These experiments suggested the *law of conservation of mass:* matter cannot be created nor destroyed. We shall see later that there is an exception both to this and to the *law of conservation of energy.* Mass can be converted to an equivalent amount of energy, and energy to an equivalent mass; more correctly, mass and energy, together, are conserved. But under most ordinary circumstances, mass changes are so slight that they escape detection.

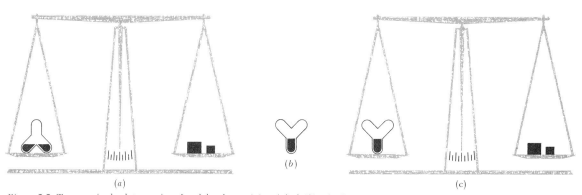

Figure 2.2. Two unmixed substances in a Landolt tube are (a) weighed, (b) mixed, reacted, and (c) reweighed. There is no change in mass.

THE COMPOSITION LAWS
FOR PURE SUBSTANCES

Next, Lavoisier constructed some equipment like that shown in Fig. 2.3. He placed pure mercury in the round flask and estimated the total air volume in the flask and bell jar. This air was confined and separated from the atmosphere by a pool of mercury at the base of the bell jar. He heated the flask for 12 days and observed two things: a red calx formed on the pure mercury in the flask, and the air volume was reduced by about one-sixth. The remaining air did not permit a candle to burn or small animals to live. Lavoisier called this remaining air "azotic," or "mephitic." It supports neither combustion nor respiration.

Lavoisier isolated the red calx, heated it to a higher temperature than used before in its preparation, and collected a gas which was given off as the red calx was converted back into mercury. The volume of gas given off was equal to the air-volume contraction when the mercury was converted to calx. This gas caused a candle to burn more rapidly than in air and also supported respiration.

This gas had exactly the same properties as a gas prepared independently by Joseph Priestley (1733–1804) in England and Carl Scheele (1742–1786) in Sweden. Lavoisier was the first to recognize it as an element and named it "oxygen." This name derives from the Greek word for acid-producer since the union of it with elements such as sulfur and phosphorus, when dissolved in water, produced *acids*, substances which taste sour and dissolve metal. From this work Lavoisier concluded that air comprises two gases, oxygen and azote (the latter gas is now called "nitrogen"). He also

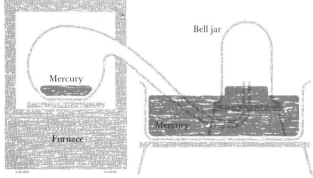

Figure 2.3. Lavoisier's experiment.

Table 2.3. The Composition of Three Calxes (Metal + Oxygen → Calx)

Original metal	Metal mass, g	Calx mass, g	Oxygen mass, g	Composition of calx % Metal	% Oxygen	Calx/metal-mass ratio
Mercury	200.6	216.6	16.0	92.61	7.39	1.080
Tin	118.7	150.7	32.0	78.8	21.2	1.270
Lead	207.2	223.2	16.0	92.83	7.17	1.077

described *combustion* and *calcination* (the formation of calx) as the union of a substance with oxygen.

Lavoisier saw that the quotient obtained by dividing the mass of the formed calx by the mass of consumed mercury was a constant number independent of how much pure mercury he started with; each time he made mercury calx, the quotient was the same. With the metal and calx of tin, the quotient had another constant value; with lead, still another constant value was obtained. Since he knew the original metal mass, the formed calx mass, and that the mass increase was due to union with oxygen, he concluded that the difference between the calx and metal was the oxygen mass. Thus the mass composition of a calx can be calculated. Examples of the composition of the calxes of three substances are given in Table 2.3, using metric mass units. In the first case the fraction of the calx which is metal is

$$\frac{200.6 \text{ g mercury}}{216.6 \text{ g calx}} = 0.9261 = 92.61\%$$

and the fraction of oxygen is

$$\frac{16.0}{216.6} = 0.0739 = 7.39\%$$

The sum of $92.61 + 7.39$ percent is 100.00 percent.

This constancy of composition is true for all substances, and Lavoisier suggested that each substance has its own characteristic set of elements, each of which is present in a definite mass percent peculiar to that substance. This is the *law of definite proportions*, or the *law of constant composition*, which was first stated by one of Lavoisier's successors, Joseph Louis Proust (1755–1826).

Jean Servais Stas (1813–1891) of Belgium established the validity of this law by preparing with great care and with very pure starting materials a number of compounds by several methods and analyzing for the mass fraction of each element. For example, he made a compound of silver and chlorine by four separate methods and found that for each 100 g of pure silver, the following masses of the compound were obtained:

132.8423 g
132.8475 g
132.848 g
132.842 g

It can be seen that, with a precision better than 1 part in 10,000, the compound had a constant composition.

For his work in chemistry, especially his careful use of mass determination as a tool of investigation, Lavoisier is often called the "father of modern chemistry."

THE LAW OF MULTIPLE PROPORTIONS

Although each compound has a definite element-mass composition, it frequently happens that a set of elements can form more than one compound. The composition of anything is defined both by a list of its components and by how much of each component there is. The determination of which elements and of how much of each are contained in a compound is called the "chemical analysis" of the compound. Several examples will illustrate this procedure.

> **Example.** Both water and hydrogen peroxide contain the two elements hydrogen and oxygen. Water and hydrogen peroxide are distinct compounds because they have different properties. Each can be analyzed by appropriate methods for its mass percent of the two elements and tabulated as in Table 2.4. (The mass percent is simply the number of grams of each element in 100 g of the

Table 2.4

Substance	g in 100 g of substance	
	Mass % H	Mass % O
Water	11.19	88.81
Hydrogen peroxide	5.93	94.07

Table 2.5

Substance	g in 100 g of substance		g O/g H	Ratio
	Mass % H	Mass % O		
Water	11.19	88.81	7.937	1.00
Hydrogen peroxide	5.93	94.07	15.9	2.00

substance.) To make water, 88.81 g of oxygen combine with 11.19 g of hydrogen; to make hydrogen peroxide, 94.07 g of oxygen and 5.93 g of hydrogen are needed.

Another way to compare the composition of water and hydrogen peroxide is to calculate the number of grams of oxygen combining with 1 g of hydrogen in each compound. Table 2.5 contains this calculation and this comparison. The values obtained by dividing the mass of oxygen by the mass of hydrogen seem to have a simple relationship; 15.9 is about twice as big as 7.937. (To confirm this, divide each by the smaller of the two, or 7.937. This gives you the ratio of the oxygen to the hydrogen in the two compounds, which we have placed in the last column of the table. This ratio is simply a comparison of these two compounds. The mass of oxygen per unit mass of hydrogen in the two compounds is in the ratio of 1 : 2.) The conclusion is that there are twice as many grams of oxygen per gram of hydrogen in hydrogen peroxide as there are in water.

Let's work out a second example.

Example. Barium can form two oxides with different oxygen contents, barium oxide and barium peroxide (which has a greater amount of oxygen). The masses of barium per unit mass of oxygen are related by a simple numerical ratio: there are twice as many grams of barium per gram of oxygen in the oxide as there are in the peroxide. Table 2.6 lists the mass percents of barium and oxygen which were obtained by analysis. From these numbers we can calculate the mass of oxygen per unit mass of barium, that is, g Ba/g O. Dividing these numbers by the smaller of the two gives the simple whole-number ratio, in this case 2.000 : 1.000.

Table 2.6

Substance	Mass % Ba	Mass % O	g Ba/g O	Ratio
Barium oxide	89.57	10.43	8.588	2.000
Barium peroxide	81.11	18.89	4.294	1.000

HYDROCARBONS. Compounds containing only the two elements carbon and hydrogen are called "hydrocarbons." The analysis of a hydrocarbon is made by burning it in an excess of oxygen, converting it completely to carbon dioxide and water. As in Fig. 2.4, a weighed sample of the hydrocarbon is placed in a furnace and burned in a stream of oxygen gas to give carbon dioxide and water. These resultant gases are then passed through a tube containing a substance which quantitatively (that is, completely) removes the water and then through a second tube which quantitatively removes the carbon dioxide. The mass of these tubes can be determined before and after the combustion so that the mass of water and carbon dioxide can be determined precisely.

> *Example.* We know that when 12 g of carbon are completely burned in oxygen to carbon dioxide, the mass of the carbon dioxide formed is 44 g. Using these values, we can calculate from the mass of the carbon dioxide produced how many grams of carbon were contained in the original hydrocarbon. Similarly, from the observation that 2.015 g of hydrogen produces 18.015 g of water, we can calculate how many grams of hydrogen were in the original hydrocarbon.
>
> In Table 2.7, using five hydrocarbons as examples, we have listed the mass percent of the two components C and H for all five hydrocarbons. As in previous examples, the ratio of carbon to hydrogen is obtained by dividing the mass of C by the mass of H, as shown in column (4). To calculate the lowest ratio, divide each

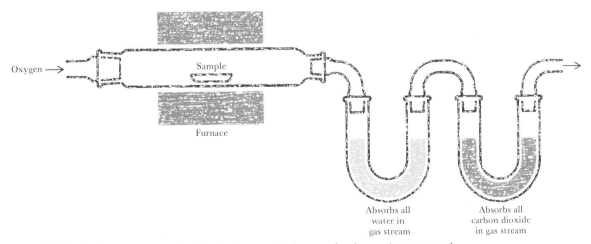

Figure 2.4. Combustion apparatus used to determine how much hydrogen and carbon are in a compound.

Table 2.7. Analyses of Five Hydrocarbons

(1) Substance	(2) %C	(3) %H	(4) g C/g H	(5) Ratio	(6) Integer
Methane	74.87	25.13	2.979	$1.000 \times 3 =$	3
Ethane	79.89	20.11	3.972	$1.333 \times 3 =$	4
Ethylene	85.63	14.37	5.959	$2.000 \times 3 =$	6
Acetylene	92.26	7.74	11.92	$4.001 \times 3 =$	12
Benzene	92.26	7.74	11.92	$4.001 \times 3 =$	12

number in column (4) by the smallest number in the column, in this case 2.979. The ratios in column (5) may all be rounded off to whole numbers, except for 1.333. To make this quantity an integer, you must multiply by 3 to give 3.999, which can be rounded off to 4. In the last column the smallest whole-number integer is shown.

The numbers in column (4) are proportional to those in column (6); that is, the grams of carbon per gram of hydrogen in the several hydrocarbons are in the ratio of the integers 3, 4, 6, 12, and 12. Acetylene and benzene have exactly the same chemical analysis. This does not mean that they are the same compound chemically or physically since, for example, acetylene is a gas and benzene is a liquid at room temperature.

Examining compounds of hydrogen and carbon, John Dalton (1766–1844) noticed these simple whole-number ratios and in 1804 first stated the *law of multiple proportions:* When two elements *A* and *B* combine to form a series of compounds, the masses of element *A* per unit mass of element *B* are in the ratios of small whole numbers.

THE ATOMIC THEORY

At 20°C it is possible to dissolve as much as 203.9 g of table sugar in 100.0 g of water, but no more than that. Any table sugar added in excess of 203.9 g will settle out of the liquid, that is, it will not dissolve. A liquid that has dissolved in it as much of a substance as possible at a given temperature is said to be *saturated* with that substance. Because it is possible to dissolve any amount of sugar in any amount of water under the saturation level, there is no limit to the *number* of solutions of different compositions we can make from sugar and water.

Table 2.8. Analyses of Five Hydrocarbons

(1) Substance	(2) g C/g H	(3) Ratio	(4) Unit of carbon
Methane	2.979	3	$2.979 \div 3 = 0.993$
Ethane	3.972	4	$3.972 \div 4 = 0.993$
Ethylene	5.959	6	$5.959 \div 6 = 0.993$
Acetylene	11.92	12	$11.92 \div 12 = 0.993$
Benzene	11.92	12	$11.92 \div 12 = 0.993$

Looking again at Table 2.7 (analyses of five hydrocarbons), we can see that the mass of carbon combined with a unit mass (such as a gram) of hydrogen does not assume any undefined ratio but that carbon combines with hydrogen in some regular manner. We can elaborate on this regularity (Table 2.8). The carbon/hydrogen mass ratios (g C/g H) and the whole-number ratios for each hydrocarbon are repeated from Table 2.7. If we divide the former by the latter, a common quotient 0.993 is obtained. It can be seen, then, that there is a basic unit, 0.993 g of carbon; methane has three ($3 \times 0.993 = 2.979$), ethane four, ethylene six, and acetylene and benzene each twelve of these basic carbon units. The same regular pattern described by the *law of multiple proportions* occurs in any series of compounds made from a given set of elements.

If carbon can be added to a fixed amount of hydrogen in *only* definite units, it is tempting for scientists to imagine that carbon is composed of discrete units, i.e., atoms. Since we can also calculate the grams of hydrogen combining with 1 g of carbon and see that the law of multiple proportions holds here as well, we can imagine hydrogen, too, as being composed of atoms. All elements, in the same way, can be so imagined.

Since the composition of a compound is independent of the amount of that compound analyzed, the law of multiple proportions is true for very small samples of compounds. Atoms, if they exist, must also be very small.

DALTON'S THEORY

Dalton, in 1803, gave the atomic theory of Democritus and others a specific form by postulating:

1. For each element there are real, distinct, and immutable atoms.
2. Atoms are indestructible.

3. For each element, the atoms are indistinguishable from one another.
4. For each element there is a unique type of atom.
5. The combining proportions mentioned in the laws of definite and multiple proportions are related to the relative masses of atoms.

Using Dalton's postulated atoms and symbols (Table 2.9), this hypothesis can be used to describe the three laws: the law of conservation of mass, the law of constant composition, and the law of multiple proportions.

DALTON'S MODEL FOR WATER. Water is formed by the combination of hydrogen and oxygen. Illustrating this by using Dalton's symbols, a single Dalton atom each of hydrogen and oxygen combining to form a single compound-atom (in our terms, molecule) of water would be depicted as follows:

$$\odot + \bigcirc \longrightarrow \odot\bigcirc$$

If 10 of the hydrogen atoms and 10 of the oxygen atoms combine, the resultant illustration would be

$$\text{(10 hydrogen atoms)} + \text{(10 oxygen atoms)} \longrightarrow \text{(10 water compound-atoms)}$$

No matter how many atoms are involved, we can see two things. First, the formation of water is described as the union of hydrogen and oxygen atoms. There is a constant number of hydrogen and oxygen atoms before and after the formation of water. Atoms are not created nor do they vanish. Therefore, there is no change in mass before and after the reaction; mass is conserved.

Second, each Dalton compound-atom of water has exactly one hydrogen atom and one oxygen atom. No matter how many or few water compound-atoms we examine, the number of hydrogen atoms will equal the number of oxygen atoms. Since all hydrogen atoms are of the same mass, and all oxygen atoms have the same mass, the mass fractions of hydrogen and oxygen in water are always constant, independent of how many water compound-atoms are analyzed. *The composition of water is constant.* In Dalton's model, a

Table 2.9. Dalton's Symbols

Substance	Symbol
Hydrogen	⊙
Oxygen	◯
Nitrogen	⊘
Carbon	●
Sulfur	⊕
Phosphorus	⊗
Copper	Ⓒ
Lead	Ⓛ
Water	⊙◯
Ammonia	⊙⊘
Olefiant gas	⊙●
Marsh gas	⊙●⊙
Carbonic oxide	●◯
Carbonic acid	◯●◯

chemical reaction is the reorganization of atoms into new associations or groupings.

DALTON'S MODELS FOR HYDROCARBONS. Marsh gas, which results from the decomposition of vegetable matter in the absence of air, was first obtained in 1776 by Volta, who collected the gas released when mud near lake shores was disturbed. Olefiant gas was produced in 1794 by treating grain alcohol, the product of grain fermentation, with sulfuric acid. These two gases are known as methane and ethylene, respectively, and their analyses are found in Table 2.7.

Since ethylene has 5.959 g of carbon for each gram of hydrogen, compared to the 2.979 g of carbon in methane, and because 5.959 is about twice 2.979, Dalton depicted these compound atoms as

 Olefiant gas (ethylene)

 Marsh gas (methane)

This is the simplest picture which describes the fact that ethylene has twice as much carbon per unit of hydrogen as does methane. (On the basis of later work, using Dalton's theory, it will be seen that ethylene actually contains two carbon atoms and four hydrogen atoms, and methane contains one carbon atom and four hydrogen atoms.)

For each atom of hydrogen, ethylene has one carbon atom and methane has one-half a carbon atom. The numbers of carbon atoms 1 and $\frac{1}{2}$ are in the ratio 2:1; so must be the ratio of the masses of carbon per gram of hydrogen. Thus, the law of multiple proportions holds for these two hydrocarbons.

It is frequently the case that an algebraic representation is helpful in stating an idea. Since, in Dalton's view, each kind of atom

Table 2.10. The Law of Multiple Proportions

| Hydrocarbon | In each compound-atom | | | | Total mass | | |
	C atoms	H atoms	Mass C	Mass H	C	H	g C/g H
First	a	b	m_C	m_H	am_C	bm_H	$\dfrac{am_C}{bm_H}$
Second	p	q	m_C	m_H	pm_C	qm_H	$\dfrac{pm_C}{qm_H}$

has a special mass, we can define

m_H = the mass of a single hydrogen atom
m_C = the mass of a single carbon atom

If we look at two hydrocarbons and suppose that the first compound-atom hydrocarbon has a carbon and b hydrogen atoms and that the second compound-atom hydrocarbon has p carbon and q hydrogen atoms, then we can compute the grams of carbon per gram of hydrogen in each, as follows.

As shown in Table 2.10, each compound-atom of the first hydrocarbon has a carbon atoms, each of which has a mass m_C. Thus, the total carbon mass in the first hydrocarbon is $a \times m_C$, or am_C. Similarly, the other terms bm_H, pm_C, and qm_H are calculated. The last column is simply the ratio of the carbon/hydrogen masses, or g C/g H. This ratio is independent of the quantity of the hydrocarbon that is analyzed. The two ratios

$$\frac{am_C}{bm_H} \quad \text{and} \quad \frac{pm_C}{qm_H}$$

are themselves in the ratio $a/b : p/q$ or

$$\frac{a/b}{p/q} = \frac{aq}{pb}$$

Since a, b, p, and q are whole numbers, then aq/pb or $aq:pb$ is also the ratio of whole numbers, as described in the law of multiple proportions.

THE SUCCESS OF DALTON'S THEORY

Because Dalton's hypothesis so elegantly and successfully describes the three experimental laws, it is called a "theory." The model of matter as being composed of atoms, as described by Dalton, is accepted as a real representation of our universe because it is successful in describing and predicting natural phenomena: we believe atoms to be real.

From the analysis of water, there are 7.937 g of oxygen for each 1.000 g of hydrogen. This must be the ratio of the masses of the individual atoms in Dalton's compound-atom for water ⊙◯.

Therefore, it is possible to determine the *relative* masses of individual atoms from the analysis of a large collection of these.

Although Dalton's theory was further developed and refined, the essential features remain and John Dalton is considered to be the author of the modern atomic theory. Dalton's symbols are too inconvenient for complicated chemistry, and in 1811 J. J. Berzelius introduced the use of one- or two-letter symbols. A table of these symbols appears on the inside of the front cover of this book.

THE BEHAVIOR OF GASES

We know two things about how gases behave. First, from Boyle's investigation (Fig. 1.1, Table 2.11), for a gas at constant temperature, when the pressure p of the gas is changed, there is an accompanying volume V change such that the product of pressure times volume is constant:

$pV = $ constant

Second, if a gas is confined within a constant volume vessel (Fig. 1.15), the pressure increases *linearly* with an increase in tem-

Table 2.11. Robert Boyle's Data for the Volume of Air†

Air volume, in., V	Confining pressure, in. Hg, p	Product, pV
12	29.12	349.5
11	31.94	351.3
10	35.31	353.1
9	39.31	353.8
8	44.19	353.5
7	50.31	352.2
6	58.81	352.9
5	70.69	353.4
4	87.88	351.5
3	117.56	352.7

† Boyle estimated the air volume by the length of the air space along the shorter leg of this J-shaped tube (Fig. 1-1). The pressure is the length of the mercury column in the longer leg plus the pressure of the atmosphere above the open, longer tube (estimated by Boyle to be equivalent to 29.12 in. of mercury). The product is seen to be rather constant. Although Boyle measured his lengths in eighths and sixteenths of an inch, the more convenient decimal equivalents are used to calculate this product.

perature (Fig. 1.16); that is, when pressure is plotted against temperature, the resulting graph is a straight line. The equation which describes the pressure-temperature relationship for a particular amount of helium in our example is

$$p = 0.0394t + 10.76$$

where the pressure p is measured in pounds per square inch (psi) and the temperature t is measured in degrees centigrade (°C).

Since the pressure becomes smaller as the temperature decreases, it is of interest to see at what temperature the pressure can be expected to be zero. We can determine this by using either the graphical or the equation representation for the dependence of pressure on temperature. The temperature range (page 31) is from −25.0 to +150.0°C, and the pressure is 9.78 to 16.67 psi. It is evident that the pressure will be zero at some temperature below −25.0°C, that is, below the lowest temperature at which a measurement was made. The segment between −25.0 and +150.0°C, comprising experimental data, is straight. If we can assume that the behavior at temperatures below −25.0°C can also be described by the same straight line, then we can extend the line segment to zero, as indicated by the dashed line in Fig. 2.5a. Extending the line beyond the experimental values is called "extrapolation." The straight line hits the horizontal axis at −273°C; for $p = 0$, the extrapolated temperature is −273°C.

As a matter of convenience we can redefine the temperature scale so that $T = 273 + t$, indicated in Fig. 2.5b. Then the line segment and its extrapolated extension goes right through the new origin ($p = 0$, $T = 0$). From the equation

$$p = 0.0394t + 10.76$$

the same information can be had. To find the temperature at which $p = 0$

$$0 = 0.0394t + 10.76$$
$$0.0394t = -10.76$$
$$t = \frac{-10.76}{0.0394} = -273°C$$

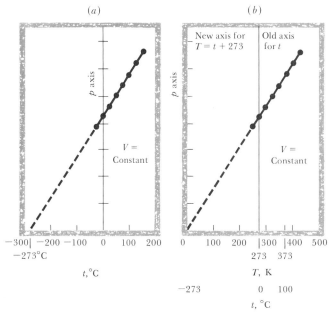

Figure 2.5. Effect of temperature on the pressure of a gas when the volume is constant. The straight line through the experimental points is extended or extrapolated to −273°C, where the extrapolated pressure is zero. The extrapolated portion is shown as a broken line.

Now, if we let

$$T = t + 273$$
$$t = T - 273$$
$$p = 0.0394(T - 273) + 10.76$$
$$= 0.0394T - 10.76 + 10.76$$
$$= 0.0394T$$

This redefined temperature scale is the *absolute*, or *kelvin*, scale mentioned in Chap. 1. From what we know about straight lines, this equation has a p-axis intercept of $p = 0$; at $T = 0$, $p = 0$. This last equation has the same slope as $p = 0.0394t + 10.76$ and contains exactly the same information but is for many purposes more practical.

The slope 0.0394 is valid only for a particular amount of gas in a vessel of certain volume: 1.308 g of helium in 10,000 cm³. For other amounts in other volumes the value of the slope will differ from 0.0394. In general, however, with a constant amount of gas in a constant volume, the equation

$$p = cT$$

where c is a constant, describes the linear p versus T relationship.

AN AIR THERMOMETER. An air thermometer can be made from a vessel containing air whose pressure can be kept constant. As the temperature increases, so does the volume of the gas. In 1702 Guillaume Amontons (1663–1705) made such a thermometer; at the boiling point of water the measured gas volume was 73 units, and at the freezing point it was $51\frac{1}{2}$ units. Amontons guessed that at some low temperature the air volume should be zero.

Air at a pressure of 1 atm does not become liquid until a temperature of $-193°C$ is reached. No matter how high the pressure, it is still necessary to cool oxygen to $-118.8°C$ and nitrogen to $-147.1°C$ to liquefy these gases. Since such low temperatures were not attainable in Amontons' time, his guess that air is a gas at all temperatures was a reasonable and fortuitous one. In his thermometer the air volume should be zero at $-240°C$. Johann Heinrich Lambert (1728–1777) made a more precise gas thermometer and estimated that at $-270°C$ the extrapolated volume should be zero.

THE BEHAVIOR OF GAS AT CONSTANT PRESSURE

Both Jacques Charles (1787) and John Dalton (1801) independently found that gases at constant pressure expanded uniformly with an increase in temperature. A graph plotting gas temperatures and corresponding volumes is a straight line if the pressure is constant for all the measurements. A typical plot of volume V versus temperature t is shown in the two equivalent graphs (Fig. 2.6).

As in the p versus t graphs (Fig. 2.5) it is found that the straight-line segment extrapolated back to $V = 0$ crosses the temperature axis at $t = -273°C$. Here, too, it is convenient to define the absolute temperature scale $T = t + 273$. The equation representation for the constant-pressure volume and temperature relationship is

$$V = bT$$

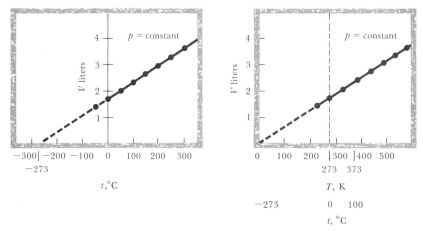

Figure 2.6. Effect of temperature changes on the volume of a gas when the pressure is kept constant. The extrapolation is shown as a broken line.

where b is some constant, depending upon how much gas is confined at constant pressure. After careful study this linear relationship was established in 1802 by Joseph Louis Gay-Lussac (1778–1850) and is known as either "Charles' law" or "Gay-Lussac's law."

The three relations for a constant mass of gas are

$pV = $ constant $\qquad$ if T is constant
$p = cT \qquad\qquad$ if V is constant
$V = bT \qquad\qquad$ if p is constant

These can be combined into one equation

$$pV = BT$$

where B is a constant for a constant mass of gas and is proportional to the amount of gas. The combined equation is consistent with the original ones:

1. If the mass and temperature of a gas are each maintained constant, so is BT and so must

$pV = BT = $ constant $\qquad$ (Boyle's law)

2. If the mass and volume are each constant, so is B/V, and since

$$p = \frac{B}{V} T = \text{constant} \times T$$

3. If the mass and pressure are constant, so is B/p, and since

$$V = \frac{B}{p} T = \text{constant} \times T \qquad \text{(Charles' law)}$$

4. If the temperature is kept constant and gas is continuously pumped into an initially empty container of constant volume, then the pressure is continuously increased. Because T and V are fixed,

$$p = \frac{T}{V} B = \text{constant} \times B$$

Both p and B are proportional to how much gas is put into the container; B is proportional, then, to the mass of the gas.

TEMPERATURE AND IDEAL GASES

The relationship, the *ideal gas equation*,

$$pV = BT$$

describes a gas quite well, provided the gas is not close to the temperature and pressure of liquefaction. For all gases at sufficiently high temperatures and sufficiently low pressures, this equation is valid. Gases for which the equation is valid are said to behave ideally.

All gases will *condense* into liquids at low enough temperatures, and liquids are *not* described by the ideal gas equation. The Celsius temperature at which pV should be zero *if the ideal gas equation were valid* can be determined by using gases which behave ideally at very low temperatures. Very precise measurements with helium and hydrogen, which do not become liquid until -268.9 and $-252.8°C$, respectively, indicate that the best extrapolated temperature value for which $pV = 0$ is $t = -273.15°C$.

THE IDEAL GAS THERMOMETER. The ideal-gas-thermometer scale is defined by the equation $pV = BT$ and the *absolute zero* 273.15°C below the freezing point of water. The freezing point of water in contact with air is reproducible to one-thousandth of a degree and is 273.150 ± 0.001 on the absolute scale. The freezing point in the absence of air, the *triple point*, is 0.01° higher but is reproducible to one ten-thousandth of a degree. The triple point 273.1600 ± 0.0001 absolute is the international temperature standard.

The pressure and volume of any gas at a pressure sufficiently low for it to behave ideally can be measured at the triple point and will have values $p = p_1$, $V = V_1$, and $T = 273.16$ K. Then the same mass of a gas can have its pressure and volume measured at a second temperature to be determined by the ideal gas thermometer; for example,

$$p = p_2 \qquad V = V_2 \qquad T = T_2$$

Because

$$p_1 V_1 = B273.16$$

and

$$p_2 V_2 = BT_2$$

dividing the second by the first equation gives

$$\frac{p_2 V_2}{p_1 V_1} = \frac{T_2}{273.16}$$

or

$$T_2 = 273.16 \, \frac{p_2 V_2}{p_1 V_1}$$

which is the second temperature on the ideal gas, or absolute, scale. The same mass of gas is used in each of the two sets of measurements of p and V, and so B is the same for the two sets and cancels in the division. *The temperature of any substance can be obtained with the ideal gas thermometer by measuring the pressure and volume of a gas at both the triple point and at the temperature of the substance.*

Example. An automobile tire is filled with air at a pressure of 26 psi and at 25°C. After being driven on, the tire temperature is 45°C and there is no change in the volume within the tire. Calculate the tire air pressure at the higher temperature.

At constant volume the pressure varies directly with the absolute temperature:

$$p = \frac{B}{V} \, T = cT$$

For two pressures p_1 and p_2 at two temperatures T_1 and T_2, all at constant volume,

$$p_1 = cT_1$$
$$p_2 = cT_2$$

and

$$\frac{p_1}{p_2} = \frac{T_1}{T_2}$$

or

$$p_1 = p_2 \times \frac{T_1}{T_2}$$

In our example, if $p_2 = 26$ psi, then

$$p_1 = 26 \times \frac{45 + 273.15}{25 + 273.15} \text{ psi}$$

$$= 26 \times \frac{318}{298} \text{ psi}$$

$$= 26 \times 1.067 \text{ psi}$$
$$= 28 \text{ psi}$$

THERMODYNAMIC TEMPERATURE AND HEAT ENGINES

Lord William Kelvin (1824–1907) defined temperature in a way that is independent of the substance in the thermometer. To look at this we must also look at heat engines. A heat engine is a device which takes heat, a form of energy, from a hot reservoir, converts part of this energy into work, and returns the rest of the energy to a cold reservoir. For example, a steam engine absorbs heat from a furnace, the steam expands, doing work, and heat leaves the engine for the cooler surroundings.

Since the temperature of a substance increases when heat is added, this increase in temperature can be used to measure the amount of heat added to that substance. A *calorie* is defined as the amount of heat needed to increase the temperature of 1 g of water from 14.5 to 15.5°C. Later we shall see how heats are measured in a *calorimeter*. In addition to the calorie, other units for measuring heat are used and will be introduced when needed.

A schematic sketch of heat engines is shown in Fig. 2.7. The temperatures of the two reservoirs are T_2 and T_1 on the ideal gas law scale. The *efficiency* of the engine is defined as the fraction of the heat absorbed from the hot reservoir that is converted into work:

$$\text{Efficiency} = \frac{W}{Q_2}$$

According to the first law of thermodynamics — the law of conservation of energy — the energy entering the engine Q_2 must equal the work done W plus the energy leaving Q_1 if the engine is not to accumulate energy, or

$$Q_2 = W + Q_1$$

From experience we know that no heat engine is ever 100 percent efficient. To be 100 percent efficient, *all* the heat Q_2 must be converted to work W

$$Q_2 = W$$
$$\text{Efficiency} = \frac{W}{Q_2} = 1 = 100\%$$
$$Q_1 = 0$$

The last equation states that no heat is returned to the cold reservoir; none of Q_2 is wasted.

The statement of the experimental evidence†

---

† The symbol $\neq$ represents not equal to; the symbol $<$ stands for less than; $>$ for greater than; $\leq$ for equal to or less than; $\geq$ for equal to or greater than.

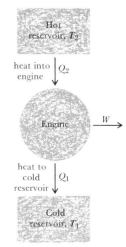

HEAT ENGINE

Figure 2.7. In a heat engine the heat Q_1 returned to a cold reservoir at T_2 is converted to work W done by the engine and heat Q_1 returned to a cold reservoir at T_1. According to the conservation of energy, $Q_2 = Q_1 + W$.

$$\text{Efficiency} = \frac{W}{Q_2} < 100\%$$

and

$$Q_1 \neq 0$$

is one form of the *second law of thermodynamics.* Another equivalent statement is that heat, which spontaneously travels from a warmer to a cooler substance, *never* spontaneously travels the reverse route — from a cooler to a warmer substance. The heat must be pumped or work must be done to accomplish this.

There is a simple relationship between the maximum efficiency of a heat engine and the reservoir temperatures T_2 and T_1 on the ideal gas scale. It is

$$\text{Efficiency}_{\text{max}} = 1 - \frac{T_1}{T_2}$$

Since

$$0 \leq T_1 \leq T_2$$
$$0 \leq \frac{T_1}{T_2} \leq 1$$

and

$$\text{Efficiency}_{\text{max}} = 1 - \frac{T_1}{T_2} \leq 1$$

Under the condition of maximum efficiency, since

$$Q_2 = W + Q_1$$
$$W = Q_2 - Q_1$$

then

$$\text{Efficiency}_{\text{max}} = \frac{W}{Q_2} = \frac{Q_2 - Q_1}{Q_2} = 1 - \frac{Q_1}{Q_2}$$
$$\text{Efficiency}_{\text{max}} = 1 - \frac{T_1}{T_2}$$

so

$$1 - \frac{Q_1}{Q_2} = 1 - \frac{T_1}{T_2}$$

and

$$\frac{Q_1}{Q_2} = \frac{T_1}{T_2}$$

Kelvin defined a *thermodynamic temperature scale* in which the temperature is proportional to the heat exchanged between a reservoir at that temperature and a heat engine operating at the maximum allowable efficiency. But from the last equation the *heat exchanged is proportional to the reservoir temperature on the ideal gas scale*. Then the thermodynamic temperature, which does not depend on the material of the heat engine, is proportional to the ideal gas temperature. By simple adjustment of that proportionality, the thermodynamic and the ideal gas temperature scales can be made identical.

It can be seen that as the cold reservoir temperature is lowered or the hot reservoir temperature is raised, the ratio T_1/T_2 becomes smaller and the efficiency $1 - T_1/T_2$ becomes larger.

> **Example.** Under a pressure of 1 atm, the boiling point of water is 100.0°C; under a pressure of 2 atm, it is 120.6°C. Under a pressure of 1 atm, the boiling point of mercury is 356.6°C. Calculate the maximum allowable efficiencies of three heat engines using for the hot reservoir (1) boiling water at 1 atm, (2) boiling water at 2 atm (i.e., superheated steam), and (3) boiling mercury at 1 atm and for the cold reservoir in each case water at 40.0°C.
>
> First the reservoir temperatures must be in the absolute temperature scale, obtained by adding 273.15° to the above Celsius temperatures:
>
> 100.0 + 273.15° = 373.2 K
> 120.6 + 273.15° = 393.8 K
> 356.6 + 273.15° = 629.8 K
> 40.0 + 273.15° = 313.2 K
>
> Then the maximum allowable efficiencies of the three engines are calculated from the equation Efficiency $= 1 - T_1/T_2$:

$$1 - \frac{313.2}{373.2} = 1 - 0.8392 = 0.1608 = 16.08\%$$

$$1 - \frac{313.2}{393.8} = 1 - 0.7952 = 0.2048 = 20.48\%$$

$$1 - \frac{313.2}{629.8} = 1 - 0.4973 = 0.5027 = 50.27\%$$

Example. Using the three heat engines just described operating at their maximum allowable efficiencies, assume that each engine takes up 1 million cal from the hot reservoir and calculate the calories converted to work and the calories returned to the cold reservoir.

Since

$$\text{Efficiency} = \frac{W}{Q_2}$$

and

$$Q_2 = W + Q_1$$

then

$$W = \text{efficiency} \times Q_2$$

and

$$\begin{aligned} Q_1 &= Q_2 - W \\ &= Q_2 - (\text{efficiency} \times Q_2) \\ &= Q_2(1 - \text{efficiency}) \end{aligned}$$

Thus for the first engine:

$$W = 0.1608 \times 10^6 = 1.608 \times 10^5 \text{ cal}$$
$$Q_1 = 10.000 \times 10^5 - 1.608 \times 10^5 = 8.392 \times 10^5 \text{ cal}$$
$$Q_2 = W + Q_1 = 1.608 \times 10^5 + 8.392 \times 10^5 = 10.000 \times 10^5 \text{ cal}$$

For the second engine:

$$W = 0.2048 \times 10^6 = 2.048 \times 10^5 \text{ cal}$$
$$Q_1 = 10.000 \times 10^5 - 2.048 \times 10^5 = 7.952 \times 10^5 \text{ cal}$$
$$Q_2 = W + Q_1 = 2.048 \times 10^5 + 7.952 \times 10^5 = 10.000 \times 10^5 \text{ cal}$$

For the third engine:

$$W = 0.5027 \times 10^6 = 5.027 \times 10^5 \text{ cal}$$
$$Q_1 = 10.000 \times 10^5 - 5.027 \times 10^5 = 4.973 \times 10^5 \text{ cal}$$
$$Q_2 = W + Q_1 = 5.027 \times 10^5 + 4.973 \times 10^5 = 10.000 \times 10^5 \text{ cal}$$

Example. Using the same three heat engines with the same three efficiencies, calculate how much heat must be taken from the hot reservoir in order for the engine to do 1 million cal of work.
From

$$\text{Efficiency} = \frac{W}{Q_2}$$

we arrive at

$$Q_2 = \frac{W}{\text{efficiency}}$$

Then the quantities desired are

$$Q_2 = \frac{10^6}{0.1608} = 6.219 \times 10^6 \text{ cal}$$

$$Q_2 = \frac{10^6}{0.2048} = 4.883 \times 10^6 \text{ cal}$$

$$Q_2 = \frac{10^6}{0.5027} = 1.989 \times 10^6 \text{ cal}$$

REFRIGERATORS AND ABSOLUTE ZERO

A refrigerator is simply a heat engine run in reverse: heat is pumped from a cold reservoir into a hot reservoir by having work done on the refrigerator engine (e.g., by putting in electric energy, Fig. 2.8). Here, as heat is removed, the temperature T_1 is allowed to decrease. As T_1 approaches absolute zero, i.e., as

$$T_1 \longrightarrow 0$$
$$\frac{T_1}{T_2} \longrightarrow 0$$

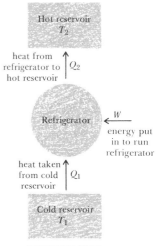

REFRIGERATOR

Figure 2.8. In a refrigerator heat Q_1 is removed from the cold reservoir if work W is done to operate the refrigerator and the combined energy
$$W + Q_1 = Q_2$$
is expelled to the hot reservoir.

$$1 - \frac{T_1}{T_2} \longrightarrow 1$$

$$\frac{W}{Q_2} = \frac{Q_2 - Q_1}{Q_2} = 1 - \frac{Q_1}{Q_2} \longrightarrow 1$$

so

$$\frac{Q_1}{Q_2} \longrightarrow 0$$

and

$$Q_1 \longrightarrow 0$$

The closer T_1 approaches absolute zero, the smaller is Q_1, the heat removed from the cold reservoir. If $T_1 = 0$, then $Q_1 = 0$ and no more heat can be removed from the cold reservoir; it cannot be cooled below $T_1 = 0$, the *absolute zero*.

We now have two interpretations of *absolute zero*. It is the extrapolated temperature at which the pressure or volume, or both, of an ideal gas should vanish if it were to remain an ideal gas down to that temperature. It is also the temperature at which no more energy can be removed from a substance. As absolute zero is approached, it becomes increasingly difficult to remove heat from a substance. It has been possible to produce temperatures down to one-millionth of a degree Celsius above absolute zero (N. Kurti, 1960). We shall look at some of the techniques used to attain such extreme cold in Chap. 9.

HEAT PUMPS AND REFRIGERATORS

We shall assume here that the optimum work and heat relationships of refrigerators depend on the two reservoir temperatures in exactly the same way they would if the refrigerator were operated in reverse—as a heat engine. However, it can be a source of confusion to speak about the efficiency of a refrigerator because when the efficiency is high (i.e., close to 100 percent) the amount of heat removed from the cold reservoir is very small (close to zero). The more efficient the refrigerator, the poorer a job it does in removing heat from the cold reservoir. A better basis of comparison for refrigerators is the *coefficient of performance*. This is the ratio of heat removed to work done:

$$\text{COP} = \frac{Q_1}{W}$$

> **Example.** How much electric energy W must be supplied to a refrigerator to pump 1 million cal ($Q_2 = 1.000 \times 10^6$ cal) from a pond at 2°C into a house at 20°C? What is the COP?
>
> The efficiency is

$$1 - \tfrac{275}{293} = 1 - 0.939 = 0.061$$

Since

$$\frac{W}{Q_2} = 0.061$$

$$W = 0.061 Q_2$$
$$= 0.061 \times 1.000 \times 10^6$$
$$= 6.1 \times 10^4 \text{ cal}$$

$$\text{COP} = \frac{Q_1}{W} = \frac{9.39 \times 10^5}{6.1 \times 10^4}$$
$$= 15$$

Because 15 cal are removed from a free source (the pond) for each calorie bought, such a heat pump is an attractive method for heating homes. Considerable research and development have been invested in attempting to make this practical.

> **Example.** An air conditioner pumps heat from a room at 20.0°C to the outdoors at 40.0°C. The air conditioner is rated as being able to pump from the room 10,000 Btu/hr (1 Btu is a British thermal unit, the amount of heat needed to raise the temperature of 1 lb of water 1°F). Calculate how much heat is expelled to the outdoors each hour.
>
> The efficiency is

$$1 - \frac{293.2}{313.2} = 1 - 0.9361 = 0.0639$$

also

$$0.0639 = \frac{W}{Q_2} = \frac{W}{W + Q_1}$$

Then

$$\frac{1}{0.0639} = \frac{W + Q_1}{W} = 1 + \frac{Q_1}{W}$$

$$15.6 = 1 + \frac{Q_1}{W}$$

$$14.6 = \frac{Q_1}{W} = \frac{10{,}000 \text{ Btu}}{W}$$

$$W = \frac{10{,}000 \text{ Btu}}{14.6} = 685 \text{ Btu}$$

$$Q_2 = 10{,}000 + 685 = 10{,}685 \text{ Btu}$$

Example. A refrigerator engine pumps heat from the freezer compartment at $-5.0°C$ to the room at $25.0°C$. To remove 10,000 cal from the freezer compartment, how many calories are added to the room and how many calories of energy are required to operate the refrigerator engine?

Using absolute temperatures, the efficiency is

$$1 - \frac{268.2}{298.2} = 1 - 0.8994 = 0.1006$$

Since

$$0.1006 = \frac{W}{Q_2}$$

$$\frac{1}{0.1006} = \frac{Q_2}{W} = \frac{W + Q_1}{W} = 1 + \frac{Q_1}{W}$$

$$9.94 = 1 + \frac{Q_1}{W}$$

$$8.94 = \frac{Q_1}{W} = \frac{10{,}000}{W}$$

and

$$W = \frac{10{,}000}{8.94} = 1190 \text{ cal}$$

Therefore, 1190 cal must be put into the refrigerator engine in the form of electric energy. And $Q_2 = 10{,}000 + 1190 = 11{,}190$ cal are expelled into the room.

> **Example.** If the refrigerator engine in the above example operates under the same conditions but with the freezer-compartment door open to the room, will the room temperature rise or fall? The refrigerator engine removes 10,000 cal from part of the room, the electric energy introduced is 1190 cal and the total 11,190 cal is put into the room. The net effect is the introduction of 1190 cal from outside the room. Therefore, the room will become warmer.

THE MOTION OF BODIES

The ideal gas equation $pV = BT$ is a law based upon experimental evidence and does not depend on whether or not the gas is composed of particles. However, it is possible to calculate the pressure exerted on a surface due to a collection of particles hitting that surface. For this calculation, some definitions concerning the motion of particles are needed.

Displacement is the change in position of a particle; it is the difference between the final and the initial positions of the particle. To completely specify a displacement, it is necessary to state both the *distance* through which the particle moves and the *direction* it travels.

Imagine a straight line drawn through the initial and final positions of a particle; this can be called the "x axis" (Fig. 2.9). Select some arbitrary point on this axis, called the "origin." The distances x_1 and x_2 measured from the origin specify the initial and final positions of the particle, in this case, a car. The displacement is $x_2 - x_1$ and is the difference in the values of x between the final and initial positions of the car. It is evident that the magnitude of the difference $x_2 - x_1$ does not depend on where the origin is arbitrarily placed. Wherever on the axis the origin is selected $x_2 - x_1$ will have exactly the same value. We write this difference as

$$\Delta x = x_2 - x_1$$

which is read "delta x." If the car moves from x_2 to x_1, its displacement is

$$\Delta x = x_1 - x_2$$

which has the same absolute value, or magnitude, as $x_2 - x_1$ but the

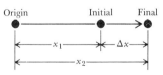

Figure 2.9. Displacement where x is the position of the moving particle.

opposite sign since the second displacement (from x_2 to x_1) moves in the direction opposite to the first displacement (from x_1 to x_2).

A quantity which must be described both by magnitude and direction is called a "vector." A car or particle may move in any number of directions. Displacement is, therefore, a vector. A quantity described by magnitude alone is a "scalar." For example, mass and length are scalars.

Velocity is the time rate at which a displacement occurs. If a displacement of Δx takes a time interval Δt, then the velocity

$$u = \frac{\Delta x}{\Delta t}$$

in the direction of the displacement. Velocity, too, is a vector. The corresponding scalar, where direction is not specified, is *speed*. A car may have a speed of 10 mph. The velocity may be positive or negative; that is, the car may be going forward at 10 mph or in reverse at 10 mph.

Acceleration is the rate at which a velocity changes. If the velocity changes from u_1 to u_2 (where $\Delta u = u_2 - u_1$) in a time interval Δt, the acceleration

$$a = \frac{\Delta u}{\Delta t}$$

Acceleration is a vector. A car's velocity may be accelerated or decelerated, depending on whether the gas pedal or brake is stepped on. The units of displacement can be meters (m); of velocity, meters per second (m/s); of acceleration, meters per second per second (m/s^2).

FORCES

When the motion of a body is changed, that is, when its velocity is not constant, a *force* is said to be exerted on that body. Also, when the shape or form of a body is altered, a force is needed to effect that alteration.

Forces may be of many types. Because the effect of any force must be described by the direction of the change, force is a vector, that is, a quantity for which direction must be specified. Some examples of forces are:

1. When a rubber band is pulled at each end, it stretches. The force causing the increase in length is said to be a *tensile force*, or *tension*.

2. When force is applied all over the surface of a body, its volume decreases. The body is compressed by a *compressive force*, or *compression*.

3. In general, when the shape or size of a body is altered, the force causing this is a *stress* and the resulting change is a *strain*. Stretching and volume contraction are strains; tension and compression are stresses.

4. The force of attraction between two bodies which depends on the masses of the bodies and on the distances between them is the *gravitational force*. Weight is such a force.

5. There is a force of attraction between north and south magnetic poles and a force of repulsion between two norths and also between two souths. This force, which depends on the magnitudes of the magnetic poles and the distance between them, is the *magnetic force*.

6. There are two types of electric charge, called "positive" and "negative" charges. Like charges are repelled by and unlike charges attracted to each other by an *electric*, or *coulombic*, *force*, after Charles Coulomb (1738–1809), who stated the formula which describes this force in terms of the magnitude of each charge and the distance between them.

7. A *contact force* is the force applied to a body by touching it. Hitting a golf ball with a club, causing the ball to move, is an example of this kind of force.

8. A ball rolling along a level surface will slow down and eventually come to rest because of the *friction*, or *frictional force*, acting on the ball by the surface.

9. An object moving through a fluid, e.g., air or water, has a retarding force acting to slow it down. This force is a *fluid frictional force*, or *viscous force*. A fluid which exerts a large viscous force on objects moving within it is said to be very viscous or to have a high *viscosity*.

WORK, KINETIC ENERGY, AND PRESSURE

When a force is required to move a body through a distance, some energy must be used to move the body. The energy is found to be equal to the product of the force times the distance and is called the "work" needed to move the body. Work is a form of energy, and both work and energy are scalars.

The energy associated with the motion of a body is the *kinetic energy*. As work is done on a body to increase its motion, the amount of work expended can appear as an increase in the kinetic energy $\frac{1}{2}mu^2$, where m is the mass and u the velocity of the body. Kinetic energy, like all energy, is a scalar quantity; that is, no direction needs to be specified when giving the quantity.

Pressure is defined as the force acting on a surface divided by the area of that surface.

NEWTON'S LAWS OF MOTION

Sir Isaac Newton (1642–1727) formulated three laws which describe the relationships among force, velocity, acceleration, and mass. The simplicity of these laws masks the genius that Newton exhibited in formulating these laws, which have remained valid through centuries of experimentation.

The first law is that a body has a constant velocity, that is, it remains at rest (zero velocity) or moves in a straight line at a constant speed, until it is acted upon by a force.

The second law is that the force F needed to effect a particular acceleration is equal to the product of the body mass m and the acceleration a:

$$F = ma$$

The larger the mass, the greater the force needed for a desired acceleration; the larger the mass, the smaller the acceleration for a particular force.

A body, such as a satellite, traveling along a circular path at constant speed changes its direction continuously although not its speed. Therefore, its velocity changes because velocity is a vector. Then, a force is needed to make a body travel on a circular path. When the direction of motion changes continuously, the force causing this change in direction must, itself, change direction continuously.

Newton's third law states that for every force there is another of the same magnitude but acting in the opposite direction. The gravitational force of the earth acting on a body, pulling the body toward the earth, is exactly balanced by the gravitational force pulling the earth toward the body. If the mass of the earth greatly exceeds the mass of the body, the acceleration of the former is much less than that of the latter. A rubber ball striking a wall with a force is acted upon by an equal force deforming it and sending it back from where it came.

THE KINETIC THEORY OF MATTER

Daniel Bernoulli (1700–1782) applied the Newtonian laws to the problem of how pressure can be described by many small particles striking a wall. Suppose a particle with mass m is traveling within a rectangular box (Fig. 2.10) whose dimensions are $a \times b \times c$ with a velocity u m/s parallel to the edge of length a. Just after collision with a wall of area $b \times c$ m², the particle moves in the opposite direction, with a velocity of $-u$ m/s. The particle moves until it hits the wall at the far end of the rectangular box, turns around, and, again, hits the first wall.

   Collisions which result in a velocity change in direction only but not in magnitude or speed are said to be "elastic collisions." We assume these particles undergo such elastic collisions. If there should be changes in speed, too, these are called "inelastic collisions." Between any two collisions of the particle at one wall, the particle must travel twice the length of the rectangular box, a distance of $2a$ m, at a speed of u m/s. The time between successive collisions at that wall is

$$2a \text{ m} \times \frac{1\text{s}}{u\text{m}} = \frac{2a}{u} \text{ s}$$

   The time-average acceleration of the particle is the velocity change $-2u$ m/s divided by the time interval $2a/u$ s, or

$$-2u \text{ m/s} \times \frac{u}{2a} \frac{1}{\text{s}} = \frac{-u^2}{a} \text{ m/s}^2$$

The average force acting on the particle is the mass times acceleration

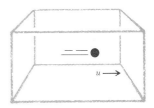
Before collision

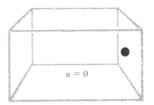
At collision

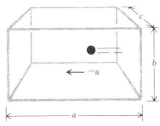
After collision

Figure 2.10. Collision of a moving particle with the wall of a container.

$$F = m\frac{-u^2}{a} = \frac{-mu^2}{a}$$

and, by Newton's third law, the average force on the wall due to the single particle hitting it once each $2a/u$ s must be

$$F = +\frac{mu^2}{a}$$

The pressure due to this single particle is the force divided by the area bc, or

$$p = \frac{mu^2}{abc}$$

If there are N particles in the solid container and if on the average one-third of these, or $N/3$, travel in the back-and-forth direction, including collision with that specified wall, the total pressure from these $N/3$ particles is

$$p = \frac{N}{3}\frac{mu^2}{abc}$$

The volume of the rectangular solid container is $V = abc$, and the kinetic energy of each particle is $mu^2/2$; and so we can rewrite the expression for total pressure:

$$p = \frac{2N}{3V}\frac{mu^2}{2}$$

$$pV = \frac{2}{3}N\frac{mu^2}{2}$$

We can compare this to the ideal gas law

$$pV = BT$$

and if we identify the absolute temperature T with the total kinetic energy of all the particles $Nmu^2/2$, with a few constants (that is, $\frac{2}{3}$ and B) included, we can say that at constant average kinetic energy, which is at constant temperature, pV is constant (Boyle's law). In this model, at absolute zero the kinetic energy is also zero and particle motion ceases.

Actually, the acceleration of each particle at collision takes place not over a time interval $2a/u$ s but over a considerably shorter period—the time the particle is in contact with the wall. Consequently the acceleration and the force are much higher for this shorter time and are zero when the particle is not at the wall. If we could measure the pressure with very sensitive instruments, we would find not a steady pressure but a discontinuous, or interrupted, series of alternating high and zero pressures.

It is the nature of systems such as this gas, comprising a great many particles, that the high-pressure impulses are rather closely spaced, and we measure, with the techniques available, an averaged-out pressure intermediate to high-pressure impulses and the no pressure between the impulses. The success of the Bernoulli derivation is another reason for believing in an atomic theory.

ROCKET PROPULSION. Although a definite force is applied to the wall by the particles colliding with that wall, the entire vessel does not move because the same force is applied to the opposite wall by particles striking it. The two forces cancel each other.

However, if the containing vessel has an opening in one wall, particles will pass through that opening instead of striking the wall (Fig. 2.11). As a result, fewer particles strike the wall with the hole than the intact wall opposite. A greater force is applied to the intact wall than to the wall with the hole. The result is that the vessel has an unbalanced force in the direction away from the hole, is accelerated, and moves in such a way that the hole is at the trailing surface. This is the description of how rockets are propelled. As gas leaves an exhaust in one direction, the rocket itself moves in the opposite direction (Fig. 2.12).

BROWNIAN MOVEMENT

The botanist Robert Brown (1773–1858) observed that pollen grains suspended in a liquid and seen under a microscope move about in a random zigzag path. Jean Perrin (1870–1942) and Albert Einstein (1879–1955) described the random motion, with abruptly changing direction, as the result of many collisions on the pollen grain by rapidly moving discrete particles, much smaller than the pollen grain and too small even to be seen by microscope.

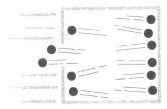

Figure 2.11. Many collisions at the right wall cause the container to move to the right as the gas molecules escape at the left-wall opening. If gas molecules could not escape, they would rebound and hit the opposite wall; hence, the container would not move.

Figure 2.12. The Apollo 14 space vehicle. The 363-foot-high rocket generated more than 7.5 million pounds of thrust as it took off. (The National Aeronautics and Space Agency)

An imbalance at any moment from all the submicroscopic particles colliding with the pollen grain results in a net unbalanced force in one direction and the acceleration of the pollen grain in that direction. The submicroscopic particles move rapidly and at random, so that at the next moment the imbalance is in another direction. Einstein was able to predict the average rate at which the pollen grain moves away from an initial position. (In Chap. 9 we shall look at the way in which substances move about or *diffuse* through space in greater mathematical detail.)

In addition to Bernoulli's description of Boyle's law and Dalton's atomic model accounting for the several mass-combination laws, the treatment of Brownian movement is a third reason that we accept the reality of atoms.

PROBLEMS

2.1. Convert the following Celsius-scale temperatures to equivalent temperatures in degrees Kelvin:

      a. 18°C   *b.* 37.2°C   *c.* −100°C   *d.* 23.98°C

2.2. Convert the following Celsius-scale temperatures to equivalent temperatures in degrees Kelvin:

      a. 19.42°C   *b.* 81°C   *c.* −124.6°C   *d.* 946°C

2.3. A gas is under a pressure of 1.00 atm when the temperature is 25°C. If the temperature is raised to 100°C and the volume is maintained constant, what is the gas pressure?

2.4. At 24.12°C a gas has a pressure of 1.000 atm. Calculate the gas pressure when the temperature is raised to 342.6°C and the volume is unchanged.

2.5. A gas at 25.07°C has a volume of 22.31 cm³. Keeping the pressure constant but lowering the temperature to the ice point 0.000°C, what will the new volume be?

2.6. At 25°C a gas has a volume 392 cm³. If the pressure is not changed but the temperature is raised to 102°C, what is the volume?

2.7. An ideal gas thermometer at the triple point contains a gas at 1.0000 atm with a volume of 106.22 cm³. At a temperature to be determined the pressure is 0.9420 atm and the volume is 164.82 cm³. What is that temperature in degrees Celsius?

2.8. The pressure and volume of helium in an ideal gas thermometer at a certain temperature are 0.1943 atm and 200.91 cm³. At the ice point these values are 0.1728 atm and 172.49 cm³. What is that first temperature in degrees Celsius?

2.9. Phosphorus forms three distinct compounds with oxygen. On a mass basis these are analyzed and found to be 56.35, 49.19, and 43.65 percent phosphorus, respectively. Demonstrate the law of multiple proportions.

2.10. Nitrogen and oxygen form six distinct compounds. These have mass percent nitrogen analyses of 63.65, 46.68, 36.86, 30.45, 25.94, and 22.59 percent. Show that the law of multiple proportions obtains here.

2.11. A heat engine operates with maximum allowed efficiency between reservoirs at 200 and 30°C. What is that efficiency?

2.12. With hot and cold reservoirs of 1000 and 10°C, what is the greatest efficiency you can expect for a heat engine operating between these temperatures?

2.13. A refrigerator operating between 100 and 10 K removes 1 cal from the colder reservoir. Assume efficiencies are at the ideal upper limit. How much work, in calories, is needed and how much heat is expelled to the warmer reservoir?

2.14. A refrigerator removes heat from a colder reservoir at 30 K and expels heat to a warmer reservoir at 60 K. Assume the performance of this refrigerator is that of an ideal heat engine working optimally between these temperatures. Calculate how much work is needed to remove 100 cal from the colder reservoir and how much heat is put into the hotter reservoir.

The Atomic Composition and the Behavior of Molecules

3

The soundness and acceptance of the atomic theory derives from the success of this theory in describing mass-combination laws, the ideal gas law, and Brownian movement. In this chapter we shall continue to develop the theory, learning more about these very small particles—molecules. We can then use this knowledge to better understand some more practical principles.

THE COMBINING VOLUME OF GASES

Henry Cavendish (1731–1810) observed that when hydrogen and oxygen gases combine to form water, the volume of hydrogen consumed is twice that of oxygen when both gases are measured at the same temperature and pressure. In 1805 Alexander von Humboldt and Gay-Lussac independently confirmed this simple observation. If the combination takes place at a temperature above the boiling point of water, the product is a gas and its volume is equal to that of the consumed hydrogen and twice that of the consumed oxygen (Fig. 3.1).

Other reactions in which gaseous reactants are converted into gaseous products show this simple behavior. Bromine, an element discovered in 1826 by Antoine-Jerome Balard, is a red-brown liquid with a very irritating odor resembling chlorine gas. When hydrogen combines with bromine gas, the volume of hydrogen used up is equal to the volume of bromine used up and each gas has half the volume of the hydrogen bromide formed (Fig. 3.2).

In 1808, after looking at many such reactions, Gay-Lussac announced his *law of combining volumes: At constant temperature and pressure, the volumes of gaseous reactants used up and gaseous products*

Hydrogen Oxygen Water vapor

Figure 3.1. Volumes of gases used and produced when water vapor is formed from its elements.

Hydrogen Bromine Hydrogen bromide

Figure 3.2. Volumes of gases used and produced when hydrogen bromide is formed from its elements.

formed are in the ratios of small whole numbers. In our examples, the hydrogen, oxygen, water-vapor volumes are in the ratio 2 : 1 : 2; the hydrogen, bromine, hydrogen bromide volumes are in the ratio 1 : 1 : 2.

AVOGADRO'S HYPOTHESIS AND MOLECULES

In Dalton's atomic model, atoms of each element combine with atoms of other elements in naturally determined ratios. When a particular compound is formed, it has a constant composition because atoms of elements join in set ratios characteristic of that compound. The numbers of atoms of each element combining to form a single compound-atom are always expressed in whole numbers, or integers. This whole-number ratio, or integer ratio, then, is set for any amount of a substance, that is, any number of its individual compound atoms.

If the numbers of atoms combining are in whole-number ratios and if the volumes of gaseous elements combining are in whole-number ratios, then a simple relationship may exist between the volume of a gas and the number of atoms contained in that volume.

In 1811, Amedeo Avogadro (1776–1856) stated his hypothesis concerning this relationship. All samples of gases with the same volume at the same pressure and temperature contain the same number of elementary particles. He called these particles "molecules."

There is an important consequence of this hypothesis. In the formation of hydrogen bromide, for example, when a volume of hydrogen combines with an equal volume of bromine to form twice that volume of product, any given number of hydrogen molecules combine with an equal number of bromine molecules to produce twice that number of product molecules. Each hydrogen molecule combines with a single bromine molecule to form two hydrogen bromide molecules. If one hydrogen molecule can form two equal hydrogen bromide molecules, then the hydrogen molecule must be divisible into two equal parts (as is the bromine molecule). A hydrogen molecule is, therefore, *not* the same as a hydrogen atom, which is indivisible. Further, because the hydrogen molecule is divided into two equal parts, it must contain an even number of

Figure 3.3. Reaction of hydrogen and bromine to form hydrogen bromide.

hydrogen atoms. The smallest even number is two, and the simplest possibility is that a hydrogen molecule contains two hydrogen atoms. We write the symbol for this molecule as H_2, in which the subscript indicates that there are two hydrogen atoms in a single molecule. Consistent with Avogadro's hypothesis, one bromine molecule can react with hydrogen to form two hydrogen bromide molecules and therefore must also contain an even number of bromine atoms. We write Br_2, similarly, as the simplest possibility.

Using the law of combining volumes, the atomic theory, and Avogadro's hypothesis, the reaction between hydrogen and bromine can be shown as in Fig. 3.3. The reaction can also be described by the equation

$$H_2 + Br_2 \longrightarrow 2HBr$$

The equation states that one molecule of hydrogen, containing two atoms of hydrogen, combines with one molecule of bromine, containing two atoms of bromine, to form two molecules of hydrogen bromide, each containing one hydrogen and one bromine atom.

In the formation of water, two volumes of hydrogen and one volume of oxygen form two volumes of water vapor. Two molecules of hydrogen and one molecule of oxygen produce two molecules of water. One molecule of oxygen, producing two water molecules, must contain an even number of oxygen atoms. The smallest possible oxygen molecule is then O_2. The formation of water may be represented by either Fig. 3.4 or

$$2H_2 + O_2 \longrightarrow 2H_2O$$

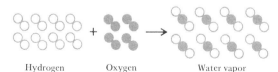

Figure 3.4. Reaction of hydrogen and oxygen to form water. This reaction does not occur spontaneously.

Two hydrogen molecules, each having two hydrogen atoms, react with one oxygen molecule that has two oxygen atoms to form two water molecules. A total of four hydrogen atoms and two oxygen atoms forms two water molecules. Each water molecule has two hydrogen atoms and one oxygen atom and is written H_2O.

MOLECULES AND ATOMS

A *molecule* is the smallest unit of a compound which has the same elemental composition as a larger amount of that compound. When the molecule is further broken up into smaller units, each having a distinct identity, these units are the *atoms* of the elements forming the compound.

Molecules which have two atoms are said to be *diatomic;* those with three, *triatomic;* with more, *polyatomic;* and those with only one, *monatomic.* The molecules H_2, O_2, and Br_2 are diatomic. Water H_2O is triatomic. Molecules of vaporized metals, for example, mercury Hg, sodium Na, and potassium K, and of gases like helium He are monatomic. This can be demonstrated experimentally. For example, two volumes of mercury vapor react with one volume of oxygen O_2. Two molecules of mercury react with one molecule, that is, with two atoms, of oxygen. One molecule of mercury reacts with one atom of oxygen. Therefore, a mercury molecule *can* contain only one mercury atom; the simplest possible mercury molecule is Hg. In the case of monatomic molecules and molecules containing a single atom of a particular kind, for example, oxygen in H_2O, the subscript 1 is omitted.

The conclusion that an element such as bromine or hydrogen contains not one but *two* atoms was first seen to be a threat to the very successful atomic theory. As conceived by Dalton, elements were the simplest basic units of matter, cannot be broken up further, and so must be atomic. After we reexamine the ideal gas law in light of Avogadro's hypothesis, we can look at other reasons for believing that bromine is an example of a diatomic molecule.

THE IDEAL GAS LAW

The ideal gas law $pV = BT$ contains the factor B. For a given amount of a gas, B remains constant as the pressure, volume, and temperature of the gas change.

Bernoulli's derivation, using Newtonian mechanics, yielded the equation

$$pV = \frac{N}{3} mu^2 = \tfrac{2}{3} N \frac{mu^2}{2}$$

We can set

$$BT = \tfrac{2}{3} N \frac{mu^2}{2}$$

Recall that B increases proportionally with an increase in the mass of the gas, that m is the mass per particle, N is the number of particles, and Nm is the total mass of all the particles in the gas. Hence the equation is consistent. We increase the mass Nm of a gas by increasing the number N of particles, and at the same time B increases. The factor B must then vary with the number of particles, and $B = Nk$, where k is another constant. Now we can replace BT with NkT and write the ideal gas law as

$$pV = NkT$$

In accord with Avogadro's hypothesis, the number of gas molecules N is determined by specifying the pressure, volume, and temperature and is independent of the nature of the gas. The constant k must be independent of the nature of the gas, too. It has the same value for all gases and for any amount of gas and for this reason is called a "universal constant." The factor B varies with the amount of gas. The form of the ideal gas law $pV = NkT$ now explicitly involves the number of molecules rather than the mass of the gas.

THE DISSOCIATION OF A DIATOMIC MOLECULE

If a small amount of bromine is placed in a vessel of constant volume and the pressure measured as the temperature is varied, the data obtained can be put in a table (Table 3.1) or plotted in a graph (Fig. 3.5).

In the graph in Fig. 3.5, at low temperatures the pressure varies linearly with temperature [$p = (Nk/V)T = $ constant $\times T$] but at about 600 or 700 K there is the beginning of a

Table 3.1. Pressure of a Sample of Bromine Gas at Constant Volume and Various Temperatures

T, K	p, atm
200	0.00502
300	0.00753
400	0.0100
500	0.0127
600	0.0158
700	0.0196
800	0.0247
900	0.0310
1000	0.0380
1100	0.0453
1200	0.0524
1300	0.0591
1400	0.0657
1500	0.0717
1600	0.0771
1700	0.0828
1800	0.0886
1900	0.0933
2000	0.0986

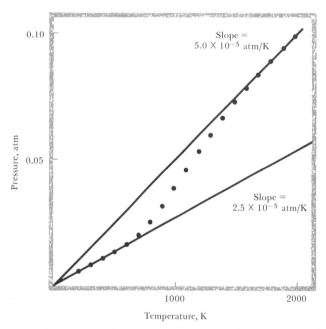

Figure 3.5. Variation of pressure with temperature of a fixed amount of bromine confined within a constant volume.

noticeable departure from linearity. At higher temperatures the experimental points lie close to the line $p = (2Nk/V)T$, which has exactly twice the slope of the first line. This second line would result if there were exactly twice as many molecules in the same volume.

Since the mass of bromine in the volume is kept constant, the simplest description of what happens is that each bromine molecule breaks up into two atoms, thus doubling the number of molecules in the volume. At higher temperatures the experimental points continue to lie on the second line. Since they show no further departure from linearity, the splitting into two atoms is the full extent of division the original bromine molecules can undergo. Our conclusion is that bromine is diatomic.

The evidence from the high-temperature division, or *dissociation*, of molecules confirms the evidence from combining volume ratios that the vapors of the metals sodium, potassium, zinc, and mercury and of the gases helium, krypton, and xenon are monatomic. Hydrogen, oxygen, nitrogen, and chlorine, as well as bromine, are diatomic.

ATOMIC AND MOLECULAR WEIGHTS

Stanislao Cannizzaro (1826–1910) developed a method for determining the relative masses of molecules and atoms based on Avogadro's hypothesis. If a set of different gases are all held at the same temperature, pressure, and volume, then the gases will have exactly the same number of molecules. It is customary to compare gases at 0°C (273.15 K) and at the pressure of 1 atm; this constant condition is referred to as "standard temperature and pressure" (STP).

In Table 3.2 a number of common gases and the corresponding densities at STP are listed. The data for the gases need not be obtained at STP; they can be taken at any pressure and temperature and converted by the ideal gas law to STP conditions. Since each density is the number of grams of gas contained in 1 liter at STP and since each liter at STP contains exactly the same number of molecules, it is evident that these densities are proportional to the individual masses of these molecules.

Hydrogen is the least dense of these gases and, in fact, is the least dense of all the elements. We also know that a molecule of hydrogen contains two atoms of hydrogen and is written H_2. The hydrogen atom is the lightest atom, and if we assign a relative weight of 1 to the hydrogen atom, then the hydrogen molecule must have a relative weight of 2; that is, its *molecular weight* is 2.

Table 3.2. Gas Densities at STP and Their Masses

Gas	Symbol	Calculated density, g/liter	Mass of 22.4 liters, g
Hydrogen	H_2	0.0899	2.015
Helium	He	0.1786	4.00
Nitrogen	N_2	1.250	28.02
Oxygen	O_2	1.428	32.00
Neon	Ne	0.9004	20.18
Chlorine	Cl_2	3.163	70.90
Argon	Ar	1.702	39.94
Krypton	Kr	3.739	83.81
Xenon	Xe	5.858	131.3
Carbon dioxide	CO_2	1.963	44.00
Ammonia	NH_3	0.7598	17.03
Methane	CH_4	0.7156	16.04
Ethane	C_2H_6	1.342	30.08
Propane	C_3H_8	1.967	44.09

Comparing the density of hydrogen to that of oxygen and knowing that oxygen is diatomic, we can see that one molecule of oxygen O_2 is 16 times as heavy as one molecule of hydrogen and that one atom of oxygen is 16 times as heavy as one atom of hydrogen.

It is also possible to determine the mass of each of the molecules in Table 3.2 *relative* to that of the hydrogen molecule. But for practical purposes, because oxygen is the most plentiful element and because it is heavier than hydrogen, it was the practice to define all atomic weights relative to the atomic weight of oxygen, which was defined originally as exactly 16.0000. If 1 liter of oxygen (STP) contains 1.428 g O_2, then 22.4 liters of O_2 (STP) will weigh 22.4×1.428 g, or 32.00 g. In the third column of Table 3.2, we have calculated the mass of each gas contained in 22.4 liters at STP. We can say that O_2 has the relative mass of 32 g; then, for example, exactly the same number of nitrogen molecules will have a mass of 28.02 g. The *molecular weight* of nitrogen is 28.02, based on the arbitrary assignment of 32 for the molecular weight of O_2.

It is important to note that 28.02 is not the mass of one molecule of nitrogen but the mass of that very large number of molecules which is contained in 22.4 liters at STP. Similarly, the numbers in the third column of Table 3.2 are molecular weights of the corresponding gases.

Table 3.3 lists atomic weights, which can be calculated from the molecular weights of the first nine entries in Table 3.2, that is, from those gases which are elements and from knowing whether the molecule is diatomic or monatomic. In addition to gas densities, there are several other techniques for obtaining molecular and atomic weights. The most precise of these makes it especially con-

Table 3.3. Atomic Weights of Some Common Elements†

Element	Symbol	Modern C standard	Older O standard
Hydrogen	H	1.0080	1.0080
Helium	He	4.0026	4.003
Nitrogen	N	14.0067	14.008
Oxygen	O	15.9994	16.000
Neon	Ne	20.179	20.183
Chlorine	Cl	35.453	35.457
Argon	Ar	39.948	39.944
Krypton	Kr	83.80	83.80
Xenon	Xe	131.30	131.30
Carbon	C	12.011	12.011

† A complete table of the atomic weights of all of the elements appears on the inside of the front cover of this book. For some elements, the change in standard *and* remeasurement of atomic weight resulted in no net change.

venient to use carbon rather than oxygen as the standard. There has been a small shift in all of the atomic weights in converting from oxygen to the new carbon standard which is now used. These are compared in Table 3.3 for several elements. Atomic weights are relative measures of mass, and using a carbon rather than an oxygen standard shifts all the atomic weight values by the same factor, keeping the ratio between any two atomic weights the same. In Chap. 5 we shall see why the carbon atomic weight is not an integer.

INDIVIDUAL ATOMIC AND MOLECULAR MASSES

The terms *atomic weight* and *molecular weight* are used for the mass of that very large number of atoms or molecules contained in 22.4 liters of gas at STP. The terms *atomic mass* and *molecular mass* are used for the mass of a single atom and a single molecule. The number of molecules contained in 22.4 liters of gas at STP has a special name; this *number* of molecules of a particular substance is said to be a *mole* of that substance. Since a mole is a definite number of molecules, we can have a mole of a substance which is a gas or a liquid or a solid. A mole of oxygen has a mass of 31.9988 g. The actual number of molecules present is very large, called "Avogadro's number" N_A, and is

$$N_A = (6.02252 \pm 0.00027) \times 10^{23}$$

We shall see later (page 216) how this number is determined.

If 1 mole of oxygen O_2 has a mass of 31.9988 g and consists of 6.02252×10^{23} molecules, then each molecule has a mass of

$$m = \frac{31.9988 \text{ g}}{1 \text{ mole}} \times \frac{1 \text{ mole}}{6.02252 \times 10^{23} \text{ molecules}}$$
$$= 5.3132 \times 10^{-23} \text{ g/molecule}$$

We say that the molecular weight M of oxygen is 31.9988 g, that the molecular mass m is 5.3132×10^{-23} g, and that $N_A m = M$, where N_A is Avogadro's number.

MOLECULAR WEIGHTS AND THE IDEAL GAS LAW

Since 1 mole of anything is a definite count, namely, 6.02252×10^{23} of that substance, it is easy to convert the number of molecules into the number of moles. If N is the number of molecules, N_A Avogadro's number, and n the number of moles, then

$$\frac{N}{N_A} = n$$

or

$$N = nN_A$$

Using this in the ideal gas law,

$$pV = NkT = nN_AkT = nRT$$

where

$$R = N_Ak = \frac{pV}{nT}$$

R is a universal constant.

The conversion from molecules to moles is no different from converting a number of eggs to dozens of eggs by dividing the former by 12 or converting sheets of paper to reams of paper by dividing by 500. Since we have defined the mole as the number of molecules of a substance contained in 22.4 liters at STP, we can evaluate the gas constant R:

$$R = \frac{pV}{nT}$$

$$= \frac{1 \text{ atm} \times 22.4 \text{ liters}}{1 \text{ mole} \times 273.15 \text{ K}}$$

$$= 0.0821 \text{ liter-atm/mole-K}$$

With this value, we can now do certain calculations describing the behavior of gases.

> ***Example.*** 1.00 g of gaseous benzene at 150°C and 1.00 atm pressure occupies 0.445 liter. Calculate the molecular weight of benzene.
>
> From the pressure, volume, and temperature data it is possible to calculate the number of moles of benzene in that 1-g sample:

$$n = \frac{pV}{RT}$$

$$= 1 \text{ atm} \times 0.445 \text{ liter} \times \frac{\text{mole-K}}{0.0821 \text{ liter-atm}} \times \frac{1}{423 \text{ K}}$$

$$= 0.0128 \text{ mole}$$

From

$$0.0128 \text{ mole benzene} = 1.00 \text{ g benzene}$$

$$1 \text{ mole benzene} = \frac{1.00 \text{ g}}{0.0128} \text{ benzene}$$

$$= 78.1 \text{ g benzene}$$

Thus the molecular weight of benzene is 78.1.

Example. 20.0 g of oxygen are contained in a vessel of 10.0-liters capacity and at 300°C. Calculate the pressure of the gas.

$$20.0 \text{ g O}_2 \times \frac{\text{mole O}_2}{32.0 \text{ g O}_2} = 0.625 \text{ mole of O}_2$$

$$p = \frac{nRT}{V}$$

$$= 0.625 \text{ mole} \times \frac{0.0821 \text{ liter-atm}}{\text{mole-K}} \times \frac{573 \text{ K}}{10.0 \text{ liters}}$$

$$= 2.94 \text{ atm}$$

THE MOTION OF MOLECULES

We can calculate the speed of gas molecules from the several equivalent forms of the ideal gas law

$$pV = \frac{N}{3} mu^2 \tag{3.1}$$

$$= NkT \tag{3.2}$$

$$= nRT \tag{3.3}$$

and the relation between the number of molecules and moles $N = N_A n$ and between molecular mass and molecular weight $M = N_A m$. In Eq. (3.1) the right-hand side can be altered:

$$N \frac{mu^2}{3} = \frac{(N/N_A)(N_A m)u^2}{3}$$

$$= \frac{nMu^2}{3}$$

and this is equal to the right-hand side of Eq. (3.3):

$$\frac{nMu^2}{3} = nRT = pV$$

from which

$$u^2 = \frac{3RT}{M}$$

and

$$u = \sqrt{\frac{3RT}{M}}$$

The speed is determined by the temperature and the molecular weight. For a given gas, the higher the temperature, the greater the speed. Comparing several gaseous substances all at the same temperature, the greater the molecular weight (and the greater the molecular mass), the slower the speed. In other words, the heavier the particle, the slower it moves. Since the speed varies inversely as the square root of the molecular weight and since the speed determines how fast a gas will escape through an opening made in the containing vessel, a method is available for estimating the molecular weight of a gas by the escape or *effusion* rate.

> **Example.** Oxygen at 1.5 atm pressure and 40°C is placed in a vessel fitted with a small orifice through which the gas can escape. The rate of escape is measured by the rate at which the oxygen pressure drops. Following this, the vessel is filled with ammonia at the same pressure and temperature and the escape rate measured. The ammonia escapes 37 percent faster than the oxygen. If the oxygen molecular weight is 32 g/mole, estimate the ammonia molecular weight.
>
> Since the temperature, volume, and initial pressure are the same for each gas, so must be the number of molecules and the number of moles. From the gas equations, if p, V, T, and n are

equal, then Mu^2 must be the same for the two gases. Then

$$M(\text{ammonia}) = M(\text{oxygen}) \left(\frac{u(\text{oxygen})}{u(\text{ammonia})} \right)^2$$

$$= 32 \left(\frac{1}{1.37} \right)^2$$

$$= 32 \times \frac{1}{1.88}$$

$$= 17$$

If a mixture of gases is escaping through a small hole, the lower-molecular-weight molecules will effuse faster. Air escaping from a balloon through a pinhole will contain more nitrogen and less oxygen than the original air within the balloon because nitrogen, being lighter, moves faster and escapes more rapidly.

CALCULATING THE SPEED OF A MOLECULE

To calculate the gas-molecule speed we use the formula

$$u = \left(\frac{3RT}{M} \right)^{\frac{1}{2}}$$

The value of R we obtained is 0.0821 liter-atm/mole-K. If we look at the dimension of the speed, it is

$$u = \left[3R \, \frac{\text{liter-atm}}{\text{mole-K}} \, T(k) \times \frac{1 \text{ mole}}{M \text{ g}} \right]^{\frac{1}{2}}$$

$$= \left(\frac{3RT}{M} \, \frac{\text{liter-atm}}{\text{g}} \right)^{\frac{1}{2}}$$

We should like the speed to have the units meters per second, and it is not obvious how the square root of liter-atmospheres per gram is related to meters per second.

The gas constant contains the dimension liter-atmosphere, which is the dimension of energy or work. A gas expanding in volume against a pressure does work which is equal to the product of the volume change and that pressure and which has the units of volume times pressure ($V \times p$). The kinetic energy of a moving body is $\frac{1}{2}mu^2$ and has the units mass times velocity squared, for example, gram meters–squared per square second. The energy in liter-atmospheres can be converted into the equivalent kinetic energy in gram meters–squared per square second, and the gas constant is

$$R = \frac{8.314 \times 10^3 \text{ g-m}^2}{\text{mole-K s}^2}$$

Now the dimension of speed is

$$u = \left[3R \, \frac{\text{g-m}^2}{\text{mole-K s}^2} \, T(\text{K}) \times \frac{1}{M} \frac{\text{mole}}{\text{g}} \right]^{\frac{1}{2}}$$

$$= \left(\frac{3RT}{M} \frac{\text{m}^2}{\text{s}^2} \right)^{\frac{1}{2}}$$

$$= \left(\frac{3RT}{M} \right)^{\frac{1}{2}} \frac{\text{m}}{\text{s}}$$

> **Example.** Calculate the speed of an oxygen molecule at 26.8°C given $M = 32.00$ g/mole and $R = 8.314 \times 10^3$ g-m²/mole-K s².
>
> $$u = \left(\frac{3 \times 8.314 \times 10^3 \text{ g-m}^2}{\text{mole-K s}^2} \times 300.0 \text{ K} \times \frac{\text{mole}}{32.00 \text{ g}} \right)^{\frac{1}{2}}$$
>
> $$= \left(\frac{3 \times 8.314 \times 3.000 \times 10^4}{3.200} \right)^{\frac{1}{2}} \frac{\text{m}}{\text{s}}$$
>
> This calculation can be done using logarithms:
>
> | Number | Logarithm |
> |---|---|
> | 3 | 0.4771 |
> | × 8.314 | +0.9198 |
> | × 3.000 | +0.4771 |
> | × 10⁴ | +4.0000 |
> | Numerator | 5.8740 |
> | ÷ 3.200 | −0.5051 |
> | | 5.3689 |
> | Square root | × ½ |
> | log $u =$ | 2.6844 |
> | $u =$ | 4.835 × 10² m/s |
>
> To convert this to miles per hour, we use the conversion
>
> 1 mile = 1.60935 km

Then

$$u = \frac{4.835 \times 10^2 \text{ m}}{\text{s}} \times \frac{\text{km}}{10^3 \text{ m}} \times \frac{1 \text{ mile}}{1.609 \text{ km}} \times \frac{60 \text{ s}}{1 \text{ min}} \times \frac{60 \text{ min}}{1 \text{ hr}}$$

$$= 1,080 \text{ mph}$$

Example. Hydrogen remains a gas at normal atmospheric pressure until it is cooled to $-252.8°C$. What is the speed of a hydrogen gas molecule at that temperature?

$$u = \left(\frac{3 \times 8.314 \times 10^3 \times 20.4}{2.00}\right)^{\frac{1}{2}} \text{ m/s}$$

$$= 5.04 \times 10^2 \text{ m/s}$$

Although the hydrogen temperature in this example is much less than the oxygen temperature in the previous example, the mass difference between the two molecules more than compensates for the low temperature; hydrogen has a high speed even at 20° above absolute zero.

MEASURING MOLECULAR SPEEDS

It should be interesting to confirm these calculated values through experiment and to corroborate the formula. Gas molecular speeds can be measured by the simple device shown in Fig. 3.6. Two wheels, each with a sector cut out, are placed on a rotatable axle.

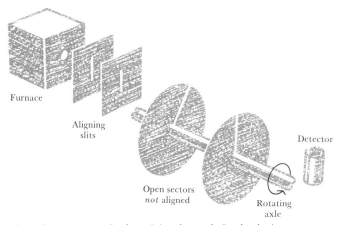

Figure 3.6. Apparatus for determining the speed of molecules in a gas.

The relative placement of the cutout sectors and the spacing between the wheels are adjustable. Gas molecules leave the source, e.g., a furnace, and pass through a set of slits, eliminating all molecules except those moving parallel to the axle. As the wheels rotate, a certain number of molecules pass through the open sector of the first wheel; the others hit the solid part of the wheel and are deflected. Those passing through the first wheel meet the second wheel and may either pass through or be deflected. Those which pass through both open sectors are counted by a detection device.

In order for a given molecule to go through both open sectors, it is necessary that the second opening be exactly in line with the moving beam of molecules as they reach the second wheel. If we know (1) the spacing between the wheels, (2) the rotation rate of the wheels, and (3) the relative angular displacement of the open sectors, we can calculate the speed of molecules hitting the detector.

Suppose the spacing between wheels is d cm, the speed of the molecules successfully passing through to the detection device is u cm/s, and the rotation rate is R turns/s. The transit time from one open sector to the other is

$$t = \frac{d \text{ cm}}{u \text{ cm/s}} = \frac{d}{u} \text{ s}$$

If the relative angular displacement of the open sectors is 90°, or a right angle, then the wheels should travel a quarter-turn in the transit time t so that the molecules can pass through to the detector. If the rotation rate R turns/s is known, we can see that in 1 s the wheel makes R turns; one turn is done in $1/R$ s; and a quarter-turn in one-fourth that time, or $1/4R$ s. This time interval must coincide with the transit time,

$$t = \frac{1}{4R} = \frac{d}{u}$$

thus

$$u = 4Rd \text{ cm/s}$$

For a given R and d, we can count how many molecules travel at a speed of u cm/s. There is a range of speeds in any sample of molecules, but the average speed is described by the equation

$$u = \left(\frac{3RT}{M}\right)^{\frac{1}{2}}$$

which confirms the kinetic model. The propagation of sound depends on molecular motion in air and is described by a similar equation with a constant of 1.4 rather than 3. Sound travels faster in hydrogen and helium than in air.

MOLECULAR WEIGHTS OF GASES†

We determined the molecular weight of benzene by using the ideal gas law. Specifically, 1.00 g of benzene under a pressure of 1.00 atm and at 150°C occupies a volume of 0.445 liter. Once the pressure, temperature, and volume are given, the *number of moles* of the substance can be precisely determined by the ideal gas law.

In our example, any substance obeying the ideal gas law, and at that pressure, temperature, and volume, must be of the amount of 0.0128 mole. From knowing that 0.0128 mole of benzene is 1.00 g of benzene, it follows that 1 mole is 78.1 g; 78.1 g is the molecular weight of benzene. For determining molecular weights, the important property of an ideal gas is that the pressure of the gas depends only on the *number of molecules*, or *number of moles*, of that gas contained in a particular volume at a particular temperature. This is Avogadro's hypothesis.

PROPERTIES OF SOLUTIONS

When we shake a mixture of fine sand and water, the result is a turbid liquid which, upon close examination, looks exactly like a collection of fine sand grains suspended in water. In time the sand will settle out, leaving clear water. Alternatively, we can force the mixture through a fine filter, which retains the sand and permits only water to pass. If we had enough patience we should discover

---

† Many substances cannot be converted to a gas without decomposition or without some chemical change. Solid sugar is converted to caramel rather than to gaseous sugar, for example. To determine the molecular weight of sugar, techniques other than gas density and gas effusion are needed.

that nothing at all happened to the individual grains of sand, other than being made wet by the water. The number of grains before mixing and after separation are exactly the same. The shape and size of each grain is also unchanged by the experience.

When we shake a mixture of sugar and water, a clear liquid results which, upon visual examination, gives no evidence of a white granular solid. No matter how long we wait, no sugar will settle out of the closed container. Nor will filtration separate the sugar from the water. We can effect a separation by heating to evaporate off the water. Crystals of sugar will form. But the number, shape, and size of these crystals will differ from the number, shape, and size of the original granules of sugar.

The sand and water mixture is a *suspension* and the sugar and water mixture is a *solution*. It is evident that the granules of sugar are broken down into extremely small pieces, invisible to the eye and, in fact, too small to be seen through a microscope. The solution is *homogeneous* throughout, which means that we cannot detect, even with a microscope, the presence of two kinds of substances in the solution. The suspension of sand and water is said to be a *heterogeneous* mixture since we can distinguish between grains of sand and water.

Solutions are made up of a *solvent*, the liquid which does the dissolving (in our example, water), and one or more *solutes*, the substances dissolved (in our example, sugar). Solutions have different properties than pure solvents. A sugar solution boils above 100°C and freezes below 0°C. In contrast, ice forms in a sand and water suspension at the normal freezing point, and steam forms at the normal boiling point. The difference between the boiling points of a solution and the pure solvent depends on the solute concentration — the more concentrated the solution, the greater the boiling-point elevation. If the solutions are not too concentrated, the boiling-point elevation varies only with concentration of solute and is independent of the nature of solute.

Two other properties of solutions that depend only on the moles of solute, not on the kind of solute, are *osmosis* and *vapor pressure*. Osmosis is the transfer of pure solvent through a membrane separating two solutions of different concentrations. Vapor pressure is the pressure of the gas which evaporates from a pure solvent or solution. Both osmosis and vapor pressure are used to determine molecular weights. The simplest property to discuss in detail is vapor pressure, which will be the subject of the next few sections. We shall look at several aspects of this property and then construct a molecular model.

THE PHENOMENON OF VAPOR PRESSURE

Suppose, as in Fig. 3.7, a liquid is placed inside a cylinder and the region above the liquid surface is evacuated. Some liquid evaporates into the free space, and the gas pressure is measured on the gauge. The entire system, liquid-vapor-piston-cylinder, is kept at a constant temperature. If the liquid is water and the temperature is 30°C, the measured pressure is 31.8 torr, where torr is the pressure of a column of mercury 1 mm high at 0°C.

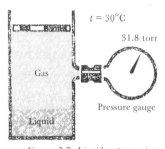

Figure 3.7. Liquid-vapor system in a cylinder equipped with a piston and pressure gauge.

If the piston is abruptly moved in, the gas volume is decreased and, from the ideal gas law, the pressure must increase to some value above 31.8 torr. However, there will be a condensation of water from the vapor to the liquid region, and, according to the ideal gas law, the pressure will drop. It will be found, again, to be 31.8 torr. Similarly, if the piston is abruptly moved up, the gas volume expands and, momentarily, the water vapor pressure drops. Very soon, however, some liquid will evaporate, adding more water to the gas region, until the gas pressure is 31.8 torr.

We say the *equilibrium vapor pressure* of water at 30°C is 31.8 torr. It is an *equilibrium* value since it is independent of the circumstances existing before the equilibrium value was established, and since it remains constant if the temperature is fixed. Table 3.4 lists the equilibrium vapor pressures for pure water at several temperatures.

If we repeat the vapor-pressure measurements over a temperature range 60 to 95°C but this time add the nonvolatile solid (i.e., a solid which does not evaporate significantly) sugar to water, we shall find again that at each temperature there is a well-defined, reproducible equilibrium vapor pressure. These data are found in the third and fourth columns of Table 3.4. The two solutions contain 818 g and 9 kg of sugar, respectively, in 1 kg of water. The latter solution is an especially viscous liquid, or syrup. The first observation that we make is that the measured equilibrium vapor pressure increases when the temperature is increased. Second, the measured vapor pressure of water decreases with the addition of sugar; the more sugar, the greater the decrease.

If we wish to see by how much the water vapor pressure is depressed, we can divide the vapor pressure over a solution by the

Table 3.4. Equilibrium Vapor Pressure of Water over Pure Water and over Two Aqueous Solutions of Sugar at Several Temperatures

t, °C	Equilibrium vapor pressure, torr		
	Pure water	818 g sugar/1 kg water	9 kg sugar/1 kg water
0	4.58		
5	6.54		
10	9.21		
15	12.8		
20	17.5		
25	23.8		
30	31.8		
35	42.2		
40	55.3		
45	71.9		
50	92.5		
55	118		
60	149	142	64.4
65	188	178	81.4
70	234	221	102.2
75	289	275	127
80	355	337	157
85	434	412	194
90	526	500	236
95	634	603	287
100	760		
105	906		
110	1,075		
115	1,268		

vapor pressure over pure water, both at the same temperature. These ratios are entered in Table 3.5.

Although the ratio for either solution over the temperature range is not exactly constant, it does not change by much—less than 1 percent in the first solution and about 4 percent in the second. The vapor-pressure lowering due to the presence of a nonvolatile solute and expressed as a fraction of the pure-water vapor pressure seems to be almost independent of temperature changes but characteristic of the nonvolatile solute concentration.

Next, we can make a graph of the three equilibrium vapor pressures over the measured temperature range. In Fig. 3.8 one heavy solid line is a graph of the vapor pressure over pure water for a range of temperatures. If we add sugar to the water, the vapor pressure is decreased and the data for the two sugar solutions are represented by the two lower curved lines in Fig. 3.8. The

Table 3.5. Relative Vapor-pressure Lowering

t °C	Ratio of solution to pure solvent vapor pressure	
	818 g sugar/1 kg water	9 kg sugar/1 kg water
60	0.953	0.432
65	0.947	0.433
70	0.944	0.437
75	0.952	0.439
80	0.949	0.442
85	0.949	0.447
90	0.951	0.449
95	0.951	0.453

horizontal line represents a pressure of 1 atm, or 760 torr. When the vapor pressure is 760 torr and the atmosphere is also at 760 torr, the liquid boils and the temperature is the boiling point of the liquid, as indicated in the figure. It is evident that a solution must

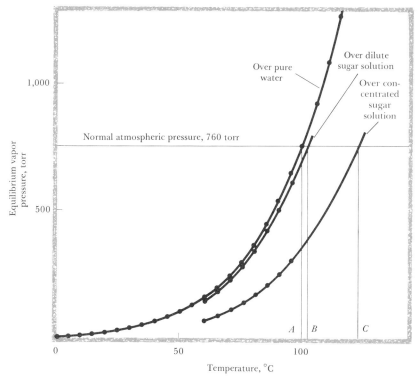

Figure 3.8. Equilibrium vapor pressures as a function of temperature where A is the boiling point of pure water, B is the boiling point of a dilute sugar solution, and C is the boiling point of a concentrated sugar solution.

be raised to a higher temperature in order for the vapor pressure of the water in the solution to be at 1 atm. Therefore, a nonvolatile solute solution boils at a higher temperature than the pure water. The two solutions boil at 101.4 and 122.8°C, respectively.

THE TEMPERATURE OF BOILING WATER

The normal boiling point of water at sea level is 100°C; that is, at 100°C the vapor pressure of water is 1 atm, the normal air pressure at sea level. At any elevation above sea level, the normal air pressure is less than 1 atm; a temperature less than 100°C is required to bring the water vapor pressure up to the air pressure of the surroundings. The boiling point of water decreases with an increase in elevation above sea level.

Salt Lake City, Utah, has an elevation of about 1,350 m above sea level and a normal atmospheric pressure of 650 torr. Pure water has a vapor pressure of 650 torr at 95.7°C. Therefore, in Salt Lake City water boils at 95.7°C. Mauna Kea, Hawaii, with an elevation of 3,981 m above sea level, has a normal atmospheric pressure of 462 torr. There, water boils at 86.6°C. At a still higher elevation of 4,800 m and with a normal atmospheric pressure of 424 torr, water boils at 84.4°C on Mont Blanc, France.

Because there is locally a continual change in the atmospheric pressure associated with meteorological changes, the boiling point also continually changes. At an atmospheric pressure of exactly 760.00 torr, the pure-water boiling point is 100.00°C. But when the ambient, or surrounding, pressure drops to 749.20 torr, the boiling point drops to 99.6°C. With an ambient pressure of 765.45, the boiling point is 100.2°C.

Since the efficiency of a steam engine can be increased by raising the boiler temperature, it is common to use pressure above the normal atmospheric value within the boiler. At a pressure of 2,250 torr (three times the normal pressure), water boils at 134°C rather than at 100°C.

RELATIVE HUMIDITY

Air does not always contain water vapor at its equilibrium-vapor-pressure value. If the water vapor pressure at 20°C is actually 14.1 torr, it is 14.1/17.5, or 0.805, of the equilibrium value. We say the *relative humidity* is 80.5 percent. If the temperature is 30°C and the relative humidity is 50 percent, the actual water vapor pressure is 31.8×0.50, or 15.9 torr. This particular water vapor pressure is the equilibrium, or saturation, value at about 18°C. If the air at 30°C and at 50 percent relative humidity is cooled to 18°C or below, we should expect some condensation of water from the vapor to the liquid state. This is the basis of rain and dew formation.

Suppose the outside weather air is at 0°C and the relative humidity is 90 percent; the actual water vapor pressure is 4.58×0.90, or 4.12 torr. When this air is drawn into a house at 25°C, this pressure is 4.12/23.8, or 0.173, of the saturation value at 25°C. The inside relative humidity is 17.3 percent. In winter it is common to have low relative humidities indoors.

EVAPORATION AND CONDENSATION

We can now look at a molecular model for the equilibrium vapor pressure of a solvent and the way it is affected by the presence of a nonvolatile solute. From the Bernoulli derivation of the ideal gas law, our concept of pressure is that of force per unit area, at the gas container's surface, which comes from the collisions of gas molecules at the surface. Pressure, then, is a measure of how many collisions per second per unit area the gas molecules make with a container's surface.

It is reasonable to assume that, as in the experiments measuring equilibrium vapor pressures over pure water and over sugar solutions, water molecules moving about in the gas strike the piston and cylinder walls and also the liquid water surface in contact with the gas. Further, we can imagine that either all or some definite fraction of those gas molecules hitting the liquid surface condense, i.e., become liquid. The rate of condensation, then, is proportional to the number of molecules reaching the liquid surface and thus proportional to the gas pressure. The rate, the number of molecules condensing per second per unit area, is

Condensation rate $= k_1 p$

where p is the gas pressure and k_1 the constant of proportionality which depends on the gas molecular speed and, ultimately, on the molecular mass and temperature of the gas.

With condensation there should be a decrease in the number of gas molecules and in the gas pressure. But the equilibrium vapor pressure does not change with time; therefore, if condensation is to occur, it must be balanced by a process in which molecules are transferred from the liquid to the gas, namely, *evaporation*. The rate of evaporation, the number of molecules escaping per second and per unit area from the liquid surface, must depend upon how many liquid molecules there are per unit area and the probability of escape for each one.

THE KINETIC THEORY OF LIQUIDS AND DYNAMIC EQUILIBRIUM

A gas is *compressible*, which means that when we apply a pressure slightly greater than the gas pressure, the gas volume is reduced. In contrast, a liquid is not very compressible. Very high pressures must be applied to reduce the liquid volume by just a small amount. A comparison of the densities of liquid and gaseous water (at STP) — 1.00 g/cm^3 and 8.04×10^{-4} g/cm^3 — indicates that the packing of gas molecules is only $1/1{,}240$ as dense as the packing of liquid molecules.

It follows, then, that a gas must be mostly empty space with large distances between molecules. On the other hand, molecules within a liquid are closely packed, and the number within a given volume or at a given surface cannot be easily changed.

Below its boiling point a liquid is stable: it does not spontaneously and abruptly fall apart and become gaseous. Forces of attraction must act between the molecules for a liquid to exist. Under the proper circumstances, a liquid can evaporate. Therefore, these attractive forces can be overcome.

We have already seen that if we identify the mean kinetic energy of gas molecules with the absolute temperature, we have a formula for calculating the molecular speed. The experiment with the rotating wheels (page 91) indicated that this is an average, or mean, speed and the molecular speeds are distributed over a range of values, with some moving very fast and others very slowly.

We can apply the same kinetic-molecular model to the description of a liquid. The absolute temperature is related to the mean molecular speed within the liquid. The range of speeds includes some fast molecules and some slow ones. If the liquid temperature is constant, the mean speed is also constant and the fractions of those molecules moving very fast and very slow must also be constant. If a significant number of high-speed molecules leaves the liquid, the average speed of those remaining is less than the average speed before removal. The temperature drops. For the liquid temperature to remain constant, some heat must flow into the liquid, making it possible for some of the slower molecules to move faster and reestablishing the mean-speed characteristic of the original temperature.

A common example of this process that can be described with the kinetic-molecular model is the cooling effect experienced when rubbing alcohol is evaporating from skin. As the faster alcohol molecules leave, the alcohol temperature drops below the skin temperature. This thermal imbalance is attended by a flow of heat from the warmer skin to the cooler alcohol. The heat flow continues until the two temperatures are equal or until all the alcohol is evaporated.

If the liquid water is maintained at a constant temperature and if a certain minimum energy is needed for escape, the fraction of molecules having this minimum energy and minimum speed will be constant. Where the number of molecules per unit area and the fraction of these with the necessary escape speed are constant, the number leaving per second and per unit area is also constant as long as the temperature is fixed. We can write

Evaporation rate $= k_2$

where k_2 is constant.

If the evaporation rate exactly equals the condensation rate,

$$k_2 = k_1 p$$

or

$$p = \frac{k_2}{k_1} = \text{constant}$$

The equilibrium vapor pressure p is equal to the ratio of two constants and is also constant provided the temperature does not change. Since we are discussing here the equilibrium vapor pressure over a pure liquid, we shall find it convenient to designate the pressure by a superscript, and so

$$p^0 = \frac{k_2}{k_1} = \text{constant}$$

In the experiments measuring vapor pressures within the cylinder and movable piston, we saw that an abrupt gas-volume contraction, by moving the piston in, was attended by an abrupt gas-pressure increase followed by a gas-pressure decrease to the equilibrium value. In our kinetic-molecular model the abrupt pressure increase is associated with an increase in collisions between gas molecules and the liquid surface. Within the smaller space, the molecules have a shorter distance to travel between collisions with the surface and, thus, can make more collisions. This increased collision rate brings about an increased condensation rate. If the condensation rate is greater than the rate at equilibrium, condensation goes faster than evaporation. There is a net transfer of molecules from gas to liquid until the gas pressure drops to the equilibrium value. Then the condensation rate equals the evaporation rate (Fig. 3.9).

The equilibrium vapor pressure p^0, then, depends on the two parameters k_1 and k_2, each of which varies with temperature. As the temperature changes, so does each k_1 and k_2 and so does the ratio k_2/k_1. It follows that p^0 also should change with temperature. It turns out that the condensation-rate constant k_1 increases much more slowly than the escape-rate constant k_2. The ratio k_2/k_1 and the pressure p^0 become larger as the temperature increases.

When the piston is abruptly moved up, the gas pressure drops below its previous equilibrium value ($p < p^0$). Confined to a larger volume, each molecule has a longer path to travel between collisions with the surfaces and thus makes less frequent collisions. The condensation rate drops below the evaporation rate, which is unaffected by the change in gas volume. There is a net transfer from liquid to gas. As this transfer continues, the gas pressure and with it the condensation rate increase until condensation again equals evaporation.

In this model, equilibrium is the consequence of balance between two dynamic processes, evaporation and condensation, and, for this reason, is called "dynamic equilibrium." Now we shall see how this model can describe the vapor-pressure lowering when a nonvolatile solute is added to the pure solvent.

The presence of a nonvolatile solute reduces the number of solvent molecules at the surface and, in turn, the evaporation rate.

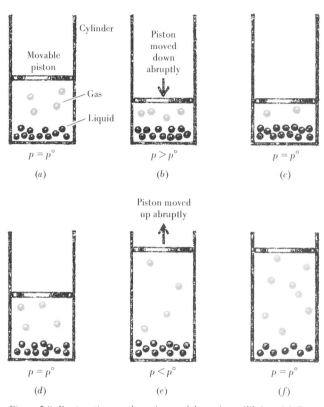

Figure 3.9. Evaporation, condensation, and dynamic equilibrium. (a) Condensation rate equals evaporation rate—dynamic equilibrium; (b) condensation rate exceeds evaporation rate—condensation; (c) condensation rate again equals evaporation rate—dynamic equilibrium; (d) condensation and evaporation rates are balanced—dynamic equilibrium; (e) condensation rate is less than evaporation rate—evaporation; (f) again, condensation equals evaporation rate—dynamic equilibrium.

When a nonvolatile solid is added to a pure liquid in dynamic equilibrium with vapor, the evaporation rate from the surface drops below the condensation rate. There is a net transfer of material to the liquid and a decrease in the vapor pressure until the condensation rate falls to a lower value equal to the smaller evaporation rate. At dynamic equilibrium, the two rates balance each other and the vapor pressure is smaller over the solution than it is over the pure solvent (Fig. 3.10).

The exact description of vapor-pressure lowering involves a method for determining solute molecular weights, which is the subject of the next section.

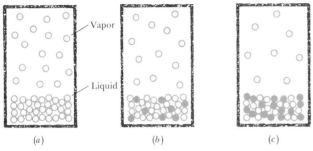

Figure 3.10. Lowering of the solvent vapor pressure by the addition of a nonvolatile solute. (a) Pure liquid water in equilibrium with water vapor; (b) one-fourth nonvolatile solute, three-fourths volatile solvent—therefore, the vapor pressure at equilibrium is three-fourths the equilibrium vapor pressure over pure solvent; (c) one-half nonvolatile solute, one-half volatile solvent—therefore, the equilibrium vapor pressure is one-half the equilibrium vapor pressure over pure solvent.

VAPOR–PRESSURE LOWERING AND SOLUTE MOLECULAR WEIGHT

The escape rate of molecules from a pure solvent is written as k_2 molecules/s-cm². When a nonvolatile solute is added, the escape rate of solvent molecules is less than this because at the surface there are not as many solvent molecules per unit area. The simplest statement of the escape rate from a solution is

Evaporation rate $= k_2 x_1$ molecules/s-cm²

where x_1 is the fraction of all the molecules at the surface which are solvent.

At one extreme, if no solute is present, all the surface molecules are solvent ($x_1 = 1$) and the evaporation rate is k_2. At the other extreme, if the solvent is completely absent, there is none at the liquid surface to escape ($x_1 = 0$) and the evaporation rate must be zero.

The vapor over a solution of nonvolatile solute contains only solvent molecules at a pressure p. The condensation rate is

$k_1 p$ molecules/s-cm²

As before, if the condensation rate exceeds the evaporation rate, there is a net transfer of material to the liquid. This is accompanied by a drop in the vapor pressure and a drop in the condensation rate. When this rate exactly equals the evaporation rate, dynamic equilibrium is established.

If the evaporation rate is greater than the condensation rate, a net transfer of solvent from solution to vapor takes place and with it an increase in the vapor pressure and in the condensation rate. The net transfer continues until the condensation rate exactly equals the evaporation rate. Again, equilibrium is attained.

At equilibrium over a solution

$$k_1 p = k_2 x_1$$

while at equilibrium over a pure solvent

$$k_1 p^0 = k_2$$

If we divide the former equation by the latter,

$$\frac{k_1 p}{k_1 p^0} = \frac{k_2 x_1}{k_2}$$

and

$$\frac{p}{p^0} = x_1$$

or

$$p = p^0 x_1$$

This equation is called "Raoult's law" in honor of Francois Marie Raoult (1830–1901), who first suggested that the equilibrium vapor pressure over a solution depends on the molecular fraction of the solution which is solvent. In other words, the equilibrium vapor pressure over a solution of volatile solvent and nonvolatile solute is equal to the pure solvent equilibrium vapor pressure times that solvent molecular fraction.

We cannot yet calculate the number of molecules of either solvent or solute in a solution. Both numbers are needed to obtain the fraction which is solvent x_1. However, we can calculate something which is just as useful, the number of moles of solvent and of solute. The fraction of the total number of moles which is solvent is exactly equal to the fraction of the total number of molecules which is solvent.

To calculate the number of moles of solute and of solvent we must know the mass and molecular weight of each. For example, by the gas density method, the molecular weight of water is 18.015 g/mole. Then 1 kg of water is

$$1,000.00 \text{ g water} \times \frac{\text{mole water}}{18.015 \text{ g water}} = 55.508 \text{ moles water}$$

When we dissolve 50.0000 g of sucrose in 1 kg of water, we have 55.508 moles of water. To calculate the moles of sugar and then the water mole fraction, we must convert the 50 g of sucrose into the number of moles of sucrose. For this we need what is not yet known to us—the molecular weight of sucrose. Conversely, if we know the mole fraction which is water and if we know how many moles of water are in a solution, we can calculate the moles of sucrose dissolved. Comparing the moles of sucrose with the grams of sucrose, we obtain the grams per mole, or the molecular weight, of sucrose.

> **Example.** At a carefully controlled temperature of $25.0000 \pm 0.0005°C$, the equilibrium vapor pressure of water over pure water is 23.756 torr, and over the solution of 50.0000 g sucrose/1000.00 g water it is 23.693 torr. From Raoult's law the water mole fraction is
>
> $$x_1 = \frac{p}{p^0}$$
> $$= \frac{23.693}{23.756}$$
> $$= 0.99735$$

This means that 99.735 percent of all the molecules in that solution are water and 0.265 percent is sucrose.

Suppose we let n_2 be the number of sucrose moles in that solution of 1,000.00 g, or 55.508 moles, of water. The total number of moles is $55.508 + n_2$. The fraction of water in the solution is

$$x_1 = \frac{55.508}{55.508 + n_2}$$

From setting

$$x_1 = \frac{55.508}{55.508 + n_2} = 0.99735$$

and inverting the second and third terms

$$\frac{55.508 + n_2}{55.508} = \frac{1}{0.99735}$$

$$\frac{55.508}{55.508} + \frac{n_2}{55.508} = 1.0027$$

$$1 + \frac{n_2}{55.508} = 1.0027$$

$$\frac{n_2}{55.508} = 0.0027$$

$$n_2 = (55.508)(0.0027)$$
$$= 0.15 \text{ mole sucrose}$$

Since

$$0.15 \text{ mole sucrose} = 50.0000 \text{ g sucrose}$$

$$1 \text{ mole sucrose} = \frac{50.0000 \text{ g sucrose}}{0.15}$$

$$= 330 \text{ g sucrose}$$

This value differs from the best experimental value of 342.30 g/mole for two reasons: The first reason involves the experimental aspects of measurement, and the other is based on theoretical considerations.

Even though the vapor pressures have been determined to five significant figures with an uncertainty of about one part in twenty-three thousand, the difference between the two pressures is not large. When 1 is subtracted from the ratio, we are left with 0.0027 with an uncertainty of 1 in the last place of a two-significant-figure value; an error of $\frac{1}{27}$, or 3.7 percent, cannot be avoided.

The correctness of this method for determining the sucrose molecular weight is based upon the validity of Raoult's law. A solution which is described by this law is said to be an *ideal solution*, just as a gas described by the ideal gas law is said to be an ideal gas. In practice not many solute-solvent systems are ideal solutions. It has been observed that the more dilute the solution is, the more accurately its vapor pressure is described by Raoult's law. (Similarly, the more dilute the gas, that is, the lower its pressure, the closer to ideal it behaves.)

A dilemma exists. The more dilute the solution, the more accurate is Raoult's law and the more valid is this method. But when the solution is more dilute, the change in the vapor pressure is less, the experimental uncertainty is greater, and the error in the calculated molecular weight is larger. At higher concentrations, the vapor-pressure lowering is greater and we may use more significant figures but Raoult's law is less valid and our inaccuracy in calculating the solvent mole fraction is larger. When precision is low, accuracy is high; when precision is high, accuracy is low. This is a common condition in the sciences.

NONIDEAL BEHAVIOR

The departure of behavior from ideality, both of the equilibrium vapor pressure from the Raoult's law prediction and of the gas pressure from the ideal-gas-law calculation, is not a pure misfortune. The discrepancy between actual, experimentally observed behavior and that predicted by simple laws can be used for learning more about the solution or the gas studied.

In the case of a nonideal gas, a smaller actual pressure, compared with what an ideal gas pressure should be, can be interpreted as a measure of a smaller effective number of gas molecules; that is, a measure of association or clustering of molecules in the gas can be obtained. Especially at temperatures or pressures near liquefaction, attraction between gas molecules can be expected to be strong, clusters likely, an effective reduction in the number of gas molecules understandable, and nonideal behavior predictable.

When a solution departs from ideal behavior, the measured equilibrium solvent vapor pressure can be either greater or smaller than that predicted by Raoult's law. When greater, we interpret this as a repulsion between the two substances. An extreme example is two immiscible liquids. The vapor pressure over such a mixture is the sum of the vapor pressures of the pure liquids. Although aniline boils at 184°C, a mixture of it and water together has a vapor pressure of 760 torr at about 98°C; both liquids boil off together. This is a common technique used to boil off liquids which decompose at their normal boiling points. When the equilibrium vapor pressure is smaller, we interpret this as a strong solute-solvent attraction, which reduces the solvent-escape rate to a value even smaller than we would expect by simply diluting the solvent concentration. Typically, such solutions liberate heat when solute and solvent are mixed. A good example of this is the mixing of concentrated sulfuric acid and water which is attended by a substantial release of energy, which may be enough to boil the water. The equilibrium water vapor pressure is only a small percent of the Raoult's law predicted value.

> **Example.** From the sucrose-water data, the equilibrium water vapor pressure over the concentrated 90 percent sugar syrup at 95°C is 286.68 torr and over water at the same temperature is 633.90 torr. Knowing the molecular weights of solute and solvent, 342.30 and 18.015 g/mole, and the respective masses, 9,000 and 1,000 g, we can calculate the solvent mole fraction:

$$9,000 \text{ g sucrose} \times \frac{\text{mole sucrose}}{342.30 \text{ g sucrose}} = 26.29 \text{ moles sucrose}$$

$$1,000 \text{ g water} \times \frac{\text{mole water}}{18.015 \text{ g water}} = 55.51 \text{ moles water}$$

> Then 26.29 moles sucrose plus 55.51 moles water are 81.80 total:

$$x_1 = \frac{55.51}{81.80} = 0.6786$$

> Ideally

$$p = 0.6786 \times 633.90 \text{ torr}$$
$$= 430.2 \text{ torr}$$

$$\frac{p \text{ (observed)}}{p \text{ (ideal)}} = \frac{286.68}{430.2} = 0.6664$$

> The measured equilibrium water vapor pressure is only 66.64 percent of the ideal value. This indicates a strong water-sucrose attraction.

BOILING–POINT ELEVATION

We have already seen that the two sugar solutions boil at temperatures above the pure-water boiling point. This is because at 100.0°C the water vapor pressure has been reduced below 760.00 torr by the diluting presence of nonvolatile solute. To increase the water vapor pressure to 760.00 torr, a higher temperature is needed. For each mole of nonvolatile solute dissolved in 1 kg of water, the boiling point increases by 0.512°C.

> **Example.** When 10.00 g of sucrose are dissolved in 100.00 g of water, the resulting solution boils at 100.15°C. What is the solute molecular weight?
>
> From the boiling-point elevation we can calculate the sucrose concentration in moles per kilogram of water:

$$0.15° \text{ elevation} \times \frac{1 \text{ mole solute/kg water}}{0.512° \text{ elevation}}$$

$$= \frac{0.15}{0.512} \frac{\text{mole solute}}{\text{kg water}} = 0.29 \text{ mole solute/kg water}$$

It is the sucrose concentration rather than the total amount of sucrose present which determines the boiling-point elevation. The concentration

$$\frac{10.00 \text{ g sucrose}}{100.00 \text{ g water}} = \frac{100.0 \text{ g sucrose}}{1,000.0 \text{ g water}}$$

Therefore

$$\frac{0.29 \text{ mole solute}}{\text{kg water}} = \frac{100.0 \text{ g sucrose}}{1,000.0 \text{ g water}}$$

$$0.29 \text{ mole solute} = 100.00 \text{ g sucrose}$$

$$1 \text{ mole} = \frac{100.0 \text{ g sucrose}}{0.29}$$

$$= 350 \text{ g sucrose}$$

FREEZING–POINT DEPRESSION

When an aqueous solution freezes so that pure ice but no solute solidifies, the freezing-point depression is 1.86°C for each mole of nonfreezing-out solute dissolved in 1 kg of water.

> **Example.** A solution of 5.000 g of sucrose in 50.00 g of water freezes at −0.544°C. From this, calculate the solute molecular weight.

$$\frac{5.000 \text{ g sucrose}}{50.00 \text{ g water}} \times \frac{1,000.0 \text{ g water}}{\text{kg water}} = \frac{100.0 \text{ g sucrose}}{\text{kg water}}$$

$$0.544° \text{ depression} \times \frac{1 \text{ mole solute/kg water}}{1.86° \text{ depression}} = \frac{0.2925 \text{ mole solute}}{\text{kg water}}$$

$$0.2925 \text{ mole solute} = 100.0 \text{ g sucrose}$$

$$1 \text{ mole} = \frac{100.0}{0.2925} \text{ g sucrose}$$

$$= 343 \text{ g sucrose}$$

The freezing-point-depression method gives the more accurate molecular weight for sucrose because the depression in freezing point 1.86°C is greater than the elevation in boiling point 0.512°C, making the former technique more sensitive by a factor of 3.6.

Although the salt content, or salinity, of seawater varies from place to place within the ocean and so, too, must the seawater freezing point, a typical freezing point is −1.89°C, or 28.6°F. As some water is converted to ice, the concentration of the remaining salt in seawater must increase. With this, the equilibrium freezing point decreases. Therefore, in our example, the first ice appears at −1.89°C but the ice and seawater mixture must be cooled further below this temperature for additional ice to form.

When enough water, in the form of ice, has been removed from a sample of seawater, the salt concentration increases until the remaining water cannot dissolve all the salt. Then both salt and water freeze out. However, this does not occur until the seawater is cooled to about −8.3°C, or 17°F.

THE EFFECT OF DENSITY ON FREEZING. An interesting feature of the cooling and freezing of water can be described by the densities of water and ice. As cold air blows over a body of water, the top layer of liquid is cooled. The volume of water contracts as it is cooled to a temperature of 3.98°C. Since the same mass occupies a smaller volume, the cooler water is denser than that warmer water beneath and sinks. Warmer water from beneath, being less dense than the colder exposed layer, rises to the surface, in turn, to be cooled by the cold air. In this way there is a continuous circulation of water until it is all cooled to the air temperature. The motion of the fluid because of density differences is called "convection."

Water has a very curious property in that the density does not increase with further cooling below 3.98°C. At this temperature water is at its maximum density and as the temperature drops from that value to the freezing point, the density actually decreases. Between 3.98°C and the freezing point, the top layer of water is both colder and less dense than the water beneath. It does not sink, and convection ceases. If the air is sufficiently cold, it will continue to carry away heat from the top layer of water. The temperature of this layer drops until the freezing point is reached, either at 0°C for pure water or below that temperature for a solution, and ice forms.

Another special property of water plays an important role. As it freezes, the ice expands so that it has a volume about 8.3 percent larger than the water. The density of ice is about 0.92 that of the liquid, and ice floats on the liquid. When ice covers the water, it serves as an insulator, separating the cold air from the water, and the further formation of ice is slowed down or even prevented. Heat flowing into the water from solid earth is sufficient to compensate for the heat carried away by cold air through the ice layer.

Because bodies of water can remain fluid even when the air temperature is considerably below the water freezing point, organisms living in the water almost always survive even in the coldest weather. There are problems, however, from excluded sunlight and a reduced oxygen supply.

As already discussed, seawater must be cooled to −8.3°C, or 17°F, before solid salt freezes out from the water. These temperatures are uncommon in seawater. Icebergs, for this reason, are almost purely water. Some unfrozen droplets of salt solution, or brine, may be entrapped during ice formation. With some special treatment icebergs can be melted for potable, that is, drinkable, water. It may be practical someday to tow a large iceberg to a coastal city which requires more fresh water than can be supplied from continental sources and thus help to solve the local water problem.

OSMOTIC PRESSURE

Consider this simple experiment. A raw potato, shown in cross section in Fig. 3.11, is prepared in this manner. First a slice is removed to make a flat surface for standing on a dish. At the top a hole is bored to a depth of 1 to 2 in. This is filled about two-thirds full with ordinary table sugar. Observe the hole and its contents over a period of hours, including a final observation the following morning. It will be seen that the sugar becomes moist and in some cases

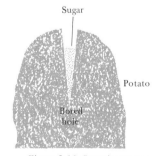

Figure 3.11. Osmotic pressure experiment using a potato slice.

the liquid fills the hole and may even run over the side of the potato.

To better study the process by which water is transferred from the potato to the sugar, we shall redesign the experiment. In Fig. 3.12 we have a thistle tube, a membrane, a sugar solution, and a beaker of pure water arranged as shown. The membrane selected is one which will allow water but not sugar to *permeate*, or pass through, it. It is called a "semipermeable membrane" and represents the surface between the potato and the sugar syrup through which water passed.

The beaker of water replaces the potato which was the source of water in our first experiment. About three-fourths of the mass of a raw potato is water, and so this feature of the experiment is not as drastic as it may seem. The membrane is secured over the wide end of the thistle tube, which is partially filled with sugar solution. Then the thistle-tube, membrane, and sugar-solution system is placed in the beaker of water.

The level of the liquid in the narrow part of the thistle tube will rise to a definite height above the surface of the water in the beaker. This height depends upon the concentration of sugar in the solution; the greater the sugar concentration, the greater the height. The liquid rises in the thistle tube because of an influx of water from the beaker through the membrane. The tendency of the water to pass through the membrane depends upon the fact that the concentration of water in the solution is less than the concentration of water in pure water.

The height of the liquid in the thistle tube exerts a hydrostatic force opposing the influx of water. When the liquid in the thistle tube reaches its final level, the hydrostatic pressure exactly balances the influx tendency. Equilibrium is established, and there is no further net transfer of water through the membrane. Solvent transfer, or *osmosis*, stops. We say that the height of the liquid in the thistle-tube column is the *osmotic pressure* of the solution inside the thistle tube. The osmotic pressure can be expressed as the height of the solution in the column above the beaker water level. But to know precisely what pressure corresponds to this height the solution density must be known. For comparison, it is more convenient to express all osmotic pressures in terms of atmospheres.

The equation which describes the dependence of the osmotic pressure π on the concentration of the solute in dilute solution is

$$\pi = cRT$$

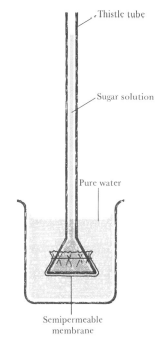

Thistle tube

Sugar solution

Pure water

Semipermeable membrane

Figure 3.12. Thistle-tube and membrane experiment to measure osmotic pressure.

where c is the number of moles of solute in 1 liter of solution and R is the ideal gas law constant 0.08205 liter-atm/K-mole. If we measure the osmotic pressure, we can calculate the solute concentration c (moles per liter). Knowing how many grams of solute per liter of solution produced this osmotic pressure, we obtain the solute molecular weight.

> **Example.** A solution containing 171.2 g sucrose/liter at 10.0°C has an osmotic pressure of 12.30 atm. From this calculate the sucrose molecular weight.
>
> $$\pi = cRT$$
>
> $$c = \frac{\pi}{RT}$$
>
> $$= 12.30 \text{ atm} \times \frac{1 \text{ K-mole}}{0.08205 \text{ liter-atm}} \times \frac{1}{283.2°}$$
>
> $$= 0.5293 \text{ mole/liter}$$
>
> Since
>
> $$0.5293 \text{ mole/liter} = 171.2 \text{ g/liter}$$
>
> $$1 \text{ mole} = \frac{171.2 \text{ g}}{0.5293}$$
>
> $$= 323.4 \text{ g}$$

An osmotic pressure of 12.30 atm is balanced by a column of the sucrose solution 127 m high. Measuring so large an osmotic pressure is thus an easier task than measuring the other colligative properties,† vapor pressure or a temperature change. However, the sucrose solution is nonideal to an extent independent of which colligative property is chosen. The four sucrose molecular weights do not agree among themselves because the extent of nonideality depends on both the temperature and concentration of the solution. The sucrose molecular weights were obtained under four different concentration and temperature conditions and cannot be expected to be in agreement.

---

† A colligative (colli gā tiv or collig′ a tiv) property depends only on the *number* of solute molecules, not on the particular solute, in a solution. Examples are boiling-point elevation, osmotic pressure, freezing-point depression, Raoult's law, gas pressure.

The membrane chosen must be nonpermeable to solute. This is a major problem in osmotic pressure work because molecules of the size of sucrose do permeate slowly, making the method inaccurate. The method is used more successfully with much larger molecules, for example, proteins, which definitely cannot pass through most membranes.

The liquid height in the thistle tube is said to be an equilibrium value because it depends exactly on the solution concentration. If the liquid height is less than the equilibrium value, water will continue to flow into the thistle tube and the column will rise. If, on the other hand, we introduce more solution into the thistle-tube column so that the liquid level is greater than the equilibrium value, the column level will drop down to the equilibrium value. This occurs with a passage of water from the solution into the beaker of water. In this process, which is called "reverse osmosis" (Fig. 3.13), the applied pressure is greater than the equilibrium os-

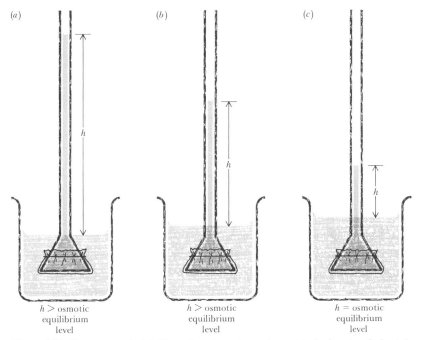

(a) (b) (c)

$h >$ osmotic
equilibrium
level

$h >$ osmotic
equilibrium
level

$h =$ osmotic
equilibrium
level

Figure 3.13. Reverse osmosis. (a) Water flows through membrane into beaker since hydrostatic pressure exceeds tendency for pure water to flow into concentrated solution from beaker; (b) water continues to flow into beaker, leaving a more concentrated solution in thistle tube; (c) hydrostatic pressure exactly balances tendency for pure water to flow into and dilute concentrated solution, and flow stops.

motic pressure and pure water is forced to leave the solution. This is a practical means of purifying water, for example, separating fresh water from seawater.

As an example, ordinary seawater at 25°C has an osmotic pressure of 26.33 atm, or 387 psi. If we were to remove one-half of the water from a sample of seawater, the concentration and hence the osmotic pressure of the resultant brine or salt solution would be twice as great: 52.66 atm, or 774 psi. Therefore, a pressure of 774 psi in a perfectly designed reverse osmotic apparatus will remove water until the salt concentration is doubled.

The osmotic-pressure phenomenon arises whenever a membrane permeable to solvent molecules and impermeable to at least some solute molecules separates two solutions of different concentrations in that solute which is not passed through the membrane. In general, solvent molecules will flow through the membrane from the less to the more concentrated solution.

OSMOTIC PRESSURE IN LIVING ORGANISMS

Osmotic-pressure considerations are of major importance in living systems. All living cells and tissues (which are collections of cells) are separated from some fluid by a cell membrane. The proper water balance between cell and fluid is governed, in part, by the osmotic pressure due to solute concentrations in the cell and in the fluid.

If a cell is placed in a high-concentration solution (i.e., in a solution having a higher concentration in that molecular species which cannot go through the membrane), the solution is said to be *hypertonic* to the cell and water passes from the cell into the solution. The cell becomes smaller and wrinkled; it becomes *crenated*. When a cell is placed in a relatively lower-concentration solution, the solution is said to be *hypotonic* to the cell and water passes from the solution into the cell. If, for example, the blood plasma lost some of its dissolved matter, it would be a more dilute solution and water would pass into the red blood cells. The red cells would then swell and even rupture (*hemolysis*).

If a cell is placed in a solution so that the nontransferable molecule concentration in the cell equals that in the solution, the solution is said to be *isotonic* to the cell and there will be no net transfer of water in either direction. Isotonic solutions can be made by dissolving 0.95 g of common table salt in 99.05 g of water. At the normal body temperature (37°C), blood has an osmotic pressure of about 7.6 atm due to all dissolved solute molecules.

Cells in the blood are carried by the fluid of the blood, the *plasma*. All other cells in our body are bathed in an *interstitial fluid*, which is separated from the plasma by the walls of blood vessels. Arteries carry blood to capillaries, where exchange of nutrients, wastes, fluids, etc., takes place between plasma and interstitial fluid. (Capillaries are very fine blood vessels that connect arteries and veins.) Veins return blood to the pump called the "heart." Most dissolved molecules can pass from plasma to interstitial fluid and back. However, there are quite large protein molecules in the plasma and interstitial fluid which do not readily cross the capillary wall. These are at a higher concentration in the plasma than in the interstitial fluid, and the osmotic pressure from this difference would ordinarily cause water to flow from interstitial fluid to blood plasma.

There is a hydrostatic pressure, the arterial blood pressure, which opposes this flow into the blood plasma and, in fact, is great enough to cause a flow in the opposite direction, from the plasma to interstitial fluid (Fig. 3.14). On the venous end of the capillary, the hydrostatic pressure is much less and water moves from interstitial fluid back to the blood by osmosis. At some intermediate neu-

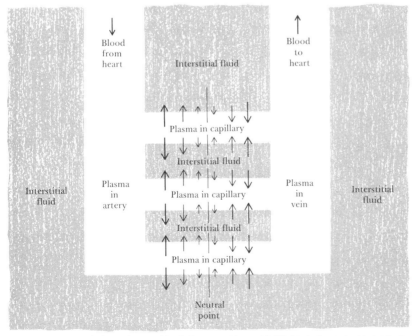

Figure 3.14. Water balance between blood plasma and interstitial fluid. At the neutral point in the capillary vessel there is no net flow in either direction.

tral point in the capillary system, the two effects exactly cancel each other. This system of fluid exchanges across the capillary was first suggested by E. H. Starling (1866–1927) and is often called the "Starling equilibrium."

This system for exchange of water from blood to interstitial fluid may become imbalanced, usually with excessive accumulation of interstitial fluid. This condition of water-swollen tissue is called "edema." Edema may be caused by several different conditions; examination of some of these may enable us to judge the correctness of our model for capillary-fluid exchanges.

Take a sharp object or fingernail and sharply draw it across your arm. After about a minute an edemous line can be observed. This injury to the capillaries allowed the plasma proteins to enter the interstitial fluid. The osmotic pressure becomes equal on each side of the capillary while the arterial pressure remains the same, resulting in a net movement into the interstitial fluid.

Normally some plasma protein "leaks" into the interstitial fluid and is removed by the lymphatic system. If the lymphatics become blocked, proteins accumulate in the interstitial fluid and edema results. Starvation causes a reduction in the plasma-protein concentration. An excessive quantity of fluid is filtered, and a reduced quantity is reabsorbed by osmosis. Edema results. In heart failure, blood accumulates in the veins and the venous blood pressure increases. There is a reduced osmotic reabsorption of water, and the tissues become edemous.

It is evident that living systems are in a careful state of adjustment and balance. Osmotic pressure plays a major role in water balance.

THE MOLECULAR FORMULA

We have already seen that for three of the simpler molecules studied, we can write oxygen as O_2, carbon dioxide as CO_2, and water as H_2O. These formulas are symbols which represent certain information. The water molecule H_2O contains two hydrogen and one oxygen atoms. A mole of water contains 2 moles of hydrogen atoms and 1 mole of oxygen atoms. A mole of water is 18.0154 g and contains 15.9994 g of oxygen and 2.0160 g of hydrogen.

Now we shall see how the molecular formula can be determined for more complicated molecules. By chemical analysis methane is 74.87 percent carbon and 25.13 percent hydrogen. The gas density molecular weight is 16.04 g/mole. A 100.00-g sample of methane contains 74.87 g C and 25.13 g H. The atomic weights of carbon and hydrogen are 12.011 and 1.0080 g/mole, respectively.

We can convert the masses of these two elements to moles:

$$74.87 \text{ g C} \times \frac{1 \text{ mole C}}{12.011 \text{ g C}} = 6.233 \text{ moles C}$$

$$25.13 \text{ g H} \times \frac{1 \text{ mole H}}{1.0080 \text{ g H}} = 24.93 \text{ moles H}$$

The 100.00-g sample of methane is an arbitrary amount. What is important is the relative mass of each element in methane, which is independent of the amount of methane selected. The number of moles of each element, 6.233 moles C and 24.93 moles H, is determined by how much methane is in the sample, that is, 100.00 g. The *relative* number of moles of each, however, is characteristic of methane itself and does not depend on how much methane is taken for analysis.

By inspection we see that there are four times as many moles of hydrogen as there are of carbon in any methane sample. If this were not immediately obvious, we would divide the larger by the smaller:

$$\frac{24.93 \text{ moles H}}{6.233 \text{ moles C}} = \frac{4.000 \text{ moles H}}{1 \text{ mole C}}$$

There are four times as many atoms of hydrogen as there are carbon atoms in any methane sample. Each molecule of methane contains hydrogen and carbon atoms in the *number* ratio of 4:1.

We can write an endless list of possible molecules with this hydrogen/carbon ratio:

$$CH_4 \qquad C_2H_8 \qquad C_3H_{12} \qquad C_4H_{16} \qquad \ldots$$

each of which has four hydrogens for each carbon atom and the same mass composition—74.87 percent C and 25.13 percent H. The molecular-weight information is used to select which one of all possible molecular formulas represents methane.

One mole of the first candidate CH_4 contains 1 mole of carbon and 4 moles of hydrogen. The elements contained in 1 mole of methane have these masses:

$$1 \text{ mole C} \times \frac{12.011 \text{ g C}}{1 \text{ mole C}} = 12.011 \text{ g C}$$

$$4 \text{ moles H} \times \frac{1.0080 \text{ g H}}{1 \text{ mole H}} = 4.0320 \text{ g H}$$

$$1 \text{ mole CH}_4 = 16.0431 \text{ g CH}_4$$

If the molecule is CH_4, its molecular weight is 16.043 g/mole.

The second possibility is C_2H_8, and its molecular weight is similarly calculated:

$$2 \text{ moles C} \times \frac{12.011 \text{ g C}}{1 \text{ mole C}} = 24.022 \text{ g C}$$

$$8 \text{ moles H} \times \frac{1.0080 \text{ g H}}{1 \text{ mole H}} = 8.0640 \text{ g H}$$

$$1 \text{ mole C}_2\text{H}_8 = 32.086 \text{ g C}_2\text{H}_8$$

This molecular weight is exactly twice the molecular weight of CH_4. Because C_2H_8 has exactly twice as many carbons and twice as many hydrogens as does CH_4, its mass and molecular weight must be twice the corresponding values for CH_4. For C_3H_{12} the molecular weight is three times 16.043, or 48.129 g/mole. For C_4H_{16}, the molecular weight is 64.172 g/mole.

From the gas-density data, methane has a molecular weight of 16.04. The correct representation must then be CH_4. These calculations can be concisely presented in tables such as the following:

(1) Element	(2) Analysis by mass, g	(3) Atomic weight	(4) Relative no. atoms	(5) Relative no. atoms
C	74.87	12.011	6.233	1.000
H	25.13	1.0080	24.93	4.000

The values in the fifth column are obtained by dividing those in the fourth column each by 6.233.

(1)	(2)	(3)	(4)
Possible molecules	Per mole of methane		Molecular weight
	g C	g H	
CH_4	12.011	4.0320	16.043
C_2H_8	24.022	8.0640	32.086
C_3H_{12}	36.033	12.0960	48.129

Experimental molecular weight is 16.04; therefore methane is CH_4.

Example. The formula and correct molecular weight of sucrose may be determined by burning a sample in oxygen and collecting and analyzing the combustion products. A 1.0000-g sample is found to contain 0.4211 g of carbon and 0.0648 g of hydrogen. Since no elements other than carbon, hydrogen, and oxygen were present during the burning of sucrose, the balance, 0.5142 g, is necessarily oxygen. (This can be confirmed by carefully measuring how much oxygen gas is consumed during the combustion. It will be found to be 0.5142 g less than if sucrose were a hydrocarbon.) By the colligative property techniques, the molecular weight of sucrose was found to be 330 g (vapor pressure), 350 g (boiling point), 343 g (freezing point), and 323.4 g (osmotic pressure). What is the formula and the correct molecular weight of sucrose?

(1)	(2)	(3)	(4)	(5)
Element	Mass in 1.000 g sucrose	Element atomic weight	Relative number of atoms	Relative number of atoms
C	0.4211	12.011	0.03506	1.091
H	0.0648	1.0080	0.0643	2.000
O	0.5142	15.9994	0.03214	1.000

The fifth column, as with methane, is obtained by dividing column (4) values by 0.03214, the smallest value in the column. The hydrogen/oxygen ratio is 2:1, as in water. Since almost all sugars, starches, and cellulose contain these two elements in this 2:1 ratio, they are collectively called "carbohydrates," where the word "hydrate" denotes a water-containing substance.

But the ratio of carbon to oxygen atoms is 1.091:1.000. This means we must do some more manipulation of numbers before the number of atoms of each element in sucrose, which must be integral, can be written down. We can reason this way. If there is one oxygen atom per sucrose molecule, there are 1.091 carbon atoms, which contradicts our theory that atoms and molecules cannot be present in other than whole-number quantities. Two oxygens in one sucrose are accompanied by 2×1.091, or 2.182, carbons; three oxygens by 3×1.091, or 3.273, carbons. Eventually, we find that 11 oxygen atoms in a single sucrose molecule must be associated with 11×1.091, or 12.001, carbon atoms. Within the precision of the analysis, this is as close to a whole number of carbon atoms as we can expect to be.

Since for each oxygen atom there are two hydrogen atoms in a sucrose molecule, for 11 oxygens there must be 22 hydrogens. A sucrose molecule *may* contain 12 carbons, 22 hydrogens, and 11 oxygens. The simplest formula is $C_{12}H_{22}O_{11}$. But equally possible are $C_{24}H_{44}O_{22}$ and $C_{36}H_{66}O_{33}$.

We can calculate the molecular weight of each possible sucrose molecule. For the simplest $C_{12}H_{22}O_{11}$:

Element	Number of atoms	Atomic weight	Mass of element
C	12	12.011	144.132
H	22	1.0080	22.176
O	11	15.9994	175.993
			342.301

Therefore the smallest possible molecular weight is 342.301 since the last column of masses adds up to 342.301. For $C_{24}H_{44}O_{22}$, the molecular weight is twice this, or 684.602.

The experimental molecular weights range from 323.4 up to 350, with an average of 337. These values agree best with the calculated molecular weight 342.301. We conclude that sucrose is $C_{12}H_{22}O_{11}$ and that its molecular weight is 342.301. The calculated molecular weight is based upon the atomic weights which are known more accurately and precisely than those obtained by the colligative methods. Refinements are made by using atomic weights, as in the sucrose *calculated* molecular weight 342.301.

CHEMICAL REACTIONS
AND BALANCED EQUATIONS

The equations representing the reaction between hydrogen and oxygen

$$2H_2 + O_2 \longrightarrow 2H_2O$$

and between hydrogen and bromine

$$H_2 + Br_2 \longrightarrow 2HBr$$

are both said to be *balanced* because the conservation of each kind of atom is observed. In the first, there are four hydrogen and two oxygen atoms both before and after the reaction; in the second, there are two hydrogens and two bromines on each side of the arrow.

Ammonia NH_3 is formed from hydrogen and nitrogen. If we write the reaction as

$$N_2 + H_2 \longrightarrow NH_3$$

we can see this is incorrect. Two nitrogen atoms react and end up as one nitrogen atom; two hydrogens become three. Neither hydrogen nor nitrogen is conserved, which violates the atomic theory.

To correctly represent the ammonia formation reaction we reason that since N_2 has two nitrogen atoms and NH_3 but one, there must be two ammonia molecules formed from each nitrogen molecule. The product is $2NH_3$. Since the product is two molecules of ammonia, each with three hydrogen atoms, six hydrogens in all are found in the product. Six hydrogens must be present in the reactant, and so we write $3H_2$. The correct representation is

$$N_2 + 3H_2 \longrightarrow 2NH_3$$

The numbers we have determined, 1, 3, and 2, are called "stoichiometric coefficients." They signify that one molecule of nitrogen and three molecules of hydrogen are used up to form two molecules of ammonia. They also signify that 1 mole of nitrogen and 3 moles of hydrogen react to form 2 moles of ammonia.

What is important is the ratio of these stoichiometric coefficients. We would be equally correct in writing the reaction as

$$\tfrac{1}{2}N_2 + \tfrac{3}{2}H_2 \longrightarrow NH_3$$

or

$$2N_2 + 6H_2 \longrightarrow 4NH_3$$

or

$$3N_2 + 9H_2 \longrightarrow 6NH_3$$

The first correct representation, with coefficients 1, 3, and 2, is, however, also the most convenient one. Usually we select for our balanced equation the representation in which the stoichiometric coefficients are all whole numbers and as small as possible.

> **Example.** Octane C_8H_{18}, the principal component in gasoline, burns to form carbon dioxide CO_2 and water H_2O if combustion is complete. Write the balanced equation.
> We start with
>
> $$C_8H_{18} + O_2 \longrightarrow CO_2 + H_2O$$
>
> Since octane has eight carbon and eighteen hydrogen atoms, there must be eight carbon dioxide and nine water molecules formed for each octane burned. We write
>
> $$C_8H_{18} + O_2 \longrightarrow 8CO_2 + 9H_2O$$
>
> The products contain $(8 \times 2) + (9 \times 1) = 16 + 9 = 25$ oxygen atoms, and so 25 oxygen atoms must be used up. Because each oxygen molecule contains two atoms, we need only $\frac{1}{2} \times 25$ or $\frac{25}{2}$ oxygen molecules. The balanced equation is
>
> $$C_8H_{18} + \tfrac{25}{2}O_2 \longrightarrow 8CO_2 + 9H_2O$$
>
> Finally, because the coefficient before oxygen is not integral, we multiply throughout by 2. The balanced equation remains balanced but now has only integral coefficients:
>
> $$2C_8H_{18} + 25O_2 \longrightarrow 16CO_2 + 18H_2O$$
>
> (Chemists use several special methods for balancing complicated equations, but almost always in this book the equations will be written already correctly balanced.)

STOICHIOMETRY

A balanced equation representing a reaction contains information about how many molecules of each reactant and product are involved. It also contains information about the number of moles of reactants and products in the reaction. From knowing the molecular weights of all reactants and products, we can deduce the masses of each reactant used up and each product formed. This information is the *stoichiometry* of the reaction.

> **Example.** Calculate the relative masses of octane and oxygen used up and of carbon dioxide and water formed in the combustion of octane.
>
> We begin by calculating the four molecular weights from the given atomic weights:
>
> C   12.011
> H   1.0080
> O   15.9994
>
> Octane C_8H_{18}:
>
> $(8 \times 12.011) + (18 \times 1.0080) = 114.232$
>
> Oxygen O_2:
>
> $(2 \times 15.9994) = 31.9988$
>
> Carbon dioxide CO_2:
>
> $(1 \times 12.011) + (2 \times 15.9994) = 44.0098$
>
> Water H_2O:
>
> $(2 \times 1.0080) + (1 \times 15.9994) = 18.0154$
>
> For our purposes, we round these numbers off to the nearest 0.01 g.
>
> Because 2 moles of octane are used up 2×114.23 g, or 228.46 g, of octane are burned. Twenty-five moles, or 25×32.00 g, which is 800.0 g, of oxygen are consumed. Similarly, 16 moles of carbon dioxide are 704.2 g; and 18 moles of water are 324.3 g. This can all be written as
>
> $$2C_8H_{18} + 25O_2 \longrightarrow 16CO_2 + 18H_2O$$
> $$2(114.23) + 25(32.00) \longrightarrow 16(44.01) + 18(18.02)$$
> $$228.46 \text{ g} + 800.0 \text{ g} \longrightarrow 704.2 \text{ g} + 324.3 \text{ g}$$
>
> This means that to burn 228.46 g of octane, 800.0 g of oxygen are required and 704.2 g of carbon dioxide and 324.3 g of

oxygen are produced. Observe that, in accord with the law of conservation of mass,

$$228.46 + 800.0 = 704.2 + 324.3$$

or

$$1,028.5 = 1,028.5$$

Example. The combustion of sulfur to sulfur dioxide is

$$S + O_2 \longrightarrow SO_2$$

When 1.000 kg of sulfur are burned, how much sulfur dioxide is formed? (The atomic weight of sulfur is 32.06, and that of oxygen is 16.00.)

The equation is balanced as written. The molecular weight of the product is

$$(1 \times 32.06) + (2 \times 16.00) = 64.06 \text{ g/mole}$$

We are not asked to calculate how much oxygen is involved, and therefore we need only the relative masses of sulfur and sulfur dioxide:

$$\begin{array}{ccc} S & + O_2 \longrightarrow & SO_2 \\ 32.06 & & 64.06 \end{array}$$

For each 32.06 g of sulfur used, 64.06 g of sulfur dioxide are formed; there is an equivalence between 32.06 g S and 64.06 g SO_2.

We can calculate for each gram of sulfur burned how many grams of sulfur dioxide are formed:

$$32.06 \text{ g S} = 64.06 \text{ g SO}_2$$

$$1.000 \text{ g S} = \frac{64.06}{32.06} \text{ g SO}_2$$

$$= 1.998 \text{ g SO}_2$$

Then, for 1.000 kg S, or 1,000 g S, there are 1,998 g SO_2 formed.

Because the numbers 32.06 and 64.06 are in the ratios of the mass of sulfur used and sulfur dioxide formed, the units can be in grams, kilograms, pounds, or any mass we wish to use. Concisely, we can write

$$1.000 \text{ kg S} \times \frac{64.06 \text{ kg SO}_2}{32.06 \text{ kg S}} = 1.998 \text{ kg SO}_2$$

Formally, the units kilograms of sulfur cancel in the multiplication.

Example. When 1 kg of sulfur is burned to sulfur dioxide, what volume of gas (at STP) is formed?

One kg of sulfur forms 1.998 kg SO_2. From the molecular weight we can convert this to moles of SO_2 and then to liters of SO_2.

$$1.998 \text{ kg SO}_2 \times \frac{10^3 \text{ g}}{1 \text{ kg}} \times \frac{1 \text{ mole SO}_2}{64.06 \text{ g SO}_2} \times \frac{22.4 \text{ liters}}{1 \text{ mole}}$$

$$= \frac{1.998 \times 22.4 \times 10^3}{64.06} \text{ liters SO}_2$$

$$= 0.699 \times 10^3 \text{ liters SO}_2$$

$$= 699 \text{ liters SO}_2$$

AIR POLLUTION AND SULFUR DIOXIDE

Ordinarily, during daylight hours, the sun's rays warm the surface of the exposed earth. This surface is warmer than the air layer immediately above it, but surface heat rises to warm this adjacent layer of air, thereby causing its temperature to rise and its volume to expand. This expansion decreases the air's density, making it less dense than the colder air which is further from the earth's surface, and the warm air rises up by convection. The colder air, in turn, sinks to the earth's surface because of its greater density, thereby causing a continuous circulation of air. This air-convection process in which heated air rises permits the accumulated air pollutants to be carried to higher regions of the atmosphere where winds can disperse the waste matter.

At night the cooled air remains cool and dense and the convection process stops, resulting in the condition called the "nocturnal inversion." Since no surface air rises, air pollutants can accumulate overnight (Fig. 3.15).

It sometimes happens, usually owing to the presence of fog or heavy smoke (or a combination of both—smog) that convection cannot take place, that is, that surface air cannot rise, so that

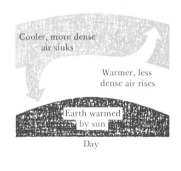

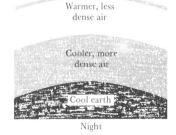

Figure 3.15. Normal daylight unstable condition; air layers move up by convection from the surface of the earth. Nocturnal inversion and occasional daytime inversion produce a stable condition; denser layer is on bottom with no convective motion of air layers.

normal cleansing of the air cannot occur. This resultant daytime inversion traps pollutants, such as industrial waste products, which build up to potentially dangerous levels (Fig. 3.15).

The Department of Health, Education, and Welfare (HEW) standard for a minimum sulfur dioxide concentration above which a serious hazard to health exists is $\frac{1}{10}$ liter SO_2 in 1 million liters of air. As a volume ratio, this is $\frac{1}{10}$ per million and is written as 0.1 parts per million or 0.1 ppm. Sulfur dioxide is only one of several major industrial waste products.

> **Example.** An industrial town of 20,000 persons is located in a valley. During a typical winter day, 200,000 kg of soft coal (about 220 tons) are burned and the waste gases are allowed to escape into the air. Before burning, this coal is by weight 2.0 percent sulfur. During a continuous inversion which confines the town's air in an area measuring 2 by 4 km and within an altitude of 0.5 km, the coal-combustion products accumulate. Let us calculate how long it should take for the sulfur dioxide level to reach the HEW minimum danger level and what the level would be after 3 days of continuous inversion. You are given the conversion factor
>
> $1 \ m^3 = 10^3$ liters
>
> Assume all gases exist at standard temperature and pressure.
> From the previous problem we know that 1 kg of sulfur will burn to produce 699 liters of sulfur dioxide at STP. From knowing how much coal is burned and what part of this burned coal is sulfur, we can calculate that in 1 day the amount of SO_2 released is
>
> $$200,000 \text{ kg coal} \times \frac{0.02 \text{ kg S}}{1 \text{ kg coal}} \times \frac{699 \text{ liters } SO_2}{1 \text{ kg S}}$$
> $$= 2 \times 10^5 \times 2 \times 10^{-2} \times 6.99 \times 10^2 \text{ liters } SO_2$$
> $$= 2.8 \times 10^6 \text{ liters } SO_2$$
>
> The gas is confined to a volume 2 by 4 by 0.5 km, or 2×10^3 by 4×10^3 by 5×10^2 m. The volume is $4 \times 10^9 \ m^3$, which is
>
> $$4 \times 10^9 \ m^3 \times \frac{10^3 \text{ liters}}{1 \ m^3} = 4 \times 10^{12} \text{ liters}$$
>
> The ratio of the two volumes is

$$\frac{2.8 \times 10^6 \text{ liters } SO_2}{4 \times 10^{12} \text{ liters air}} = \frac{0.7 \text{ liters } SO_2}{10^6 \text{ liters air}}$$

In 1 day the sulfur dioxide concentration is seven times the HEW standard. Assuming the air was free of this pollutant just as the inversion layer formed, if the level reaches 0.7 ppm in 24 hr, it takes 3.5 hr to reach 0.1 ppm. After 3 continuous days of inversion and accumulation, the level will reach 2.1 ppm.

In Chap. 11 we shall discuss this and other aspects of air pollution in more detail.

DESALINATION OF SEAWATER

Of the necessities of life, reasonably pure air and clean water head the list. Water is needed for agricultural, industrial, and municipal purposes. The last includes drinking, cooking, and cleaning. In 1970, with a population of 205 million in the United States, the average daily water consumption was 330 billion gal, an average of 1,600 gal per person per day. By far, the greater proportion was used in agriculture and industry, but each household nevertheless used nearly 100 gal/day for each member.

With increases in population and in industrial and agricultural production, the consumption of water increases continuously. But the growing need for good-quality water is aggravated by the pollution of our water resources by industrial and municipal wastes. Acute water shortages have already occurred, mostly in the New York City area and in the southwestern United States, but future critical larger-scale water shortages are not unlikely.

Because three-fourths of the earth's surface is covered by seawater and because a large fraction of the world's population lives near the seas and oceans, a great deal of research has been concerned with obtaining fresh water from the ocean.

Because of its high salinity, seawater is not potable. If it should be swallowed, two phenomena occur which intensify rather than quench thirst. Since it is hypertonic to normal body fluids, there is an osmotic transfer of water from the body fluids across the intestine wall into the seawater in the intestines; diarrhea results. At the same time, salt passes in the opposite direction—from intestinal seawater into the body fluids. Osmosis of water out of the body fluids and movement of salt into the body fluids results in a higher than normal salt concentration in the blood. The kidney attempts to excrete this excess salt. However, water is needed for the salt excretion, and additional water is lost in the elimination of salty urine. A person drinking quantities of seawater would become severely dehydrated. Of course, this is not likely to happen; ocean water has a vile and bitter taste.

In our previous discussion of physical properties, we examined several methods of desalination, or the recovery of pure water from a salt solution. These methods were: (1) the evaporation of water from a solution of water and some less volatile solute, (2) the freezing out of water from a solution of several components, and (3) the osmotic and reverse osmotic passage of water through a semipermeable membrane.

DESALINATION BY DISTILLATION

Pure water can be driven off, or evaporated, from a salt solution by supplying heat to convert the liquid water to water vapor. If, after collecting the vapor, the water vapor is then cooled to a temperature where the vapor pressure is greater than equilibrium, condensation from vapor to pure liquid water occurs. The combination of evaporation and condensation is called "distillation." The pure water thus obtained is called the "distillate" or *condensate*. It contains none of the nonvolatile solids present in the original solution.

The equipment used for distillation comprises an *evaporator*, in which the liquid is heated to boiling, and *condenser*, in which the vapor is cooled to condensation. Collectively, this equipment is called a "still." A typical still is diagrammed in Fig. 3.16.

Heat is added to the evaporator liquid by the steam passing through a coiled pipe. The pipe is bent this way to increase the surface area on which the steam can be condensed to liquid. The steam leaving the evaporator is cooler than that entering and, in part, has condensed. When the evaporator liquid boils, its vapor passes through another coiled pipe, contained in the condenser and shaped for better heat transfer. The condenser is cooled by water flowing in cold and leaving warmer, taking with it some heat supplied by the vapor from the evaporator. This vapor condenses to pure water as it cools within the condenser.

Simple distillation to obtain potable water from seawater is expensive for two reasons. As pure water is removed, the seawater

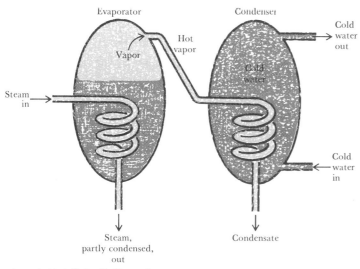

Figure 3.16. Still for distilling salt water.

becomes more concentrated and must be boiled at higher temperatures. An insoluble scale forms and adheres to the inner surface of the evaporator. This cuts down the rate of heat transfer to the evaporator liquid and also corrodes the equipment. In addition, the heat carried off by the condenser cooling water is wasted energy.

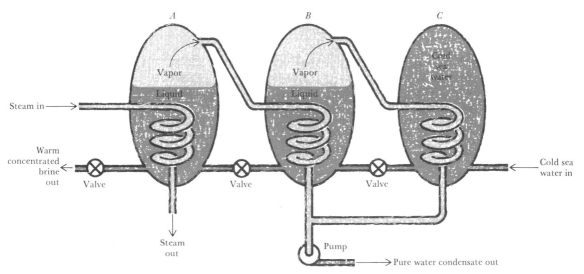

Figure 3.17. Two-stage still. First still—evaporator A and condenser B; second still—evaporator B and condenser C.

Two stills can be operated in tandem, as illustrated in Fig. 3.17. As before, steam heats the liquid contents of tank *A* and the vapor is cooled and condensed in tank *B*. But the cooling liquid in tank *B* is, in turn, heated and can be evaporated and condensed in tank *C*. The vapors which condense in tanks *B* and *C* are pumped from their respective evaporators *A* and *B*. The pump, by lowering the pressure within *A* and *B*, lowers the boiling points of the seawater.

The cooling fluid enters tank *C* from the sea and, when warm enough, is passed to tank *B*. Here it is heated more by the condensing vapor from *A* and, in part, is evaporated into *C* while the rest (which is still liquid) goes into evaporator *A*. As the liquid seawater goes from *C* to *B* to *A*, it becomes progressively warmer and more concentrated. From *A* it is returned to the ocean.

Since the cooling liquid in the condensers is not immediately discharged from the still, part of the heat it contains is used for further evaporation. Operating under reduced pressures, the evaporators are less subject to scale deposit and corrosion than in the simple still. The three tanks constituting two stills are called a "two-stage still." Greater savings in maintenance and heating costs can be made by using multistage stills.

DISTILLATION DESALINATION PLANTS CURRENTLY IN USE. Until 1950 Kuwait, an oil-rich but water-poor nation, was unable to supply itself with water from its own resources, and from 1925 to 1950 water was imported by boat from Iraq, a distance of 100 mi. In 1950 the Emir of Kuwait commissioned the construction of a multistage distillation plant. By 1960, about 16 million gal of drinkable water were produced each day, at a cost of about 70 cents/1,000 gal. The energy, or heat, needed to bring about the boiling under reduced pressure came from the combustion of petroleum, a plentiful and free fuel.

On September 2, 1958, five demonstration plants for desalinating seawater were authorized by the Eighty-fifth Congress. One of these was built at Point Loma, Calif., for the city of San Diego. It comprised 36 stages and produced 1.25 million gal/day.

Seawater was admitted to the condenser of the thirty-sixth stage at 12 to 21°C and was heated by vapor from stage 35 to about 28°C. The tank of stage 36 was at a pressure of about 28 torr, low enough to permit boiling. The first stage was at a temperature of 90.5°C and under 525 torr pressure. After being operated several years, it was dismantled and reassembled at the U.S. Naval Base, Guantanamo Bay, Cuba.

By using efficiently designed stills, the cost of water by distillation has been reduced in each succeeding year. The still on Gibraltar, installed in 1942, produced water at about $1.50/1,000 gal, and the one on the isle of Guernsey, at about 55 cents/1,000 gal. For comparison, the city of Chicago, Ill., treats the water taken from Lake Michigan at a cost of 2 cents/1,000 gal, pumps it at 1.6 cents/1,000 gal, and sells it to consumers at an average price of 22 cents/1,000 gal.

Further economies can be introduced by coupling desalination distillations and other operations. The town of Hempstead, N.Y., burns its own refuse as fuel to produce 500,000 gal of water and generate 2,500 kw of electricity each day. The metal magnesium and bromine are by-products produced from concentrated brine, along with potable water. Seawater is the principal source of magnesium in the United States.

DESALINATION BY FREEZING

Although for large operations desalination by distillation is the cheapest way of obtaining potable water, for lesser amounts freezing appears to be competitive with distillation. No scale forms, and the temperature is low enough so the corrosion and special maintenance problems of the distillation process are avoided. Freezing is also more versatile in that it can be used to process polluted water. The most serious problem is that of washing out entrapped liquid salt solution within the formed ice. The energy needed to freeze seawater can be recovered as the ice melts to cool more seawater or in conjunction with frozen-food preparation.

DESALINATION BY OSMOSIS

Reverse osmosis, also called "pressure osmosis" or "hyperfiltration," depends on the availability of membranes strong enough to withstand the high pressures involved. Recently, they have been made available on a commercial scale.

An interesting suggestion for obtaining small amounts of potable water (for example, for survival kits in lifeboats) has been made (Fig. 3.18). A large bag made of semipermeable membrane contains a little concentrated fruit juice and sugar solution. In an

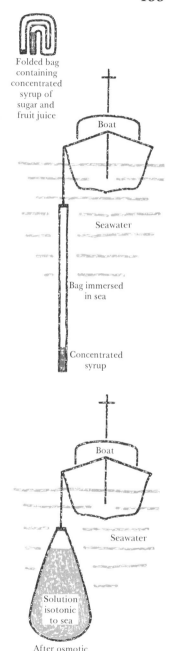

Folded bag containing concentrated syrup of sugar and fruit juice

Boat

Seawater

Bag immersed in sea

Concentrated syrup

Boat

Seawater

Solution isotonic to sea

After osmotic equilibrium

Figure 3.18. Fresh water from the sea by osmosis.

emergency, the folded bag is lowered into the ocean. By normal osmosis, pure water passes through the membrane bag from the sea into the syrup, which then becomes dilute. At osmotic equilibrium the bag contains a large amount of potable water and the nutrients initially present.

PROBLEMS

3.1. Calculate the average velocity of a water molecule after it has escaped from boiling liquid water.

3.2. Air, which is about $\frac{4}{5}$ nitrogen and $\frac{1}{5}$ oxygen, has an average molecular weight of 28.9. Phosgene ($COCl_2$) is a very poisonous gas sometimes used in warfare. What is its molecular weight? Do you expect phosgene to be more or less dense than air?

3.3. In the human body, sucrose is oxidized to water and carbon dioxide:

$$C_{12}H_{22}O_{11} + 12O_2 \longrightarrow 11H_2O + 12CO_2$$

When 100 g (about eight tablespoons) of sucrose are consumed, how much water is produced?

3.4. Referring to Prob. 3.3, how many liters of carbon dioxide at STP are produced from that much sucrose?

3.5. In the following reactions, which are balanced as written?

a. The oxidation of glucose:

$$C_6H_{12}O_6 + 5O_2 \longrightarrow 6CO_2 + 5H_2O$$

b. The oxidation of benzene:

$$2C_6H_6 + 15O_2 \longrightarrow 12CO_2 + 6H_2O$$

c. The reaction between zinc Zn and hydrochloric acid HCl to make zinc chloride $ZnCl_2$ and hydrogen H_2:

$$Zn + HCl \longrightarrow ZnCl_2 + H_2$$

Light and Electromagnetic Radiation

The atomic composition of a substance can be represented by a molecular formula, e.g., methane by CH_4. Although such formulas contain important information about the represented substances, several substances often have the same formula. The three hydrocarbons which have the same formula C_5H_{12} but different properties are shown in Table 4.1. Substances with the same molecular formula and molecular weight but with different properties are *isomers* of each other. Thus, the three hydrocarbons in Table 4.1 are pentane isomers.

More information than the molecular formula is needed to distinguish among several isomers. We must know the geometric arrangement, or structure, of molecules—the way in which atoms are joined together within a molecule. From this information, we can determine if there is a relationship between the atomic arrangement within any molecule and the behavior of that molecule, that is, if the form determines the function of a molecule.

Scientists use many tools to investigate molecular structure, the most important of which involve the interaction of light with substances. To understand how this is done we must first know something about light. Also, because the structure of a molecule depends on the character of its constituent atoms, we ought to know something about atoms. In this chapter we shall study light. In Chap. 5 we shall look at the nature of atoms.

Table 4.1. Three Isomers of Pentane C_5H_{12}

Substance	Molecular weight	Density, g/cm³	Melting point, °C	Boiling point, °C
Isopentane	72.15	0.621	−160.5	28
Normal pentane	72.15	0.626	−131	36.2
Neopentane	72.15	0.613	−20	9.5

THOMAS YOUNG'S
DOUBLE–SLIT EXPERIMENT

In the eighteenth century the majority of scientists believed that light consisted of very small, fast-moving particles, or corpuscles. This *corpuscular theory* of light was accepted mainly because it was advocated by one of the greatest scientists of all times, Sir Isaac Newton (1642–1727). The reflection of light was thought to be the rebounding of these corpuscles from a reflecting surface.

Christian Huygens (1629–1695) and Robert Hooke (1635–1703) suggested (unconvincingly to most of their contemporaries) an alternative theory. In their view certain phenomena of light (for example, the color patterns of a thin oil film on water and the bending, or diffraction, of light as it passes sharp-edged objects) can be described by a vibrating-wave model. But it was not until 1803, when Thomas Young (1773–1829) reported a few simple experiments, that confidence developed in a *wave theory* of light.

In his most famous experiment, Young passed light from a source through a single slit, then through two closely spaced parallel slits, and then onto a screen (Fig. 4.1). He saw a series of parallel alternating light and dark regions. When one of the pair of parallel slits was blocked, the light going through the single open slit of the pair illuminated the screen in a simple way, with the central light and dark regions now uniformly bright. When this experiment is done with light of a single color and two slits, the fringes are especially sharp. An actual photograph is reproduced in the front endpaper.

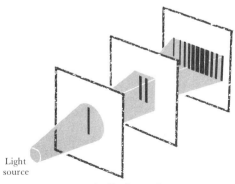

Light source

Figure 4.1. Young's double-slit experiment.

THE PERIODIC BEHAVIOR OF LIGHT

When a small stone is dropped into water, a concentric and growing series of rings forms on the water surface—an alternation of crests and troughs. By looking at just one point on the water surface, e.g., by observing a cork float loosely anchored, one will see that point rise and fall repeatedly. A phenomenon which displays a repeated pattern is said to go through *cycles* or to have *periodic* behavior.

When two small stones are dropped at the same time but at different places into water, two sets of rings form, one from each point of entry of the stone into water. As the two sets grow, meet, and cross over, an observer will see a new feature. Where a crest from each set comes together, the result is a crest much higher than either of the original crests; where two troughs come together, the result is a trough deeper than either of the original troughs. But when a trough from one set meets a crest from the other, the result is neither as deep as the former nor as high as the latter. If the original trough is as deep as the original crest is high, the result is neither a crest nor a trough: the water (and the cork float) remains stationary. The crest and trough seem to have destroyed each other.

Young reasoned that the dark regions seen in his double-slit experiment must be the consequence of canceling of light coming from each of the two slits like waves from two stones. Light, like the water waves described above, must have a periodic behavior. Some property of the light goes through cycles, increasing and decreasing in a repeated manner. Although at this point we do not know what aspect of light has this periodic, or wave, property, we can define certain features of waves of any kind.

In Fig. 4.2 an unspecified property is plotted on the vertical axis—positive above and negative below the horizontal axis. The horizontal axis can represent either of two things. If an observer at one instant of time were able to measure the variation in property with position in space, his results would be represented as shown. The spatial distance from one trough to the next, from one crest to the next, or from any point to the next corresponding point is the distance of one *cycle*, or one *vibration*. It is called the "wavelength" of the periodic phenomenon and has the dimension of distance.

If, on the other hand, the observer remained at a fixed position and measured the variations in this property over a period of

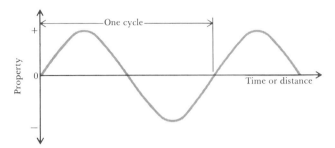

Figure 4.2. Simple graph showing variation of a property in a periodic, or repeated, way as a factor such as the time or distance (or both) changes. Many properties are periodic. The amplitude is the height of the crest, the maximum value for the property.

time, his results would be represented by the same figure. The time interval between one crest and the next passing his point of observation is the *period* of the vibration and has the dimension "seconds per cycle." The number of crests which pass him each second is the *frequency* of the vibration and is stated in cycles per second (cps) or in hertz (Hz). One Hz, which is 1 cps, is named in honor of Heinrich Hertz (1857–1894), who investigated the nature of vibrating light waves. Since the period is the seconds per cycle and the frequency is the cycles per second, it follows that one is the reciprocal of the other.

> **Example.** The note A_3 on the conventional piano has a frequency of 220.00 cps. What is its period T?

$$T = \frac{1}{220.00 \text{ cps}}$$
$$= 4.5455 \times 10^{-3} \text{ seconds per cycle}$$

If an observer can count the number of vibrations, or cycles, that pass him each second and if he can also measure the wavelength of each cycle, he can calculate how fast a wave travels past him. Wavelength is commonly represented by the symbol λ (lambda) and frequency by ν (nu). Then

$$\lambda \, \frac{\text{m}}{\text{cycle}} \times \nu \, \frac{\text{cycles}}{\text{s}} = c \, \frac{\text{m}}{\text{s}}$$

The product of wavelength times frequency is the speed of the wave.

INTERFERENCE AND
WAVE PROPERTIES

In assuming that light has a wavelike character, we have supposed that some aspect of light assumes positive and negative values in regular alternation, either in time (if position is fixed) or in space (if time is fixed). When two different waves come together, we can imagine that the positive and negative values of the varying property are added in an algebraic way—with the usual rules of adding signed values.

The combination of two waves is illustrated first when crests are added to crests, with enhanced maximum heights and minimum depths. This *constructive interference* is illustrated in Fig. 4.3a.

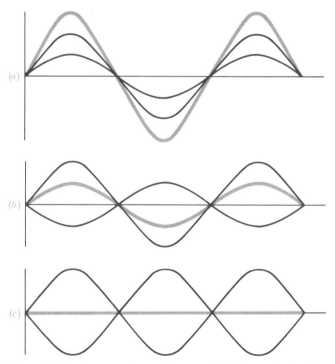

Figure 4.3. Interference and wave phenomena. (a) Constructive interference—the combination of amplitudes (thin lines) produce an enhanced resultant amplitude (heavy line); (b) destructive interference—when the amplitudes are directed oppositely, the resultant amplitude is smaller; (c) complete destructive interference—when the amplitudes exactly cancel each other, a resultant amplitude of zero, or a flat line, is observed.

When the waves are combined crest to trough, the result is a decrease in magnitude of both crest and trough and is an example of *destructive interference* (Fig. 4.3*b*). If the crest height of one wave exactly equals the trough depth of the other, there should be complete destruction of the vibration (Fig. 4.3*c*).

So successful were Young and workers who followed him in describing an impressive set of properties of light that the wave theory remained the dominant model of light until the end of the nineteenth century. Work by Heinrich Hertz, Albert Einstein (1879–1955), and Arthur Compton (1892–1962) indicated that some properties of light can more easily be understood using the particle, or corpuscular, theory. In Compton's studies of the interaction between light and very small particles, light behaves as if it has mass† and thus as if it is a particle.

Our present view of light is that it is neither solely waves nor solely particles. When we measure certain properties of light or use light in certain ways, as in Young's double-slit experiment, the description of what happens can more simply be stated in terms of a wave theory. We say light has wavelike properties. Under other circumstances the description of what we observe can be given more conveniently by using a corpuscular theory; light has particle properties. We speak of the wave-particle *dual nature* of light.

DIFFRACTION AND INTERFERENCE

Using the wave-theory model of light, we can describe Young's observations. Young first passed light through a single slit to ensure that light reaching both of the pair of slits was in phase, i.e., that maxima occurred simultaneously at the two slits (Fig. 4.4). In Fig. 4.4*a* the light from two slits travels to the screen along two paths of equal distance, and at the screen a maximum from one coincides with a maximum from the second. There is constructive interference, and that place on the screen is bright. In Fig. 4.4*b*, the light from the upper slit has to travel a greater distance (by an amount equal to one-half the wavelength) than does light from the lower slit. The two waves meet at the screen *out of phase*, there is destructive interference, and a dark region results. Further down the screen in Fig. 4.4*c*, the path difference from slit to screen is exactly

---

† To be more precise, light particles exist only when they move; at rest the mass cannot be detected.

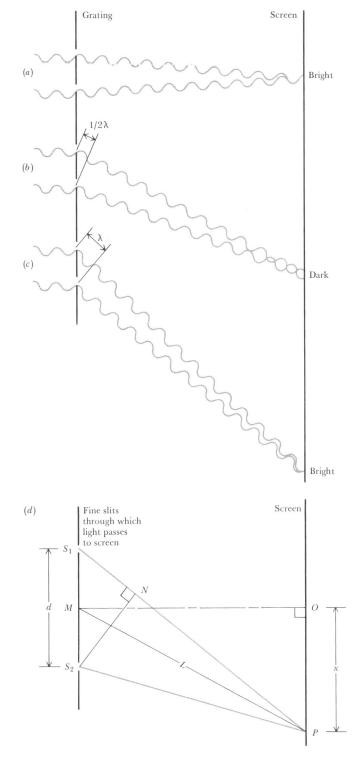

Figure 4.4. Bright and dark regions produced on a screen by the interference of light coming from a pair of slits. (a) Same grating-to-screen distances; (b) grating-to-screen distances differ by one-half a wavelength ($\lambda/2$); (c) grating-to-screen distances differ by a full wave-length (λ). (d) Rules for constructive and destructive interference. Right triangles $S_1 S_2 N$ and MOP are simi-lar, and corresponding sides proportional: $S_1 N/d = x/L$ and $x = (L/d)S_1 N$. Because of the great slit-to-screen distance, compared to d, the distances NP and $S_2 P$ are very nearly equal. The path length difference to P from the two slits is just $S_1 N$. If $S_1 N = \lambda$, 2λ, 3λ, or any whole-number multiple of the wavelength $m\lambda$, there is construc-tive interference. If $S_1 N = \frac{1}{2}\lambda$, $\frac{3}{2}\lambda$, $\frac{5}{2}\lambda$, etc., there is destructive interference. In general, where $S_1 N = m\lambda/2$, as m takes on alternating even and odd integer values there will be alternatingly constructive and destructive interference.

one wavelength, there is constructive interference, and a bright
region.

In addition to serving as good evidence for the wave theory
of light, diffraction and interference can be put to important use in
scientific investigation. By combining the equations of Fig. 4.4d

$$x = \frac{L}{d} S_1 N$$

and

$$S_1 N = m \frac{\lambda}{2}$$

we have

$$x = m \frac{\lambda L}{2d}$$

Bright fringes occur when $m = 0, 2, 4, 6$, etc. The number of bright
fringes can be found by counting. Among the four parameters x, λ,
d, and L, if any three are known, the fourth can be calculated.

If the incident light consists of a mixture of waves of many
wavelengths, the diffraction can be used to isolate light of a particular wavelength. The position on the screen of maximum light intensity depends on the wavelength of light. When light consists of
many wavelengths, there are no sharp and narrow regions of
brightness. Rather, the light extends over broad areas of the
screen. At each screen position the brightness is due to constructive
interference for a special wavelength. A rainbow effect can be seen
on the screen, with blue-violet closer to the slits, extending through
green and yellow to red, which is the most remote from the slits. To
each wavelength there is a corresponding color. What we see as
white light can be broken up or *dispersed* into its constituent colors.

If a small hole is cut into the screen at any point, light of that
wavelength which has a maximum there can pass through. This
light, which comprises a single or narrow range of wavelengths, is
said to be "monochromatic." Such a device which is a source of
monochromatic light is called a "monochromator." If d, L, and x
are known, the wavelengths of pure monochromatic light or of a
mixture can be calculated.

Example. A green-blue monochromatic light from a mercury-vapor lamp passes through two slits spaced 0.5 mm apart and falls on a screen 1 m from the slits. The distance between the first and second bright fringe is 1 mm. What is the wavelength of the light?

From the equation

$$\lambda = x \frac{d}{L}$$

$x = 1 \text{ mm} = 10^{-3} \text{ m}$
$d = 0.5 \text{ mm} = 5 \times 10^{-4} \text{ m}$
$L = 1 \text{ m}$

Thus

$$\lambda = \frac{(10^{-3} \text{ m}) (5 \times 10^{-4} \text{ m})}{1 \text{ m}}$$

$$= 5 \times 10^{-7} \text{ m}$$

If the incident monochromatic wavelength is known and the slits-to-screen and fringe-to-fringe distances on the screen measured, the spacing between the parallel slits can be calculated. This is the basis of an extremely valuable technique used by scientists to investigate molecular structure (Figs. 5.25–5.28).

Although by knowing λ, x, and d we should be able to calculate L and thus use diffraction as a means of measuring length or distance, there is a better method available, which will be discussed in the next section. From the diffraction equation

$$x = m \frac{\lambda L}{2d}$$

we can see that the smaller the slit spacing d, the larger the fringe-to-fringe distance x and the more precise our measurements.

Although the double-slit equipment of Young can be used, much sharper fringes and much finer measurements can be had if not two but a larger number of slits are used. The set of slits, which must be very closely and evenly spaced and parallel, is called a "diffraction grating." The task of constructing such a grating was first accomplished by Henry Rowland (1848–1901), who invented a remarkably fine screw of constant pitch. To this a diamond ruling point was attached. Turning the screw by a small and precisely controlled amount, the diamond point was advanced through a small and precisely controlled distance. In this manner Rowland's engine was used to cut 14,438 parallel and evenly spaced lines per in. and to make gratings up to 6 in. wide with spacing between the lines of 1.693×10^{-6} m, or 1.693μ.

Such gratings are quite expensive. Less costly ones can be made by casting a soft plastic substance on the ruled glass or metal surface and peeling this replica grating off. Plastic is subject to shrinkage and distortion and so is not as precise as the original ruled grating. If the ruled lines are cut through, the grating can be used as in Young's original experiment, with light passing through each slit. Otherwise light can be reflected from the ruled surface, and the same dark and light fringed areas are also seen. This latter technique is more common.

THE SPEED OF LIGHT

To measure the speed at which light moves we must know the time required for it to travel through a definite distance. This task is both one of the most important and one of the most difficult in all science. The speed of light is a fundamental property of our universe, appears frequently in scientific theories, and is the upper limit of speed known to man. Measuring it is a formidable job because of its very large magnitude.

A FIRST APPROXIMATION. One of the first attempts at obtaining a value for the speed of light was made by Galileo Galilei (1564–1642). Two men, each with a covered lantern, were stationed at night on two hilltops separated by about 2 miles. One person uncovered his lantern and noted the time. The other, upon seeing the first lantern's light, immediately uncovered his. Then the first person noted the time he saw the second light. The interval was the time it took light to go from the first to the second hill and then back to the first. From knowing back-and-forth distance (about 4 miles) and the time, the speed was calculated.

Although the reasoning was correct, there are enormous errors involved. Because light travels at about 186,000 mi/s, the time for traveling 4 miles is only 4/186,000 or 2×10^{-5} s. But the time needed to recognize a light signal and to remove the second lantern cover was about $\frac{1}{2}$ s; compared to such speeds, human reflexes are quite slow. Any variation in this experimentally unavoidable time delay far exceeds the actual travel time of light. If light travels at a measurable speed, a finer technique than this was needed.

FINER MEASUREMENTS OF THE SPEED OF LIGHT. In 1676 the Danish astronomer Olaf Römer (1644–1710) measured the time it took for one of Jupiter's moons to make a complete revolution about the planet. Five months later, observing what seemed to be a delay in this moon's circuit, he correctly attributed this delay to light having to travel a longer distance from Jupiter to earth than it did during his original measurement. This is the first evidence for believing that light travels at a measurable speed. Römer's method gives a speed-of-light value close to that accepted today.

The first value for light's speed through terrestrial methods was obtained in 1849 by Armand Hippolyte Louis Fizeau (1819–1896), who made his measurements entirely within the city of Paris. His equipment, shown in Fig. 4.5*a*, comprised a light

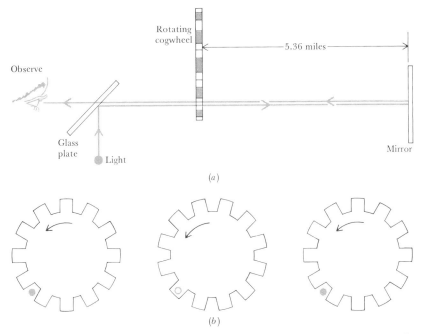

Figure 4.5. In Fizeau's experiment the reflected flash can be seen only if the opening rotates into the line of sight between the observer and mirror. From the rotation speed of the cogwheel, the time difference between two successively aligned openings can be calculated. This is also the time of travel of the light from the cogwheel to mirror and back. Knowing the distance of travel of the light, Fizeau calculated the speed of light. (a) Light beam chopped into discrete flashes by rotating cogwheel; (b) light flash passes through opening to mirror. Returning light flash is blocked by cogwheel tooth. When the cogwheel speed is adjusted, the returning flash is seen through the opening.

source, sending a light beam to a glass plate which reflects the light between the teeth of a cogwheel to a mirror 5.36 miles away. An observer behind the cogwheel views the light reflected from the distant mirror.

Using improved techniques and making measurements of the speed in an evacuated pipe 1 mile long, Albert A. Michelson (1852–1931) and his coworkers obtained a value of $c = 2.99774(\pm0.00011) \times 10^8$ m/s. The very best measurements to date indicate a value of

$$c = 2.997925\,(\pm0.000001) \times 10^8 \text{ m/s}$$

As defined in the previous section, the wavelength λ is the distance a propagating wave moves in one complete oscillation. The frequency ν is the number of oscillations made each second by the wave. Then

$$\lambda\,\frac{\text{m}}{\text{oscillation}} \times \nu\,\frac{\text{oscillations}}{\text{s}} = c\,\frac{\text{m}}{\text{s}}$$

and the product $\lambda\nu$ is the speed c. If we measure the wavelength of light, as with a diffraction grating, we can calculate the corresponding frequency.

> **Example.** What is the frequency of a blue-green light with a wavelength of 5×10^{-7} m?
>
> $$\nu = \frac{c}{\lambda}$$
>
> $$= \frac{2.997925 \times 10^8 \text{ m/s}}{5 \times 10^{-7} \text{ m/oscillation}}$$
>
> $$= 6 \times 10^{14} \text{ oscillations/s}$$

It follows from this relationship, the shorter the wavelength, the higher the frequency.

THE INTERFEROMETER

Among the uses to which the interference of light waves can be put, of special interest is the instrument developed by Albert A. Mi-

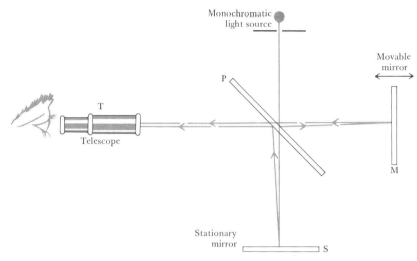

Figure 4.6. Michelson's interferometer. Light travels by two paths from source through P (plate which reflects part of incident light and transmits the rest) to telescope: source—P-M-P-T and source—P-S-P-T. As the mirror M is moved, the difference in length between the two paths, alternatingly, equals an integral and a whole number plus one-half number of wavelengths of the monochromatic light used. Alternatingly, the viewer sees bright and dark fringes from the consequent constructive and destructive interference. The distance through which mirror M is moved can be calculated from the number of dark and bright fringes observed and the known light wavelength.

chelson for measuring lengths with precisions and accuracies that are truly remarkable. The instrument is called an "interferometer." A simple version is illustrated in Fig. 4.6.

    With refinements on Michelson's original interferometer, it is possible to establish distances with a precision of a few percent of the wavelength used. Automatic equipment can be used for counting the number of dark or bright fringes appearing at the telescope.

    Michelson and René Benoît (1844–1922), the director of the International Bureau of Weights and Measures at Sèvres, measured the standard meter in units of light from a cadmium lamp. Using three separate monochromatic wavelengths, they found the results listed in Table 4.2. Today this is the way our standard of length is defined. The monochromatic light used is emitted by a specially designed krypton lamp, at the triple point of nitrogen (−210°C, see Chap. 2). The meter is defined as exactly

$1 \text{ m} = 1,650,763.73 \text{ } \lambda$

Table 4.2. Measurement of the Standard Meter

Color	Wavelengths/1 standard m	m/1 wavelength
Red	1,553,163.5	6.4384722×10^{-7}
Green	1,966,249.7	5.0858240×10^{-7}
Blue	2,083,372.1	4.7999107×10^{-7}

of this particular monochromatic light. There is an accuracy of a few parts in 100 million which is better by a factor of 10 than that obtained using the older metal-bar standard. It is very probable that even more accurate standards will be developed to replace this krypton standard.

> *Example.* What is the wavelength of the above krypton emission lamp?
> Since

$$1{,}650{,}763.73 \ \lambda = 1 \ m$$

$$1\lambda = \frac{1}{1{,}650{,}763.73} \ m$$

$$= \frac{1}{1.65076373 \times 10^6} \ m$$

$$= 0.605780211 \times 10^{-6} \ m$$

$$= 605.780211 \ nm$$

ELECTRIFICATION

Many of the properties of light can be described by wave behavior without specifying exactly what it is that changes in a cyclic, or periodic, manner. However, the nature of the interaction between light and matter can be better understood with a more intimate analysis of light. In addition, this knowledge of light possibly can indicate what light has in common with other phenomena and, if it does, can simplify our picture of the physical universe. Scientists are curious about light for two important reasons: to understand light for understanding's sake and, having this information, to more fully exploit the tool—light—in investigating the universe. Later we shall see how light is used to examine molecules.

After a great deal of research, it turns out that light is a special case of more general phenomena which have been observed for over two millenia: electricity and magnetism. We shall now see how the identification between light and electricity and magnetism was established.

THE FIRST EXPERIMENTS WITH ELECTRICITY

There are a number of familiar and related phenomena: the slight shock felt at touching a metal object after walking across a carpeted floor on a dry winter day, the sticking of a child's balloon on a wall after having been rubbed on clothing (also on a dry day), the clinging together of laundry made from different fabrics after having been dried by machine, and the sparks seen when these clothing are pulled apart in the dark. To Thales of Miletus (640–546 B.C.) has been attributed the first observation of electricity—that when amber, a yellow-brown fossil resin, is rubbed with a cloth, other small objects, for example, hair, are attracted to it. This is often called the "amber effect," but it is also called the "electric effect" (because the Greek word for amber is *elektron*). An object so treated to attract small objects to it is said to be *electrified* or to be *electrically charged*. A very simple set of experiments can be done to illustrate some features of the amber, or electric, effect (Fig. 4.7).

A successful description of the electric effect was proposed in 1751 by Benjamin Franklin (1706–1790) and is called the "one fluid" theory. Ordinarily, when a body is not electrified, there is within it a definite normal amount of an indestructible "electric fluid." The body is *neutral*, or *uncharged*. But upon being rubbed by a second body of another substance there is a transfer of fluid from one body to the other. Then, one body has an excess of fluid and the other a deficit; the former is said to be *positively* charged and the latter *negatively* charged. Any two positively charged bodies mutually repel each other. Any two negatively charged bodies also repel each other. But two bodies *oppositely* charged mutually attract each other. Since a magnet or lodestone had no effect on such a charged body, this *electric charging* was not the same as magnetism.

In our saran and glass experiment (Fig. 4.7) we have no way of knowing which material—glass or saran—has an excess charge after the two have been in contact. But, *quite arbitrarily*, we can assign to saran, after its having been rubbed on glass, the quality of being negatively charged. With this arbitrary assignment of negative charge to saran, glass must be positively charged since it is attracted to the saran. Figure 4.7*a* to *d* illustrates the observed

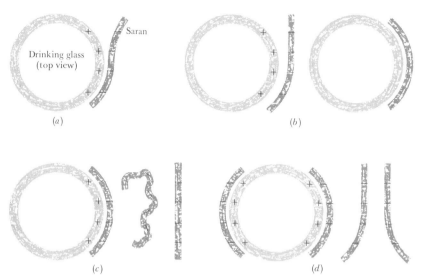

Figure 4.7. Electrification of glass and saran. For convenience we can use an ordinary clean drinking glass and two pieces of either cellophane or some other plastic wrapping material, such as saran, each about 5 by 10 cm. (a) When one of the transparent sheets is wrapped tightly around the glass and then partially peeled off, we see that it has a tendency to return to the glass; it is attracted to the glass. (b) Removing it completely but holding it close to the glass, we see the sheet attracted to the glass. (c) Removing the piece completely, crumpling it with our fingers into a tight ball, and then holding it loosely on our palm, we see it abruptly unfold. (d) Finally, putting both pieces on the glass, peeling them off one in each hand, and bringing them together, we see they repel each other. We notice that both the attraction between the sheet and glass and the repulsion between the two sheets act without the two bodies touching. The electric forces act over a distance.

phenomena in terms of the one-fluid theory and positive and negatively charged bodies.

The spontaneous opening up and unfolding of an electrified piece of saran can be described by the presence of a negative charge distributed over the entire sheet. When crumpled up into a ball, the contiguous parts of the same sheet, having the same kind of charge, repel each other. This self-repulsion of parts of a charged sheet can serve as the basis of a simple device for measuring the magnitude of an electric charge—the electroscope (Fig. 4.8).

Stephen Gray (1696–1736) first classified bodies into *conductors*, along which electric charges travel easily, and *insulators*, or *nonconductors*, along which charges move with more difficulty, that is, more slowly. Of the common metals, silver is the best electric conductor and copper the next best.

If a glass rod is rubbed with saran sheet and the charge on the rod carefully transferred to an electroscope, we can measure

Figure 4.8. Gold-leaf electroscope. (a) No charge — gold leaves hang down; (b) small charge — leaves diverge because of repulsion between like charges; (c) greater charge — more repulsion and greater divergence.

the divergence of the leaf halves. Then, after discharging the electroscope by means of a conductor, we can transfer to it the charge on the saran. The divergence will be seen to be *exactly* as great as before. Arbitrarily we have called the charge on the glass "positive" and the charge on the saran "negative." In the language of Franklin's one-fluid theory, the excess of electric fluid in the glass is *exactly* the deficit of electric fluid in the saran; the total amount of this fluid remains constant. When we create a positive charge, at the same time we create a negative charge of equal magnitude. All our experience indicates that we cannot create by itself an electric charge of one sign—we cannot alter the total charge of our universe. This is the *law of conservation of electric charge* and is equivalent to the indestructible character of Franklin's electric fluid.

ELECTRIFICATION BY INDUCTION

In addition to the attraction which exists between oppositely charged bodies, there is also the attraction an electrified body has for other small bodies, such as hairs, small bits of paper, tiny pins. This force, which acts between one charged and one neutral body and which was originally used to decide if a body (such as amber) was electrified, can be described by the one-fluid theory.

William Gilbert (1544–1603), physician to Queen Elizabeth I and an early investigator of electrical phenomena, built an instrument called a "versorium" to detect electrification. A simple version of this can be made from a drinking straw, pin, and cork (Fig. 4.9) to demonstrate how this electric fluid can be induced to move within a body. When the needle has positively and negatively

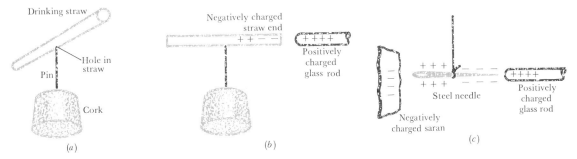

Figure 4.9. Polarization by induction. (a) Simple versorium, a balanced straw free to turn. (b) When a charged object is brought close to one end of the straw, the straw rotates so that end is as close as possible to the rod. This can be described by a charge separation within the straw and a migration of the negative charge toward the positive glass rod. (c) When a conductor material, such as a steel needle, is used, a positively charged glass rod will attract one end of the needle. Within a conductor there is a greater migration of charge. The far end of the needle is positively charged because it strongly attracts a negatively charged piece of saran.

charged ends, it is said to be *electrically polarized* and to be an *electric dipole*. When the charged glass rod is removed, by mutual attraction the positive and negative charges within the needle and within the straw are redistributed uniformly throughout the needle or straw. The polarization vanishes.

To further test the separation of charges within a metal conductor by the proximity of a charged body, two identical and touching metal spheres can be used instead of the steel needle (Fig. 4.10). Since the polarization and charging of the two spheres are *induced* by the presence of the external charged body, the processes are called "polarization by induction" and "charging by induction."

QUALITATIVE AND QUANTITATIVE LAWS. The statements that oppositely charged bodies attract and like-charged bodies repel are examples of *qualitative descriptions*. Qualitative descriptions tell us something about the physical behavior of matter. But scientists are interested in more detailed and more exact descriptions. A law, i.e., a summary of some experimentally observed phenomena, which often can be stated in the form of an equation, gives us this additional information. These kinds of laws are called "quantitative laws."

In an earlier example, the observation that when we increase the pressure on a gas its volume decreases is a qualitative description. Boyle's law, which tells us the relationship between the applied pressure and the volume, is a quantitative description of the same observations. Boyle's law tells us much more about this same behavior.

COULOMB'S LAW

A quantitative law describing repulsive and attractive forces between charged bodies, called "electrostatic forces," was discovered in 1785 by Charles Augustin Coulomb (1736–1806), who invented a torsion balance which is a modification of Gilbert's versorium. Instead of permitting the horizontal rigid rod to rotate freely about a pin, Coulomb suspended it by a wire (Fig. 4.11*a*). When the twisting force is removed, the wire and rigid rod return to their original condition and position. The greater the applied force, the greater the torsion, or twisting, of the wire. In Coulomb's balance extremely small forces produced measurable twisting of the wire and rotation of the rigid rod. By knowing the amount or

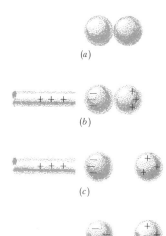

(a)

(b)

(c)

(d)

Figure 4.10. Charging by induction. (a) Two uncharged metal spheres are in contact; (b) when an electrified body is brought near one sphere, there is a polarization of charge; (c) if the two spheres are separated in the presence of the external electrified body, the separated charges cannot recombine; (d) when the external body is removed, the two spheres remain charged.

angle of wire twist, which is measured by the angle of rotation of the rigid rod, he was able to calculate the force producing the twist.

At one end of the rigid rod Coulomb fastened a ball made from pith (a very light wood) and at the other a paper ball as a counterweight to keep the rigid rod balanced and in a horizontal direction. An identical pith ball was charged and then touched to the ball on the rod, thus transferring some of the electric charge to it. Because of repulsion between the two like-charged balls and because the second ball was held in a fixed place, the first ball moved away and with it the rigid rod, twisting the suspending wire (Fig. 4.11*b*). From the angle of twist Coulomb knew the electrostatic repulsive force. He also measured the distance *r* between the two charged balls.

When he moved the second ball closer to the first, the wire twisted more and a different ball-to-ball separation and torsion force were measured. By varying the placement of the second charged ball and making a series of torsion force and ball-to-ball distances, Coulomb noticed a very simple relationship between the two. For twice the separation, the force was one-fourth; for three times the separation, the force was one-ninth; and for half the separation, the force was four times the first wire-twisting force. In general, the force *f* varied inversely as the square of the distance *r*, or

$$f \propto \frac{1}{r^2} \tag{4.1}$$

where the symbol $\propto$ is read as "is proportional to" or "varies as." By induction Coulomb was able to charge the two balls oppositely. The electrostatic attractive force also varied inversely with the square of the distance.

Finally, Coulomb learned how the electrostatic force depended on the magnitude of electric charge on each ball. When the charged pith ball was touched to the initially neutral ball, he supposed that since the two balls were identical in substance and size, the total charge ought to be equally distributed between the two. If the charged ball had a charge Q, then after contact each ought to have a charge $Q/2$. By touching either of the two charged balls with a third identical ball, he removed one-half of its charge. Thus the charge on either ball can be reduced from Q to $Q/2$, then to $Q/4$, $Q/8$, and so on.

Coulomb measured the force between the two balls with dif-

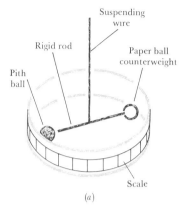

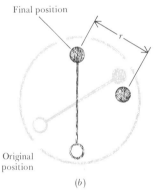

Figure 4.11. Coulomb's experiments investigating the forces between electrically charged bodies. (a) Coulomb's torsion balance in which a force is needed to rotate the rigid rod and to twist the wire; (b) top view of torsion balance after introduction of a pith ball, and rotation caused by electric force.

ferent known fractions of the initial charge on each, all at the same separation. The force was found to vary as the product of the charges on each ball. If Q_1 and Q_2 are the charges on the first and second balls,

$$f \propto Q_1 Q_2 \tag{4.2}$$

Combining both (4.1) and (4.2), the electrostatic force is

$$f \propto \frac{Q_1 Q_2}{r^2} \tag{4.3}$$

or

$$f = k \frac{Q_1 Q_2}{r^2} \qquad \text{(Coulomb's law)}$$

where k is a constant of proportionality which depends on the medium in which the two charged bodies are placed. When Q_1 and Q_2 have the same sign, the product $Q_1 Q_2$ is positive ($Q_1 Q_2 > 0$) and the repulsive force is positive. When Q_1 and Q_2 have opposite signs, $Q_1 Q_2 < 0$ and $f < 0$. *The attractive and repulsive forces act in opposite directions.*

Coulomb's law is one of the basic laws in electricity and magnetism. In honor of its discoverer, the unit of electric charge is the coulomb (coul). Formerly the coulomb was defined as the charge which must be on each of two bodies separated by 1 m and in a vacuum so that the electrostatic force between them is 8.9878×10^9 newtons (N). One N is the force which will increase the velocity of a 1-kg mass by 1 m/s². Thus, the coulomb was defined completely in terms of the meter, kilogram, and second. A newer definition of the coulomb, discussed later in this chapter, uses a more convenient technique of measurement but is also in terms of the same three basic units.

THE INVERSE SQUARE LAWS

The electrostatic force described by Coulomb's law is one of three forces which vary inversely as the square of the distance. Using a torsion balance, Coulomb also discovered that the force between two magnets has the same inverse-square dependence.

Henry Cavendish (1731–1810) also established the inverse-square nature of electrostatic forces, but his results were not made available until 1879 when James Clerk Maxwell published Cavendish's notes. One set of investigations which Cavendish made known in 1798 was his experiments with the force between masses—the gravitational force.

LAW OF GRAVITATIONAL ATTRACTION

Sir Isaac Newton was able to describe the motion of the planets about the sun with his general laws of motion and by assuming that the gravitational force f between two masses m_1 and m_2 at a distance r apart is

$$f = G\,\frac{m_1 m_2}{r^2}$$

where G is the gravitational constant.

Cavendish was able to confirm Newton's gravitation law and to determine the constant G by measuring the force between known masses at known separations. In his torsion balance, he fastened a weight at each end of the horizontal rod. In rotating the rod and weights from their rest position by applying a force, the suspending wire was twisted. When the applying force was removed, the wire untwisted and returned the rod and weights to the rest position. But because these were in motion, they continued past the rest position, twisting the wire in the opposite direction. The wire again untwisted and, because of the motion of rod and weights, retwisted in the original direction. The twisting, untwisting, and retwisting of the wire continued while Cavendish measured the time for a complete cycle, a period of about 15 min. Then Cavendish repeated the oscillations of the torsion balance when other larger weights were placed near the weights on the horizontal arm. The gravitational force acting between these weights affected the period of oscillation.

Because these forces were only about one fifty-millionth of the gravitational force between each weight and the earth, it was necessary to avoid any extraneous forces such as drafts. Cavendish observed the oscillations and measured their periods from a distance through a telescope by fixing a mirror on the suspending wire. As the wire twisted, the mirror turned and light was reflected by the mirror at an angle which oscillated in step with the torsion balance's motion.

Cavendish corroborated the form of Newton's law of gravitation and obtained a value of $G = 6.71 \times 10^{-11}$ N-m²/kg². Today the best value for the constant is 6.668×10^{-11} N-m²/kg². The gravitational force between two masses each 1 kg at a separation of 1 m is 6.668×10^{-11} N.

The force between a body and the earth is the weight of that body. Once Cavendish knew the value of G, he was able to estimate the mass of the earth by knowing the gravitational force between it and a small body of known mass. The separation is the distance from the midpoint of the earth to the midpoint of the small body.

> **Example.** The weight of a 1.00-kg mass at the surface of the earth, a distance of 6.4×10^6 m from the earth's center, is 9.80665 N. What is the mass of the earth?
>
> If we represent the weight by f, the earth's mass by m_1, the second mass by $m_2 = 1.00$ kg, and the distance between the two masses by $r = 6.4 \times 10^6$ m, then

$$m_1 = \frac{fr^2}{Gm_2}$$

$$= \frac{(9.80665 \text{ N})(6.4 \times 10^6 \text{ m})^2}{(6.668 \times 10^{-11} \text{ N-m}^2/\text{kg}^2)(1.00 \text{ kg})}$$

$$= 6.0 \times 10^{24} \text{ kg}$$

The best value, to date, is 5.975×10^{24} kg.

> **Example.** If the 1.00-kg mass is moved to a distance of 6.4×10^6 m *above* the surface of the earth, how much will it weigh?
>
> The mass is now 12.8×10^6 m away from the center of the earth, or twice as far as before. The weight is one-fourth the former value, or 2.45166 N.

FIELDS

The physical world as we conceive it progressively appears more complicated with continued scientific investigation. There is a compensating simplification, at the same time, because behavior in two seemingly different phenomena can often be described by the same model. In this chapter we are tracing the historical development of three converging ideas—electricity, magnetism, and light—to the point where they can be described by a single comprehensive theory. Another kind of simplification is possible when the mathematical descriptions of two phenomena have identical forms. Then a common vocabulary and similar concepts can be used to describe the two phenomena. The laws describing electrostatic and gravitational forces, both being inverse-square laws, are of this type.

In Fig. 4.12 we consider three aspects of the inverse-square-force laws. The first two are attractive forces between two masses and between opposite charges. The third is a repulsive force between like charges. (Thus far, no example of a repulsive gravitational force has been discovered.) One body (mass m_1 or charge Q_1) is kept in a fixed position; the other (mass m_2 or charge $-Q_2$ or $+Q_2$) can be moved so that the distance r between the two masses or charges can vary.

Although the force, either attractive or repulsive, cannot be observed until two masses or two charged bodies are present, it is convenient to talk about the *gravitational* or *electric fields* about an isolated mass or charged body. For example, we can say a *gravitational field* exists about the fixed mass m_1; this means that a second mass within this field will be acted upon by the force of gravity. The magnitude of the gravitational field is defined as the force that would be exerted on a 1-kg mass. Because this force depends on the distance from the central mass m_1, the magnitude of the gravitational field also must depend on this distance. We designate this magnitude by the symbol g, the force (in newtons) per mass (in kilograms). Because newtons per kilograms have the units meters per second per second, g is called the "acceleration due to gravity," which is the amount by which a kilogram mass is accelerated in a gravitation field of magnitude g.

If instead of a 1-kg mass we place a 2-kg mass within the same gravitation field, the force on it will be twice as much. Because the force needed to accelerate a mass is equal to the product of the mass and the acceleration, twice the mass requires twice the force. Thus the 2-kg mass will be accelerated by exactly the same amount as the 1-kg mass if the gravitational field for each is the same. From this, g is the amount by which any mass is accelerated in a gravitational field of magnitude g. Simply, if we know the value of g and the mass m_2 in a field of that magnitude, the force on that mass is $f = m_2 g$.

In an analogous manner we can also talk about an electric field existing about a central fixed charge $+Q_1$. Any other charge Q_2 placed within this field, i.e., near the central charge, will have an

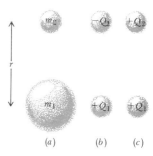

Figure 4.12. Inverse square laws and gravitational and electric fields.

$(a)\ f = G\,\dfrac{m_1 m_2}{r^2} = m_2 g$

— *gravitational force of attraction;*

$(b)\ f = -\dfrac{kQ_1 Q_2}{r^2} = -EQ_2$

— *electrostatic force of attraction;*

$(c)\ f = \dfrac{kQ_1 Q_2}{r^2} = EQ_2$

— *electrostatic force of repulsion.*

electrostatic force acting on it. Depending on the signs of the two charges, this force can be attractive or repulsive. Thus the electric field has a direction associated with it. A field, like a force, is a *vector*.

Most of our immediate experience with gravitational fields is with movement around the surface of the earth and with distances that are small compared to the earth's radius. We can expect the acceleration due to gravity to decrease somewhat with an increase in elevation. At sea level $g = 9.80665$ m/s²; and at an elevation of 500 m, $g = 9.80511$ m/s². Since the mass-to-mass distance at sea level is the earth's radius (6.4×10^6 m), the difference in elevation of 500 m is a relative change of only $(5 \times 10^2)/(6.4 \times 10^6)$, or 7.8×10^{-5}, less than one in ten thousand. Relative variations in $1/r^2$ are even smaller. For most calculations we can use the sea-level value.

With electrostatic forces, distances through which charges move are almost always relatively large, and so the assumption of a relatively constant field E is not reasonable.

POTENTIAL ENERGY AND POTENTIAL

The work that must be done, i.e., the energy that must be expended, to move something through a distance against a force resisting the motion is the product of the force and the distance through which the force acts. For a gravitational force $f = m_2 g$, if the mass m_2 is lifted a small distance h (Fig. 4.13a) away from the center of the earth, the work needed for this is $W = m_2 gh$. The work can be written in this simple way because g is essentially constant. If the distance h were not very small compared to the distances r_A and r_B from the earth's center to the initial and final positions of the mass m_2, we would have to consider g as varying and the equation for work W would be a little more complicated.

Since the gravitational force on m_2 is directed toward m_1, moving the mass m_2 from A to B requires energy, or work. If the mass m_2 now at B is released, it will spontaneously move toward m_1 under the influence of the gravitational field. When it gets to A, it will have exactly the energy, in the form of kinetic energy of motion, that was used to move it originally from A to B. If this were not so, there would be a violation of the first law of thermodynamics—the law of conservation of energy.

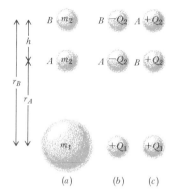

Figure 4.13. Work, potential energy, and attractive and repulsive forces. (a) Gravitational force of attraction acts to move masses toward each other. Work is required to move the masses apart, e.g., to move m_2 from A to B. This mass is said to have a higher potential energy at B than at A by an amount equal to the work done in moving it from A to B. The gravitational potential, which is the potential energy when m_2 is a unit mass, is also higher at B than at A. (b) Under an electrostatic attractive force, work is needed to move charge $-Q_2$ away from charge $+Q_1$. The charge $-Q_2$ has a higher potential energy at B than at A by an amount equal to the work done in moving $-Q_2$ from A to B. (c) Under a repulsive electrostatic force, work is needed to move charge $+Q_2$ closer to charge $+Q_1$. The charge $+Q_2$ has a higher potential energy at B than at A by an amount equal to the work done in moving $+Q_2$ from A to B.

Because the work done in the move from A to B is recoverable in the reverse move from B to A, the energy is stored rather than lost. We call that stored energy the "potential energy." We say the mass m_2 has more potential energy at position B than it did at position A by the amount m_2gh, or the gravitation potential energy difference of mass m_2 between positions A and B is m_2gh. Commonly, gravitational potential energy is measured relative to the value at sea level so that the potential energy at sea level arbitrarily is taken to be zero.

Just as we used the force per kilogram as the magnitude of gravitational field, its acceleration g, we can also speak about the potential energy per kilogram, which is called the "gravitational potential." The gravitational potential between two points in a gravitational field is the work needed to move 1 kg from the first point to the second. The convenience of the potential is that if we know the work required to move 1 kg between two points, we can immediately calculate the work needed to move m_2 kg between the same two points; it is exactly m_2 times as much as was needed for 1 kg.

When the mass m_2 at position B is released, it travels spontaneously toward mass m_1. We can describe what happens in three equivalent ways: First we can say that a gravitational force of attraction exists between the two masses. Second, we can say a gravitational field exists around and directed toward mass m_1; any other mass, for example, m_2, within this field will be impelled to move in the direction of the field. Third, the gravitational potential at B is higher than at some point, for example, A, closer to the central mass m_1. Any mass free to do so will spontaneously move from a point of higher gravitational potential to a point of lower potential.

An exactly analogous set of descriptions can be used for the electrostatic force, with the added feature of repulsive as well as attractive forces. Figure 4.13b and c illustrates the two forces: an attractive force between the unlike charges $+Q_1$ and $-Q_2$ and a repulsive force between the like charges of $+Q_1$ and $+Q_2$. In both cases moving the second body, with charge $-Q_2$ or $+Q_2$, from point A to point B, requires work against the electrostatic force opposing this motion. The force per unit charge, the electric field, acts against this motion. The work done in moving from A to B can be recovered if the body with charge $-Q_2$ or $+Q_2$ at B is released; it will spontaneously move toward A. As in the gravitational case there is a strict conservation of energy. The energy used to move the second charged body from A to B is stored as electric potential energy. We say that the difference in potential energy of charge

$-Q_2$ or $+Q_2$ at points A and B is exactly equal to the work needed to move that charge between those points.

The work needed to move a charge of 1 coul from A to B is the potential energy per coulomb or the *electrical potential difference* between points A and B. The value of the concept of electric potential is that if we know the work needed to move a single coulomb of charge from A to B, we can calculate immediately how much work is needed to move any number of coulombs between the same two points.

A body with an electric charge and in an electric field will move *spontaneously* from a point of higher electric potential to a point of lower electric potential. A mass falling spontaneously in a gravitational field can be a source of work, or energy, e.g., a waterfall turning a water wheel. Similarly, a charged body moving spontaneously to a lower electric potential can be a source of work, or energy. From the potential difference and the magnitude of the charge moved by the electric field between two points, we can calculate the work obtained in the spontaneous charge motion.

The unit of work, or energy, is the *joule*, named in honor of James Joule. It is the work done when a force of 1 N displaces something a distance of 1 m. Since the newton was defined in terms of the kilogram, meter, and second, so is the joule. The unit of charge, the coulomb, was also originally defined in the same units.

The electric potential difference between two points in an electric field was defined as the work needed to move a 1-coul charge from one to another point; it is the energy per charge and has the units joules per coulomb. In honor of Count Alessandro Volta (1745–1827), an investigator whose work will be discussed in the next section, 1 joule/coul is called a "volt" (V). The work needed to move 1 coul between two locations whose electric potentials differ by 1 V is 1 joule. The work needed to move any charge between any two positions with known electric potential difference is simply the product of the charge, in coulombs, times the potential difference, in volts, and has the units joules. Conversely, if a known charge spontaneously moves from a high electric potential to a lower potential, the work it can do (in joules) is the product of the charge and the voltage decrease.

VOLTAIC CELLS

In 1775–1778, Volta invented an electrophorus, a device which accumulates electric charge by rubbing two dissimilar materials, and built one of the earliest electroscopes. He also analyzed marsh gas, isolated and identified methane from it, and invented a lamp for burning the gas. However, his best-known and probably most important discovery was a process that replaced the older and less convenient amber-effect methods of generating electric charges.

In 1786, Luigi Galvani (1737–1798), a biology professor, observed the contraction of a frog's muscle when an electric spark was in the vicinity and also when two unlike metal objects touched the muscle. Galvani interpreted this contraction as an electrical process within the muscle itself. Volta, however, after his own investigations, came to the conclusion that what is essential for the generation of an electric field and the transport of charge is simply the presence of two dissimilar metals in a conducting liquid. He confirmed his opinion by constructing (in 1800) a series, or battery, of cells—cups filled with salt water into each of which were immersed one copper and one zinc strip. The cups were connected in series—the zinc from one to the copper of the other—by a conductor, as shown in Fig. 4.14.

An electric potential difference, or voltage, was created between the end, or terminal, strips. As with fields and potential differences created by the amber effect, electric charges could be made to travel along conductors from the terminal of higher electric potential to that of lower potential. In every way testable, the two phenomena—the amber effect and Volta's cell—involved exactly the same thing: the transfer of the electric "fluid." In the arbitrary manner of assigning signs† to the terminals of Volta's cell, called simply a "voltaic cell," or an "electric battery," the copper terminal has a positive charge and the zinc terminal a negative charge. With a conductor connecting them, positive charges move toward the zinc and negative charges toward the copper.

† That is, consistent with the amber-effect arbitrary sign assignment.

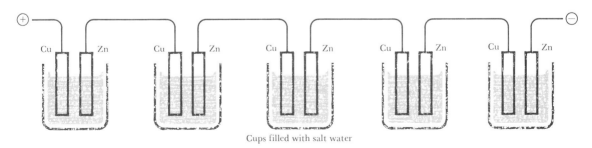

Cups filled with salt water

Figure 4.14. Volta's battery of cells producing an electric potential difference. One copper strip and one zinc strip are immersed in each cup.

The travel of electric charge along a conductor is called an electric "current." It was the accepted convention to say that current flows in the conductor from the positive to the negative terminal, i.e., that the direction of the current is the direction in which positive electricity would travel. However, what actually travels in a metal conductor is negative electricity, which flows from the negative to the positive terminal. Thus the current and the negative electric charge travel in opposite directions; at times this is a source of confusion.

Although the potential difference between the two terminals is created with a single cup, or cell, Volta found that the potential difference increased as he added more cells in series.

ELECTROLYSIS

When Volta announced his discovery, one of the first uses to which it was put was the decomposition of water by Nicholson and Carlisle. A platinum wire was connected to each terminal of a voltaic cell, or battery, and the free ends were immersed in water (Fig. 4.15). Gas bubbles formed around each of the two free platinum ends. Upon analysis it was found that hydrogen collected at the wire joined to the zinc, or negative, terminal and oxygen collected at the copper, or positive, terminal. Further, twice as much hydrogen as oxygen was formed. With Cavendish's formation, or synthesis, of water from two volumes of hydrogen and one volume of oxygen, this is more evidence in support of the elemental composition of water.

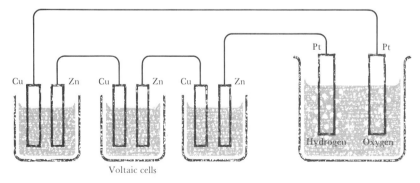

Figure 4.15. The electrolysis of water, decomposed into its elements hydrogen and oxygen, using a voltaic cell, was first demonstrated in 1800 by William Nicholson (1753–1816) and Sir Antony Carlisle (1768–1840).

The decomposition of a substance by passing an electric charge through it is called "electrolysis." The metal wires—in this case, platinum—leading into the liquid are called "electrodes," as are the copper and zinc strips in Volta's cells.

Although the argument between Galvani and Volta was resolved in favor of the latter with the construction of the first electric battery, it was discovered much later that the contraction of muscles does involve the transfer of electric charges. Both of them were correct.

MAGNETISM FROM ELECTRICITY

Lucretius [97(?)–54 B.C.] mentioned the existence of a certain mineral ore, a lodestone, found near Magnesia (today the Turkish city Manisa). Lodestone, now called "magnetite," which is composed principally of iron and oxygen, attracts objects made from iron. Also, from the name of that city, the word "magnet" and the names of two metals—"magnesium" and "manganese"—are derived.

With the convenience of electric batteries, opportunities to do research in electrical phenomena were opened to many more workers. One of the fundamental discoveries, which established a relationship between magnetism and electricity, was made in 1819 by Hans Christian Oersted (1777–1851). When the terminal strips of a voltaic battery are connected by a wire conductor, a continuous, or uninterrupted, path is provided for the travel of electricity; then, the *circuit* is said to be *closed* and a current can flow. Otherwise, if there is a break in the continuous path through the voltaic cell and wire, the circuit is said to be *open* and no current can flow.

Oersted observed that when a current flowed through a wire conductor, the needle of a nearby magnetic compass moved from its ordinary north-south alignment to a position nearly perpendicular to the wire. In some manner the system comprising the battery, connecting wire, closed circuit, and electric current was able to exert a force on a magnetized needle. To systematically investigate the phenomenon he discovered, Oersted used a series of 20 copper-zinc voltaic cells (about 15 V), the terminals of which were connected by a conductor wire. Before the circuit was closed, he arranged the wire to be in the north-south direction parallel to and either directly over or under his compass needle (Fig. 4.16a). When the circuit was closed, the needle pointed in the east-west direction

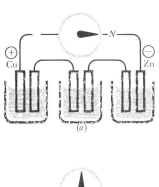

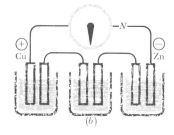

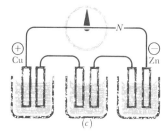

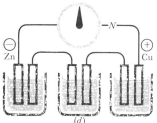

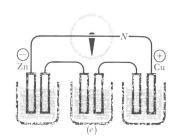

Figure 4.16. Oersted's discovery of a magnetic field accompanying an electric current. (a) Circuit is open, and compass needle points toward the earth's north pole; (b) circuit is closed, and wire is below the compass needle, which now points east; (c) circuit is closed, and wire is above the compass needle, which points west; (d) circuit is closed, but the electric flow is in the reverse direction and the wire is below the needle, which points west; (e) circuit is closed, the flow is still in the reverse direction, and the wire is above the needle, which points east.

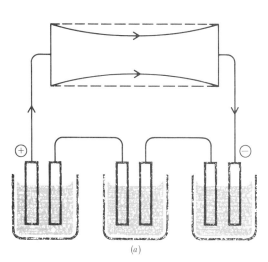

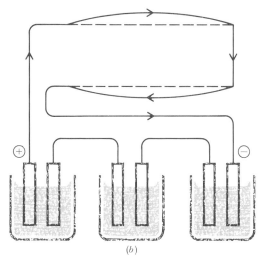

Figure 4.17. Ampère's discovery of the magnetic attraction and repulsion between two parallel wires carrying an electric current. Dashed lines indicate that wires were initially parallel before the circuit was closed. (a) When current flows in the same direction in each of two initially parallel wires, an attractive force bends the two wires toward each other; (b) if the current flows in opposite directions in each of two initially parallel wires, a repulsive force bends the two wires away from each other.

(Fig. 4.16b). Depending on the placement of the wire, the needle end marked "north" pointed either east or west (Fig. 4.16c, d, e).

The same effect was seen with connecting conductor wires of platinum, gold, silver, copper, iron, lead, and tin and with a column of liquid mercury; it did not depend on the nature of the wire conductor. The effect on the magnetic needle was not destroyed by interposing between the wire and the needle glass, wood, water, resin, stone, or metals other than iron; however, when iron was interposed between the wire and needle, the needle deflection away from the earth's north pole was lessened. Neither a needle of copper nor one made from glass was deflected in Oersted's experiments; the effect was seen only with magnetic materials, i.e., with substances on which a magnetic force acts.

Finally, Oersted demonstrated that this effect did not occur with just a static electric charge. It was not the same as the electrostatic force; what was required was a *moving* electric charge, an electric current. Associated with a moving electric charge, therefore, is a magnetic field, the existence of which can be demonstrated by the field's effect on materials affected by magnets or lodestones.

If a current through a wire is attended by a magnetic field, then when an electric current travels through each of two nearby wires, their magnetic fields ought to produce either an attractive or a repulsive force between the wires. André Marie Ampère (1775–1836) discovered just such an effect in 1820. When current flowed in the same direction in two parallel wires, these wires were attracted to each other; when the current was reversed in direction in one wire, the wires were repelled by each other (Fig. 4.17). When the two wires were not parallel but perpendicular, there was no force. The magnetic fields accompanying an electric current can be oriented, as can two magnets, either for repulsion or for attraction.

Both Oersted's and Ampère's magnetic field discoveries can be represented in a simple way. As in Fig. 4.18b, we can imagine a current flowing through a wire and, perpendicular to this, a plane in which we can place small test magnets. These are used to ascertain the presence and direction of the magnetic field. The direction of the field, indicated by the arrow, is the direction the "north" end of a magnet will point in the field. The *right-hand rule* is a convenient device for remembering the direction of the field: If the wire is held in the right hand with the thumb pointing in the direction of the current, the curled fingers point in the direction of the magnetic field (Fig. 4.18a).

It can be seen that Oersted's observations are described by this rule. If the current is reversed, so must be the right-hand

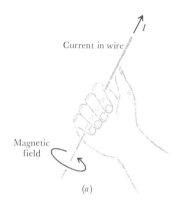

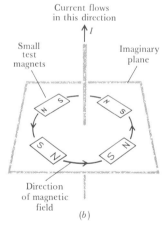

Figure 4.18. Right-hand rule and the direction of the magnetic field which accompanies an electric current. Small test magnets are aligned so as to indicate the presence and direction of the magnetic field accompanying the electric current. An imaginary plane is drawn perpendicular to the direction of current.

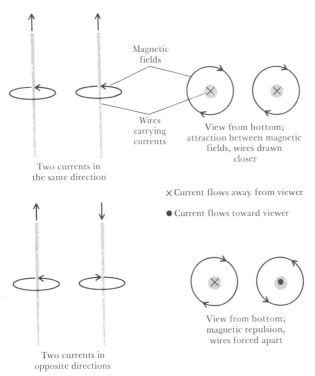

Figure 4.19. Ampère's discovery and the right-hand rule. Viewed along the wires, the magnetic fields are aligned for attraction if the currents flow in the same direction, and for repulsion if in opposite directions.

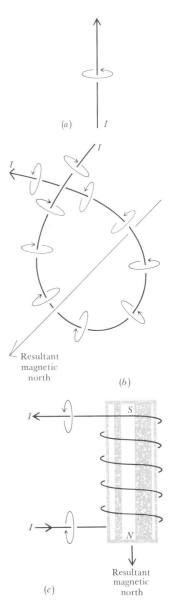

Figure 4.20. (a) Magnetic field about a current i traveling in a straight-wire conductor; (b) magnetic field about a current traveling in a simple single-loop wire; (c) magnetic field in a series of loops — a solenoid.

thumb and thus the direction of the curled fingers. Depending on whether the wire is above or below the compass needle, the "north" end can point east or west. Ampère's discovery and the right-hand rule are illustrated in Fig. 4.19.

ELECTROMAGNETS AND MAGNETIZATION

If a magnetic field is associated with a moving electric charge in the circular way described by the right-hand rule, a simple twisting of the conductor wire can enhance this field.

Within a single loop (Fig. 4.20) the magnetic fields about each point of the wire are all pointed in the same direction. In effect, the front face of the plane of the loop is magnetic north, and the reverse face is magnetic south. A series of loops, obtained by wrapping a wire around a pencil or pipe, has the form of a *helix*, or *solenoid*. The greater the number of turns in the solenoid, the stronger the magnetic field. If an iron bar is slipped within the solenoid while a current flows, the magnetic field is further enhanced.

This arrangement of solenoid and iron core is called an "electromagnet." Although both electromagnets and lodestones produce magnetic fields which are indistinguishable from each other, there is an important difference. When the current of the electromagnet is stopped, the magnetism vanishes or drops to a much smaller level; for this reason it is called a "temporary magnet." A lodestone, on the other hand, is said to be a "permanent magnet."

Certain materials, the best known of which is iron, are attracted to a magnet and are themselves magnetized. That is, an iron bar in contact with a lodestone or with an electromagnet can also pick up iron particles (Fig. 4.21). If the bar remains in contact with the magnet or lodestone for a sufficiently long time, then, after separation, it will be permanently magnetized. Iron and other materials which can be permanently magnetized are called "ferromagnetic" materials. The word "permanent" must be qualified since the field around a permanent magnet does decrease in time and can be destroyed completely and quickly if the magnet is heated.

There are two kinds of magnetic poles—north and south—just as there are two kinds of electricity—positive and negative. For each, the rules of likes repelling and unlikes attracting apply. But there is an important difference between inducing magnetism in an iron bar and inducing an electrification in the same

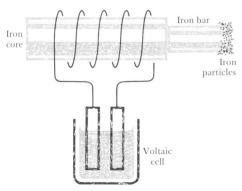

Figure 4.21. Iron bar is magnetized by being in contact with a magnet.

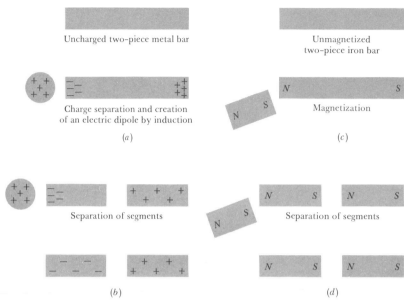

Figure 4.22. (a) When a positive charge is brought near the two contiguous pieces of the conducting iron bar, a negative charge develops at the near end and a positive charge at the far end. (b) If the two pieces are separated, the pieces will remain charged—one negatively and the other positively. It is possible to isolate one from the other, as did Coulomb in his investigations of electrostatic forces. (c) We can repeat the same process but use a magnet in place of the positive charge and for a time sufficiently long to induce permanent magnetism. (d) The two-segment iron bar develops a north and south end, but when it is separated, each segment now has a north and a south end. If we were to cut each segment into many pieces, each small piece would also have a north and south end. We can never isolate a north pole from a south pole; one is always accompanied by the other. We say the magnets, for this reason, are dipoles rather than monopoles. Electric charge, in contrast, can be isolated as electric monopoles.

bar. Consider again the induction experiment, this time in a conductor cut into two pieces (Fig. 4.22).

Forces between magnets appear to have somewhat more complicated behavior than those between electric charges. But this is only because between any two magnetic dipoles there are four forces to be considered: two attractive forces and two repulsive forces (Fig. 4.23). As in Coulomb's law, the force depends inversely on the square of the separation distance. Force also depends on the product of the magnitudes of the magnetic poles.

In 1877 Rowland transferred an electric charge produced by the amber effect to an insulated disk and observed a weak magnetic effect when the disk was rotated at high speeds. It was comparable in magnitude to the magnetic effect expected with a weak current flowing under the same conditions. This was additional evi-

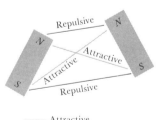

Figure 4.23. Because magnets always exist as dipoles, four forces must be considered in the interaction between two magnets: two attractive forces between opposite poles and two repulsive forces between like poles.

dence which established the equivalence between electricity produced by the amber effect and by voltaic cells.

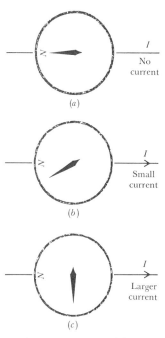

Figure 4.24. Compass deflections. (a) No electric current; the compass needle points toward earth's magnetic north pole; (b) a small current: the compass is influenced by the magnetic fields of both the earth and the current; (c) a larger current: the needle is influenced mostly by the field of this current, which is much greater than the field of the earth.

ELECTRICAL MEASUREMENTS AND UNITS

In Oersted's experiment the magnetic compass needle was either parallel or perpendicular to the wire, depending on the absence or presence of a current in the wire. There was a competition between the earth's magnetic field and the magnetic field discovered by Oersted. If the latter field were about equal in magnitude to the earth's field, the needle would be neither parallel nor perpendicular to the wire but at some intermediate angle (Fig. 4.24). Thus the magnetic compass can be used to measure the magnitude of electric current.

A *galvanometer* is simply a more effectively designed instrument for measuring the amount of current flowing through a wire. The wire can be looped in the form of a solenoid and held in place by springs between the poles of a permanent magnet. Within the solenoid is placed an iron core, and attached to this is a pointer (Fig. 4.25).

With no current passing through the coil, there is no magnetic force to act against the restraining springs. The position on the scale opposite the pointer can be marked zero. When a current flows, the produced electromagnetic field has its north near the north of the permanent magnet. The repulsive force acts against the spring, and the coil and pointer move to a new position along the scale. The larger the current, the stronger the produced electromagnetic field and the greater the pointer deflection.

If the current flows in the opposite direction, the north and south poles of the solenoid and core are reversed. Then the north of the permanent magnet and the south of the electromagnet attract each other. The core and needle move in the opposite direction. A galvanometer can measure not only the magnitude but also the direction of an electric current.

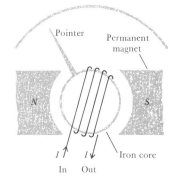

Figure 4.25. The galvanometer pointer deflection reflects the magnitude of current.

THE AMPERE AND OTHER UNITS OF ELECTRICITY. The unit of electric current is the ampere (A). When 1 A flows through each of two straight, parallel, very long and very fine wires placed 1 m apart and in a vacuum, the force, first discovered by Ampère, is 2×10^{-7} N for each meter of wire length. When a galvanometer is designed especially for measuring current and its scale is calibrated in amperes, it is called an "ammeter."

Originally the unit of electric charge—the coulomb—was defined by the electrostatic force acting between two charged bodies. The unit of current was first defined as 1 A, equal to a flow of 1 coul/s. It is more convenient to precisely and accurately measure current than charge, and so the ampere is defined as in the previous paragraph. Now, however, the coulomb is defined as the amount of electricity which is transported in 1 s when a current of 1 A flows.

The unit of work, or energy—the *joule*—has already been defined to be the effort required to move something with a force of 1 N through a distance of 1 m. In performing work against electrostatic forces, the *volt* was also defined as the potential difference through which a charge of 1 coul can be moved with the expenditure of 1 joule in energy.

The rate at which energy is produced or used is *power*, and the unit of power is the *watt* (W), which is the rate of 1 joule/s. This unit is named in honor of James Watt (1736–1819), the inventor of the steam engine. Conversely, the energy consumed is the product of the power (energy per unit time) and the time: a joule is 1 W-s. Electric energy is sold in units of kilowatt-hours.

> **Example.** If a 75-W bulb is plugged into a 110-V circuit, calculate the current through the bulb filament, the coulombs transported in one day, and the energy used up in one day.
>
> Since
>
> $$1 \text{ joule} = 1 \text{ coul-V}$$
> $$1 \text{ joule/s} = 1 \text{ coul-V/s}$$
> $$1 \text{ W} = 1 \text{ A-V}$$
> $$1 \text{ A} = 1 \text{ W/V}$$

then

$$75 \text{ W}/110 \text{ V} = 0.68 \text{ A}$$

The charge transported is

$$0.68 \text{ A} \times 1 \text{ day} \times \frac{24 \text{ hr}}{1 \text{ day}} \times \frac{60 \text{ min}}{1 \text{ hr}} \times \frac{60 \text{ s}}{1 \text{ min}}$$
$$= 0.68 \times 24 \times 60 \times 60 \text{ A-s}$$
$$= 5.9 \times 10^4 \text{ coul}$$

The energy used is

$$75 \text{ W} \times 24 \text{ hr} = 1,800 \text{ W-hr}$$
$$= 1.8 \text{ kW-hr}$$

Electric charge flows spontaneously along a conductor from a higher potential to a lower potential. This potential difference may be provided by a voltaic cell. Georg Simon Ohm (1789–1854) observed that for any conductor, the greater the potential difference or voltage, the higher the electric current which results from this voltage. He discovered a very simple relationship between current I and voltage V for metal conductors: the current was proportional to the voltage. Doubling the voltage produced twice the amperage. This can be represented by the formula

$$I = KV$$

where the constant of proportionality K is called the "conductance" of the conductor along which the current flows.

For practical reasons, the formula is most often written as

$$I = \frac{V}{R}$$

The factor in the denominator R is the resistance of the conductor. Resistance is the reciprocal of the conductance: $R = 1/K$. Both this formula and the previous one are called "Ohm's law."

The unit of electric resistance is the *ohm* (Ω), defined as the resistance between two points of a conductor when a potential difference of 1 V produces a current of 1 A. From the second and more common form of Ohm's law

$$R = \frac{V}{I}$$

Example. Calculate the resistance of the bulb filament in the last example.

$$R = \frac{110 \text{ V}}{0.68 \text{ A}}$$
$$= 160 \ \Omega$$

If a galvanometer is designed so that the current flows through a wire whose resistance is known, the current and the pointer deflection are proportional to the applied voltage. Such an instrument is called a "voltmeter."

The resistance of a material can be determined by measuring the current and potential difference between two ends of that material and then using Ohm's law. The resistance of any material depends on its temperature, shape, and size. In general, long thin wires have higher resistances than short thick wires of the same material. But, comparing wires of the same dimensions, the resistance also depends on the material. Metals and all conductors, in general, have low resistances and high conductances; nonmetals and all nonconductors have high resistances and low conductances. At room temperature silver and copper have the least resistances of all materials.

Example. Compare the resistances of the three metals, silver, copper, and lead, by measuring for each the current flowing in a wire 0.100 m long and 1.00×10^{-6} m^2 in cross-sectional area when a potential difference of 1.00 V is applied at the wire ends. Experimentally, the currents are found to be 629, 581, and 45 A respectively.

Since the same shape and size of wire is used for each metal, differences in resistances depend only on the nature of the metal. The following table lists the experimental values for current and voltage and the calculated resistances.

Substance	Potential difference, V	Current, A	Resistance, Ω
Silver	1.00	629	1.59×10^{-3}
Copper	1.00	581	1.72×10^{-3}
Lead	1.00	45	2.2×10^{-2}

From this, copper has a resistance 1.08 times as great and lead has resistance 14 times as great as silver of the same shape and size.

Example. Calculate the resistances of two nonconductors, or insulators, wood and amber, using the same technique.

Because these materials conduct so poorly, the currents passing through samples of wood or amber of the shape and size of wire are extremely small. To obtain measurable currents, high voltages must be applied to samples which are very thin and have large areas, i.e., to sheets rather than to wires. Using 100,000 V and sheets which are 1.00×10^{-2} mm thick and 1.00×10^{-2} m² in area, the measured currents across the thin sheets are 0.33 A for wood and 2.0×10^{-7} A for amber. The measured and calculated values are tabulated as follows:

Substance	Potential difference, V	Current, A	Resistance, Ω
Wood	100,000	0.33	3.0×10^5
Amber	100,000	2.0×10^{-7}	5.0×10^{11}

In spite of the conducting advantage of sheets over wires, these two insulators have much higher resistances than the three metals used in the previous experiment.

Example. Compare the resistances of the five materials mentioned in the previous examples, assuming the wood and amber are made in the shape and size of the metal wires.

The resistance is directly proportional to the length of the material and inversely proportional to the cross-sectional area. Increasing the length from a sheet thickness of 1.00×10^{-2} mm or 1.00×10^{-5} m to a wire length of 0.100 m will increase the resistance by $0.100/1.00 \times 10^{-5}$, or by a factor of 10^4. Reducing the area from a sheet of 1.00×10^{-2} m² to a wire of 1.00×10^{-6} m² will also increase the resistance by $1.00 \times 10^{-2}/1.00 \times 10^{-6}$, or by a factor of 10^4. Together, the resistance of a wood or an amber wire is $10^4 \times 10^4 = 10^8$ times as great as the resistance of the wood or amber sheet, or 3.0×10^{13} and 5.0×10^{19} Ω, respectively. (See the table below.) Resistance is an important physical property for distinguishing among and identifying substances.

Substance	Resistance, Ω
Silver	1.59×10^{-3}
Copper	1.72×10^{-3}
Lead	2.2×10^{-2}
Wood	3.0×10^{13}
Amber	5.0×10^{19}

ELECTRICITY FROM MAGNETISM

Since a magnetic field attends a moving electric charge, it is natural to ask if the converse is also possible—to produce an electric current from a magnetic field. Investigations confirming this conjecture were conducted by three contemporary scientists independently of each other—an Estonian-Russian, an American, and an Englishman. They were Heinrich Lenz (1804–1865) in St. Petersburg, Joseph Henry (1797–1878) in Albany, N.Y., and Michael Faraday (1791–1867) in London. We shall look at four of their experiments (Figs. 4.26 to 4.29).

The separate wire solenoids are carefully wrapped about a wooden spool in such a way that at no point do they touch each other (Fig. 4.26). The ends of one solenoid are linked to the plates of a voltaic cell, making what is called the "primary circuit"; the ends of the other solenoid are connected to a galvanometer, making the "secondary circuit."

While a current flows through the primary circuit and also when this circuit is open (as when one plate is lifted out of the cell), the galvanometer needle is stationary—there is no current in the secondary circuit. However, when the primary circuit is just closed (as when the plate is just returned to the voltaic cell), there is a transient galvanometer needle deflection. There must be, for this short time, a secondary current. When the primary circuit is opened (as during the actual lifting of one plate from the cell liquid), the galvanometer needle momentarily swings in the other direction.

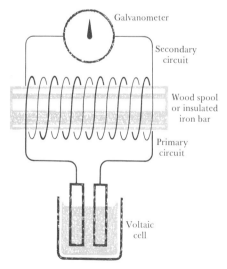

Figure 4.26. Current flows in the secondary circuit only when the primary circuit current just begins or just stops.

There is a transient secondary current but in the direction opposite to the former transient current. A secondary current is produced only when the primary current is begun or interrupted or when the primary current is changing.

The same behavior is seen, but with a stronger transient secondary current, when an iron core is used in place of the wooden spool. Here it is necessary to insulate the two solenoids from each other and from the conductor iron core. The transient secondary current can be described in terms of the magnetic field of the primary current.

When the primary circuit is open, there is no primary current and no magnetic field in the solenoid; we can say the magnetic field is zero. Just as the circuit is closed, a primary current flows and the magnetic field *increases* from zero to some steady value, which depends on the solenoid and current. It is during this *increase* in the magnetic field that a secondary current is detected. Once the magnetic field reaches this steady value, it does not change and no secondary current flows. Opening the primary circuit stops its current, and the magnetic field changes, dropping to zero. During this *decrease* a secondary current flows in the opposite direction. After this, with the primary circuit open, there is no primary current, no magnetic field, and no secondary current.

Next, suppose we take two long wire conductors and, after folding the middle part of each into a zigzag pattern, fasten each to a stiff wooden board (Fig. 4.27). The ends of one wire are connected to a voltaic cell to form the primary circuit. A galvanometer is connected to the ends of the other wire to form the secondary circuit. A steady current flows through the primary circuit and is accompanied by a steady magnetic field.

When the second board is moved either toward or away from the first one (but without the two wires touching), the galvanometer needle moves in one direction or the other. A secondary current develops as the secondary wire *moves through* the magnetic field of the primary current. The direction of this current depends on the direction of relative motion of the wire and the magnetic field.

A V-shaped magnet can be made from two bar magnets, joining the north end of one to the south end of the other (Fig. 4.28). A solenoid is wrapped around a magnetizable iron core, and each end of the wire is connected to a galvanometer. When the two free ends of the two bar magnets are touched to the iron core, a transient current is detected by the galvanometer. At the moment

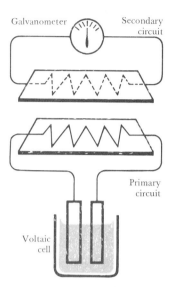

Figure 4.27. Secondary current flows when there is relative motion between the pair of wires.

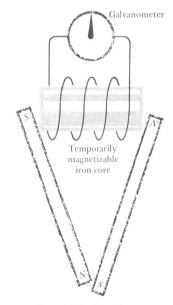

Figure 4.28. Secondary current flows only when the bar magnets just touch or are just removed from the iron core.

of touching, the iron core, originally not magnetized, becomes a magnet. During the time its magnetic field builds up from zero, a transient current flows; it ceases to flow when the field reaches its steady value. When the bar magnets are removed, the magnetic field of the core drops to zero. During this drop, a transient current flows in the opposite direction. In this third experiment there was no primary current involved—only the *change* in the magnetic field about the core and solenoid.

Finally (Fig. 4.29), if the ends of a long wire are joined to a galvanometer and part of this wire is moved about a magnetic field, a current flows in the wire. The direction of the current, as before, depends on the direction of motion of the wire. When the wire is stationary, no current flows.

There is one feature common to these four demonstrations: a wire conductor in a *changing* magnetic field. It does not matter how this feature is brought about: by starting or interrupting an electromagnet, magnetizing or demagnetizing an iron bar, moving a wire in the magnetic field of a current, or moving a wire in the magnetic field of a permanent magnet (or an electromagnet, too). A current is induced to flow in the wire when this feature is present. This process of creating an electric current—actually creating a potential difference which causes a current to flow—is called "induction."

When an induced current flows, it is accompanied by its own magnetic field. Thus there are two magnetic fields: the initial one inducing the current and an induced field. Lenz found that the induced magnetic field is always directed so that a repulsive force acts between the two fields. Work is required, then, to create an in-

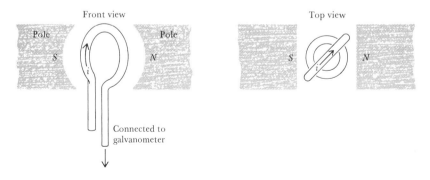

Figure 4.29. Current flows when the loop moves in the magnetic field.

duced voltage and current. Because the induced voltage can be used to do work, i.e., to move an electric charge, Lenz's observation is consistent with the first law of thermodynamics; otherwise electric energy could be created with no expenditure of energy, or work.

ELECTRIC GENERATORS AND MOTORS

The induction effect, using a changing magnetic field, offers a third means (in addition to the amber effect and the voltaic cell) for producing, or generating, an electric current. This method, in fact, is the cheapest, and therefore it is the principal source of electricity today. The work (described by Lenz's law) necessary to move a wire loop conductor within a magnetic field is supplied by the least expensive energy source available.

In hydroelectric generation, falling water can turn a water wheel and, with it, attached wire loops. Alternatively, a heat engine run by burning the fossil fuels coal and oil can do this work. Unfortunately, much of the cheap fuel used for this contains large amounts of sulfur, and the by-products of such fuel consumption include major air pollutants. In a later chapter we shall look at the use of nuclear reactors to supply the energy to run electric generators.

A practical combination of a magnetic field and movable wire loops is called an electric "generator." For the continuous generation of electricity the loops are not merely moved back and forth but are rotated. Because of the direction (described by Lenz's law) of the induced magnetic field accompanying the generated current, an interesting feature attends this current. Consider a single loop placed between magnetic poles as in Fig. 4.30. Although either a permanent or an electromagnet can be used, for simplicity we shall consider only the permanent magnet. To avoid twisting of wires as the loop is rotated, the simple arrangement shown can be used. One end of the wire loop slides around a conductor ring, and the other end rotates within this ring. Wires can be joined to this end and to the conductor ring for the withdrawal and consumption of the generated electric energy.

As the loop rotates, the direction of the current generated regularly alternates. This periodic alternation is shown in Fig. 4.31.

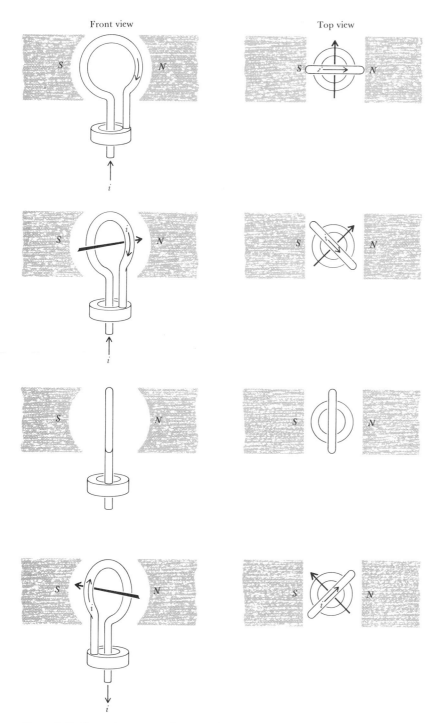

Front view   Top view

Figure 4.30. Alternating current. The current direction changes so the induced magnetic field of the loop interacts with the permanent magnetic field to always require work to turn the loop (Lenz's law). Heavy arrow points to induced magnetic north pole.

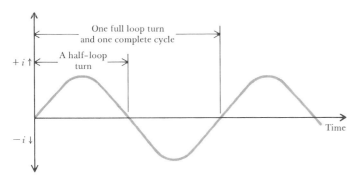

Figure 4.31. Regular periodic variation in the electric current generated by a loop turning continuously in a magnetic field.

There is a continuous rather than an abrupt change in the current. The electric current thus produced is called "alternating current" (ac). The electric potential difference which causes this alternating direction must also change in a similar way. During one-half of the loop turn, one generator terminal has a higher potential than the other terminal; during the second half, the same terminal has a lower potential than the other. In contrast, the current from a voltaic cell is always in one direction and is called "direct current" (dc).

It is possible to obtain direct current from a generator by replacing the conductor ring with a terminal that is designed differently. However, alternating current can be more efficiently transmitted over long distances. For most purposes, then, this is what is used—for lighting, heating, electric motors, etc. In the United States, 110 V is the common potential difference available for home use; and the ac frequency is 60 cps (maintained at this value with remarkable constancy).

An electric *motor* converts electric energy into mechanical work. In design and principle it resembles a generator, except that alternating current from a generator is passed through the loops. The magnetic field, alternating in direction with the alternating current, is repelled continuously by the magnetic field of the permanent magnet. The loops are continuously rotated. A shaft connected to the loops rotates with them and thus can perform work. It is possible to design a motor using direct current and even one that may use either alternating or direct current.

> *Example.* The cost of 1 kW-hr of electric energy made by a generator is about 5 cents. A 1-W flashlight bulb uses two 1.6-V dry cells

(for a total of 3.2 V) at a total cost of 40 cents. If the bulb can shine for 200 hr, calculate the cost per kilowatt-hour using a voltaic cell. The total energy consumed is

$$1 \text{ W} \times 200 \text{ hr} = 200 \text{ W-hr}$$
$$= 0.200 \text{ kW-hr}$$

The cost per kilowatt-hour is:

$$40 \text{ cents}/0.200 \text{ kW-hr} = \$2.00/\text{kW-hr}$$

This is 40 times as expensive as using generator electricity.

MAXWELL'S EQUATIONS

James Clerk Maxwell (1831–1879) recognized a pattern of behavior in all the electrical and magnetic phenomena known at that time. With only four equations he was able to quantitatively describe the results of Coulomb, Oersted, Ampère, Ohm, Faraday, and Lenz. Maxwell's equations were said to be complete in that there was nothing in the way of electric and magnetic force fields that they did not account for. They are one of the very greatest creative efforts of science.

The equations themselves can be combined to describe the way in which magnetic and electric fields change with time and position in space. These equations have exactly the form of equations used to describe *transverse wave motion*, where something is propagated in space and is attended by an oscillation perpendicular to the propagation. The vibrating violin string, for example, is described in this way.

A *wave equation* includes as one of its parameters the speed of propagation. In Maxwell's equation the speed of propagation of the electric and magnetic oscillations can be obtained experimentally from the two definitions of electric force. From Coulomb's original description,

$$\text{Force (N)} = \frac{\left(\frac{8.9875 \times 10^9 \text{ N-m}^2}{\text{coul}^2}\right)(Q_1 Q_2 \text{ coul}^2)}{r^2 \text{m}^2}$$

According to Ampère's discovery,

$$\text{Force/length}\left(\frac{N}{m}\right) = \left(\frac{2 \times 10^{-7} \text{ N}}{A^2}\right) \frac{I^2}{r}\left(\frac{A^2}{m}\right)$$

In Maxwell's formulation the numerical constants must be combined in this way to obtain the square of the speed of propagation:

$$v^2 = \frac{8.9878 \times 10^9 \text{ N-m}^2}{\text{coul}^2} \times \frac{A^2}{10^{-7} \text{ N}}$$
$$= 8.9878 \times 10^{16} \text{ m}^2/\text{s}^2$$
$$v = 2.9979 \times 10^8 \text{ m/s}$$

This is exactly the measured value of the speed of light.

Maxwell's equations lead to the prediction of an oscillating electric field which is attended by an oscillating magnetic field. One can equally correctly speak of an oscillating magnetic field which is attended by an oscillating electric field. A change in one field can induce the presence of the other field. The pair of oscillating fields can be pictured as in Fig. 4.32; they are known as "electromagnetic radiation," or "electromagnetic waves." Since the waves travel at the speed of light, the suspicion exists that light and electromagnetic waves have something in common.

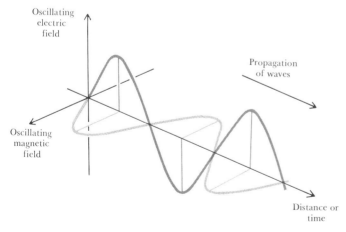

Figure 4.32. Electromagnetic radiation consists of an electric field oscillating periodically in one plane. Here a magnetic field is oscillating periodically perpendicular to and synchronously with the electric field; also, both the electric and magnetic fields are oscillating in directions perpendicular to the direction of motion of the radiation.

THE BEHAVIOR OF
ELECTROMAGNETIC WAVES

Joseph Henry (1797–1878) showed that when two metal spheres are heavily and oppositely charged with electricity and then brought close to each other, a visible spark jumps across the gap from one sphere to the other. The spheres are still oppositely charged but now in reverse; the sphere that was originally positive is now negative and vice versa. This charge imbalance results in another spark, which jumps in the opposite direction. An alternating series of back-and-forth jumping of sparks continues until the charges leak away. From the magnitude of the charge, sphere size, and air resistance, it is possible to estimate the charging and discharging rate.

To test Maxwell's equations, Heinrich Hertz (1857–1894) built an oscillating spark generator. In one typical experiment he fastened a highly polished brass sphere to the ends of each of two brass rods. The rods were aligned as in Fig. 4.33*a*, and the spheres were about 3 mm apart. The rods were joined to a solenoid, so that the system, solenoid, joining wires, brass rods, spheres, and air gap, formed the secondary circuit. The primary circuit comprised another solenoid wound around the first one and

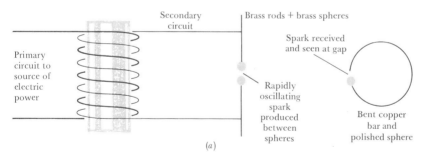

Figure 4.33. (a) Hertz's demonstration of the production and propagation of an electromagnetic signal. (b) When a metal sheet was placed 0.33 m behind the original oscillating spark, the transmitted signal was found to be stronger and could be detected at a greater distance. The enhancement of the transmitted signal strength depended sensitively on the distance between the metal sheet and the signal-generating brass spheres. (c) If the signal is a wave, and if the sheet is exactly one-half wavelength away, the reflected and direct signals combine by constructive interference to produce a stronger signal for the detector. (d) If the sheet is held three-eighths of a wavelength from the generating gap, the reflected and direct signals are not perfectly in phase; the combined amplitude is less here than in (c), and the received signal is weaker at the detector. (e) at $\frac{1}{4}$ wavelength complete destructive interference.

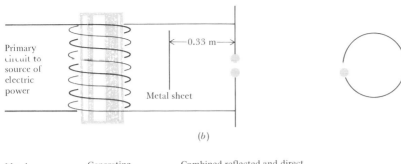

Primary
circuit to
source of
electric
power

0.33 m

Metal sheet

(b)

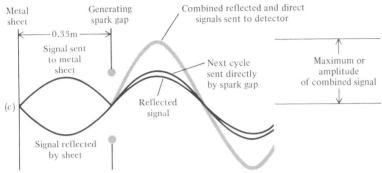

Metal
sheet

Generating
spark gap

Combined reflected and direct
signals sent to detector

0.33m

Signal sent
to metal
sheet

Next cycle
sent directly
by spark gap

Maximum or
amplitude
of combined signal

Reflected
signal

(c)

Signal reflected
by sheet

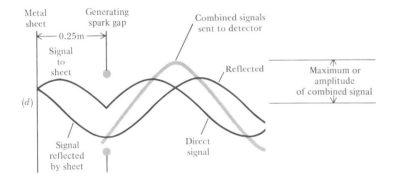

Metal
sheet

Generating
spark gap

Combined signals
sent to detector

0.25m

Signal
to
sheet

Reflected

Maximum or
amplitude
of combined signal

(d)

Signal
reflected
by sheet

Direct
signal

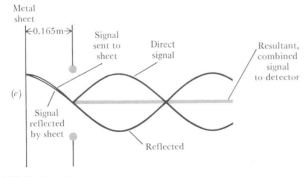

Metal
sheet

0.165m

Signal
sent to
sheet

Direct
signal

Resultant,
combined
signal
to detector

(e)

Signal
reflected
by sheet

Reflected

Figure 4.33 (Continued).

a voltaic battery. Since air has a high resistance, a very large potential difference was needed to force an electric current, i.e., a spark, across the gap. This was provided by induction between the two solenoids. Hertz estimated that the oscillating spark produced had a frequency of about 1 billion/s.

In the same room Hertz placed a copper bar curved into an almost complete circle. At one end of the bar he placed a polished sphere; the other end was sharpened to a point. The distance from the point to the sphere was a few hundredths of a millimeter. When an oscillating spark was moving in the first part of his apparatus, i.e., between the two brass spheres, he detected a spark in the second gap, i.e., between the point and sphere. Even though the two pieces of equipment were about 2 m apart, an electric effect had traveled across the laboratory room from one gap to the other—from the sender to the receiver, or *detector*. (This was the first broadcast of electromagnetic radiation.)

The maximum distance through which the electric effect could be transmitted from the sender to the detector was increased by using polished metal sheets. Hertz found that if he placed such a sheet behind the original spark sender (Fig. 4.33*b*), the detector gap would show a spark if the path from the sender to it was 5 or 6 m. However, the increased distance of electric-effect transmission depended quite sensitively on where the metal sheet was placed. In one of his experiments, for example, Hertz found that if the sheet was exactly 0.33 m behind the original spark, the effect could be detected at a maximum distance. He reasoned that this was exactly one-half the wavelength of the electromagnetic wave. In traveling to the sheet and back to the original gap, a distance of 0.33×2, or 0.66 m, the electromagnetic wave made one complete oscillation and joined a wave just beginning at the gap. The two waves, in phase, travel to the detector with an enhanced intensity because of constructive interference (Fig. 4.33*c*). Thus, combined this way, electromagnetic waves can be detected at a greater distance.

From the wavelength thus measured and his estimation of the spark frequency, Hertz was able to arrive at an estimated value of the speed of propagation of the electromagnetic wave. This was approximately equal to the speed of light. More modern techniques of measuring spark frequency improve this agreement.

Polished metal sheets placed around the laboratory improved the distance of detection and thus the intensity of the transmitted signal, provided the sheets were in certain positions. Hertz found that, as with light reflected by a mirror, the incident and

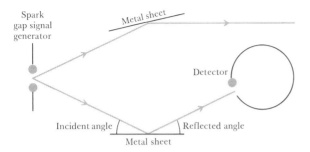

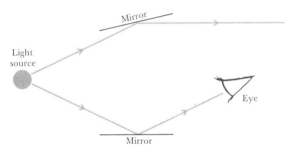

Figure 4.34. Both ordinary light and Hertz's electromagnetic signal are reflected by surfaces so that the incident and reflected angles are equal.

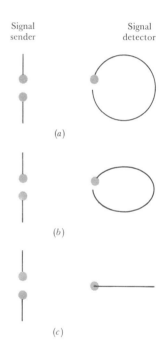

reflected angles for the electromagnetic signal are equal (Fig. 4.34). By placing metal objects between the sender and detector, the signal was blocked from reaching the detector. Hertz showed that the signal travels in straight lines, as does light. Objects opaque to the electromagnetic waves cast shadows over the detector. Insulators, e.g., wood, did not stop the signals and were transparent to the transmission of this effect.

For these studies Hertz had the sender and detector aligned in the same plane (Fig. 4.35). In rotating the detector to remove this alignment, he saw that the signal was not detected, or reduced in intensity. This means that the vibrating electromagnetic signal put out by the sender was traveling as Maxwell's equation predicted and as illustrated in Fig. 4.36. The curve describing the electric field variation is planar; i.e., it lies entirely in a plane. The corresponding magnetic field variation curve is also planar, and its plane is perpendicular to the electric field curve. Hertz's waves are said to be "plane-polarized."

Figure 4.35. Hertz's experiment demonstrates the plane-polarized nature of his generated electromagnetic radiation. (a) When the detector and sender are aligned, a strong signal is received at the detector; (b) a weaker signal is seen when the detector is rotated; (c) no signal is received at the detector when it is rotated by 90° from the first alignment.

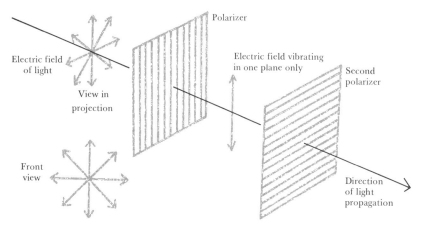

Figure 4.36. Plane polarization of light.

When an ordinary lightbulb is seen through two Polaroid sheets, such as two lenses from Polaroid sunglasses, the intensity of light reaching the eyes can be reduced to a very low level by rotating one of the sheets. This effect can be described most simply by considering light to be one form of electromagnetic radiation, as the work of Hertz and Maxwell indicates. As in Fig. 4.36 (for convenience only the electric field is drawn), the electric field oscillations coming from the bulb filament are in all possible planes. Passing through the first Polaroid sheet, all the oscillations except those in one plane are filtered out. The first sheet passes light oscillating in one plane only. If the second sheet is aligned parallel to the first, the polarized light will pass through it, too; if it is aligned perpendicular to the first, it will be filtered out and no light will pass through. Thus both ordinary light and Hertz's electromagnetic waves can be polarized.

When light passes from one medium into another, where the speed of light differs in each, the light is seen to be bent, or *refracted*. For example, a stick placed partly in water appears bent. Like a grating, a glass prism breaks up, or disperses, light into its various constituent colors because the various wavelengths of light are refracted to different extents. Hertz constructed a huge prism (not from glass but from an asphalt material) which weighed about 1,200 lb. He showed that his electromagnetic signal was refracted by it.

Thus, in addition to having a common speed, both light and electromagnetic waves can be reflected, refracted, and polarized and exhibit interference. Light is, then, an electromagnetic radiation with frequencies and wavelengths detectable by the human eye.

THE ELECTROMAGNETIC SPECTRUM

The wavelength or frequency, or both, of any electromagnetic radiation can be measured. It is convenient to arrange the entire known collection of electromagnetic radiation in a table according to wavelength or frequency. This arrangement is called the electromagnetic "spectrum" (Fig. 4.37). Since the values vary over a wide range, the spectrum has a logarithmic scale.

The visible spectrum, which we can see with our eyes, has the wavelengths from about 3.60×10^{-7} to 8.00×10^{-7} m, that is, from 360 to 800 nm. The corresponding frequencies are 8.3×10^{14} and 3.8×10^{14} Hz. The shortest visible wavelength and highest frequency is violet. That part of the spectrum adjacent to the visible with higher frequencies is called the "ultraviolet" (uv) region. The low-frequency end of the visible spectrum is red, and adjacent to it is the "infrared" (ir) spectrum.

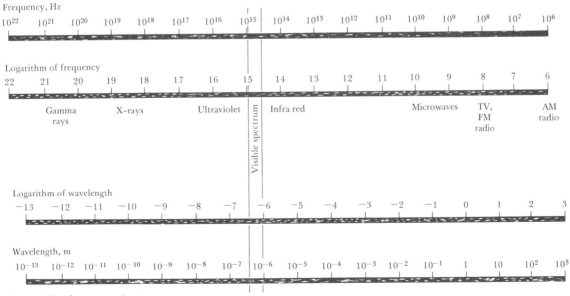

Figure 4.37. Electromagnetic spectrum.

Example. A radio station has a frequency of 810 kc. What is the wavelength of its broadcast electromagnetic radiation?

The frequency is

$$\nu = 810 \times 10^3 \text{ Hz}$$

and so its wavelength is

$$\lambda = \frac{c}{\nu}$$
$$= \frac{3.00 \times 10^8 \text{ m/s}}{810 \times 10^3/\text{s}}$$
$$= 370 \text{ m}$$

Example. A radar unit uses radiation of 3.00×10^{-2} m wavelength. What frequency is this?

$$\nu = \frac{c}{\lambda}$$
$$= \frac{3.00 \times 10^8 \text{ m/s}}{3.00 \times 10^{-2} \text{ m}}$$
$$= 1.00 \times 10^{10} \text{ Hz}$$

PROBLEMS

4.1. On Sunday, June 20, 1969, Neil A. Armstrong, the first human being to step on the surface of the moon, spoke to listeners on earth. How long did it take for the radio message to travel the 3.8×10^8 m from moon to earth?

4.2. What is the travel time for a photon to go from the sun to the earth, a distance of 1.49×10^{11} m?

4.3. One light year is the distance light travels in 1 yr. The nearest star, Proxima Centauri, is about 4 light years away from the earth. How far is that in kilometers?

4.4. In 1866, Mahlon Loomis sent the first message by wireless telegraph between two high mountain peaks 18 miles, or 29 km, apart. How long did it take for the signal to go this far?

4.5. The earth has a mass of 5.98×10^{27} g and a mean radius of 6.38×10^6 m. For the moon, the corresponding values are 7.34×10^{25} g and 1.74×10^6 m. Estimate the weight of a 160-lb man at the surface of the moon.

4.6. Jupiter has a mass of 1.88×10^{30} g and a radius of 7.14×10^7 m. How much will a 160-lb man weigh on the surface of this planet?

The Structure of Atoms

Now we shall resume our exploration into the fine structure of matter. Atoms are much too small to be seen or to be individually manipulated and studied. The specimens scientists can study ordinarily contain an enormous number of atoms. With what we know about physical phenomena such as temperature, electricity, and light, we can see how matter behaves when a specimen is heated, when an electric current is passed through it, and when light of various wavelengths interacts with it. We shall look for patterns of regularity in the behavior of matter, and we shall try to organize and thus simplify these patterns. Then we shall see what can be inferred about the constituent atoms themselves.

PROUT'S HYPOTHESIS

The atomic weights of some of the more common elements are: hydrogen, 1.0; carbon, 12.0; nitrogen, 14.0; oxygen, 16.0; sodium, 23.0; potassium, 39.1; aluminum, 27.0; and sulfur, 32.1. These values are all very close to whole numbers. In 1815 William Prout (1785–1850) proposed that all atomic weights are exact multiples of the atomic weight for hydrogen and that hydrogen is the primary substance from which all other elements and thus all matter are made.

The molecular weight of chlorine gas was found experimentally to be 70.90. When chlorine gas was shown to be diatomic, the atomic weight *of the element* was seen to be one-half this, or 35.45. Since Dalton's postulate of indivisible atoms was well corroborated by experimental evidence, the notion that a chlorine atom was made up of $35\frac{1}{2}$ hydrogen atoms was not acceptable. For this reason, Prout's hypothesis was dropped. It became acceptable in a modified form when the structure of atoms was more clearly understood.

SPECIFIC HEAT

If 5 g of aluminum are heated to 100.0°C and dropped into 10 g of water at 0.0°C, the temperature of both water and metal after a short while will be about 9.7°C. Heat is transferred from the warmer metal to the cooler liquid; the temperature of metal drops and that of water increases until they are equal. Then there is no further heat transfer because heat is transferred only when a temperature difference exists. With no heat transfer and with no other energy changes, the temperatures remain constant (Fig. 5-1*a*).

When the same experiment is done with 5 g of gold, the final temperature is about 1.5°C (Fig. 5.1*b*). We conclude that at 100.0°C the gold carries with it less heat than an equal mass of aluminum at the same temperature. If we wish to heat the two metal samples from room temperature to some higher temperature, we shall find that it requires much less energy, or heat, to warm up gold than it does an equal mass of aluminum.

To describe these observations as simply as possible we define the *specific heat* of a substance as the amount of heat needed to increase the temperature of a 1-g sample of that substance by

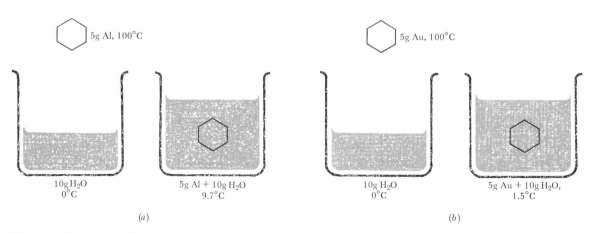

Figure 5.1. Aluminum metal at 100°C contains more heat than the same mass of gold at the same temperature. The greater the specific heat, the less the temperature changes for a given amount of heat added or removed. (a) 5 g of aluminum are added to 10 g of water (note the temperature change); (b) 5 g of gold are added to 10 g of water [compare the temperature change with (a)].

1°C. The specific heat of water is defined as 1 cal/° g. The specific heats of other substances are then expressed in terms of this calorie and are measured by transferring heat between the substances and, usually, water. For convenience, materials other than water are sometimes used as references.

The technique of measuring heat leaving the warmer substance, by the increase in temperature of the cooler water, is called "calorimetry." The apparatus used—a container of water, precise thermometers, and auxiliary equipment, all insulated against heat either leaving or entering from the surroundings—is called a "calorimeter."

> **Example.** From the information given, calculate the specific heats of aluminum and gold.
>
> For the aluminum calorimetry, 10.0 g of water, with a specific heat of 1.00 cal/°-g, have a temperature increase of $9.7 - 0.0°C = 9.7°C$. This is a total heat gain of
>
> $$10.0 \text{ g} \times \frac{1.00 \text{ cal}}{°\text{-g}} \times 9.7° = 97 \text{ cal}$$
>
> Assume the calorimetry was done carefully so that exactly 97 cal left the aluminum. We shall let c cal/°-g be the unknown aluminum specific heat. The temperature change is
>
> $$100.0 - 9.7 = 90.3°C$$
>
> Then
>
> $$5.0 \text{ g} \times \frac{c \text{ cal}}{°\text{-g}} \times 90.3° = 97 \text{ cal}$$
>
> Thus
>
> $$c = 97 \text{ cal} \times \frac{1}{5.0 \text{ g}} \times \frac{1}{90.3°}$$
>
> $$= \frac{97}{450} \frac{\text{cal}}{°\text{-g}}$$
>
> $$= 0.22 \text{ cal}/°\text{-g}$$
>
> For the gold calorimetry, the water temperature increase is 1.5°C and the metal decrease is 98.5°C. The heat gained by the water is

$$10.0 \text{ g} \times \frac{1.00 \text{ cal}}{°\text{-g}} \times 1.5° = 15 \text{ cal}$$

The heat lost by the metal is

$$5.0 \text{ g} \times \frac{c \text{ cal}}{°\text{-g}} \times 98.5° = 15 \text{ cal}$$

Then

$$c = 15 \text{ cal} \times \frac{1}{5.0 \text{ g} \times 98.5°}$$
$$= \frac{15 \text{ cal}}{490°\text{-g}}$$
$$= 0.031 \text{ cal}/°\text{-g}$$

Information about specific heats enables us to describe many heat transfer phenomena.

> **Example.** A 10.0-g cube of aluminum at 80.0°C is placed in contact with a 16.0-g cube of gold at 20.0°C. If the two cubes are within an insulated container so that heat neither enters nor leaves, what is the final temperature of the two cubes?
>
> The final temperature for each cube will be represented by t°C. Then the temperature decrease of the aluminum is $80.0 - t$°C, and the heat loss (to the gold) is

$$10.0 \text{ g} \times \frac{0.22 \text{ cal}}{°\text{-g}} \times (80.0 - t)°$$

> The temperature increase of the gold is $t - 20.0$°C, and its heat gain (from the aluminum) is

$$16.0 \text{ g} \times \frac{0.031 \text{ cal}}{°\text{-g}} \times (t - 20.0)°$$

> Since there is no other energy involved, in accord with the first law of thermodynamics the two heats must be equal.

$$10.0 \times 0.22 \times (80.0 - t) \text{ cal} = 16.0 \times 0.031 \times (t - 20.0) \text{ cal}$$
$$(180 - 2.2t) = 0.50t - 9.9$$
$$190 = 2.7t$$
$$t = 70°C$$

The specific heats of all substances change with temperature. As absolute zero is approached all specific heat values also approach zero, but at room temperatures and for all but very precise work, we can assume the specific heat to be constant. If the specific heat is known over a wide temperature range, then we are able to calculate the heat, or energy, needed to bring the substance from one temperature to another. In this way chemists are able to calculate the energy of a substance at any temperature, above the energy of the same substance at absolute zero. From this, they are able to describe quite well the heats and energies involved in chemical and physical changes. This important branch of science is called "thermochemistry." Later, in Chap. 6, we shall discuss thermochemistry in greater detail.

Except for a very small number of substances, water has the highest known specific heat. This has an extremely important significance in describing weather or meteorological processes. Because of water's large specific heat, great amounts of heat can be added to or taken from bodies of water with relatively small temperature changes. During a warm day, water heats up more slowly than a contiguous mass of land. At night, water cools off more slowly than the land. As a result, water has a stabilizing effect on the regional temperature.

THERMAL POLLUTION

An important aspect of cooling by heat transfer in water is thermal pollution. Water from rivers or lakes is often diverted into power plants, particularly nuclear power plants, to cool the reactors and then returned to the main stream. The warmer water then heats the main stream water, causing an increase in temperature. If this increase is large enough, the fish and plants natural to the water cannot survive the change. Less oxygen is dissolved in warm water than in cold, and the fish may suffocate (Fig. 5.2).

THE LAW OF DULONG AND PETIT

A metal is an element that is solid at room temperature, shiny, a good conductor of electricity and heat, able to be hammered without breaking (i.e., malleable) and to be drawn out (ductile), and

Figure 5.2. Thermal pollution. Downstream from a thermonuclear power plant, thousands of dead fish evidence the destructive effects of heating a natural water source.

opaque. Examples are copper, iron, zinc, silver, gold, and sodium; mercury, an exceptional metal, is liquid at room temperature.

An interesting aspect of the specific heat of metals was first pointed out in 1819 by Pierre Louis Dulong (1785–1838) and Alexis Therese Petit (1791–1820). In Table 5.1 some common metals are listed, together with their specific heats, their atomic weights, and the product of these—the *molar heat capacity* (called often the "atomic heat") in calories per degree per mole. For metals, the molar heat capacity at room temperature or above is about 6 cal/°-mole. This law is useful in estimating the atomic weights of metals.

> **Example.** The specific heat of palladium is 0.057 cal/°-g. Estimate its atomic weight M from this.
>
> The molar heat capacity is approximately 6.1 cal/°-mole.

Table 5.1. Specific Heats of Metals

Metal	Specific heat, cal/°-g	Atomic weight, g/mole	Molar heat capacity, cal/°-mole
Aluminum	0.22	26.98	5.9
Cadmium	0.055	112.4	6.2
Copper	0.092	63.54	5.8
Gold	0.031	197.0	6.1
Iron	0.11	55.85	6.1
Lead	0.031	207.2	6.4
Magnesium	0.25	24.31	6.1
Platinum	0.032	195.1	6.2
Silver	0.056	107.9	6.0
Tin	0.054	118.7	6.4
Tungsten	0.034	183.8	6.2
Zinc	0.092	65.37	6.0
Average			6.1

Thus

$$\frac{0.057 \text{ cal}}{°\text{-g}} \times M \approx \frac{6.1 \text{ cal}}{°\text{-mole}}$$

$$M \approx \frac{6.1}{0.057} \frac{\text{g}}{\text{mole}}$$

$$= 110 \text{ g/mole}$$

The symbol $\approx$ is read "is approximately equal to." The atomic weight of this metal is 106.4. (See inside the front cover of this book.)

FARADAY'S LAWS OF ELECTROLYSIS

Pure water is a poor conductor. But if a little common salt, an acid, or an alkali is added, the resulting solution will conduct electricity. Such a substance which, when added to water, increases the electrical conductivity is called an "electrolyte" (Fig. 5.3).

When a direct current is passed through such a solution, gas collects at each of the two wires, or *electrodes*, dipping into the water solution. Water is broken down, or *electrolyzed*, into its constituent elements hydrogen and oxygen. Faraday was able to collect the two

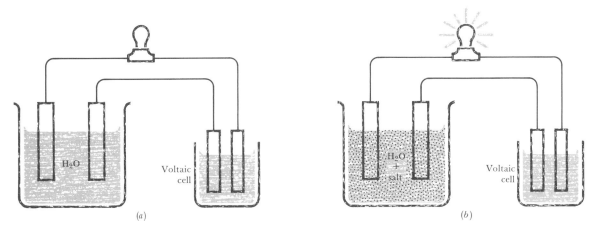

Figure 5.3. (a) Water is a poor conductor — hence there is no current to light the bulb; (b) when a salt, acid, or alkali is added to the water, the solution is a good conductor and a current flows.

gases separately and measure the amounts generated by this *electrolysis* process. Using an ammeter, he measured the current (coulombs per second), and using a clock, he measured the length of time of the electrolysis. The product of these is the number of coulombs passing through the circuit, including the solution (Fig. 5.4).

Faraday found that the amount of hydrogen and oxygen collected at each electrode depended only on the quantity of electricity passed and was independent of the electrolyte added, the nature of the electrodes, the duration of the electrolysis, and the current flowing. Faraday's *first law of electrolysis* (1834) states that the

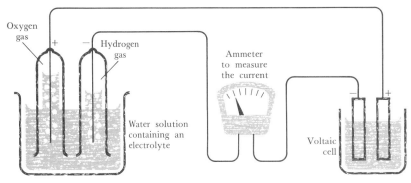

Figure 5.4. Electrolysis of water. The amount of gas collected is exactly proportional to the quantity of electricity passed through the water solution.

amount of material decomposed by this process varies exactly proportionally to the total quantity of electricity passed. Doubling the coulombs used will double the mass of material collected at either electrode.

Faraday suggested that the collected gases can be used to measure the number of coulombs passing through a circuit. An instrument that does this is called a "coulometer." He also introduced several of the terms that are still used in discussing electrolysis and electrolytes. The electrode which is positively charged is called the "anode"; the negative electrode is the "cathode." A negatively charged species in the solution which moves toward the positively charged anode is called an "anion." The positively charged species traveling to the negative cathode is a "cation." Collectively, cations and anions are called "ions." (This word is derived from a Greek word meaning *to travel,* or *to go.*) Since the electrolytes do convert water into a conductor, they must, in some manner, give rise to the charged ions that, when moving, are part of the electric current.

Several electrolyses can be performed in series (Fig. 5.5). Faraday found that the masses of materials collected at the several electrodes were always in a fixed ratio. For each gram of hydrogen collected, he always obtained 8 g of oxygen, 36 g of chlorine, 125 g of iodine, 104 g of lead, and 58 g of tin. In Fig. 5.5, for each gram of hydrogen, we should collect, in addition, 23 g of sodium and 32

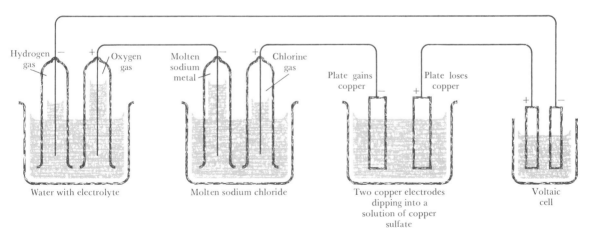

Figure 5.5. Three electrolyses operating in series. The amount of copper loss by one plate is exactly equal to the mass of copper gain by the other plate. The amounts of hydrogen, oxygen, and chlorine gases and metallic sodium collected and the amount of copper either collected or lost all increase with the number of coulombs of electricity passing through this closed circuit. But the masses of these several gases and metals transformed are always in the same ratio.

Table 5.2. Comparison of Equivalent Weights and Atomic Weights

Element	Equivalent weight	Atomic weight	Atomic weight Equivalent weight
Hydrogen	1.0080	1.0080	1
Oxygen	7.9997	15.9994	2
Chlorine	17.726	35.453	2
Iodine	63.452	126.9045	2
Lead	103.6	207.2	2
Tin	59.35	118.69	2
Copper	31.773	63.546	2
Silver	107.868	107.868	1
Aluminum	8.9938	26.9815	3

g of copper. Faraday stated that in a series of electrolyses 8 g of oxygen, 36 g of chlorine, 125 g of iodine, and 104 g of lead are each *equivalent* to 1 g of hydrogen. His *second law of electrolysis* states that in a series of electrolyses the masses of materials transferred to or from electrodes are always in the ratios of their *equivalent weights*.

If we compare the equivalent weights with the corresponding atomic weights using modern values for each, we can see an interesting thing (Table 5.2). The equivalent weight of any element is either equal to the atomic weight or one-half or one-third of the atomic weight. For some elements the fraction is one-fourth or one-sixth.

Faraday remarked that although nothing was then known about the structure of atoms, *it seemed quite likely that this structure would have an electrical character*. The best experimental value for the quantity of electricity needed to obtain 1 equivalent of a substance through electrolysis is 96,487.0 coul/equivalent. This number is called the "Faraday constant" and is represented by $\mathscr{F}$.

> **Example.** In an electroplating process a current of 5.00 A flows for 180.0 min. How much copper will plate out on the cathode?
>
> The quantity of electricity passed through the circuit is

$$5.00 \text{ A} \times \frac{1 \text{ coul}}{1 \text{ A-s}} \times 180.0 \text{ min} \times \frac{60 \text{ s}}{1 \text{ min}} = 54,000 \text{ coul}$$

$$= 5.40 \times 10^4 \text{ coul}$$

This can be converted to equivalents of copper:

$$5.40 \times 10^4 \text{ coul} \times \frac{\text{equivalent}}{96,487.0 \text{ coul}} = 0.560 \text{ equivalents}$$

In turn, this is converted to the mass of copper plated out:

$$0.560 \text{ equivalents} \times \frac{31.77 \text{ g}}{\text{equivalent}} = 17.8 \text{ g copper}$$

FUEL CELLS

In a voltaic cell a chemical process produces electric energy. This production continues as long as material is available for the chemical process to occur. Ultimately batteries do run down. They can be recharged by supplying electric energy to undo the chemical change, which can again be allowed to occur with a second production of electric energy. Thus, batteries can be used to store electric energy and for this reason are called "storage batteries."

The recharging of a battery is one example of electrolysis—causing a chemical change by the addition of electric energy. Thus an electrolysis can be considered the reverse of the normal operation of a voltaic cell. If the electrolysis of water produces hydrogen and oxygen, then the recombination of these gases to form water can be made to occur with the production of electric energy.

The production of hydrogen and oxygen can be made a continuous, or uninterrupted, process by supplying water to the electrodes as fast as it is decomposed. The reverse, the formation of water, can also be made continuous by supplying the two gases continuously from tanks. Therefore, the production of electric energy by this method can be continuous. This operation, the prolonged operation of a voltaic cell by supplying the needed substances continuously, is an alternative method to recharging. Such a device is called a "fuel cell" (Fig. 5.6).

Fuel cells have certain advantages over other energy-producing methods. The production rate can be easily controlled by varying the resistance of the circuit. If the resistance is very large, no current flows and no energy is delivered. The fuel cell is not limited in its efficiency in the way a heat engine is; it can be operated at moderate temperatures. There are no moving parts to wear out or to waste energy by friction. The fuel consumption can be complete and the products formed harmless; thus there is no contribution to air pollution. On the other hand, sometimes several hundred cells are needed to supply the energy needed. Also, the current is direct rather than alternating, and this is not always convenient.

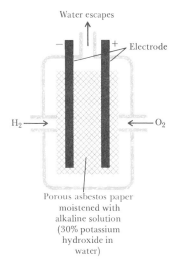

Figure 5.6. The reverse of water electrolysis. In the fuel cell, the elements are fed in continuously to form water and electric energy. A delicate balance is required between water production and evaporation to maintain the liquid level constant within the cell.

Figure 5.7. Fuel cells are used to provide electric power for the Apollo spacecraft. (National Aeronautics and Space Administration.)

One commercial version of a fuel cell uses 36 such cells to supply 500 W of power at 28 V. Using hydrogen and air as fuel, it operates at 65°C. At this temperature the water loss by evaporation balances the formation of water from its elements. In spacecraft where this amount of energy is needed for up to a month, the fuel cell at present is the lightest and most convenient form of packaged energy (Fig. 5.7).

PERIODIC PROPERTIES OF THE ELEMENTS

By the middle of the nineteenth century about sixty elements were known to exist. They were distinguishable by each having a unique set of physical and chemical properties. Almost all are solids at room temperature. Those which are gases have low atomic weights. Only two elements are liquid at room temperature: bromine and mercury.

It is possible to group elements according to their chemical properties. For example, the metals lithium, sodium, potassium, rubidium, and cesium react violently with water to produce hydrogen gas, while forming solutions which are bitter in taste and slippery. The solutions are said to be *basic*, or *alkaline*, and these elements are called the "alkali metals."

Another set of metallic elements—magnesium, calcium, strontium, and barium—react much more slowly with water but also form basic solutions. They are called the "alkaline earths." Neither they nor the alkali metals are ever found *free*, i.e., in the pure state, in nature. They are prepared usually by the electrolysis of molten salts of these metals. Both in aqueous solution and in the molten state the metals are collected at the negative electrode and thus are cations.

Of the four elements fluorine, chlorine, bromine, and iodine, the first two are gases, bromine is a liquid, and iodine is a solid. With hydrogen, each forms a diatomic molecule which dissolves in water to form a sour-tasting, metal-corroding solution called an "acid." Because the compounds they form (with, for example, the alkali metals and alkaline earths) resemble each other, for simplicity they can be studied together. Collectively, the four elements are called the "halogens." In compounds the halogens are typically anions.

To see what other similarities exist among the elements we can list some of their physical properties: their melting and boiling points, densities, and atomic weights (see Table 5.3). For convenience we shall consider only the first 32 elements (when listed in order of increasing atomic weights) known around 1850 and use modern values for the physical properties. For comparison of normally gaseous elements, the densities used are for the liquid state.

It is evident that the melting and boiling points and the densities do not simply increase as the atomic weight gets larger. To see better what sort of variation does occur in these properties a graph is useful (Fig. 5.8). The properties shown periodically increase and decrease as the atomic weight increases. Chemically, carbon and silicon behave similarly; they are on corresponding parts of two adjacent graph peaks. The alkali metals lithium, sodium, potassium, and rubidium are all found at or near the graph troughs.

If all the known elements are used in such graphical representations, this periodic variation in these and other physical properties will be seen to be a general phenomenon. The *periodic law* states, in summary, that *the properties of the elements vary in a periodic*

Table 5.3. Some Physical Properties of the Lighter Elements

Name	Symbol	Atomic weight	Melting point, °C	Boiling point, °C	Density, g/cm³
Hydrogen	H	1.01	−259	−253	0.071
Lithium	Li	6.94	180.5	1365	0.53
Beryllium	Be	9.01	1283	2484	1.85
Boron	B	10.8	2180	3660	2.34
Carbon	C	12.0	3727	4830	2.26
Nitrogen	N	14.0	−210	−196	0.81
Oxygen	O	16.0	−219	−183	1.14
Fluorine	F	19.0	−220	−188	1.11
Sodium	Na	23.0	97.8	903.7	0.97
Magnesium	Mg	24.3	649	1105	1.74
Aluminum	Al	27.0	662	2467	2.70
Silicon	Si	28.1	1412	3167	2.33
Phosphorus	P	31.0	44	277	1.82
Sulfur	S	32.1	115	445	2.07
Chlorine	Cl	35.5	−101	−35	1.56
Potassium	K	39.1	63.2	774	0.86
Calcium	Ca	40.0	838	1440	1.55
Titanium	Ti	47.9	1668	3260	4.51
Vanadium	V	50.9	1900	3450	6.10
Chromium	Cr	52.0	1880	2665	7.19
Manganese	Mn	54.9	1245	2150	7.43
Iron	Fe	55.8	1536	2872	7.86
Nickel	Ni	58.7	1450	2730	8.90
Cobalt	Co	58.9	1495	2900	8.90
Copper	Cu	63.5	1080	2600	8.96
Zinc	Zn	65.4	420	906	7.14
Arsenic	As	74.9	613	817	5.72
Selenium	Se	79.0	217	685	4.79
Bromine	Br	79.9	−7.2	59.8	3.12
Rubidium	Rb	85.5	39	688	1.53
Strontium	Sr	87.6	768	1380	2.60
Zirconium	Zr	91.2	1855	4474	6.49

way with an increase in the atomic weight. This law was first stated in 1869 by Dimitri Ivanovitch Mendeleev (1834–1907) and independently by Lothar Meyer (1830–1895).

THE PERIODIC TABLE

Once the periodic law was recognized as an important general description of the relationship between an element's atomic weight and its set of properties, the next logical step was to arrange all the elements in a table with similar elements listed near each other.

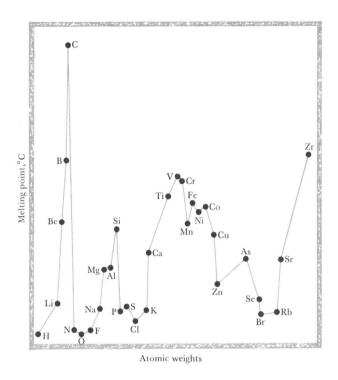

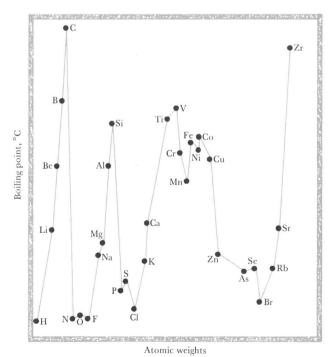

Figure 5.8. *Variation of melting points and boiling points with atomic weight.*

One form of such a *periodic table* is bound as an endpaper at the back of this book. All the known elements found in nature or man-made are listed. Within each box is the name, symbol, atomic weight, and ordinal number for each element. The last is simply the counted position, beginning with hydrogen, the lightest element, as number 1. This is called the "atomic number." For example, the element whose atomic number is 82 is lead, represented by Pb with an atomic weight of 207.2.

Notice the elements in the column labeled "Group I*A*"; they are the alkali metals. Those in group VII*A* are the halogens. In a like manner any column contains elements which have similar properties. Such an arrangement makes the study of chemistry much easier. It will also prove to have a close relationship to the actual structure of atoms.

It should be observed that there are two special rows of elements: numbers 58 to 71 and 90 to 105. These follow lanthanum (number 57) and actinium (number 89) and thus are called the "lanthanide" and "actinide" series. The 14 elements of the lanthanide series all have quite similar properties and for this reason can be separated only with difficulty. In the actinide series the elements beyond uranium (92), the *transuranic* elements, are all man-made. The 101st element is named after the man who designed the first periodic table.

An examination of each atomic weight and number indicates that there are a few cases where the atomic weight *does not* increase: from element 18 to 19, from 27 to 28, from 52 to 53, and within the actinide series. Consider the elements 52 and 53. Iodine has all the properties of a halogen, and tellurium has those of group VI*A*. They are placed where they are on the basis of their properties other than their atomic weights. The same has been done with the other elements. This indicates that the atomic number is a more important and more fundamental property of an element than is its atomic weight. The *periodic law* can be more exactly stated: *the properties of the elements vary in a periodic way with an increase in the atomic number.*

VALENCE

One criterion used by Mendeleev to arrange the elements within a periodic table and thus assign atomic numbers is the manner in which elements form compounds. For example, compounds of the alkali metals and halogens (groups I*A* and VII*A*) are HF, KBr,

NaCl, RbI, CsCl, and so on. They have the general feature of one atom from each group and can be represented by MX, where M is the metal and X the halogen atom.

Using group IIA and VIIA elements, the compounds are $MgBr_2$, $SrCl_2$, BaF_2, and so on, and are represented by MX_2, indicating one alkaline earth and two halogen atoms. With groups IIIA and VIIA, the compounds $AlCl_3$, BF_3, $GaBr_3$, $TlCl_3$, and so on, are represented by MX_3. Groups IVA and VIIA form, in general, MX_4 (for example, SiF_4, $SnCl_4$, and CI_4). Group VIA instead of VIIA forms with IA elements (for example, H_2O), compounds of the form M_2O; with IIA elements (for example, MgS) MO compounds; with IIIA elements (for example, Al_2O_3) compounds with the general formula M_2O_3; and with IVA (for example, SiO_2) the compounds MO_2.

In discussing the combining capacity of elements, chemists use the term *valence*. This can be defined for convenience as the number of hydrogen or chlorine atoms with which an element combines. In our examples, sodium in NaCl has a valence of 1; magnesium in $MgBr_2$ has a valence of 2 (since bromine and chlorine behave similarly here); thallium in $TlCl_3$, a valence of 3; and tin in $SnCl_4$, 4. In the compound H_2O, oxygen has a valence of 2; in H_2S, sulfur also is 2; in NH_3, nitrogen is 3.

Both magnesium and calcium have valences of 2, as in $MgCl_2$ and $CaCl_2$. Each metal forms a compound with oxygen, namely, MgO and CaO; oxygen also has a valence of 2. But there is no stable compound MgCa, and so we must further designate the valence with an algebraic sign $+$ or $-$. Since the alkali and alkaline earth metals in compounds are cations, we say sodium has a valence of $+1$; calcium, $+2$. Because the halogens in compounds often are anions in solution or in the molten state, we say chlorine and the others have a valence of -1.

Now the rules of combining atoms are obvious. The valences of the atoms used must add up to zero in the formed compound. This corresponds to an electrically neutral molecule just as the positive or negative sign of the valences corresponds to an electrically positive or negative charge, as in ions.

Given what we know about the valence signs, we can combine elements with confidence to form simple molecules. We know that LiCl is stable but not $LiCl_2$ or $LiCl_3$. We know H_2O, Al_2O_3, and MgS are plausible but not H_2O_3, AlO, or Mg_2S. Unfortunately there are three serious matters which prevent us, with simple valence rules, from predicting all possible molecules. First, under diverse conditions, many elements have several valences. The compounds NH_3, N_2O, N_2O_3, NO, and N_2O_5 are all known. In these compounds, with valences of $+1$ for hydrogen and -2 for oxygen, nitrogen has valences of -3, $+1$, $+3$, $+2$, and $+5$. In CuCl and $CuCl_2$, both known to exist, copper has valences of $+1$ and $+2$.

Second, many important compounds of carbon must be considered in more detail before the simple valence concept can be used. The hydrocarbons CH_4, C_2H_6, C_3H_8, C_4H_{10}, and so on suggest that carbon has a valence of -4, -3, $-2\frac{2}{3}$, $-2\frac{1}{2}$, and so on. However, when we study the structure of molecules, we shall see that carbon's true valence is almost always 4.

Third, diatomic molecules such as O_2, N_2, H_2, Cl_2, and F_2 and triatomic molecules such as ozone O_3, all of which are stable, are not predictable from our simple valence rules. More general valence rules will be discussed after the structure of atoms is developed. Then we shall know exactly what it is that holds together the atoms within a molecule.

THE PREDICTION OF NEW ELEMENTS

When Mendeleev prepared his first periodic table, not all the elements were then known. For example, germanium (atomic number 32) was not discovered until 17 years after his table was made. However, Mendeleev was able to suggest many of the properties of this element, which, placed just below silicon, he called "eka-silicon" and represented by Es. Its chemical and physical properties can be guessed from those of the elements above and below it in the same group—silicon and tin. In addition to eka-silicon (germanium), Mendeleev also described and predicted the existence of several other elements, among them eka-boron and eka-aluminum. These were discovered in 1875 and 1879 and named "scandium" and "gallium," respectively.

DISCHARGE TUBES

Because, ordinarily, gases are poor conductors, high voltages are required for a current to flow through them. For example, nearly 12,000 V are needed to have a spark travel through 1 cm of air from one slender needle electrode to another. Interestingly enough, as the air pressure is reduced, so is the voltage necessary to cause a spark.

The flow of electric current through gases at low pressures was studied in detail during the latter half of the nineteenth century. From these investigations an impressive body of information developed. The equipment used, the *discharge tube*, was made from glass; it was fitted with two metal electrodes (a negatively charged cathode and a positive anode) across which large potential differences could be imposed, and, with the proper pumps and valves, it could be filled with any gas at various pressures (Fig. 5.9).

Once the discharge tube is prepared and filled and the voltage increased to the level where a current, the spark, travels through the gas, an emanation, or ray, is seen to travel from the cathode to the anode; for this reason, it is called the "cathode ray." The properties of this ray were determined by simple experiments.

If the anode is made from a screen, part of the cathode ray passes through it and strikes the end of the glass tube, causing this to glow. When a small object is placed within the tube and in the path of the cathode ray, a sharp outline or shadow is seen at the glowing tube end. Thus, the cathode ray travels in a straight path. If a small paddle wheel is put in the ray, it turns, indicating the emanation has a momentum it can transfer to the wheel.

The cathode ray, which travels toward the positive anode, is deflected by either a magnetic or an electric field acting across its path. The way in which the ray is deflected is exactly the way in which a negatively charged particle is expected to behave in magnetic and in electric fields. Finally, if the ray strikes an insulator

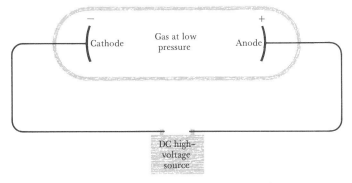

Figure 5.9. When the gas within the discharge tube is subjected to a high voltage, an emanation or ray is seen to travel from the cathode to the anode and, characteristic of charged particles, this ray is deflected by electric or magnetic fields. If the anode is a screen the ray hits the tube past the anode and causes a glow.

which is then brought to an electroscope, the insulator is found to be negatively charged.

At the end of the nineteenth century, a body of evidence was being collected which indicated that cathode rays were collections of negative charges moving through the gas. However, the contrary belief—that cathode rays were a form of electromagnetic radiation—was also popular. The resolution of this matter will be discussed soon.

THE DISCOVERY OF X-RAYS

In 1895 Wilhelm Roentgen (1845–1923) noticed that when cathode rays struck the anode, there was an apparent reflection from the anode surface (Fig. 5.10). The seemingly reflected ray was able to pass through cardboard, wood, paper, thin metal sheets, and other matter opaque to visible and ultraviolet light. Although produced by cathode rays, this new ray differed in that it was not affected by electromagnetic fields. Roentgen named his phenomenon "x-rays."

The penetration of x-rays is greatest in the lighter elements and least in those of high atomic weight. For this reason, lead is used as a shield against x-rays. Soon after Roentgen's discovery, it was observed that bone is less transparent than muscle to x-rays and that a sharp shadow of bone structure can be formed on photographic film. Almost immediately, the x-ray became a major diagnostic tool for physicians. For his work Roentgen received the first Nobel Prize in Physics in 1901.

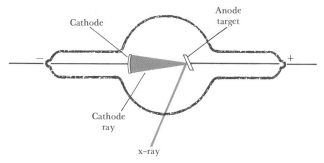

Figure 5.10. The production of an x-ray in a discharge tube. This ray is not deflected by electric or magnetic fields, thus is different from the cathode ray in Figure 5.9.

X-rays proved to be a new example of electromagnetic radiation. The wavelengths were found to depend on the material of the anode and to be much shorter than anything known up to that time. If the anode, or target, is copper, one wavelength observed is 1.54×10^{-10} m, compared to 4×10^{-7} m for violet light. Thus, x-rays are shorter than visible light by a factor of 3,000. Because the length 10^{-10} m occurs so often in x-ray work, it has a special name—the angstrom unit (Å). The copper x-ray wavelength is 1.54 Å. This unit is in honor of Anders Jonas Ångstrom (1814–1874), who made careful measurements of the wavelengths of light from the sun.

NATURAL RADIOACTIVITY

The glow produced when a cathode ray hits the glass wall of a discharge tube has a color which depends on the composition of the glass. This glow is called "fluorescence" if it is present only when the cathode ray falls on the glass and "phosphorescence" if the glow persists after the cathode ray is stopped.

Henri Becquerel (1852–1908), while investigating the fluorescence of certain compounds of the metal uranium, observed that a penetrating radiation was given off which, like Roentgen's x-rays, passed through opaque materials. He discovered this in 1896 when a uranium-containing sample was placed near a photographic plate wrapped in a lightproof package and the plate was found to be exposed.

It was the element uranium which was responsible for this radiation, whether pure or in a combined state. Later thorium, too, displayed the same type of behavior, emitting an unusually penetrating radiation. In 1898 Marie (1867–1934) and Pierre Curie (1859–1906) found two new elements in uranium ore, both of which were more active than uranium in emitting penetrating radiation. They called these "polonium" and "radium."

RUTHERFORD'S EXPERIMENTS. In 1899 Ernest Rutherford (1871–1937) found that the radiation emitted by these *radioactive elements* actually consisted of two parts. One, which was easily absorbed by matter, he called "alpha rays" α; the other, which was more penetrating, was designated "beta rays" β. The following year Paul-Ulrich Villard (1860–1934) noticed a third emanation more penetrating than these, which was called "gamma rays," γ.

Rutherford next investigated the way in which the three rays were affected by electromagnetic fields. An element with this property of radiation emission (called a radioactive element) was placed within a lead container so that a narrow beam of the rays, passing through an electric or a magnetic field, struck a photographic film (Fig. 5.11). The film was exposed in three places. Because the alpha rays were deflected toward the negative plate of the electric field, they were considered to be positively charged. The beta rays, which were deflected toward the positive plate, must be negatively charged. The gamma rays, in not deviating, must be uncharged or must be moving very fast through the field.

Later it was found that alpha particles are actually positively charged helium ions, that beta rays are identical to cathode rays, and that gamma rays are very short electromagnetic radiations with about the wavelength of x-rays. Since beta rays are deflected through a larger displacement than the alpha particles by the same field, it followed that if both are particles, beta particles are either much lighter or more heavily charged than alpha particles.

The significance of radioactivity in some naturally occurring elements is that *these atoms are not indestructible and can be divided into more fundamental constituents,* contrary to the original atomic theory. To learn more about the makeup of atoms it was necessary to identify all the rays emitted by radioactive elements. Cathode or beta rays had to be analyzed in more detail to decide whether they are waves or particles and, if particles, to determine their mass and charge.

THE ELECTRON

In 1897 Joseph J. Thomson (1856–1940) performed an experiment which established the particle nature of cathode rays and thus of beta rays (since they are the same). It was already known that cathode rays are deflected by electric and magnetic fields. The magnitude of the deflection can be measured by the position of the fluorescent glow spot at the tube end. The behavior of the cathode ray can be described most simply in terms of a stream of negatively charged particles. From an analysis of the electric and magnetic field strengths and the shift in the glow spot, Thomson was able to calculate the ratio of the mass to charge of the individual particles constituting the cathode ray.

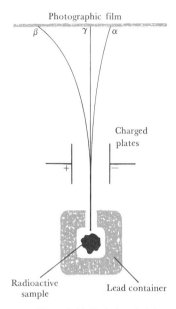

Figure 5.11. Emission of alpha, beta, and gamma rays by a radioactive element. Because the gamma ray shows no deflection in the electric field, either it has no charge or it moves so fast or is too massive to be measurably affected by the field. In either case, other experiments are needed to determine its nature.

Ray	Electric charge	Magnitude of deflection	Relative mass
α	Positive	Small	Heavy
β	Negative	Large	Light
γ	Neutral	None	?

In his tube Thomson used air, hydrogen, and carbon dioxide, and for his cathode material, aluminum and platinum. The mass-to-charge ratio was the same in all cases. Therefore, *the particle he discovered was not peculiar to any of these gases or metals but found in all substances.* Because the particles were able to travel fairly far in a gas before being deflected by collision with gas molecules, it followed that they were quite small. This fundamental negatively charged particle was called the "electron." For his major discovery Thomson received the Nobel Prize in Physics in 1906. Using the best data available today, the mass-to-charge ratio is

$$\frac{m}{e} = 5.68571 \times 10^{-9} \text{ g/coul}$$

THE CHARGE ON THE ELECTRON

Thomson's result was the ratio of two important numbers: the mass of the electron and its electric charge. Both must be known to describe this particle. If either the mass or the charge is determined, the other can be derived using Thomson's ratio. It is easier, here, to manipulate an electric charge than a mass. Robert A. Millikan (1865–1953) developed a technique for measuring the charge on fine oil droplets and from this calculated the electron's charge in 1909.

It was known that radiation from radioactive elements passing through air will convert neutral nitrogen and oxygen molecules into either cations or anions. This *ionizing radiation* thus changes a good insulator, air, into a conductor. Measuring gas conductivity is a common technique for detecting radioactivity.

An ordinary atomizer used for dispensing perfume or medicine as a fine spray can disperse oil into very small droplets. These, in falling through ionized air, will collide with nitrogen and oxygen ions and, occasionally, will absorb an ion, acquiring the electric charge. Once charged, the oil droplet can be controlled by an electric field.

Millikan was able to time the descent of an oil drop under the influence of gravity alone and then, applying an electric field, the ascent of the same drop (Fig. 5.12). A smaller drop falls more slowly than a larger one. From the speed of descent it is possible to calculate the drop size. The time for the drop to fall a definite distance remained constant during a typical experiment of $4\frac{1}{2}$ hr, indicating the oil-drop size remains constant.

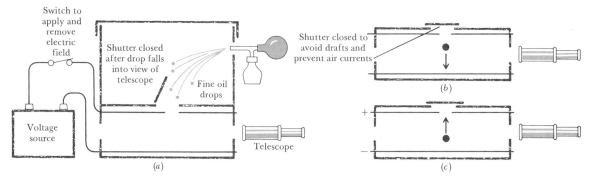

Figure 5.12. Millikan's oil-drop experiment to determine the fundamental electric charge. (a) Apparatus; (b) no electric field—drop falls; (c) electric field forces the negatively charged drop to rise.

The time for ascent when the field is applied depends on the magnitude of charge on the drop. The greater the charge, the greater the force, the higher the velocity, and the shorter the time needed for the drop to rise through the definite distance. From time to time Millikan observed that the time for ascent changed, sometimes increasing and sometimes decreasing. This was interpreted to happen when the negative charge on the oil drop decreased or increased by absorbing a positive or negative ion, respectively.

By considering the forces acting on the moving oil drop—gravity, friction, and, when the field is applied, electric—Millikan was able to calculate the actual charges on the drop during a series of ascents and descents. Some of his data are presented in Table 5.4. Although he used another unit for electric charge, in this table his values have been converted to coulombs.

The first column lists the various times it took for the drop to fall the measured distance of 10.10 mm. At the end of each descent, the field was applied, and the drop time for ascending the same distance is given in the second column. Looking at the first four ascent times, they are seen to decrease each time. The ascent speed will increase if the drop charge becomes more negative. Between the fourth and fifth ascent, a positive charge evidently was added to the drop (or a negative charge lost). The third column gives Millikan's computed electric charges based on the descent and ascent times. Observe that in the 12 charges listed there are only 6 distinct values.

In order to see if these drop charges have any special relation we can proceed as we did in arriving at the law of multiple

Table 5.4. Millikan's Calculation of the Fundamental Electric Charge

(1) Time for descent when electric field is absent, s	(2) Time for ascent with field applied, s	(3) Charge on drop, coul	(4) Difference between successive values	(5) Multiple of fundamental charge	(6) Fundamental charge = drop charge ÷ multiple factor
22.8	29.0	1.150×10^{-18}		7	1.643×10^{-19}
			$+1.66 \times 10^{-19}$		
22.0	21.8	1.316		8	1.645
			$+1.64$		
22.2	17.3	1.482		9	1.647
			$+1.66$		
22.0	14.2	1.648		10	1.648
			-1.67		
22.7	21.5	1.316		8	1.645
			$+6.56$		
22.9	11.0	1.972		12	1.643
			-4.90		
22.4	17.4	1.482		9	1.647
			$+1.66$		
22.8	14.3	1.648		10	1.648
			$+1.51$		
22.8	12.2	1.799		11	1.635
			-1.51		
22.8	14.1	1.648		10	1.648
			-1.66		
22.8	17.1	1.482		9	1.647
			$+4.90$		
22.8	10.9	1.972		12	1.643

Distance for ascent and descent = 10.10 mm
Distance between plates = 16.00 mm
Voltage when field applied = 7930 V

Average of these 12 = 1.645×10^{-19} coul
Average of 40 values = 1.640×10^{-19} coul
Best value today = 1.60210×10^{-19} coul
± 0.00002

proportions. Alternatively, we can write down in column (4) the differences in successive calculated drop charges. Here we see something simple, interesting, and possibly important. Using all 40 values calculated by Millikan, the *differences* in charges, which are due to the drop taking up an anion or cation, are all either about 1.64×10^{-19} coul, three times that $(3 \times 1.64 \times 10^{-19} = 4.92 \times 10^{-19} \approx 4.90 \times 10^{-19}$ coul), four times that $(4 \times 1.64 \times 10^{-19} = 6.56 \times 10^{-19}$ coul), or very close to some other whole-number multiple of 1.64×10^{-19} coul. If the oil drop picks up or loses only 1.64×10^{-19} coul or some integral multiple of this, then 1.64×10^{-19} coul may be the fundamental charge. If so, *electric charge can be added or subtracted only in these units.*

The total charge first measured on the drop is 1.150×10^{-18} coul. If the unit charge is 1.64×10^{-19} coul, dividing the former by the latter will give the number of units on the drop; that number is 7.01. Similarly, the second measurement indicates the drop to have

$$\frac{1.316 \times 10^{-18}}{1.64 \times 10^{-19}} = 8.02 \text{ units}$$

For each of Millikan's 40 measured drop charges, the number of units of this fundamental 1.64×10^{-19} coul charge is always very close to a whole number. In column (5) we write the integer thus obtained. Now, dividing the drop charge [column (3)] by the number of units [column (5)], the quotient is the charge per unit [column (6)]. It is seen that the number is remarkably constant. Using all of his data, Millikan found this to be 1.640×10^{-19} coul. The best value for this available today is

$$1.60210(\pm 0.00002) \times 10^{-19} \text{ coul}$$

This number is the minimum positive or negative charge of an individual ion of nitrogen or oxygen. Since a beta ray, which is a collection of electrons, can produce anions from neutral molecules, it is logical to conceive of an anion as a molecule or atom with one or more extra electrons. Similarly a cation is a molecule or atom with one or more electrons removed. *The fundamental charge is then the charge of a single electron.* Combining this with Thomson's ratio, we have the *mass of the electron:*

$$\frac{5.68571 \times 10^{-9} \text{ g}}{\text{coul}} \times 1.60210 \times 10^{-19} \text{ coul} = 9.10908 \times 10^{-28} \text{ g}$$

Since all elements can be converted into ions, all atoms have electrons. In some cases the electrons are spontaneously ejected in the form of beta rays. In others, they are pulled off by a strong electric field, as in cathode rays or in electrolysis. Chemical reactions, ionizing radiation, and high temperatures are still other means of adding or removing electrons from atoms and molecules.

CALCULATION OF AVOGADRO'S NUMBER

When discussing Faraday's laws of electrolysis, it was mentioned that 96,487.0 coul of electric charge are transferred when 1 mole of the ions silver, hydrogen, chloride, etc., are made from or converted to the corresponding neutral species. Our understanding of ionization and electrolysis is that electrons are added and removed in integral numbers.

We now have enough information to calculate Avogadro's number—the number of ions, atoms, or molecules in 1 mole of any substance. When 1 mole of silver is electrolyzed, 96,487.0 coul, that is, 1 $\mathscr{F}$, of charge are involved at the electrode process. If one electron is transferred to convert a silver atom to a silver ion or an ion to an atom, then 1 mole of electrons is required for 1 mole of silver ions or atoms. From the charge on a single electron we can compute how many electrons are needed for a total charge of 1 $\mathscr{F}$. This is also the number of silver atoms in 1 mole of silver, i.e., Avogadro's number.

Example. From the experimental data given, calculate Avogadro's number N_A:

$$N_A = \frac{96{,}487.0 \text{ coul}}{\text{mole}} \times \frac{\text{single Ag atom or ion}}{1.60210 \times 10^{-19} \text{ coul}}$$

$$= \frac{9.64870 \times 10^4}{1.60210 \times 10^{-19}}$$

$$= 6.02252 \times 10^{23}/\text{mole}$$

THE SIZE OF INDIVIDUAL ATOMS AND MOLECULES

Although Avogadro's number is arbitrary in the sense that the definition of the mole is also arbitrary, nevertheless we are now able to estimate the size of individual atoms and molecules.

Example. The molecular weight of hydrogen H_2 is 2.0160. From this calculate the mass of a single hydrogen atom.

One mole of hydrogen atoms has a mass of 1.0080 g and contains 6.02252×10^{23} atoms. Then one atom has a mass of

$$\frac{1.0080 \text{ g}}{6.02252 \times 10^{23} \text{ atoms}} = 1.6737 \times 10^{-24} \text{ g}$$

This is over 1,800 times the mass of a single electron. Either a hydrogen atom contains many electrons, or its mass is due to something else more massive than an electron.

Example. Water has a density of 1.000 g/cm³ and a molecular weight of 18.02. From this calculate the molecular volume and estimate the distance of a single molecule from one end to the other.

In the liquid state water cannot be compressed into a smaller volume without using a great amount of energy. While a gas, which is easily compressible, has open spaces between molecules, a liquid has molecules that touch each other. The size of one molecule can be calculated if we know how many molecules are in a given volume. The volume of 1 mole of water is

$$\frac{18.02 \text{ g}}{\text{mole}} \times \frac{\text{cm}^3}{1.000 \text{ g}} = \frac{18.02 \text{ cm}^3}{\text{mole}}$$

and the volume of a single molecule is

$$\frac{18.02 \text{ cm}^3}{\text{mole}} \times \frac{\text{mole}}{6.023 \times 10^{23} \text{ molecules}}$$
$$= 2.992 \times 10^{-23} \text{ cm}^3/\text{molecule}$$

If we assume the simplest shape, a cubic molecule, this volume is the edge of a cube raised to the third power. The edge e is the volume to the one-third power:

$$
\begin{aligned}
e &= (2.992 \times 10^{-23} \text{ cm}^3)^{\frac{1}{3}} \\
&= (29.92 \times 10^{-24} \text{ cm}^3)^{\frac{1}{3}} \\
&= 3.104 \times 10^{-8} \text{ cm} \\
&= 3.104 \text{ Å}
\end{aligned}
$$

Since we know nothing about the shape of molecules or about how they are packed in the liquid (how much molecule and how much free space), we shall say that the molecule is about 3 Å across.

SCATTERING EXPERIMENTS

Electrons can be removed from all elements. Therefore, all atoms must contain one or more electrons. Because atoms are electrically neutral, each atom must also contain a positively charged part to balance the electron's negative charge. The next problem is to determine how these charges are found within the atom. Since atoms are so small, it is necessary to use probes of comparable size to learn something about atomic structure. Alpha particles, which are the cations He^{2+} and which can be obtained from natural radioactive elements, serve this purpose.† (See Chap. 8, Prob. 4.)

† He^{2+} is an atom of helium from which two electrons have been removed.

The way in which collisions occur tells us something about the colliding bodies. For example, if a hard rubber ball is thrown against a wall, it will bounce back; but if it hits a lighter body—a table-tennis ball or a sheet of paper—it will continue in the same direction. We can tell by these simple experiments whether that which is struck is much heavier or lighter than the hard rubber ball.

In the same manner, a stream of alpha particles can be directed against a collection of atoms. The atoms against which the alpha particles strike and are *scattered* are said to be the "target atoms." These are most conveniently handled in the form of a very thin metal foil. By hammering a metal sample of known mass and volume into a foil of measured area, it is possible to calculate the thickness of this foil. From knowing the average or effective size of each atom, the number of atoms in this thickness can be determined.

> **Example.** Gold has a density of 19.3 g/cm³ and an atomic weight of 197.0. If a 1.00-mg sample is hammered into a thin foil of uniform thickness and with an area of 5.18 cm² on one face, estimate the thickness of the foil.
>
> The density of gold is independent of its shape or form. Because the mass does not change, the volume of the foil is calculated:
>
> $$V = \frac{1.00 \text{ mg}}{19.3 \text{ g/cm}^3}$$
>
> $$= \frac{1.00 \times 10^{-3} \text{ g}}{19.3 \text{ g/cm}^3}$$
>
> $$= 5.18 \times 10^{-5} \text{ cm}^3$$
>
> This volume is also equal to the area A times the thickness h, and so $V = Ah$ and
>
> $$h = \frac{V}{A}$$
>
> $$= \frac{5.18 \times 10^{-5} \text{ cm}^3}{5.18 \text{ cm}^2}$$
>
> $$= 1.00 \times 10^{-5} \text{ cm}$$

> **Example.** How many gold atoms are found along this distance? Unlike a gas, a solid cannot easily be compressed into a smaller volume. We consider the atoms in a solid state to be in contact

with each other. If we know how many atoms are contained in a definite volume, we can calculate the size of each atom.

Imagine a cube of gold 1.00 cm on each edge. In this volume are contained

$$1.00 \text{ cm}^3 \times \frac{19.3 \text{ g}}{\text{cm}^3} \times \frac{6.02 \times 10^{23} \text{ atoms}}{197.0 \text{ g}} = 5.98 \times 10^{22} \text{ atoms}$$

Then along one edge are contained

$$(5.98 \times 10^{22})^{\frac{1}{3}} = (59.8 \times 10^{21})^{\frac{1}{3}} = 3.91 \times 10^7 \text{ gold atoms}$$

If this many are in a 1.00-cm path of gold, then

$$1.00 \times 10^{-5} \text{ cm} \times \frac{3.91 \times 10^7 \text{ gold atoms}}{1.00 \text{ cm}} = 391 \text{ gold atoms}$$

is the number of gold atoms an alpha particle must pass if it travels completely through the gold foil.

With his coworkers Hans Geiger (1882–1947) and Ernest Marsden (1889–1970), Rutherford learned much of what is known today about atomic structure through these scattering experiments. Alpha particles from polonium or some other convenient radioactive element were directed against a thin foil target (Fig. 5.13*a*). Around the foil was a fluorescent screen which sparkles or *scintillates* when struck by an alpha particle. Just by counting the scintillations at different positions on the screen, the

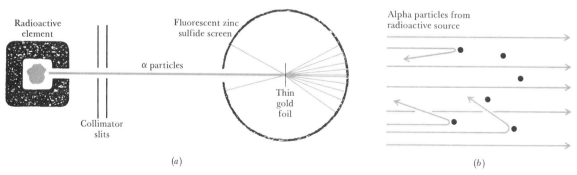

Figure 5.13. Rutherford's scattering experiment. (a) The scattering of alpha particles by a thin gold-metal foil and measurement of the angles of scattering. A very small number of alpha particles are scattered back by almost 180°. Most alpha particles go through foil with only a small deflection. (b) Since most alpha particles pass through with little or no deflection, the massive part of each gold atom must present a very small target area.

scattering behavior of the alpha-particle gold collisions was determined.

The observations were quite remarkable. Almost all the alpha particles went right through the foil with only small scattering angles, i.e., with only slight deflections away from the straight path. In spite of traveling through several hundred gold atoms, most alpha particles were not affected very much.

Because alpha particles and helium atoms are more than 7,000 times as massive as electrons, it was not expected that alpha particles would be deflected much from a straight path upon hitting electrons. A gold atom, which is about 50 times as massive as an alpha particle, has most of its mass in a form other than electrons. This structure within the gold atom could be expected to scatter through wide angles the relatively light alpha particles. Since almost all the alpha particles are not scattered very much in traveling through the foil, they must miss rather than collide with the massive part of each gold atom. The conclusion follows immediately: most of the mass of a gold atom is contained in a very small volume—much smaller than the overall effective atomic volume. Most of the atom is open space.

Less than one scintillation out of each ten thousand was seen at angles greater than 90°. But of the alpha particles which were scattered this much, a significant fraction of these were turned completely around and sent back to the emission source. When an alpha particle did make a rare collision with the nonelectron part of a gold atom, it was strongly repulsed. Rutherford remarked he was as surprised as if an artillery shell fired against a piece of tissue paper had been occasionally reflected back toward the gun.

In order to describe what happened, Rutherford calculated that the massive part of the gold atom must be about 10^{-14} m, compared with 10^{-10} m, or 1 Å, for the overall atomic size. Also this massive part, to which he gave the name "nucleus," must be positively charged. The small size means each gold atom presents a very small target area to the alpha particles; collisions will be infrequent. Coulomb's law states that the repulsive force between like charges varies as the inverse of the square of the distance. A very small nucleus provides for a very close approach by some alpha particles and consequently a very strong repulsion.

The Rutherford atom, described in 1911, comprises a nucleus which is positive, which contains almost all the mass of the atom, and which has about one ten-thousandth the diameter of the entire atom, and negatively charged electrons distributed in the space about the nucleus in sufficient number to maintain an electrically neutral atom. The scattering experiment, using this model, is described in Fig. 5.13*b*.

Further analysis of scattering data permitted an estimation of the actual nuclear charge. In 1913 the gold atom nucleus was found to have a positive charge 78 ± 4 times the magnitude of the negative electron charge. In copper, the nuclear charge was 29.3 ± 0.5 times the electron charge. In the periodic table, gold has a position number, or atomic number, of 79 and copper has one of 29. To within the precision of these results, *the atomic number is the positive charge on the nucleus and also the number of electrons in the neutral atom.*

ISOTOPES

In his discharge tube studies, Thomson found that no matter with what gas his tube was filled or of what metal his cathode was made, the cathode ray was always exactly the same. It could always be described as a collection of electrons because the measured mass-to-charge ratio had always the same value.

In addition, another ray was also observed moving away from the positively charged anode toward the cathode. This ray must bear a positive charge and for this reason is called the "positive ray." From measuring the ray deflection in electromagnetic fields, the mass-to-charge ratio of the constituent particles can be obtained. Unlike the electrons of the cathode ray, the mass-to-charge ratios depend on which gas is contained in the tube.

The mass of an individual atom or molecule can be calculated from the atomic or molecular weight and Avogadro's number. Comparing this with the mass-to-charge ratio, the charge on the positive-ray particle can be calculated. In every case, the charge is either exactly the charge of the electron or twice, three times, or some other whole-number multiple of that charge and positive in sign. Conversely, from examining the trajectory of the positively charged particle moving in a magnetic field, the mass-to-charge ratio and then the particle mass can be determined.

We have already seen that a magnetic field exerts a force on a moving electric charge. When a charged particle moves within such a field, the force acts perpendicularly to the direction of motion of the particle. This results in a curved path for the charged

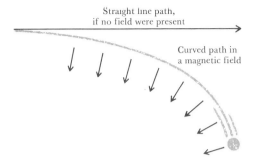

Straight line path,
if no field were present

Curved path in
a magnetic field

Figure 5.14. Influence of a magnetic field on the direction of motion of an electrically charged particle. The short arrows indicate that magnetic field acts always in a direction perpendicular to motion of charged particle, whose path continuously changes.

particle (Fig. 5.14). The measure of curvature of a path is the radius of the circle that can be drawn through the curved path. A greater curvature corresponds to a circle with a smaller radius, and a smaller curvature to a larger radius. A straight line, which has no curvature, can be thought of as a circle with an infinitely large radius (Fig. 5.15*a*).

The amount by which a charged particle moving in a magnetic field is deflected from a straight line depends on the magnitude of both the electric charge Q and the particle mass m. The deflecting force varies directly with the charge. The greater the charge, the more the deflection and the smaller the radius of curvature. For a given force the greater the mass, the less the deviation from a straight path and the larger the radius of curvature. The radius of curvature R varies as m/Q.

In 1913 J. J. Thomson directed positive rays formed from several kinds of atoms and molecules along a path within a magnetic field. He found, as expected, that the radii of curvature did depend on the molecular and atomic masses. For the gas neon, which has an atomic weight of 20.183, an especially interesting discovery was made. A schematic diagram of an apparatus, more modern than that used originally by Thomson, is shown in Fig. 5.15*b*. When a positive ray consisting of neon cations is curved by the magnetic field and strikes a photographic plate, two spots are found. Thomson found that one spot corresponded to an atom with a mass 20 times that of a single hydrogen atom and the second one to an atom with a mass 22 times the hydrogen mass. Thus, not all neon atoms are identical in all respects.

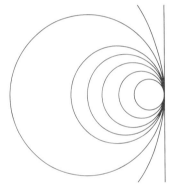

Figure 5.15a. Radius of curvature as a measure of the degree of turning in a curved path. A family of circles with various radii. Curvature is the extent to which a curved path, e.g., the arc of a circle, departs from a straight line. As the circle radius increases, the arc bends less. In the limit, a straight line can be considered the arc of a circle of infinitely large radius; it has no curvature. At the other extreme, the smaller the circle radius, the more the arc is bent from linearity. In general, the smaller the radius of curvature, the greater is the curvature.

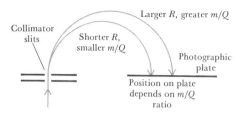

Figure 5.15b. Mass spectrograph. A magnetic field acts across the plane of this paper; one pole is above and the second is below this page. The radius of curvature R is larger for greater masses m since a given force diverts larger masses less and smaller masses more. The radius of curvature decreases as the charge Q increases since the diverting force depends on the magnitude of the charge. A high degree of bending (a small radius of curvature) is found for a small mass and a large charge; less bending (a larger radius of curvature) results from a greater mass and a smaller charge. If the detection is done by electrical means rather than by a photographic plate, the instrument is a mass spectrometer.

During the same year, Frederick Soddy (1877–1956) and other workers were able to isolate samples of certain radioactive elements which had identical chemical properties but not identical physical properties. For example, two samples of thorium, each purified but obtained from different sources, were found to have different radioactive emission rates. In another instance, the metal lead was found in samples of uranium ore and also in thorium ore. Chemically these elements are both lead, but the atomic weight of lead obtained in uranium ore is less than that found in thorium. Because all samples of thorium are chemically the same, they belong in the same place in the periodic table. The same is true of the several samples of lead. Soddy suggested the name *isotope* (from the Greek *iso*, "the same," and *topos*, "place") for the varieties of atoms of a single element.

With Thomson's apparatus, called a "mass spectrograph," and with improved versions of this, it was soon established that isotopes are found among almost all the elements. The two neon isotopes Thomson found can be represented by the symbols Ne^{20} and Ne^{22}. These are read as "neon-twenty" and "neon-twenty-two." More completely, the symbols $_{10}Ne^{20}$ and $_{10}Ne^{22}$ are used. The lower number is the atomic number, and the upper number is the atomic mass relative to the hydrogen atom.

PARTICLES WITHIN TIIE NUCLEUS

With the exception of the lightest element, hydrogen, it is possible to remove more than one electron from an atom. For example, helium can be ionized to either He^+ or He^{2+} when one or two electrons are removed from the neutral atom He. Similarly, neutral lithium can be ionized to Li^+, Li^{2+}, or Li^{3+} by removing one, two, or three electrons.

In all cases, the positive charge is some whole-number multiple of the electron charge. This is what we should expect since the atoms were originally neutral. When a whole number of electrons is removed, the residual positive charge must be equal to the removed negative charge but opposite in sign.

With hydrogen only the singly ionized cation H^+ is found. This means that a hydrogen atom contains only one electron. The nucleus of the hydrogen atom contains a positive charge equal in magnitude to that of the fundamental charge on the electron. All other nuclei have positive charges—some whole-number multiple of the charge on the hydrogen nucleus.

Rutherford suggested that the hydrogen nucleus is the prototype of all the positive charges found in atomic nuclei. The hydrogen nucleus, i.e., the cation H^+, is a second fundamental particle, which he called the "proton." The charge on any nucleus is determined by how many protons are contained there. The significance of the *atomic number* is now obvious. It is *the number of protons in the nucleus.*

The mass of the proton, the hydrogen cation H^+, can be precisely determined with a mass spectrograph. The best modern value for the proton mass is

$$m_p = 1.67252(\pm 0.00003) \times 10^{-24} \text{ g}$$

This is considerably more than the electron mass m_e. The ratio of the two masses is

$$\frac{m_p}{m_e} = \frac{1.67252 \times 10^{-24}}{9.10908 \times 10^{-28}}$$
$$= 1836.10$$

(Within the precision to which this ratio is known, it is also the number $6\pi^5 = 1,836.12 \ldots$, where π is the circle circumference/diameter ratio 3.14159. . . . However, whether this agreement is fortuitous or results from some property of these fundamental particles is not yet known.)

Although the hydrogen atom has a single proton and one electron, heavier atoms have something more in their nuclei. With Thomson's discovery, neon has two isotopes $_{10}Ne^{20}$ and $_{10}Ne^{22}$. The former has an atomic weight almost exactly 20 times and the latter 22 times the mass of a proton. Yet both isotopes have the same nuclear charge +10 and must both have 10 protons.

Rutherford, in 1920, suggested a simple way in which two nuclei can have the same charge but different masses. He invented the *neutron,* an electrically neutral particle with a mass approximately equal to that of a proton. In the case of the two neon isotopes, $_{10}Ne^{20}$ has 10 protons and 10 neutrons, $_{10}Ne^{22}$ has 10 protons and 12 neutrons, and each has 10 electrons. Rutherford also guessed that a neutron might be a combined proton and electron pair.

It is a general property of the elements, as an examination of the periodic table indicates, that the atomic weight is about twice the atomic number. Gold, for example, has a mass 197 times that of the proton and, with an atomic number of 79, contains 79 protons. With 118 neutrons too, both the nuclear charge and mass of a gold atom can be satisfactorily described.

When the metallic element beryllium is bombarded with alpha particles, something unusual happens. A new particle is emitted by beryllium which penetrates solid materials more deeply than protons or electrons. In 1932, James Chadwick showed that the new emission phenomenon can be most simply described in terms of a neutral particle with about the mass of a proton — Rutherford's neutron. Since it is uncharged, it is not subject to the same coulombic repulsive forces which act on charged particles traveling through matter.

The mass of the neutron m_n, according to the best available data, is

$$m_n = 1.67482(\pm 0.00003) \times 10^{-24} \text{ g}$$

This is greater than the sum of the electron and proton masses. The neutron is considered to be a fundamental particle rather than an electron-proton pair. The fundamental particles are listed in Table 5.5. In the three symbols used, the lower symbol, or subscript, is the electric charge in units of the proton charge. The

Table 5.5. The Three Fundamental Particles

Particle name	Particle mass, g	Particle mass relative to proton	Particle charge relative to proton	Particle symbol
Electron	9.10908×10^{-28}	$\dfrac{1}{1,836} \cong 0$	-1	$_{-1}e^0$
Proton	1.67252×10^{-24}	1	1	$_1p^1$
Neutron	1.67482×10^{-24}	$1.00138 \cong 1$	0	$_0n^1$

upper symbol, or superscript, is the mass number relative to the proton mass. Because protons and neutrons are present in atomic nuclei, collectively they are called "nucleons."

ATOMIC WEIGHTS

Of the several techniques used today for determining atomic and molecular weights, mass spectrography and mass spectrometry are the most precise. By 1926 several atomic masses were known to a precision within one ten-thousandth of the proton mass.

Because most of the work done today using mass spectrographs involves carbon-containing molecules, it has become convenient to employ carbon rather than oxygen or hydrogen as the standard for all atomic weights. Carbon has six known isotopes, but only two are both stable and found in nature: $_6C^{12}$ and $_6C^{13}$. The other four are radioactive. Almost 99 percent of naturally occurring carbon is $_6C^{12}$; it is this isotope which is the standard. Arbitrarily, but conveniently, the atomic weight of $_6C^{12}$ is taken to be exactly 12.00000. From a precise mass spectrograph, the mass of the second isotope is 1.0836125 greater than this. The atomic weight of $_6C^{13}$ is 12.00000×1.083612, or 13.00335.

All the carbon found in nature, whether pure (i.e., diamond or graphite) or in compounds (e.g., benzene and octane), is a mixture of the two isotopes $_6C^{12}$ and $_6C^{13}$. The atomic weight of natural carbon, then, is neither 12.00000 nor 13.00335 but some average of these two pure-isotope atomic weights. Because there is much more $_6C^{12}$ than $_6C^{13}$, we expect the former isotope to be counted more, i.e., to be more heavily weighted, in the averaging process and the average atomic weight to be closer to 12 than to 13.

From a mass spectrograph it is possible to obtain a count of the relative numbers of each isotope present in any element. In carbon the lighter isotope in a typical specimen is 98.889 percent abundant, and the heavier isotope is 1.111 percent abundant.

Example. Calculate the average atomic weight of carbon from the above data.

The averaging process must count the lighter isotope 98.889 times for each 1.111 times the heavier isotope is counted. It does not matter if we take 98,889 of the former and 1,111 atoms of the latter, or 9.8889×10^{23} and 1.111×10^{22}, respectively, or any other two sets of numbers. The only restriction is that the ratios must be as 98.889:1.111. We can obtain the average, then, by adding 12.00000 a total of 98,889 times and 13.00335 for 1,111 times and dividing that grand sum by $98,889 + 1,111 = 100,000$. It is easier to multiply 12.00000 by 98,889, add to this the product of 13.00335 and 1,111, and then divide the sum of these two products by $98,889 + 1,111 = 100,000$.

Atomic weight × number taken = contribution to average

$$12.00000 \times 98,889 = 1,186,668 \qquad {}_6C^{12} \text{ isotope}$$

$$13.00335 \times \frac{1,111}{100,000} = \frac{14,447}{1,201,115} \qquad {}_6C^{13} \text{ isotope}$$

$$\text{Average} = \frac{1,201,115}{100,000}$$

$$= 12.01115$$

A somewhat more elegant procedure can be used if we realize that the division by 100,000 can be performed either on the large number 1,201,115 or individually on the two numbers 98,889 and 1,111. The result is the same:

Atomic weight × number of each = contribution to average

$$12.00000 \times 0.98889 = 11.86668 \qquad {}_6C^{12} \text{ isotope}$$

$$13.00335 \times \frac{0.01111}{1.00000} = \frac{0.14447}{12.01115} \qquad {}_6C^{13} \text{ isotope}$$

Observe that the numbers of each isotope are really the fraction of each. The average is obtained in one less step.

It can now be seen why Prout's hypothesis was rejected and can now be modified and reaccepted. Chlorine has two stable isotopes ${}_{17}Cl^{35}$ and ${}_{17}Cl^{37}$, with abundances of about 75.5 and 24.5 percent and isotopic atomic weights of 34.96885 and 36.96590. The average atomic weight is 35.453. But each isotope has an atomic weight which is almost an integral multiple of the hydrogen atomic weight. Each isotope has a whole number of nucleons.

The atomic weights of pure isotopes are very close to but not exactly whole-number multiples of hydrogen's atomic weight, nor are they in the ratio of integers. The fine discrepancy, in the case of $_6C^{13}$, between exactly 13 and 13.00335 will be accounted for in Chap. 8 when we study the nucleus.

THE INADEQUACY OF CLASSICAL PHYSICS

The investigations of Thomson and Rutherford lead to an atomic model with a positive nucleus surrounded by one or more negative electrons. A coulombic force of attraction acts between the oppositely charged particles. If no other forces were present, the electrons would move toward the nucleus and, in effect, the atom would collapse to a volume much smaller than that it is known to occupy.

An analogous force, gravity, acts between the sun and each planet. To prevent a planet from moving into the sun there must be another force reacting against gravity. To see what this is we can consider what might happen if gravity were abruptly turned off as the earth revolves about the sun. Suddenly the force which acts to move the earth toward the sun would be absent. If there were no force acting on the earth, there could be no acceleration. The velocity of the earth could not change; the earth would move in a straight line. We might expect the earth to fly off, on a path tangent to its orbit, in a straight line away from the sun (Fig. 5.16).

Now we can see why it is that the earth does not move toward the sun: Inherent to its orbital motion is an inertia which, as it does to all moving masses, tends to keep the earth on a straight-line path, traveling away from the sun. Since the orbit is a stable one, this inertial force must exactly balance the attractive gravitational force. Because gravity acts to pull the earth toward the center of the solar system, it is sometimes called a "centripetal," or center-seeking, force. The opposing force, which arises from the orbital motion, is called a "centrifugal," or center-escaping, force.

If we use the solar system as a model for the atom, the electrons should move in orbits about the central nucleus. Because the electron has a negative charge, an electrostatic force of attraction will act to move it toward the nucleus. The electron also has a mass, and the resulting centrifugal force acts to move it away from the

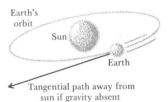

Earth's orbit

Sun

Earth

Tangential path away from sun if gravity absent

Figure 5.16. The earth in a stable orbit about the sun.

nucleus. With the proper orbit sizes and speeds, the two forces exactly balance each other.

However, because the electron has an electric charge, a very serious objection must be made against this model. As an electron moves in its orbit, the direction of its motion changes continuously. The velocity, which is a vector defined by both speed and direction, must also change continuously. The electron, then, is accelerated.

It was one of the great features of Maxwell's equations that an accelerated electric charge must lose energy in the form of electromagnetic radiation. This is what Hertz demonstrated in sending a charge rapidly back and forth in his famous spark-gap experiments establishing the validity of Maxwell's equations. Up to the end of the nineteenth century, the electromagnetic laws described perfectly all electric and magnetic phenomena.

Since the charged electron is accelerated, it should radiate energy. The electron possesses a certain amount of potential energy when it is at a fixed distance from the nucleus. If the electron-nucleus distance is increased, work must be done to overcome the electrostatic force of attraction; the electron has a higher energy. When this distance is decreased, the potential energy is less and energy is lost. We should expect the electron to spiral in toward the nucleus as it radiates energy. Eventually all electrons should touch their nuclei. The atom should collapse to a much smaller volume, about the size of the nucleus. Electrons so close to a nucleus should be subject to enormous attractive forces, and ionization should be considerably more difficult.

It was not possible to reconcile the experimental evidence of Thomson and Rutherford with the set of physical laws known at the end of the nineteenth century. Either these laws, known today as *classical physics*, were not as universal as once thought, or the careful work of Rutherford and Thomson was not properly understood. During the period when these men and their coworkers were learning about atomic structure, a major revolution was occurring in physics. From this our modern model of the atom developed.

THERMAL RADIATION

The first phenomenon which could not be described by the laws of classical physics was not concerned immediately with Rutherford's atom but rather with the manner in which hot bodies emit light. We

know from experience that objects glow when hot. A typical substance, for example, charcoal or iron, becomes a dull faint red
when its temperature is about 600°C. As the temperature is raised,
it progressively becomes brighter—red-orange, yellow, and then
white-hot (about 1400°C). At still higher temperatures the substance becomes blue-violet and is so bright as to be painful to
look at.

To see better what happens, the emitted light can be analyzed. The word *analysis* means a separation into constituents; this
literally is what is done to the light. The instrument used for the
analysis is called a "spectrophotometer," or "spectrograph" (Fig.
5.17). Once dispersed, the light can be examined at each
wavelength for brightness. This is done by a *detector*. In some instruments the dispersed light falls on a photographic plate, a different wavelength at each plate position. If the brightness is not
uniform for all wavelengths, the exposure will vary along the plate.
From the plate darkening the intensity of light at any wavelength
can be obtained.

In place of a photographic plate other types of detectors are
used. Sunlight falling on a surface will warm it; and the more intense the sunlight, the warmer the surface. A special detector surface can be exposed to dispersed light at various wavelengths and
the temperature rise of the detector recorded. The higher the temperature, the more intense the light at that particular wavelength.
A third kind of detector converts the light to an electric current

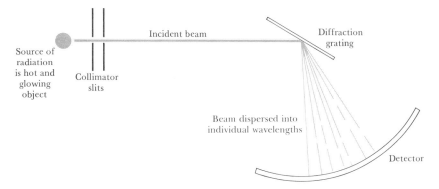

*Figure 5.17. Grating spectrograph. Light emitted from a hot glowing substance passes through one
or more slits. The beam thus formed is then either reflected from a diffraction grating or passed
through a prism. In either case, the amount by which the light is bent depends on its wavelength.
Thus the emitted light is dispersed, or broken up into its constituent wavelengths.*

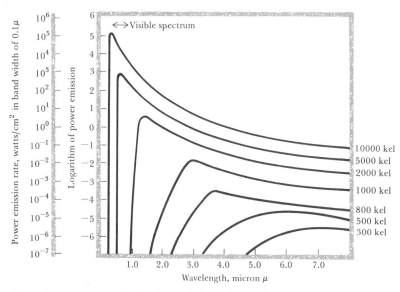

Figure 5.18. Spectral distribution of emitted blackbody-radiation power shown at seven temperatures. The wavelength of maximum power emission decreases as the temperature increases, as described by the Wien displacement law.

using a photoelectric cell. This usually consists of the metal cesium, which is known to ionize when light falls on it.

In any case, the record of emission intensities at the various wavelengths is plotted on a graph. The curve of this plot is called a "spectrum." It turns out that at any fixed temperature the emission spectra are similar for all solid bodies, but that body which emits most is also the one which reflects incident light the least. Ideally, a body which reflects no incident light at all and emits the very most is called a "blackbody." In Fig. 5.18 typical *blackbody emission spectra* are plotted for several temperatures.

We can see that at all temperatures a smooth curve describes the emission rate versus wavelength. No matter how narrow a wavelength range we examine, we shall find some emission at all wavelengths; there are no wavelengths at which radiation is absent. This type of emission, represented in the graph (Fig. 5.18), is said to be *continuous radiation* for this reason.

At any fixed wavelength we can see that the higher the temperature, the more energy emitted per second. Also, at each temperature there is a characteristic wavelength at which a maximum power emission is recorded. From the graph we can estimate the wavelength of maximum emission λ_{max} for each temperature T

Table 5.6. Wavelengths of Maximum Radiation Emission Rates from a Blackbody at Various Temperatures

T, K	Wavelength, λ max μ	Product, $T\lambda$ max, $(K \cdot \mu)$
500	6.0	3,000
800	3.8	3,040
1000	3.0	3,000
2000	1.5	3,000
3000	1.0	3,000
5000	0.6	3,000
10,000	0.3	3,000

shown; these values are listed in Table 5.6. The product $T\lambda_{max}$ is seen to be nearly constant. The constancy of this product was suggested in 1893 by Wilhelm Wien (1864–1928). This was later verified experimentally and is now called "Wien's displacement law." Today the best value for the constant is

$$T\lambda_{max} = 2897.8 \text{ K} \cdot \mu$$

At low temperatures the total emission is weak and most of that is in the far infrared spectrum invisible to the human eye. At 1000 K there is a little radiation at the red end of the visible spectrum and much less at other colors. At this temperature the object appears dimly red. At 3000 K there is quite a bit of radiation along the entire visible spectrum, but, in addition, there is too much ultraviolet to permit viewing with the unprotected eye. As the temperature increases the color of the emitting blackbody appears to change in the direction from the red to the violet end of the visible spectrum in accord with the displacement law. Wien's displacement law is used to estimate the surface temperature of hot remote bodies, such as the sun and stars (see Probs. 5.13 and 5.14 at the end of this chapter).

The total energy emitted by a blackbody in 1 s can be obtained by adding together the emissions at all wavelengths. In 1879 Josef Stefan (1835–1893) discovered that this total power P depended on the absolute temperature of the emitter T raised to the fourth power

$$P = \sigma T^4$$

In 1884, Ludwig Boltzmann (1844–1906) showed that this fourth-power law could be deduced from thermodynamics and classical physics. The Stefan-Boltzmann law means that when the absolute temperature is doubled, the power is increased by a factor of $2^4 = 16$. The best value for the Stefan-Boltzmann constant is

$$\sigma = 5.6697 \times 10^{-8} \text{ W/m}^2\text{-K}^4$$

If we know the area and temperature of a blackbody, we can calculate the emission power. Conversely if we can measure both its area and power radiated, we can calculate the temperature of a body. In Prob. 5.15 this law is used to obtain the surface temperature of the sun.

Several attempts were made to describe the shape of the blackbody-radiation spectrum using classical physics. But no combination of the laws formulated by Newton, Maxwell, and others could predict what was firmly established experimentally.

THE QUANTUM THEORY

In 1900, Max Planck (1858–1947) modified the accepted rules of physics in a small but very important way. What he did created a major revolution in physics and, with it, a profound change in all of natural science.

According to classical physics, the energy of any system can be varied by either adding to or taking from the system any amount of energy. Thus, within limits, the system can have any arbitrary amount of energy. For example, it requires work to set a cello string in vibration. The frequency at which it vibrates is governed by the mass and elasticity of the string. The loudness, or intensity level, is the rate at which energy is transferred from vibrating string to the air and then to the listeners' ears. The more work done by the cellist with his bow, the greater the energy added to the string and the louder the sound. The energy of the vibrating string varies with the square of the amplitude. With the restriction that the amplitude is not so large as to break the string, the string can have *any energy* from none, when it is not vibrating, up to the amplitude just less than its breaking limit. We say the energies allowed to the cello string have a *continuous* set of values. By this we mean that between any two vibrating string energies, no matter how close in value, another energy can be found for the vibrating string.

The energy of an electromagnetic radiation, according to Maxwell's equations, depends on the square of the amplitude of the electric or magnetic field. Here, too, there is no restriction on what energies such radiation may have. For monochromatic light, for example, the energy can have any value in a continuous set of possible energies.

First, Planck looked at some of the unsuccessful attempts to describe blackbody radiation using classical physics. Then he found an equation which did describe the radiation. This was an empirical equation, based on no theory and derived from no known model, but it described perfectly the shape of the blackbody spectrum and its dependence on temperature.

Next Planck assumed a simple model. A hot, glowing blackbody comprises a large number of oscillators—the electrical parts of each atom in vibration. Rather than permitting each oscillator to have any energy, he postulated that an oscillator's energy was restricted to only certain *discrete* values.

The *distinction between discrete and continuous*, which is the important feature of Planck's model, is illustrated by the distinction between integers and real numbers. Between any two real numbers we can always find an intermediate number. For example, between 1.5906 and 1.5907 we can select 1.59063. There is no limit to the fineness of real numbers; real numbers are a *continuous* set. On the other hand, there are no integers intermediate to 6 and 7; the integers are a *discrete* set.

If an oscillator's energy is confined to a discrete set of values, then energy can be added or removed from an oscillator in amounts exactly equal to the difference between the original and final energies. How much energy is added or removed is strictly governed by the discrete set of permitted energies. The Latin word for "how much" is *quantum*. Planck called the definite amount of energy added or removed a "quantum of energy."

The way in which energy is transferred to or from an oscillator in a blackbody is by the absorption or emission of electromagnetic radiation. Planck postulated that the energy ϵ of a quantum was exactly proportional to the frequency of the radiation ν

$$\epsilon = h\nu$$

where h, a universal constant for electromagnetic radiation of all frequencies, is now called "Planck's constant."

With this model Planck deduced a radiation equation of exactly the form of his empirical equation. The emitted intensity depended on the frequency, or wavelength, the temperature, and two universal constants. The first universal constant is h, Planck's constant; the second is k, Boltzmann's constant, which really is the ideal gas constant divided by Avogadro's number R/N_A. By adjusting the values of h and k, Planck was able to make his blackbody-radiation formula describe perfectly the measured emitted powers at all temperatures and wavelengths. The constants had the values

$h = 6.55 \times 10^{-34}$ joule-s
$k = 1.346 \times 10^{-23}$ joules/K

In going outside classical physics Planck succeeded in precisely describing blackbody radiation in all its details—the shape of the spectrum, the displacement law, and the fourth-power law. With the introduction of the quantum concept, it was possible to develop models for other phenomena not consistent with classical physics.

THE PHOTOELECTRIC EFFECT

One way to remove an electron from an atom is to supply the energy needed for ionization in the form of light. In the *photoelectric effect,* when light shines on some surfaces, electrons are ejected, leaving a positively charged substance. The details of this process, however, were so curious as to make it inexplicable in terms of classical physics.

For each element there is a limiting lower frequency of the incident light below which no electrons are ejected. For example, a clean lithium surface will show the photoelectric effect only if the incident light contains some radiation whose frequency is 5.56×10^{14} Hz or greater. This is equivalent to light of wavelength 540 nm or shorter. No matter how intense or bright a beam of light may be, if there is no radiation present whose frequency is at the threshold 5.56×10^{14} Hz or higher, no electrons are ejected. But, no matter how faint the incident light may be, if some of it has a frequency above the threshold, some electrons are removed. Thus light with a lot of energy, none of it at 5.56×10^{14} Hz or greater, cannot accomplish what light with a little energy, some of it at the threshold or greater, can do. Clearly, the photoelectric effect requires more than just energy; it requires energy in a special form.

In 1905, Albert Einstein (1879–1955) suggested a very simple model to account for the effect. Rather than use the well-established wave description of light, he found it more convenient to consider light, as Newton had, to be a collection of particles. Each particle, called a "photon," or "light quantum," has a definite frequency ν and, in accord with Planck's theory, a definite energy $\epsilon = h\nu$. Monochromatic light comprises only photons of a single frequency; polychromatic light, photons of many frequencies. The intensity of light at any frequency, or wavelength, depends on how many photons there are having that frequency.

Suppose ν_0 is the threshold frequency for the photoelectric effect. For lithium, $\nu_0 = 5.56 \times 10^{14}$ Hz. The minimum energy required to remove an electron is $h\nu_0$. When a photon strikes an atom, one of two things can happen. If the photon energy $h\nu$ is less than the threshold energy for ionization $h\nu_0$, the photon is rejected. If the photon energy is equal to or greater than the threshold energy, the photon is accepted, ionization occurs, and any photon energy in excess of the threshold $h\nu_0$ is converted into the kinetic energy of the ejected electron.

If this excess energy beyond that needed for ionization is W, then

$$W = h\nu - h\nu_0$$

and

$$h\nu = h\nu_0 + W$$

or

$$\nu = \nu_0 + \frac{1}{h} W$$

This is Einstein's photoelectric equation. If the photon has exactly the required energy, $h\nu = h\nu_0$ and $\nu = \nu_0$. Then $W = 0$, and there is no excess for the removed electron, which cannot travel away. If ν is slightly greater than ν_0, the surplus appears as a small velocity of escape of the electron from the metal surface. When ν is much greater than ν_0, the surplus and the escape velocity are both big.

The Einstein equation suggests a simple test for the correctness of the model. Monochromatic light can be shined on a clean metal surface and the escape velocity and escape kinetic energy measured. This can be done for several frequencies of monochromatic light. The escape, or surplus, kinetic energy can be measured by seeing what opposing voltage is needed to stop a current of the ejected electrons from reaching a detector. A higher escape velocity, which means a higher surplus energy, requires a greater stopping voltage.

There were several confirmations of Einstein's model. Of special importance is the work of R. A. Millikan in 1916. In plotting the incident monochromatic-light frequency ν on one axis against the escape kinetic energy W on the other axis, he obtained, as Einstein predicted, a straight line. The intercept is the threshold frequency ν_0, and the slope is $1/h$. Each metal had its own characteristic threshold, but all metals had a common slope. Millikan's value for Planck's constant was $h = 6.589 \times 10^{-34}$ joule-s, which agrees well with Planck's own value. The best value for this universal constant, obtained from several different kinds of experiments, is

$$h = 6.62559(\pm 0.00016) \times 10^{-34} \text{ joule-s}$$

THE QUANTUM THEORY OF ATOMIC HEATS

The atomic heat of a metal (which is the heat needed to raise the temperature of 1 mole of that metal by 1°C) is always about 6 cal/°C-mole if the metal is at room temperature. This law (of Dulong and Petit) fails at lower temperatures. As the metal is cooled the atomic heat decreases, and when the temperature is very close to absolute zero, the atomic heat is very nearly zero. There was no satisfactory classical model to describe this phenomenon.

Einstein, in 1907, created a model for atomic heats which, like Planck's blackbody-radiation theory and Einstein's own treatment of the photoelectric effect, also employed the notion of discrete energy levels and quanta. A solid metal is imagined to be a collection of vibrating atoms. Classically, there is no restriction on the energy of oscillation for any atom, but Einstein imposed a quantum restriction on the possible energies of oscillation the atoms may have. Only a discrete set of energies is available to the vibrating atoms. We say the vibrational energy is *quantized*.

Ludwig Boltzmann a pioneer in applying statistical methods

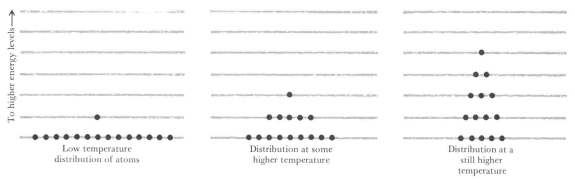

Figure 5.19. Boltzmann-type distribution of allowed energy levels at several temperatures. Atoms, molecules, or particles, in general, are represented by small black circles, and discrete permitted energies for these molecules are indicated by the set of horizontal line segments, with the lowest energy level at the bottom and the highest at the top. As the temperature is increased, the Boltzmann distribution shifts to some more particles at the higher energies (but in the Boltzmann distribution there are never more particles at any level than there are at any lower level). A discrete amount of energy is needed to move any particle from one energy level to a higher level. The total energy required to effect the distribution at a higher temperature can be calculated from knowing how many particles must be moved and how much energy is needed for each. This total energy divided by the temperature increase is the energy per degree. From this, the atomic heat in calories per degree per mole can be calculated.

to a very large number of atoms or molecules, showed that there is a most probable arrangement of atoms or molecules among possible energies. There are always fewer molecules at the higher energies and more at the lower energies. The inequality in distribution is less at higher temperatures and more at lower temperatures.

Einstein combined this Boltzmann-type distribution with his model of quantized vibrational energies (Fig. 5.19). Einstein's simple model predicted the general way in which atomic heat varies with temperature. It also showed why the law of Dulong and Petit was a general description for all metals at the higher temperatures.

With three solid achievements for the quantum theory—blackbody radiation, the photoelectric effect, and atomic heats—scientists were now ready to see if the idea of discrete energies and quanta could be used to describe atomic structure.

THE ATOMIC HYDROGEN SPECTRUM

When the light source in a grating spectrograph (Fig. 5.17) is a low-pressure hydrogen-gas-filled discharge tube instead of a hot glowing solid, the spectrum is as shown in Fig. 5.20. Unlike the continuous blackbody emission spectrum where some light of every wavelength is found, the hydrogen spectrum is *discrete*.

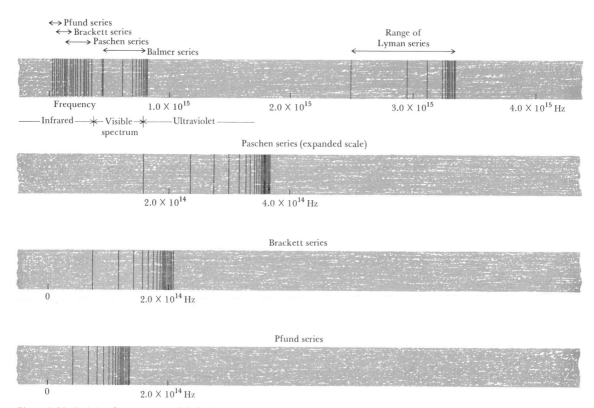

Figure 5.20. Emission-line spectrum of the hydrogen atom.

The simplest spectrum obtained is when the gas is at a very low pressure and an elevated temperature, conditions under which almost all the hydrogen is dissociated into atoms. Figure 5.20 is an *atomic emission spectrum.* Each line is where one discrete wavelength, or frequency, present in the emitted light is diffracted by the grating onto a distinct place on the detector. Each line, then, is the slit image on a photographic plate detector due to one emitted wavelength of light. For this reason, Fig. 5.20 is also called a "line spectrum."

The part of the line spectrum which is in the ultraviolet region is called the "Lyman series" in honor of Theodore Lyman (1874–1954), who investigated this portion of the atomic emission spectrum. Other parts are named in honor of the men who discovered and measured these segments of the entire spectrum. Each series has the same general pattern: a set of lines, the separation

Table 5.7. High-frequency Limit to Each Series in the Atomic Hydrogen Emission Spectrum

Series	High-frequency limit, Hz	Ratio of Lyman to other limits	High-frequency limit in terms of Lyman-series limit
Lyman	3.28805×10^{15}	1.0000	Lyman limit
Balmer	0.822013×10^{15}	4.0000	$\frac{1}{4} \times$ Lyman limit
Paschen	0.365338×10^{15}	9.0000	$\frac{1}{9} \times$ Lyman limit
Brackett	0.205504×10^{15}	15.9999	$\frac{1}{16} \times$ Lyman limit
Pfund	0.131522×10^{15}	25.0000	$\frac{1}{25} \times$ Lyman limit

between which is larger at the low-frequency end and vanishingly small at the high-frequency end.

Table 5.7 lists the high-frequency limit to each of the five series. These limits have a simple relationship: each is equal to the Lyman-series high-frequency limit divided by an integer which is also a perfect square. In general, the high-frequency limit is $(3.28805 \times 10^{15})/n^2$ Hz, where n is some integer. For the Lyman series, $n = 1$; for the Balmer series, $n = 2$; for the Paschen series, $n = 3$; for the Brackett series, $n = 4$; and for the Pfund series, $n = 5$.

Next, we can look at the frequencies within any series. Table 5.8 gives the first five and the limiting frequencies for the Lyman series. Each frequency is a simple rational fraction of the limiting frequency. Further, each rational fraction has a perfect square in its denominator and a numerator which is one less than the denominator. In short, each fraction has the form $1 - 1/n^2$. The Lyman-series frequencies are described by

$$\nu = 3.28805 \times 10^{15} \left(1 - \frac{1}{n^2}\right) \text{ Hz}$$

Table 5.8. Lyman-series Frequencies†

Frequency, Hz	Frequency divided by limiting frequency	Ratio expressed as a fraction	Ratio in the form $1 - 1/n^2$
2.46609×10^{15}	0.750016	$\frac{3}{4} = 0.75$	$1 - \frac{1}{4}$
2.92244×10^{15}	0.888806	$\frac{8}{9} = 0.888\ldots$	$1 - \frac{1}{9}$
3.08257×10^{15}	0.937507	$\frac{15}{16} = 0.9375$	$1 - \frac{1}{16}$
3.15651×10^{15}	0.959995	$\frac{24}{25} = 0.96$	$1 - \frac{1}{25}$
3.19670×10^{15}	0.972218	$\frac{35}{36} = 0.97222\ldots$	$1 - \frac{1}{36}$
$\ldots$	$\ldots$	$\ldots$	$\ldots$
3.28805×10^{15}	1.00000	$\frac{1}{1} = 1.00000$	$1 - 0$

† Frequencies of the Lyman series have the general form $\nu = 3.28805 \times 10^{15} (1 - 1/n^2)$ Hz.

where $n = 2, 3, 4, \ldots$.

Because the frequency 3.28805×10^{15} Hz or some exact fraction of it appears so often in the atomic hydrogen spectrum, it is represented by the symbol R in honor of Johannes Robert Rydberg (1854–1919), who first obtained a general formula for these spectral lines. The Rydberg constant is often written as R_H to indicate the numerical value is for atomic hydrogen alone. For other elements the Rydberg constant has other values. The Lyman-series frequencies can be represented by

$$\nu = R \left(1 - \frac{1}{n^2} \right)$$

where $n \geq 2$.

The Balmer series, in the visible part of the spectrum, was found to have the general form

$$\nu = R \left(\frac{1}{4} - \frac{1}{n^2} \right)$$

where $n \geq 3$. As n increases without limit, $1/n^2$ approaches zero and the frequency approaches $R/4$, which is the high-frequency Balmer-series limit. Similarly, the Paschen series, in the infrared, is described by

$$\nu = R \left(\frac{1}{9} - \frac{1}{n^2} \right)$$

where $n \geq 4$. In general, the entire atomic hydrogen spectrum is described by the formula

$$\nu = R \left(\frac{1}{n_1^2} - \frac{1}{n_2^2} \right)$$

where $n_2 \geq n_1$ and both are integers.

Because the general formula contains two terms, one of which is subtracted from the other, Walter Ritz (1878–1909) suggested a combination principle. All frequencies found in the emission spectrum arise from differences in terms of the general form R/n^2. Thus, the difference between the terms R/n_1^2 and R/n_2^2 corresponds to the frequency

$$\nu_{1,2} = \frac{R}{n_1{}^2} - \frac{R}{n_2{}^2}$$

$$= R \left(\frac{1}{n_1{}^2} - \frac{1}{n_2{}^2} \right)$$

The significance of these spectral terms was not made clear until the quantum theory provided a basis for atomic structure.

THE BOHR ATOM

Thus far, the constituents of the simplest atom were known: one electron and one proton. From the emission spectrum and from Planck's equation $\epsilon = h\nu$, something was known about the energies emitted by atomic hydrogen. But the model of a planetary electron in orbit about a central proton could not be stable according to the laws of classical physics. A charged particle whose velocity is not constant must lose energy by radiation; the electron must spiral in toward the proton nucleus if classical physics is descriptive of phenomena on the atomic scale.

Niels Bohr (1885–1962) assumed that, in fact, classical physics must be modified for atomic-sized systems. He postulated that an electron could travel in an orbit without radiating energy. Only a discrete number of orbits was possible for the electron, and each circular orbit was characterized by a radius and an electron speed. The electron had a different energy for each possible orbit. When the electron moves from one possible orbit to another, it must go from one energy level to another. If it goes to a lower energy, it emits radiation of frequency ν, which is related to the energy level difference $\Delta\epsilon$ by

$$\Delta\epsilon = h\nu$$

When the electron goes to a higher-energy orbit, the atom must take up the energy needed, which is the difference in energy between the lower and higher values $\Delta\epsilon$, in the form of radiation. The frequency of this absorbed radiation is calculated from the same equation.

Bohr applied classical physics to the electron-proton electrostatic attractive force and to the electron centrifugal force, but he imposed a condition on which orbits may be permitted. The angular momentum of the electron is the product of the electron mass m, the speed v, and the orbit radius r. This must be some whole-number multiple of $h/2\pi$:

$$mvr = \frac{nh}{2\pi} \qquad n = 1, 2, 3, \ldots$$

The number n is called the principal "quantum number."

From this he found the only allowed orbits must have radii defined by

$$r = \frac{n^2 h^2}{4\pi^2 e^2 m_e}$$

where e is the electron charge, m_e its mass, and h Planck's constant, all of which had been experimentally determined. The first Bohr radius, where $n = 1$, is

$$r_1 = 5.29 \times 10^{-11} m$$
$$= 0.529 \text{ Å}$$

The diameter of a hydrogen atom is then 1.06 Å, which is in excellent agreement with the atomic size calculated by other means. The successive orbit radii increase as the square of the principal quantum number. Thus the second orbit has a radius $r_2 = 2^2 r_1 = 4 \times 0.529$ Å $= 2.12$ Å. The third orbit's radius is $9r_1$, or 4.76 Å.

The energy of the atom due to the electron orbital motion is the sum of two contributions: the kinetic energy of the electron traveling about its orbit and the electrostatic potential energy between the electron and proton. Bohr found the energy E of an allowed orbit to be

$$E = \frac{-2\pi^2 e^4 m_e}{n^2 h^2}$$

The minus sign has this significance: It requires work to separate an electron from the nucleus. Thus a removed electron has a higher energy than one closer to the nucleus. If the energy of a completely removed electron is assigned the arbitrary value zero, the energy of an electron in an allowed orbit must be less than this; it must be a negative number.

Consider two different orbits with principal quantum numbers n_1 and n_2. The respective energies are

$$E_1 = \frac{-2\pi^2 e^4 m_e}{h^2} \frac{1}{n_1{}^2}$$

and

$$E_2 = \frac{-2\pi^2 e^4 m_e}{h^2} \frac{1}{n_2{}^2}$$

If the electron goes from the first to the second allowed orbit, from n_1 to n_2, the energy difference is

$$\Delta E = E_2 - E_1$$
$$= \frac{2\pi^2 e^4 m_e}{h^2} \left(\frac{1}{n_1{}^2} - \frac{1}{n_2{}^2} \right)$$

If this difference is positive, energy must be absorbed; if negative, energy is emitted. In either case

$$\Delta E = h\nu$$

and so

$$\nu = \frac{\Delta E}{h}$$
$$= \frac{2\pi^2 e^4 m_e}{h^3} \left(\frac{1}{n_1{}^2} - \frac{1}{n_2{}^2} \right)$$
$$= R \left(\frac{1}{n_1{}^2} - \frac{1}{n_2{}^2} \right)$$

The set of constants represented by R were all experimentally known to Bohr. Thus Bohr was able to calculate a value for this set of constants and compare it with the experimentally obtained Rydberg constant.

The best data available in 1913, the year Bohr announced his theory, gave a calculated Rydberg constant of

3.2880570×10^{15} Hz

The best experimental value for that constant was

3.2880696 × 10^{15} Hz

The agreement was better than 1 part in 300,000. To that precision Bohr was able to predict the entire atomic hydrogen spectrum. The simple Bohr picture is represented in Fig. 5.21.

QUANTUM MECHANICS

Bohr's atomic model worked quite well for the hydrogen atom, the helium ion He$^+$, and the lithium ion Li^{2+}. In each case the atom or ion contains but a single electron. But this simple picture failed for atoms or ions with two or more electrons. It was possible to modify some of the details of Bohr's theory to get better agreement with observation, but the addenda resulted in a model much less elegant than Bohr's original concept. What was needed was a more general theory which, in the case of a single-electron atom or ion, would reduce to Bohr's model but which was also adequate for more complicated atoms.

At the end of 1925 Werner Heisenberg (b. 1901) arrived at a set of rules enabling him to calculate the frequencies and relative intensities of spectral lines. A few months later Erwin Schrödinger (1887–1961) postulated a wave equation (the kind of equation used to describe wave motion) which he used to solve the hydrogen atom problem. In addition he was able to describe quite precisely the behavior of systems much too complicated for Bohr's theory. Schrödinger then showed that his method and Heisenberg's were, in principle, the same. Their theory, which involves considerably more mathematics than the older quantum theory, is called either "quantum mechanics" or "wave mechanics." A modern approach to quantum mechanics is to begin, not with a picture model of what a system looks like or how it functions, but rather with certain general rules. All the information about a system, an atom, for example, is presumed to be contained in a collection of mathematical symbols representing the important parameters—mass, distance, time, etc. This collection of symbols is called a "wave function," or "state function," for the system.

In order to determine some property of the system, its state function is manipulated, or operated on, in a special way. For each

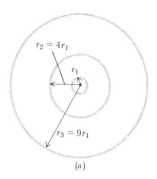

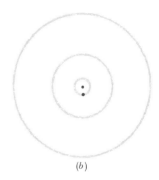

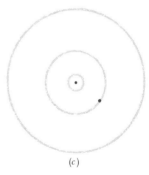

Figure 5.21. Bohr hydrogen atom. (a) The first three orbits. On this scale the nucleus, which is 10^{-5} r$_1$, is too small to be seen. (b) The electron is in the lowest allowed orbit. (c) The electron is in the next higher orbit. Energy in the form of light is absorbed to accomplish the transfer of the electron from the lowest to the next highest orbit.

property there is a definite kind of operation, and the result of this operation is a number which is the value of desired property.

Under this arrangement, the property can have only one of a discrete set of possible values; the property is quantized as in the older quantum theory. Or the best that can be done in many cases is to obtain an entire collection of all the possible property values. From this the average, or most likely, value of the property is calculated, and it is this averaged value which is compared with the experimentally measured property.

The wave function depends on one or more variables, such as time and position. For each value of the time or position variables, the wave function may be evaluated, and, in general, the wave function changes in value with a change in these variables. Max Born (1882–1970) suggested a physical interpretation of the wave function. Its value at any position multiplied by itself is a measure of the probability of finding a particle described by the wave function at that position. Thus a property of the particle, in this case its position, is known in a statistical sense. We can know the chance of finding it at various positions, and from this we can determine that position where it is most likely to be found. The Born interpretation indicates a significant difference between what we can know about a system using classical physics and what we can know using quantum mechanics.

In Fig. 5.22a the relative probability of finding an electron at a distance r from the nucleus is plotted on the vertical axis against the electron-nucleus distance on the horizontal axis. This probability is for an electron in the lowest-allowed energy in a hydrogen atom. The first Bohr radius, at $r = 0.529$ Å, is indicated and coincides with the electron-nucleus distance of greatest probability. In the newer quantum mechanical description the electron does not spend all its time at a fixed distance from the nucleus. There is some chance of finding it at any distance from the nucleus, except right at the center of the nucleus. Alternatively, we might say the electron is spread over the entire space, but more of it is found at the Bohr radius than elsewhere.

In the Bohr model the electron moves along a circular orbit and is thus located in a two-dimensional plane, which includes the circumference of the circle. The quantum mechanical model, in contrast, permits the electron to move about in three dimensions. It is difficult to represent the newer model satisfactorily with a drawing, but in Fig. 5.22 both the first Bohr circular orbit and the quantum mechanical spatial distribution are shown.

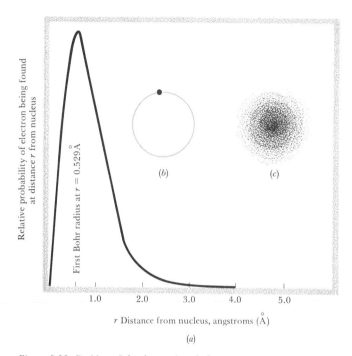

Relative probability of electron being found at distance r from nucleus

First Bohr radius at $r = 0.529 \text{Å}$

(b)

(c)

1.0   2.0   3.0   4.0   5.0

r Distance from nucleus, angstroms (Å)

(a)

Figure 5.22. Position of the electron in a hydrogen atom in its lowest energy state, according to quantum mechanics. (a) Relative chance of finding an electron at various distances from the nucleus of a hydrogen atom in its lowest allowed energy; (b) first Bohr circular orbit for electron in hydrogen atom; (c) first shell for electron in hydrogen atom is a three-dimensional distribution.

An analogy with an ideal onion may be useful. Imagine a spherical onion consisting of a large number of concentric, thin spherical layers. Each layer can be peeled off in turn, leaving a perfect sphere. Each layer also contains some record of how often a single electron can be found in it, i.e., what fraction of the time the electron spends in that particular layer. Alternatively, we can see what fraction of the single electron is in that layer. Suppose the onion has an initial radius of about 10 Å. The first layer is removed, analyzed, and found only very rarely to have had the electron within it. The electron had an equal chance to be anywhere within this layer. The next layer is similarly removed and examined, and it is found to have had the electron within it more often than the first layer; also, there was no place within this layer preferred most by the electron.

As each layer, in turn, is removed, we find increasingly more evidence of the electron having been there, and the evidence is uniform throughout each layer. When the onion is reduced to a radius of 0.529 Å, the evidence is greatest; layers removed after this show less evidence. Finally at the very core, or heart, of the onion, we find the electron was never present. Of all the layers removed, the one with radius 0.529 Å contained the electron most often.

With the newer quantum mechanics we speak, not of an orbit, but rather of a *shell*. There are a number of permitted shells, each with its characteristic energy. Within each shell the electron is allowed to move about in three dimensions. Beyond the first shell, quantum mechanics predicts the higher shells to be broken up into subshells. The description, in the form of an equation, which tells us how a single electron spends its time in space within a shell or subshell is called an "orbital." The quantum mechanical picture is not amenable to as simple a graphic representation as the Bohr model, but because it works better, we accept it as a truer model.

THE CONSTRUCTION PRINCIPLE

As a logical consequence of quantum mechanics, the permitted energy levels for electrons in an atom can be calculated. These computations for atoms other than hydrogen are so lengthy that only with high-speed computers are values for the allowed energies obtained. But there are certain features common to all atoms which can be presented without these computations.

Where the original Bohr hydrogen atom used only a single quantum number n, the principal quantum number, it is necessary to specify four quantum numbers for each electron to fully describe the atom by quantum mechanics. Of the four quantum numbers needed for each electron, the first and most important is the *total*, or *principal, quantum* number, which can have only positive integral values $n = 1, 2, 3, 4, \ldots$. It is this number which largely fixes the energy level of the orbital. We sometimes call these levels the first, second, third, etc., shells, or the K, L, M, etc., shells about the atom nucleus.

The second quantum number is l, which has possible values $0, 1, 2, \ldots, n - 1$. Thus in the third, or M, shell, $n = 3$ and l may have the values 0, 1, or 2; there are three subshells. The third quantum number m has for each l the possible integral values from $-l$ to $+l$. For $n = 3$ and $l = 2$, m can have the values $-2, -1, 0, +1$, or $+2$. The fourth quantum number is called the "spin" and can have either of two values, $+\frac{1}{2}$ or $-\frac{1}{2}$. (It has this name because certain small spectral energy differences were attributed to the magnetic field associated with a spinning negatively charged electron, oriented either in the direction of an external magnetic field or opposed to that field.) The possible combinations of the four quantum numbers for the first three shells are listed in Table 5.9.

PAULI'S EXCLUSION PRINCIPLE. The exclusion principle of Wolfgang Pauli (1900–1958), when applied to electrons in atoms, states that no two electrons in a single atom may have the same four quantum numbers. In Table 5.9 observe that in the first shell, $n = 1, l = 0, m = 0$, and the spin can be either $+\frac{1}{2}$ or $-\frac{1}{2}$; this shell can have no more than two electrons. The second shell can contain, at the most, eight electrons; the third shell can have a maximum of eighteen electrons.

The orbital designation s is used for $l = 0$; p, for $l = 1$; d, for $l = 2$; f, for $l = 3$; and g, for $l = 4$. The numerical coefficient before the letter is the principal quantum number; thus, $3s$ designates $n = 3, l = 0$. The letters s, p, d, and f stand for sharp, principal, diffuse, and fundamental and were originally applied to spectral lines before quantum mechanics was invented.

Table 5.9. Four Quantum Numbers and Their Possible Combinations for the First Three Shells

Shell	n	l	m	Spin	Orbital designation	Total electrons in shell
K	1	0	0	$\pm\frac{1}{2}$	$1s$	2
L	2	0	0	$\pm\frac{1}{2}$	$2s$	
		1	-1	$\pm\frac{1}{2}$	$2p$	
		1	0	$\pm\frac{1}{2}$	$2p$	
		1	1	$\pm\frac{1}{2}$	$2p$	8
M	3	0	0	$\pm\frac{1}{2}$	$3s$	
		1	-1	$\pm\frac{1}{2}$	$3p$	
		1	0	$\pm\frac{1}{2}$	$3p$	
		1	1	$\pm\frac{1}{2}$	$3p$	
		2	-2	$\pm\frac{1}{2}$	$3d$	
		2	-1	$\pm\frac{1}{2}$	$3d$	
		2	0	$\pm\frac{1}{2}$	$3d$	
		2	1	$\pm\frac{1}{2}$	$3d$	
		2	2	$\pm\frac{1}{2}$	$3d$	18

The 1s subshell has the lowest energy. In increasing order, the next are 2s, 2p, 3s, 3p, 4s, 3d, 4p, 5s, and so on (as in Fig. 5.23). Bohr suggested the way in which electrons are added to the shells and orbitals as the atomic number is increased: Electrons are placed one by one into the lowest available energy level. This manner of building up the atoms is called the "construction," or "aufbau," principle. It states that the only difference between an atom and one with the next highest atomic number is that the nuclear charge is increased by one and one more electron is added. The other electrons are found in the same orbitals in both atoms. Table 5.10 lists the n and l quantum numbers for the first 20 elements. The exponents in the designation column are equal to the number of electrons in a subshell. Nitrogen, for example, has the designation $1s^2 2s^2 2p^3$. There are two electrons in the 1s subshell ($n = 1$, $l = 0$), two in the 2s subshell ($n = 2$, $l = 0$), and three in $2p$ ($n = 2$, $l = 1$)—a total of seven electrons in all.

Here is a theoretical basis to the periodicity of elements. Consider the inert gases He, Ne, and Ar. The configurations are that of a completed K shell, a completed L shell, and the completed

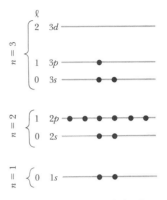

Figure 5.23. Normal aluminum atom and its 13 electrons in their subshells ($1s^2 2s^2 2p^6 3s^2 3p$).

Table 5.10. Electronic Structure of the First 20 Elements

Element	Atomic number	Shell n → Subshell l →	1 0	2 0	2 1	3 0	3 1	3 2	4 0	Designation
H	1		1							$1s$
He	2		2							$1s^2$
Li	3		2	1						$1s^2 2s$
Be	4		2	2						$1s^2 2s^2$
B	5		2	2	1					$1s^2 2s^2 2p$
C	6		2	2	2					$1s^2 2s^2 2p^2$
N	7		2	2	3					$1s^2 2s^2 2p^3$
O	8		2	2	4					$1s^2 2s^2 2p^4$
F	9		2	2	5					$1s^2 2s^2 2p^5$
Ne	10		2	2	6					$1s^2 2s^2 2p^6$
Na	11		2	2	6	1				$1s^2 2s^2 2p^6 3s$
Mg	12		2	2	6	2				$1s^2 2s^2 2p^6 3s^2$
Al	13		2	2	6	2	1			$1s^2 2s^2 2p^6 3s^2 3p$
Si	14		2	2	6	2	2			$1s^2 2s^2 2p^6 3s^2 3p^2$
P	15		2	2	6	2	3			$1s^2 2s^2 2p^6 3s^2 3p^3$
S	16		2	2	6	2	4			$1s^2 2s^2 2p^6 3s^2 3p^4$
Cl	17		2	2	6	2	5			$1s^2 2s^2 2p^6 3s^2 3p^5$
Ar	18		2	2	6	2	6			$1s^2 2s^2 2p^6 3s^2 3p^6$
K	19		2	2	6	2	6	1		$1s^2 2s^2 2p^6 3s^2 3p^6 4s$
Ca	20		2	2	6	2	6	2		$1s^2 2s^2 2p^6 3s^2 3p^6 4s^2$

first two subshells within the *M* shell, respectively. Hydrogen and the alkali metals H, Li, Na, and K all have completed subshells and then one electron beginning a new *s* subshell. It is the set of electrons in the incomplete subshells which most determines the chemical behavior of the elements.

ELECTRON TRANSITIONS AND SPECTRA

The assignment of electrons to their proper subshells is illustrated for aluminum in Fig. 5.23. From Table 5.10, the designation $1s^2 2s^2 2p^6 3s^2 3p$ instructs us to place the 13 electrons in their proper subshells. From the emitted spectral lines, the difference in energy between the levels can be calculated, and from this the energy of each level can be determined.

The arrangement of electrons shown in Fig. 5.23 is the arrangement of lowest energy, the *ground state*. If an electron is raised from the 2*p* to the 3*d* level, work is required and the new arrangement has a higher energy. The atom is said to be in an *excited state*. The work needed is the difference in energy $\Delta\epsilon$ between an electron in the 2*p* and 3*d* levels. The energy equivalent of this work is obtained by absorbing electromagnetic radiation of frequency $\nu = \Delta\epsilon/h$.

In Fig. 5.24 the energy levels of a hypothetical element are drawn and several general electron transitions are represented. *Absorption*, indicated by the arrows a_1, a_2, a_3, and a_4, permits the transition of an electron at 3*d* to the levels 4*p*, 4*f*, 5*p*, and 5*f* provided, of course, these final levels are not filled. The lengths of the arrows correspond to the energies needed and, thus, to the frequencies of light absorbed. Transition a_1 requires the least energy of the four absorptions shown; its absorption is at the lowest frequency and longest wavelength.

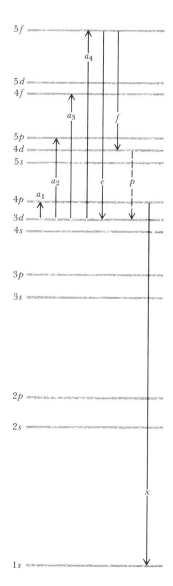

Figure 5.24. Electronic transitions. Transitions can only occur if the receiving subshell is not filled. Observe the energy overlap of subshells; 4s is lower than 3d; and so on. Absorption of radiation and elevation of electron to permitted higher energy level are shown by a_1, a_2, a_3, a_4. Emission of radiation and dropping of electron to permitted lower energy level is e; the wavelength emitted by e is exactly the wavelength absorbed by a_4. Emission of radiation of less energy than absorbed in a_4 is shown by f — lower frequency and longer wavelength emitted than absorbed (fluorescence). Emission in a transition which occurs much more slowly than others (phosphorescence) is shown by line p. If a K-shell electron is knocked out by a high-speed electron (as in a cathode-ray discharge tube), another electron, here from the 4p subshell, can drop down into the created 1s vacancy; this high-energy decrease is accompanied by a high-frequency, short-wavelength emission — an x-ray, x.

When light of all wavelengths is passed through this hypothetical element, wavelengths corresponding to the processes a_1, a_2, a_3, and a_4 will be absorbed; the transmitted light will be weaker in these wavelengths. The wavelengths *not absorbed* are what we see, and these determine the color of the substance doing the absorption. The spectrum of the transmitted light is called an "absorption spectrum." Since each substance has its own energy values for the various levels, the set of allowed absorption wavelengths is unique for each substance. The absorption spectrum is a convenient tool for identifying a substance.

Some electron transitions may be delayed and not occur immediately. Thus, in a collection of many atoms, process p, for example, can continue to occur long after the radiation which elevated electrons to $4d$ is removed. This delayed emission is called "phosphorescence."

When a high-speed electron collides with an atom, as in a cathode-ray discharge tube, an electron in the atom can be knocked out of its position. Another electron in a higher energy level can drop into this created vacancy, emitting a high-frequency, or short, wavelength. Thus, in process x, an x-ray is emitted.

We discussed earlier the standard of length, using light emitted by an isotope of krypton $_{36}Kr^{86}$. Its normal electron structure is $1s^2 2s^2 2p^6 3s^2 3p^6 3d^{10} 4s^2 4p^6$, and the levels $5p$ and $6d$ are completely vacant. If one $4p$ electron is promoted to the $5p$ level, the atom is in an *excited* state; i.e., it has more energy than in the lowest, or *ground*, state. When this excited $_{36}Kr^{86}$ is kept at the triple point of nitrogen $-210°C$ and the $5p$ electron absorbs just enough energy to go to the $6d$ level, that radiation is the standard of length. Either this absorbed light or the emitted light in the reverse process $6d$ to $5p$ has a wavelength such that 1,650,763.73 wavelengths total 1 m exactly. This special wavelength is the international standard which determines the size of the meter.

The standard of time is also a precisely defined transition, with a frequency and wavelength in that region used for television and radar signals. With a grating spectrograph capable of distinguishing wavelength differences as fine as a few tenths of an angstrom unit, individual spectral lines are seen to actually be multiplet lines closely spaced and therefore of nearly the same energy.

Pauli attributed these minute energy differences to the nucleus. Isotopes of the same element have slightly different energies. Furthermore, the nucleus has a spin and with this a magnetic moment which can assume one of a discrete set of orientations with respect to an external magnetic field. The transitions between these allowed nuclear magnetic energies have frequencies of about 10^{10} Hz and wavelengths of about 3 cm. The isotope $_{55}Cs^{133}$ has one transition especially convenient for the standard of time. A signal source which supplies exactly the proper frequency to cause this transition emits an oscillation which defines the second. Thus 1 s is the time for 9,192,631,770 cycles produced as described.

THE DUALITY OF LIGHT

With Einstein's description of the photoelectric effect, using the particle-like photon model, the old question was reopened: Is light particlelike or wavelike? When x-rays were discovered, the wave theory of light was so well established that tests were devised especially to demonstrate that these, too, were waves.

Max von Laue (1879–1960) and coworkers, in 1912, directed an x-ray beam through a crystal of zinc sulfide and then onto a photographic plate. When developed, the plate had a series of regularly spaced spots where exposed by the x-rays. These resembled the bright regions produced when light passes through a diffraction grating. This can be described by considering the atoms of the zinc sulfide crystal to be in a regular array, serving as a diffraction grating for the wavelike x-rays. Both the spacing between the atoms and the x-ray wavelength were estimated to be about 1 Å, or 10^{-10} m.

William Henry Bragg (1862–1942) recognized the Laue spots as reflections from a series of parallel planes within the crystal (Fig. 5.25). If the plate at position P shows a strong reflection, this is caused by constructive interference between the two beams meeting at P; if there is no exposure at P, the interference is destructive. For the former case, the path difference must be some whole number of wavelengths of the x-ray. In the simplest case, this can be a single wavelength λ or $AN + NB = \lambda$. Provided the source and plate are symmetrically placed with respect to the reflecting crystal, the two right triangles AMN and BMN are congruent; distance $AN = NB$. It follows, then, that $2(AN) = \lambda$ and $AN = \lambda/2$.

The angle θ at which the beam meets the crystal and at which it leaves by reflection is also the angle AMN and angle BMN. Thus the angle of incidence and angle of reflection determine the shape of

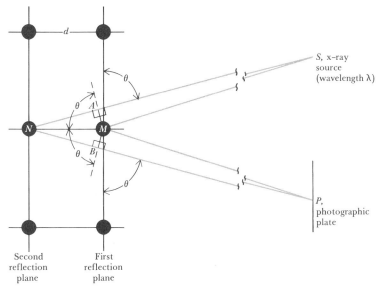

Figure 5.25. Bragg relationship for constructive interference and reflection by parallel planes in a crystal. If a bright spot is found at P, there must have been constructive interference; that is, the path difference is some whole-number multiple of x-ray wavelengths; in the simplest case, AN + NB = λ = 2AN.

the right triangle *AMN* because the sum of the three angles in a triangle must equal 180°. In a right triangle, one angle is 90°; therefore the other two acute angles must also add up to 90°. If one of these angles is known, the other may be completely determined.

Bragg suggested that, knowing the angle θ, at which a strong reflection appears on the plate, the spacing between planes d can be calculated. All right triangles which have the same acute angle θ are similar in shape. Thus the ratio of any two corresponding sides is the same for all such similar triangles no matter how large their actual size. In particular, the ratio AN/d is fixed once θ is known. That means $\lambda/2d$ is known, and if λ is known, d follows immediately.

As the path difference $AN + NB$ assumes the values λ, 2λ, 3λ, . . . , the angle for strong reflection assumes different calculable values. With his son, William Lawrence Bragg (1890–1971), W. H. Bragg, in 1913, confirmed his suggestion. They found the strongly reflected spots to be placed on the plate at angles consistent with this picture. This distance d was obtained.

A regular two-dimensional array of atoms is drawn in Fig. 5.26, and through these are drawn several families of parallel

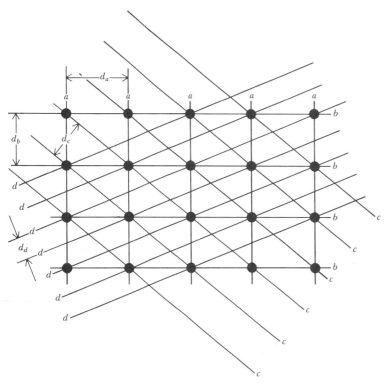

Figure 5.26. Through a regular array of atoms many series of parallel planes may be drawn. In this figure the series labeled a, b, c, and d each has a unique spacing between neighboring parallel planes d_a, d_b, d_c, and d_d, respectively. From the Bragg relationship the entire set of spacings between all possible reflection planes is determined. Using these spacings, the actual position of each atom can be deduced within the regular crystalline array.

planes (actually, parallel lines in this figure). By rotating the crystal, the various planes can do the constructively interfering reflections and the set of plane spacings d_a, d_b, d_c, and so on, can be obtained. Since these spacings depend on the spacings between the atoms themselves, it is possible to deduce the regular arrangement of atoms within the crystal. *X-ray diffraction* is a major tool in determining the atomic arrangement within crystals and molecules (Fig. 5.27).

The work of von Laue and the Braggs was strong evidence that x-rays are a wave phenomenon. (It was also evidence of the repeated, regular array of atoms within crystals.) But a careful measurement of the wavelengths of the incident and reflected x-rays indicated a slight but significant lengthening of the wave-

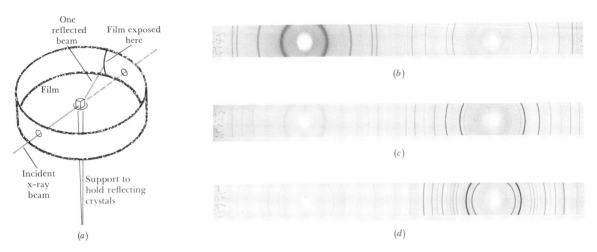

Figure 5.27. X-ray diffraction. (a) Because an x-ray beam is reflected strongly at an angle determined by a particular lattice spacing within the crystal, different crystal structures produce distinct patterns. These patterns are obtained using (b) silver, (c) sodium chloride, and (d) quartz. (See Figs. 6.9, 7.2, and 7.14 for the deduced structural models for these three substances.) The reflection from a single crystal is a set of distinct spots; from many crystals (e.g., a powder) the spots are found along arcs.

length. Arthur H. Compton (1892–1962) considered the x-ray to give up some of its energy to an electron upon the collision of an x-ray photon with the electron. If its energy $\epsilon = h\nu$ decreases, its frequency must become less and its wavelength longer. The decrease in photon energy is equal to the increase in the electron energy, according to the law of conservation of energy.

There is also a law of conservation of momentum. Momentum, which is the product of mass times velocity, is a vector. Compton assumed the photon to have a mass m and the speed-of-light velocity c. According to Einstein's relativity theory, a particle with a mass m has an energy mc^2. The quantum energy can be set equal to the relativity energy:

$$mc^2 = h\nu$$

and

$$mc = \frac{h\nu}{c}$$

so

$$mc = \frac{h}{\lambda}$$

The momentum of a photon *mc* is inversely proportional to its wavelength; as the photon momentum decreases, its wavelength increases. The loss in momentum of the photon was found to be exactly equal to the gain in momentum of the electron.

The *Compton effect*, in 1923, indicated that a photon interacting with an electron can be considered as a particle with a definite mass and momentum, as well as a wave phenomenon. Unlike other masses, however, the photon can only exist when it moves at the velocity of light. The mass of a photon is so small (see Prob. 5.16 at the end of this chapter) that ordinarily it is undetectable.

THE DUALITY OF MATTER

In 1924 Prince Louis Victor de Broglie (b. 1892) proposed an astonishing property for matter in motion. By analogy with the momentum of a photon $mc = h/\lambda$, de Broglie considered a mass *m* with velocity *v* to have a wavelength λ given by $mv = h/\lambda$. The de Broglie wavelength

$$\lambda = \frac{h}{mv}$$

can be calculated if the mass and velocity are known.

Because electrons can be accelerated to high and measurable velocities, these particles seemed to be good candidates to test de Broglie's hypothesis. In 1927 George P. Thomson (b. 1892), son of J. J. Thomson, repeated the Laue experiment, passing a stream of fast electrons (instead of an x-ray beam) through a thin crystalline metal foil and then onto a detector. He observed a diffraction pattern, like Laue's, indicating interference and, hence, some wave phenomenon. Because high-speed electrons do produce x-rays which have wave properties, it was necessary to rule out this possibility as the source of constructive and destructive interference. In a magnetic field, the entire diffraction pattern shifted. X-rays are not moved by a magnetic field, but electrons are. Thus it was a stream of electrons which moved to the detector and which was capable of interference.

At the same time, Clinton Davisson (1881–1958) and Lester Germer (b. 1896) reflected an electron beam from a nickel crystal and obtained a Bragg-like series of strong reflections. From the angles of reflection and the spacing between planes within the crystal (obtained by x-ray diffraction), they were able to calculate the wavelength of these electrons as 1.65 Å. Knowing the speed and mass of the electrons, they calculated the wavelength from de Broglie's formula and obtained 1.66 Å under these experimental conditions. (At other speeds, the wavelength has other values.)

Example. At what speed must an electron travel to have a de Broglie wavelength of 1.66 Å?

Planck's constant $h = 6.62559 \times 10^{-34}$ joule-s in many cases can be more conveniently handled in units of erg-seconds. Because a joule is the kinetic energy of 1 kg moving at 1 m/s (1 joule = 1 kg $\times$ m²/s²) and an erg is the kinetic energy of 1 g at 1 cm/s (1 erg = 1 g $\times$ cm²/s²), 1 joule = 10^7 ergs and

$$h = 6.62559 \times 10^{-27} \text{ erg-s}$$

Substituting into

$$v = \frac{h}{m\lambda}$$

$$= \frac{6.62559 \times 10^{-27} \text{ erg-s}}{9.109 \times 10^{-28} \text{ g} \times 1.66 \times 10^{-8} \text{ cm}}$$

$$= 4.38 \times 10^8 \frac{\text{erg} \times \text{s}}{\text{g} \times \text{cm}}$$

$$= 4.38 \times 10^8 \frac{(\text{g} \times \text{cm}^2/\text{s}^2) \text{ s}}{\text{g} \times \text{cm}}$$

$$= 4.38 \times 10^8 \text{ cm/s}$$
$$= 4.38 \times 10^6 \text{ m/s}$$

Example. What is the wavelength of an automobile of mass 1,500 kg moving 30 m/s?

The de Broglie wavelength

$$\lambda = \frac{h}{mv}$$

$$= \frac{6.63 \times 10^{-34} \text{ joule} \times \text{s}}{1.50 \times 10^3 \text{ kg} \times 30 \text{ m/s}}$$

$$= 1.5 \times 10^{-38} \frac{\text{joule} \times \text{s}^2}{\text{kg} \times \text{m}}$$

$$= 1.5 \times 10^{-38} \frac{(\text{kg} \times \text{m}^2/\text{s}^2) \text{ s}^2}{\text{kg} \times \text{m}}$$

$$= 1.5 \times 10^{-38} \text{ m}$$
$$= 1.5 \times 10^{-28} \text{ Å}$$

The proton, in comparison, has a diameter of about 10^{-15} m, or 10^{-5} Å.

Here is the clue to why the wave nature of matter escaped detection until de Broglie announced his formula. The wave nature of light was established only by interference—when light reacts with slits or a grating whose spacing is comparable in size to the length of light. For familiar objects moving at their usual speeds, the wavelength is so short and the frequency so high that nothing yet discovered in nature has a comparable size.

Because the voltage used to accelerate an electron can be precisely varied, it is possible to control the speed and the de Broglie wavelength. If this is about 1 or 2 Å, an electron beam can be used for diffraction studies just as x-rays are. Since x-rays penetrate deeply into matter, the information from x-ray diffraction concerns the interior of a crystal. For very thin specimens, x-rays pass right through without being reflected and diffracted. Electrons, in contrast, are not very penetrating and can be used to study the atomic arrangement of thin films.

Atoms, molecules, and neutrons can also be used in diffraction for special purposes. Hydrogen, for example, penetrates matter only slightly; mostly it is reflected by the surface layer. This type of diffraction is used for learning how atoms are arranged on the surface of a substance. Because surface atoms do not have the same environment as the interior atoms, their geometric design and many of their properties differ from the bulk of the crystal.

The size limit of what we can see using a microscope is governed by the wavelength of light used—several thousand angstroms. Using ultraviolet light, the best magnification available is about 3,000. In an *electron microscope,* an electron beam with a de Broglie wavelength of about 0.1 Å, a magnification of about 100,000, is possible. With a light microscope objects no smaller than 1000 Å can be seen or photographed, but with the electron microscope the present practical limit is about 2 Å. Pictures of very large molecules or clusters of molecules, such as individual viruses and, recently, large atoms (uranium and thorium), can be obtained by this technique.

THE UNCERTAINTY AND COMPLEMENTARITY PRINCIPLES

In 1927 Heisenberg pointed out an unusual implication of the new quantum mechanics. From the set of possible values for some property (position of a particle, for example), an average, or most probable, position is calculated. The differences, or deviations, between this average and each of the possible values can, in turn, be averaged. This *average of the deviations* is a measure of the certainty, or precision, to which we can know the experimental value of the position. The less the range of possible position values, the more likely the calculated averaged position will agree with what we measure.

In a similar way, the momentum of the same particle has a precision. What follows from quantum mechanics is that the more precisely we know the position, the less precisely we can know the momentum; and, conversely, the more precisely we measure the particle momentum, the less we are certain about its position.

There is a simple physical description of Heisenberg's discovery. Suppose we wish to measure the position of an atom whose diameter is known to be about 1 Å. We could use light with a wavelength of 5000 Å, but this is equivalent to using a 1-mile-long chain to measure the position of something 1 ft in size—what we use affects the precision to which we can measure. Clearly, radiation of a shorter wavelength will enable us to specify the position more exactly. But, as the wavelength becomes shorter, the frequency ν, energy $\epsilon = h\nu$, and momentum $mc = h/\lambda$ of the light increase. In the process of measurement with this light, more energy and more momentum can be transferred to the atom. Thus, the finer our position measurement becomes, the more the motion and momentum of the atoms are altered. Also, under the rules of quantum mechanics, the amount by which momentum of the atom is changed cannot be determined. In summary, if we wish to know the position of an atom quite precisely, we must use light of a short wavelength, but this light, with its high energy and momentum, alters the motion of the atom in an undetermined way.

Heisenberg's *uncertainty principle* has the form of an inequality. If Δx is the uncertainty in position x and Δp the uncertainty in momentum $p = mv$, the product of these is equal to or greater than Planck's constant divided by 2π:

$$(\Delta x)(\Delta p) \geqslant \frac{h}{2\pi}$$
$$\geqslant 1.055 \times 10^{-27} \text{ erg-s}$$

Notice that the units position times momentum is energy times time. Another combination of properties subject to the uncertainty principle is energy and time. To measure the time of an event in a system we use a clock. For atomic systems the clock we use is the oscillation of electromagnetic radiation. The higher the frequency, the more finely we can time the event, but at the same time we disturb the energy of the system. The more precisely we measure the time, the less precisely we can know the energy of the system.

In 1928 Bohr offered as a *principle of complementarity* an important difference between phenomena on an atomic or subatomic scale and those of the usual macroscopic size. We cannot hope to describe the former, microscopic, events with the same completeness we have become accustomed to using in classical physics. This is a law of nature and is not because of some remediably imprecise technique of our experiments. Some properties are mutually exclusive: one or the other, but not both, can be known with high precision. Thus, if an experiment is designed to measure the wavelength of an object, no information will be revealed about the object as a particle.

Using Heisenberg's inequality, we can look at an important difference between microscopic and macroscopic systems.

> **Example.** If a 10.0-g mass is found to be moving in a straight line at 5.00 cm/s so that the momentum 50.0 g-cm/s is known to 1 percent, what restriction is placed on the precision to which we can know the position of this moving mass?
>
> The precision of the momentum is $\Delta p = 0.50$ g-cm/s, and from
>
> $$(\Delta x)(\Delta p) \geq 1.055 \times 10^{-27} \text{ erg-s}$$
>
> it follows that
>
> $$\Delta x \geq \frac{1.055 \times 10^{-27} \text{ erg-s}}{0.50 \text{ g-cm/s}}$$
> $$\geq 2.1 \times 10^{-27} \text{ cm}$$
> $$\geq 2.1 \times 10^{-19} \text{ Å}$$
>
> The restriction on the precision of the position imposed by the uncertainty principle is so much smaller than the finest measuring instrument we can use that it is, in fact, no restriction at all.
>
> **Example.** An electron with a de Broglie wavelength of 1.66 Å has a velocity of 4.38×10^8 cm/s and a momentum known to 1 per-

cent in precision. What is the limiting uncertainty-principle precision imposed on the position of this electron?

The momentum is

$$p = mv = 9.11 \times 10^{-28} \text{ g} \times 4.38 \times 10^8 \text{ cm/s}$$
$$= 3.99 \times 10^{-19} \text{ g} \times \text{cm/s}$$

One percent of this is

$$\Delta p = 3.99 \times 10^{-21} \text{ g} \times \text{cm/s}$$

Then

$$\Delta x \geq \frac{1.055 \times 10^{-27} \text{ erg} \times \text{s}}{3.99 \times 10^{-21} \text{ g} \times \text{cm/s}}$$
$$\geq 2.65 \times 10^{-7} \text{ cm}$$
$$\geq 26 \text{ Å}$$

This imprecision is considerable on an atomic scale, being almost thirty times as big as a hydrogen atom.

CLASSICAL AND QUANTUM PHYSICS

Classical physics was developed to account for events occurring on a macroscopic scale. Distances, masses, forces, and times were all of the size of direct human experience. The remarkable success of this physics in describing, with precision, so wide a variety of natural phenomena pointed to the correctness of a doctrine of *determinism*.

All natural events, in this view, occur according to natural laws. If we look carefully enough, we can discover these laws and shall be able to describe all natural processes. For example, if we know part of the trajectory of a mass, i.e., the position and motion of its travel over a short path, we can define the forces acting on it. With this knowledge we can predict its future motion and also determine its past motion. Its trajectory is determined from established physical laws. As another example, if we know, for one instant of time, the positions, velocities, and masses of each of a collection of bodies colliding with each other, we shall be able to calculate the positions and velocities at any future or past time.

On the atomic scale and for smaller particles also, classical physics, being inadequate, was replaced with quantum mechanics, which was more successful. There is a region between the atomic scale and ordinary dimensions where one might expect the two kinds of physics either to both apply or to contradict each other. The former is the case.

Bohr's *correspondence principle* states that the laws of quantum mechanics become the laws of classical mechanics as the scale, or size, of events changes from the very small to the macroscopic. For example, the quantum energies are discrete for an electron in an atom, as described originally by Bohr. This energy varies as $1/n^2$, where n is the principle quantum number. As in Fig. 5.24, as n increases in value, the difference between any level and the next higher level becomes smaller. In the limit of very large n, the allowed energy levels are so close together that differences are no longer detectable. As we measure the energy of a very-high-quantum-number electron, it appears to take on any of a continuous set of values. It is what it is measured to be: all energies are possible. The quantum condition and discrete energies vanish.

We have also seen that the de Broglie wavelength is not measurable on the macroscopic scale. The Heisenberg uncertainty principle presents no real restriction on our measuring both position and momentum as precisely as we can provided we measure macroscopic objects and normal events. Because quantum mechanics reduces to classical mechanics, and because quantum mechanics cannot be deduced from classical physics, quantum mechanics is the more general theory.

Determinism still can be said to operate in quantum mechanics, but it is of a different sort. Because of the uncertainty principle and the probability interpretation of Born, the information obtained is inherently statistical. One can determine the general properties of a large collection of particles, or a macroscopic system, as before. But for an individual small particle (an electron, for example), we can speak only in a statistical way about its behavior. Evidently it is futile to calculate a well-defined trajectory for a single electron. Its path is blurred because we cannot know with fine precision both where it is and how it moves.

PROBLEMS

5.1. A 100-W bulb is used to heat up 1,000 g of water. Assume that all the electric energy is converted to heat. How long will it be before the water, initially at 20°C, will be at its boiling point? Use the equivalence 1 cal = 4.18 joules.

5.2. Using the same amount of electric energy to plate-out alkali metals from molten alkali chlorides, will the mass of sodium so obtained be greater, equal to, or less than the mass of potassium?

5.3. What is the frequency of the copper x-ray of wavelength 1.54 Å?

5.4. A square sheet of aluminum foil measuring 15.0 by 15.0 cm has a mass of 1.72 g and a density of 2.70 g/cm³. What is the thickness of this foil?

5.5. Calculate the mass and estimate the diameter of a single aluminum atom.

5.6. Assume that a proton has a diameter of 10^{-13} cm and the hydrogen atom a Bohr diameter of 10^{-8} Å. If you wish to construct a scale model of this Bohr atom and plan to use a wooden ball 1 in. in diameter for the nucleus, how far from the nucleus must the electron orbit be placed?

5.7. For each state there are record low and high temperatures. Do you expect the maximum recorded temperatures in North Dakota, South Dakota, Nebraska, and Kansas to be greater or lower than those recorded in Florida, Alabama, Mississippi, and Louisiana? (This information is listed, e.g., in the "New York Times Encyclopedic Almanac.")

5.8. Do you think the coldest recorded temperature is lower in Nevada or in Rhode Island?

5.9. If you live near a large lake or at the coastline, and if temperatures are reported for communities both close to and farther from the water, where are the higher temperatures in the summer? the lower temperatures in the winter?

5.10. Suppose, after a snowfall, some ashes are dumped on part of the fresh snow. Which will melt faster in the sunlight—the clean snow or the dirty snow?

5.11. Why do people in tropical countries wear white clothing?

5.12. If your house had a black roof, what could you do to keep the house cooler in the summertime? Having done this, what effect would this have during the winter?

5.13. The sun's emission spectrum is presented in arbitrary intensity units for part of the visible spectrum, which contains the wavelength of maximum emission:

Å	Relative intensity
3800	0.46
4000	0.53
4200	0.72
4400	0.79
4500	0.83
4600	0.85
4700	0.86
4800	0.86
4900	0.85
5000	0.84
5200	0.79
5500	0.76
6000	0.72

Using the Wien displacement law, what is the estimated surface temperature of the sun?

5.14. A distant star has an emission spectrum with a maximum energy output at 1450 Å. What is its surface temperature?

5.15. The solar constant is the amount of solar energy received per minute per square centimeter by the earth, which has an average value of about 8.36 joules/min-cm² or 2.00 cal/min-cm². From the earth-to-sun distance of 1.50×10^{11} m and the sun's radius 1.61×10^8 m, it is possible to calculate the emission rate from the sun's surface; this is 6.3×10^7 W/m². Using the Stefan-Boltzmann formula, calculate the sun's surface temperature.

5.16. What wavelength and frequency must a photon have so that its mass is equal to the mass of an electron? Is this in the visible part of the spectrum? How does it compare with a typical x-ray photon?

A *chemical bond* exists between atoms within a molecule, holding the atoms together. In this chapter we shall apply the concepts and techniques discussed in Chaps. 4 and 5 to learn some things about these bonds. In Chap. 7 we shall look at the geometric shapes of both simple and complicated molecules, but in Chap. 6 we shall be concerned with the way in which atoms and molecules stay together.

STRUCTURAL FORMULAS AND VALENCE

The development of the periodic table involved a *chemical basis* for cataloging the elements into groups. This chemical basis is called the "valence," the chemists' measure of the bonding capacity of an atom. For our purposes we define the valence of an atom as the number of atoms of hydrogen or of chlorine that are bonded to the central atom.

The simplest way to *represent* a bond is by a line joining the atomic symbols, such as hydrogen chloride H—Cl. Hydrogen and chlorine each have a valence of 1. In the same way aluminum chloride $AlCl_3$ can be represented by

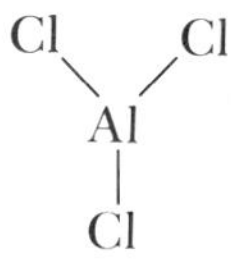

Cl   Cl
  \\   /
   Al
   |
   Cl

Because chlorine has a single valence, a single line is drawn from each Cl symbol; aluminum, with a valence of 3, has three lines from the symbol Al. The three chlorines are placed symmetrically about aluminum only for convenience; this diagram does not tell us anything about the molecule in three dimensions. In Chap. 7 we shall see how the atoms within a molecule are situated when viewed in three dimensions.

Methane CH_4 contains one carbon, with a valence of 4, and four hydrogens, each with a single valence. Its *structural formula* is

$$
\begin{array}{c}
\text{H} \\
| \\
\text{H}-\text{C}-\text{H} \\
| \\
\text{H}
\end{array}
$$

The number of bonds from each atom, represented by the number of lines from each symbol, is equal to the valence of that atom.

The structural formula contains more information than the basic molecular formula. For methane, the symbol CH_4 tells us there are one carbon and four hydrogen atoms in each molecule. The structural formula contains the additional information that each of the four hydrogen atoms is bonded directly to the carbon and that there are no direct hydrogen-hydrogen bonds.

IONIC COMPOUNDS

Compounds can be classified as electrolytes or nonelectrolytes, depending upon whether or not water solutions of these compounds conduct electricity (Chap. 5). Many electrolytes are also good conductors in the pure molten state. Such compounds are said to be *ionic;* that is, their atoms are present as cations and anions. It is convenient to look at the bonding within ionic compounds first.

The alkali metals and the alkaline earths combine with the halogens and oxygen to form ionic compounds. Some examples are KF, Na_2O, $CaCl_2$, and SrO. From our definition of valence and the observation that oxygen usually has a valence of 2, we can say that in these compounds potassium, sodium, fluorine, and chlorine each have a valence of 1; calcium, strontium, and oxygen each have a valence of 2. If we wish to use structural formulas, the four compounds can be represented as in Fig. 6.1. The number of lines drawn from each symbol is the valence of that element in these compounds.

$$\text{K}-\text{F}$$

$$\text{Na}-\text{O}-\text{Na}$$

$$\text{Cl}-\text{Ca}-\text{Cl}$$

$$\text{Sr}=\text{O}$$

Figure 6.1. Structural-formula representation of four ionic compounds. The number of lines drawn from each symbol represents the valence of that element.

Either in water solution or in the pure molten state, the alkali and alkaline earth metals, e.g., K and Sr, all form cations. The amount of positive electric charge on each cation can be determined by electrolysis and Faraday's second law. When 1 $\mathscr{F}$ of electricity, equal in charge to 1 mole of electrons, is passed through molten potassium fluoride, the product formed is 1 mole of potassium metal. When 1 $\mathscr{F}$ of electricity is passed through molten calcium chloride, only $\frac{1}{2}$ mole of calcium metal is formed. This means that one electron is enough to satisfy, or neutralize, the charge of potassium; however, since only one-half as much calcium forms with the same amount of electricity, the charge of a calcium ion requires twice as many electrons as potassium to be neutralized.

Our conclusion is that a potassium ion has a charge of $+1$ and a calcium ion has a charge of $+2$. The potassium cation is written K^+; the calcium ion is written Ca^{2+}. From the same electrolysis data we can determine the chloride ion to be Cl^-. This is consistent with the double positive charge on the calcium ion and the neutrality of calcium chloride $CaCl_2$. Similarly, the other ions can be written as Na^+, Sr^{2+}, F^-, and O^{2-}.

The two oxides Na_2O and SrO are both electrolytes, but in water each reacts to form a hydroxide: sodium hydroxide $NaOH$ and strontium hydroxide $Sr(OH)_2$. This happens because oxygen anion O^{2-} is not stable in water. The alkali and alkaline earth metal oxides are all *hygroscopic;* that is, they all have a high affinity for moisture in the air. Upon exposure they soon absorb enough of the air's humidity to form syrups and solutions.

In the pure solid state these four ionic compounds are represented in Fig. 6.2. In each case the net charge is zero because the positive and negative charges cancel each other. The valence is equal to the electric charge, which is simply the number of electrons gained or lost in forming the anions or cations.

In an ionic compound, the electrostatic force of attraction between oppositely charged ions gives rise to a bond between these ions. Many of the properties of such compounds are explained by the strong forces (coulombic) acting between electric charges over small distances. The energy needed to overcome these forces in converting solid sodium chloride to a vapor is reflected in the high boiling point of this compound—about 1400°C. This electrostatic model for bonding does not, however, account for all the properties of ionic compounds. Of particular interest is the magnitude of charge on each ion. How do sodium and potassium lose only one

$$K^+ \quad F^-$$

$$Na^+ \quad O^{2-} \quad Na^+$$

$$Cl^- \quad Ca^{2+} \quad Cl^-$$

$$Sr^{2+} \quad O^{2-}$$

Figure 6.2. Four ionic compounds shown as cations and anions. In each compound the positive charges exactly balance the negative charges; the net charge is zero.

electron, and calcium and strontium lose two? How is oxygen a divalent anion and fluoride and chloride univalent anions? We shall answer these questions in this chapter.

Table 6.1 lists the electron structures of the atoms and ions present in the four ionic compounds and, for a reason soon to be clear, the electron structures of the first four inert gases. The inert gases were all isolated and characterized during the years 1894–1898. They were considered to be remarkable because they do not form compounds readily with any elements. It was not until 1962 that Neil Bartlett (b. 1932) prepared the first compound, xenon hexafluoroplatinate $XePtF_6$.

From Table 6.1 we can see that in each inert gas all the subshells which have at least one electron are completely filled. Each s orbital subshell can contain two electrons—one with a spin of $+\frac{1}{2}$ and the other with $-\frac{1}{2}$. In the inert gases, all s orbital subshells contain two paired-off electrons with opposed spins. Similarly each

Table 6.1. Electron Structure of Some Inert Gases, and Atoms and Ions Forming Ionic Compounds

Atom or ion	Electron structure	
	Explicit form	**Abbreviated form**
He	$1s^2$	(He)
Ne	$1s^2 2s^2 2p^6$	(Ne)
Ar	$1s^2 2s^2 2p^6 3s^2 3p^6$	(Ar)
Kr	$1s^2 2s^2 2p^6 3s^2 3p^6 3d^{10} 4s^2 4p^6$	(Kr)
Na	$1s^2 2s^2 2p^6 3s$	(Ne)$3s$
K	$1s^2 2s^2 2p^6 3s^2 3p^6 4s$	(Ar)$4s$
Ca	$1s^2 2s^2 2p^6 3s^2 3p^6 4s^2$	(Ar)$4s^2$
Sr	$1s^2 2s^2 2p^6 3s^2 3p^6 3d^{10} 4s^2 4p^6 5s^2$	(Kr)$5s^2$
Na^+	$1s^2 2s^2 2p^6$	(Ne)
K^+	$1s^2 2s^2 2p^6 3s^2 3p^6$	(Ar)
Ca^{2+}	$1s^2 2s^2 2p^6 3s^2 3p^6$	(Ar)
Sr^{2+}	$1s^2 2s^2 2p^6 3s^2 3p^6 3d^{10} 4s^2 4p^6$	(Kr)
O	$1s^2 2s^2 2p^4$	(He)$2s^2 2p^4$
F	$1s^2 2s^2 2p^5$	(He)$2s^2 2p^5$
Cl	$1s^2 2s^2 2p^6 3s^2 3p^5$	(Ne)$3s^2 3p^5$
O^{2-}	$1s^2 2s^2 2p^6$	(Ne)
F^-	$1s^2 2s^2 2p^6$	(Ne)
Cl^-	$1s^2 2s^2 2p^6 3s^2 3p^6$	(Ar)

p orbital subshell can contain six electrons—three of one spin and three with the opposite spin. In the inert gases, the *p* orbital subshells all contain the full count of six paired-off electrons. In a like manner the *d* orbital subshells can contain 10 paired-off electrons; krypton has a completed 3*d* subshell. The important features common to all inert gases are the completely filled subshells and, especially, the outermost eight electrons—two in the *s* orbital and six in the *p* orbitals.

Looking at sodium and potassium, we see they both have the electron structure of an inert gas plus an additional single *s* orbital electron. This is characteristic of the alkali metals. The alkaline earth metals have two *s* orbital electrons in addition to those found in an inert gas. When the ions are formed, sodium loses a single electron and has the electron structure of the inert gas that appears just before it in the periodic table, neon. The nuclear charge is not changed, and the ion has a single positive charge. In the same way, potassium loses one electron, and calcium and strontium each two electrons. The remaining electrons in the atom have a structure identical to that of the inert gas that appears just before the metal in the periodic table. (Refer to the periodic table inside the back cover of this book.)

However, when oxygen, fluorine, and chlorine gain one or two electrons, they have the structure of the inert gases following them in the periodic table. The rule follows immediately: An atom loses or gains just enough electrons to form an ion whose electron structure is that of the nearest inert gas. In the ionic compound which is formed, all the electrons are paired off. In the great majority of all known compounds the total number of electrons is even and the electrons are paired off by spins. Molecules which have an odd number of electrons and thus at least one unpaired electron are ordinarily quite reactive and unstable.

From an examination of the electron structures of the atoms and ions in Table 6.1, we can see that the chemical reaction of atoms forming an ionic compound is simply the transfer of one or more electrons from the metal to the nonmetal (Fig. 6.3). This electron transfer is according to the rules that the subshells involved are either completely emptied (in the case of the metals) or completely filled (in the case of the nonmetals). The only electrons concerned in the chemical reaction are those in the outermost subshells; these electrons are called the "valence electrons." Other electrons are not affected by the reaction.

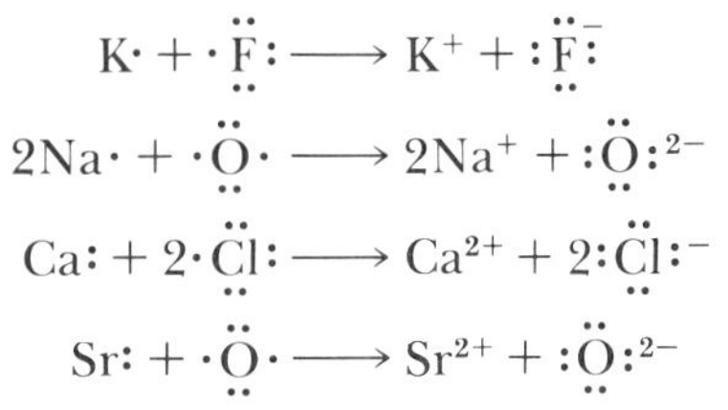

$$\text{K} \cdot + \cdot \overset{\displaystyle ..}{\underset{\displaystyle ..}{\text{F}}} : \longrightarrow \text{K}^+ + : \overset{\displaystyle ..}{\underset{\displaystyle ..}{\text{F}}} :^-$$

$$2\text{Na} \cdot + \cdot \overset{\displaystyle ..}{\underset{\displaystyle ..}{\text{O}}} \cdot \longrightarrow 2\text{Na}^+ + : \overset{\displaystyle ..}{\underset{\displaystyle ..}{\text{O}}} :^{2-}$$

$$\text{Ca} : + 2 \cdot \overset{\displaystyle ..}{\underset{\displaystyle ..}{\text{Cl}}} : \longrightarrow \text{Ca}^{2+} + 2 : \overset{\displaystyle ..}{\underset{\displaystyle ..}{\text{Cl}}} :^-$$

$$\text{Sr} : + \cdot \overset{\displaystyle ..}{\underset{\displaystyle ..}{\text{O}}} \cdot \longrightarrow \text{Sr}^{2+} + : \overset{\displaystyle ..}{\underset{\displaystyle ..}{\text{O}}} :^{2-}$$

Figure 6.3. Reaction between two elements to form an ionic compound.

There is one valence electron in the alkali metals potassium and sodium; there are two in the alkaline earths calcium and strontium, seven in the halogens fluorine and chlorine, and six in oxygen. The dots about each atom or ion represent valence electrons.

NONIONIC COMPOUNDS

Of the more than three million compounds which have been isolated, synthesized, and characterized, most are not ionic. Let us look at four such compounds to see whether or not the same general rule of completed subshells is valid.

The fluorine molecule is diatomic: F_2. Fluorine is the most reactive element. For this reason there are few materials which can be used as containers for elemental fluorine. Although nonionic, it reacts vigorously with water, producing oxygen and hydrofluoric acid HF, which is a fair electrolyte. Imagine the formation of the diatomic molecule from two atoms of fluorine F, each with the electron structure $1s^2 2s^2 2p^5$. Only the $2s$ and $2p$ electrons are involved in the formation (Fig. 6.4a). When the two atoms are brought together, the odd unpaired valence electron in each is paired off. This pair, shared by both fluorine atoms, forms the bond that holds the molecule together. The effect of sharing is that each fluorine atom has eight electrons in its second shell. In the structural formula, the bond is represented by a line drawn between the atomic symbols.

Oxygen atoms have six electrons in their second shell, two less than the maximum number. In Fig. 6.4b, the bond between the two atoms is formed by sharing two pairs of electrons. Each oxygen atom in the diatomic molecule has eight second-shell electrons.

At room temperature hydrogen chloride is a nonionic gas which is an electrolyte, since it breaks up into ions, or *ionizes*, in

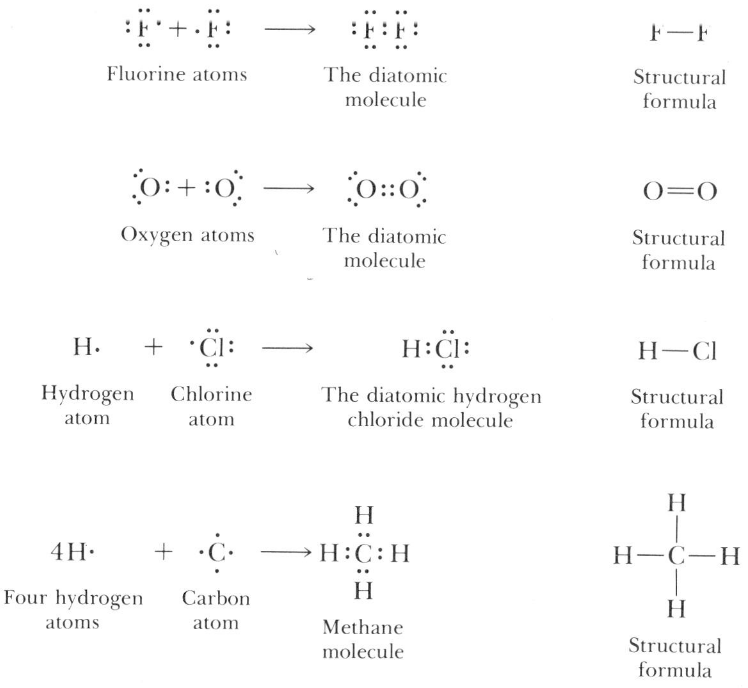

$$:\ddot{F}\cdot + \cdot\ddot{F}: \longrightarrow :\ddot{F}:\ddot{F}: \qquad\qquad F\!-\!F$$

Fluorine atoms   The diatomic   Structural
molecule   formula

$$:\ddot{O}: + :\ddot{O}: \longrightarrow :\ddot{O}::\ddot{O}: \qquad\qquad O\!=\!O$$

Oxygen atoms   The diatomic   Structural
molecule   formula

$$H\cdot \quad + \quad \cdot\ddot{Cl}: \longrightarrow H:\ddot{Cl}: \qquad\qquad H\!-\!Cl$$

Hydrogen   Chlorine   The diatomic hydrogen   Structural
atom   atom   chloride molecule   formula

$$
4H\cdot \quad + \quad \cdot\dot{C}\cdot \longrightarrow
\begin{matrix} & H & \\ & \overset{\cdot}{\underset{\cdot\cdot}{}} & \\ H: & C & :H \\ & H & \end{matrix}
\qquad\qquad
\begin{matrix} & H & \\ & | & \\ H\!-\! & C & \!-\!H \\ & | & \\ & H & \end{matrix}
$$

Four hydrogen   Carbon   Methane   Structural
atoms   atom   molecule   formula

Figure 6.4. Sharing of electrons in nonionic compounds.

water. The hydrogen atom has a single $1s$ electron; the first shell, however, can contain two electrons. A chlorine atom has seven electrons in, and needs one more electron to complete, its second shell (Fig. 6.4c).

Methane can be formed from one carbon and four hydrogen atoms as in Fig. 6.4d. In the methane molecule each hydrogen has two electrons in its first shell and each carbon has eight electrons in its second shell.

Ionic compounds are formed by a transfer of one or more electrons from one atom to another. In nonionic compounds the bond is made by a pair of electrons, one contributed by each atom and *shared* by both atoms; this is a *covalent bond*. Most of the compounds known today contain carbon. Because these compounds were first thought to be synthesized only within living organisms, they are called "organic compounds." Most organic compounds are nonionic, that is, covalent. Many ionic bonds have some covalent character, too. On page 278 we shall see how to estimate the relative ionic and covalent contributions to a chemical bond.

The modern concept of the covalent bond is based on quantum mechanics; unlike the electrostatic ionic bond, the covalent bond has no simple classical basis. Although the paired-electron idea is more clearly represented in the dot-type formula, it is more convenient, most of the time, to use the simpler formula and model in which a line stands for the pair of electrons making the bond.

HYDROCARBONS

There is a large group of organic molecules which contain only the elements carbon and hydrogen; these are called the "hydrocarbons."

SATURATED HYDROCARBONS

Within this group of hydrocarbons there are series of related molecules, one of which is shown just below. By inspection we can see that the number of hydrogen atoms is two more than twice the number of carbon atoms in each molecule of the series. The general formula is C_nH_{2n+2} where $n = 1, 2, 3, \ldots$.

Methane	CH_4
Ethane	C_2H_6
Propane	C_3H_8
Butane	C_4H_{10}
Pentane	C_5H_{12}
Hexane	C_6H_{14}
Heptane	C_7H_{16}
Octane	C_8H_{18}
. . .	. . .

At room temperature the first four members of this series are gases; from pentane ($n = 5$) through heptadecane ($n = 17$), they are liquid; and octadecane ($n = 18$) and higher members are waxy solids. Because the solids have little affinity for other molecules under ordinary circumstances, they are slow to react. For this reason the solids are collectively known as "paraffin waxes" (from the Latin, *parum*, "little," and *affinis*, "connected or linked"). The entire series is called the "paraffin series."

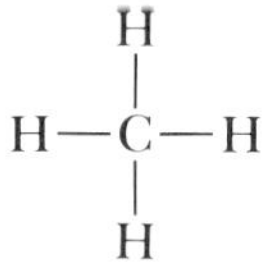

H
|
H—C—H
|
H

Methane CH$_4$

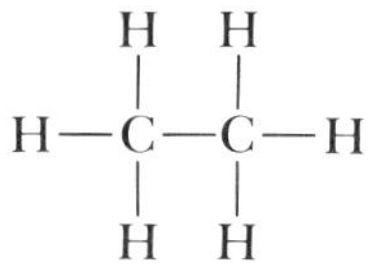

H H
| |
H—C—C—H
| |
H H

Ethane C$_2$H$_6$,
or CH$_3$CH$_3$

H H H
| | |
H—C—C—C—H
| | |
H H H

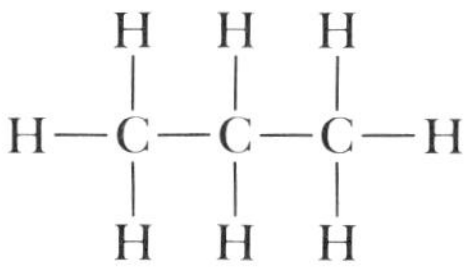

Propane C$_3$H$_8$,
or CH$_3$CH$_2$CH$_3$

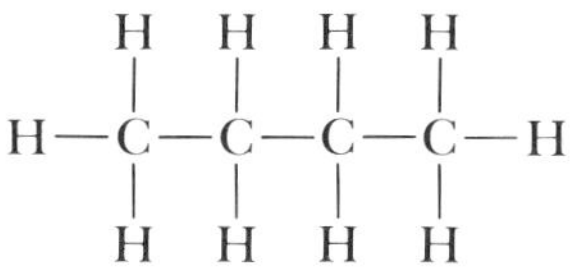

H H H H
| | | |
H—C—C—C—C—H
| | | |
H H H H

Butane C$_4$H$_{10}$,
or CH$_3$CH$_2$CH$_2$CH$_3$

Figure 6.5. Structural formulas for four paraffin hydrocarbons; each carbon has a valence of 4, and each hydrogen has a valence of 1.

If we apply simple valence theory to the paraffin molecules and assign hydrogen a valence of +1, carbon in methane should have a valence of −4; in ethane, −3; and in the others, noninteger valences. The paraffin molecules are all quite similar to each other. (Separating the paraffins is fairly difficult and is a principal operation of the petroleum industry. Octane, the main constituent of gasoline, is separated from the other hydrocarbons by a complex distillation procedure.) The simplest way to satisfy the necessary conditions (a valence of 4 for carbon and 1 for hydrogen, and a gradually changing set of properties) is for carbon atoms to be joined to each other (Fig. 6.5).

Each paraffin molecule has a compact formula which indicates the details of the structural formula (Fig. 6.5). For example, propane C$_3$H$_8$ can be written as CH$_3$CH$_2$CH$_3$. This form can be set on a single line and, like the structural formula, indicates how the eight hydrogens are connected to the three carbon atoms. In going from one member of the paraffin series to the next, a CH$_2$ unit is added. As the paraffin molecule becomes larger, each CH$_2$ added constitutes a relatively less significant change in terms of the weight of the entire molecule. The difference in properties between one and the next larger paraffin, therefore, is relatively small.

Paraffins can be very large. One of the very largest paraffins is polyethylene, the common synthetic material of which some plastics are made. Polyethylene can have as many as a half-million carbon atoms in a chain. This property—carbon bonded to carbon—is one reason why carbon is found in so many compounds. In the case of polyethylene, the difference in properties between polyethylenes of slightly different molecular weights is so slight that it is impossible to obtain a sample of this substance that is entirely of one size or molecular weight. A sample whose range of molecular weights is spread over 1 percent of an average molecular size is considered rather pure.

From Fig. 6.5 it is evident that no more hydrogens can be added to these paraffin hydrocarbons without changing the valences of carbon or hydrogen. Paraffin compounds are said to be *saturated* for this reason.

UNSATURATED HYDROCARBONS

Another hydrocarbon series is the *olefins*, of which ethylene C_2H_4 is the first member; the general formula for the olefins is C_nH_{2n}. A third hydrocarbon series is the *acetylenes*, of which acetylene C_2H_2 is the first member; the general formula for the acetylenes is C_nH_{2n-2}.

When acetylene is exposed to hydrogen, under the proper conditions, it will form ethylene:

$$C_2H_2 \ + \ H_2 \ \longrightarrow \ C_2H_4$$
Acetylene   Hydrogen   Ethylene

Ethylene can combine with more hydrogen to form ethane:

$$C_2H_4 \ + \ H_2 \ \longrightarrow C_2H_6$$
Ethylene   Hydrogen   Ethane

Because acetylene and ethylene can react with and take up more hydrogen, these compounds are said to be *unsaturated* molecules.

The reactions in which hydrogen is added to unsaturated molecules are called "hydrogenations." Chemists picture the hydrogenation of acetylene and ethylene as

$$CH\equiv CH + \ H_2 \ \longrightarrow CH_2=CH_2$$
Acetylene   Hydrogen   Ethylene

and

$$CH_2=CH_2 + \ H_2 \ \longrightarrow CH_3CH_3$$
Ethylene   Hydrogen   Ethane

The two unsaturated molecules in Fig. 6.6 have structural formulas with more than just a single bond between the carbons. Ethylene, as well as all members of the olefin series, contains a *double bond*. Acetylenes all contain a *triple bond*. This way of representing the extent of unsaturation maintains the convenient carbon and hydrogen valences.

In ethylene each carbon shares four electrons (two pairs and two bonds) with its carbon neighbor and two electrons with each of the two neighboring hydrogens. Each carbon has a total of eight electrons in its outermost, or second, shell; thus each carbon has the neon electron structure. Each hydrogen has two electrons in its outermost, or first, shell; this is the helium electron structure.

For acetylene, a carbon atom shares six electrons (three pairs and three bonds) with the other carbon and two electrons with the closer hydrogen. Again, each carbon atom has its second shell filled.

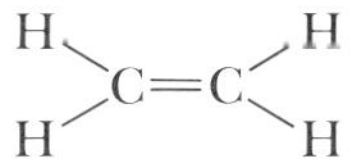

Ethylene C_2H_4, or $CH_2{=}CH_2$

Acetylene
C_2H_2, or $CH{\equiv}CH$

Figure 6.6. Structural formulas for ethylene and acetylene; each carbon has a valence of 4, and each hydrogen has a valence of 1.

ELECTRONEGATIVITY

Whether a bond between two atoms is ionic or covalent depends upon where the two paired-off electrons forming that bond are located. If each atom contributes an electron so that the pair is more or less equally shared, the bond is covalent. But if one atom loses an electron so that the pair is very far from it and very close to the other atom, the bond is ionic.

Because the position of the shared electron pair is hardly ever exactly at the midpoint between the two bonded atoms, few bonds are completely covalent. And because a removed electron never fully escapes the influence of a cation, none are entirely ionic. Chemists have devised a quantitative method of predicting the relative covalent and ionic natures of any bond from the two atoms at the bond.

The *electronegativity* of a particular atom is a measure of the probability that an electron will be located near that atom. The electronegativity can be estimated experimentally by measuring (1) the energy needed to remove an electron from an atom and (2) the energy released when an electron is added to an atom. The electronegativities of some atoms are given in Table 6.2. ·

Fluorine is the most electronegative of the elements. Whenever a bond is formed between fluorine and any other atom, the bond electrons will be closer to the fluorine than to the other atom.

Table 6.2. Electronegativities of Some Atoms

Atom	Electronegativity
F	4.0
O	3.5
Cl	3.0
N	3.0
Br	2.8
C	2.5
I	2.5
S	2.5
H	2.1
Ag	1.9
Al	1.5
Mg	1.2
Li	1.0
Na	0.9
K	0.8
Cs	0.7

Sodium has a low electronegativity and will form bonds with most other atoms in which the bond electrons will be closer to the other atom. If two atoms, for example, carbon and iodine, have approximately the same electronegativities, the electrons will show a preference for neither. The difference in atomic electronegativities is a measure of the unequal spatial distribution of electrons in a bond formed between those atoms (Table 6.3).

The designation *covalent* refers to a bond with a small difference in electronegativity between the two atoms; that is, the electrons are distributed more or less equally between the two atoms. *Polar* means there is a significant difference in electronegativities between the two atoms; for example, the electrons are distributed closer to the oxygen in O—H and to the fluorine in C—F. Polar molecules have an unequal electric charge distribution, so that one end of the polar molecule is negatively charged and the other end

Table 6.3. Bond Types and Electronegativities

Bonds	Electronegativity difference	Type
C—H	$2.5 - 2.1 = 0.4$	Covalent
C—Cl	$3.0 - 2.5 = 0.5$	Covalent
O—H	$3.5 - 2.1 = 1.4$	Polar
C—F	$4.0 - 2.5 = 1.5$	Polar
Na—F	$4.0 - 0.9 = 3.1$	Ionic

is positively charged. The molecule is an *electric dipole*. Water, with two O—H bonds, is the most common polar molecule; most of its important properties depend upon its polar character.

Ionic means that the electronegativity difference between two atoms is sufficiently large to effect a complete transfer of the electron; in our example, the electron from the sodium atom is transferred to the fluorine atom, producing a positively charged sodium ion Na^+ and a negatively charged fluoride ion F^-.

POLAR LIQUIDS AND SOLUBILITY

When a salt such as sodium chloride dissolves in water, a sodium ion Na^+ is immediately surrounded by water molecules oriented so that the negative ends of these polar molecules are directed toward the positive sodium ion. Similarly, the chloride ion Cl^- is surrounded by water molecules oriented so that the positive ends of the molecules point toward the Cl^- ion (Fig. 6.7). Salts are much less soluble in nonpolar solvents because the *solvation* of ions by

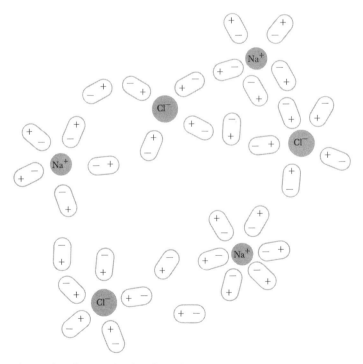

Figure 6.7. Solvation of Na^+ and Cl^-, by polar water molecules.

polar molecules is an important step in the overall solution process and is absent in nonpolar solvents.

The *solubility* of a substance is the maximum amount of that substance which will dissolve in a specified solvent at a given temperature. The solubility of an ionic compound depends on the competition between two interactions. One is the electrostatic force holding the ionic compound together; the other is the attractive force between the electric dipole of the polar solvent molecule and the compound's cations and anions. When the former is greater, the solubility is less; when the latter is greater, the solubility is more.

The ionic compounds made from the alkali metals and the halides are all quite soluble in water. The coulombic electrostatic force contributing to the bonding between cations and anions depends on the product of two single charges. In comparison, a compound such as calcium sulfate $CaSO_4$ or $Ca^{2+}SO_4^{2-}$ is very insoluble in water, which can be attributed, in part, to the larger coulombic attractive force between two doubly charged ions.

The compound silver chloride is also water-insoluble. From Table 6.3 the electronegativity difference between the two atoms is 1.1 $(3.0 - 1.9)$. Silver chloride is polar rather than ionic and is not expected to behave as sodium chloride does in water.

Water H_2O is a liquid at room temperature while hydrogen sulfide H_2S is a gas. The main reason for this difference is that the electronegativity of oxygen is 3.5 while that of sulfur is 2.5. Thus, the electronegativity difference of the O—H bond is 1.4, and that of S—H is 0.4. Therefore, water is a polar molecule. There is a strong interaction between the negative end of one water molecule and the positive end of another water molecule. Water is said to consist of associations, or clusters, of molecules and is an *associated liquid*. This effect is absent in hydrogen sulfide. Each H_2S molecule is more independent of its neighbors; for this reason, H_2S is a gas at room temperature.

ACIDS, BASES, AND SALTS

The molecule hydrogen chloride HCl is polar but dissolves in water to form the hydrated hydrogen ion, usually written as $H^+(H_2O)$, H_3O^+, or $H^+(aq)$ (the last for *aqueous*). In solution this HCl is hydrochloric acid.

An *acid* can be defined as a substance which, when dissolved in water, will yield a hydrogen ion or proton H^+. A *base*, for example, sodium hydroxide NaOH, when dissolved in water, yields an ion, in this case, the hydroxide ion OH^-, which combines with a hydrogen ion and removes it from solution. In this example

$$OH^- + H^+ \longrightarrow H_2O$$

A more general, yet simpler, definition for an acid and a base was proposed in 1923 by Johannes N. Brönsted (1879–1947) and Thomas M. Lowry (1874–1936). The Brönsted-Lowry *acid* is a *proton donor;* a *base* is a *proton acceptor.*

When a base reacts with an acid in water, we have these steps.

The solution of the base:

$$NaOH \longrightarrow Na^+(aq) + OH^- \tag{6.1}$$

The solution of the acid:

$$HCl \longrightarrow H^+(aq) + Cl^- \tag{6.2}$$

The reaction

$$Na^+(aq) + OH^- + H^+(aq) + Cl^- \longrightarrow$$
$$H_2O + Na^+(aq) + Cl^- \tag{6.3}$$

or

$$OH^- + H^+(aq) \longrightarrow H_2O$$

The *neutralization* of an acid by a base (or a base by an acid) is simply the combination of hydrogen ions with the base; the unreacted ions (here, Na^+ and Cl^-) form a *salt.* In general, the reaction between an acid and a base produces a salt. The inverse is not general; salts can be made in ways other than by neutralization.

The *strength* of an acid or base is a measure of how much of it *ionizes*, or dissociates into ions, when dissolved in a solvent. Strong acids, for example, hydrochloric acid HCl, sulfuric acid H_2SO_4, and nitric acid HNO_3, are almost completely dissociated in water. Weak acids, for example, acetic acid (the principal component of vinegar), citric acid (found in citrus fruits), and ascorbic acid (vitamin C) are found in water as mostly the undissociated, or un-ionized, neutral molecule.

All the hydroxides of the alkali metals and the alkaline earths, for example, NaOH and Ca(OH)$_2$, are strong bases in water because they completely dissociate to provide the hydroxide ion OH$^-$. This, as do all bases, combines with the acid proton H$^+$ to form water. Ammonia NH$_3$ in water forms ammonium hydroxide NH$_4$OH, which is not completely ionized. Ammonia is a weaker base. Another common base is the carbonate ion CO$_3{}^{2-}$; it accepts a proton to become the bicarbonate ion HCO$_3{}^-$, which can act as a base under some conditions. After accepting a second proton, HCO$_3{}^-$ becomes carbonic acid H$_2$CO$_3$.

THE IONIZATION OF ACIDS, BASES, AND WATER

Acetic acid C$_2$H$_4$O$_2$ can donate a single proton from each molecule. Therefore, knowing nothing more about its structure, we can write the formula as HC$_2$H$_3$O$_2$. In water the ionization can be represented as a proton transfer:

$$HC_2H_3O_2 + H_2O \longrightarrow H_3O^+ + C_2H_3O_2{}^-$$

Acetic acid	Water	Hydronium ion	Acetate ion
(acid)	(base)		

Acetic acid, donating a proton, is the acid; water, accepting the proton, is the base. The products are the hydrated proton, called the "hydronium ion" H$_3$O$^+$, and the acetate ion C$_2$H$_3$O$_2{}^-$. This ionization involves breaking the bond joining the proton to the rest of the acid molecule, which is the acetate ion, and forming a bond linking the proton to a water molecule, thus producing a hydronium ion.

Because ions conduct electricity and undissociated molecules do not, it is possible to know just how many ions are present in a solution by measuring the electrical conductivity of that solution. If a solution at 25°C contains 1.000 mole/liter of acetic acid, there is 0.996 mole/liter of the undissociated molecule and 0.0042 mole/liter each of the hydronium and acetate ions. Only 0.42 percent of the acid is ionized. In comparison, a solution of hydrochloric acid containing 1.000 mole/liter is completely ionized. The hydronium and chloride ions have concentrations of 1.000

mole/liter, and there is no undissociated hydrogen chloride HCl.

A more dilute acetic acid solution, with a concentration of 0.00100 mole/liter, will be found to have hydronium and acetate ions each at a concentration of 1.24×10^{-4} moles/liter. The undissociated acid concentration is 8.76×10^{-4} moles/liter. At this concentration the acid is 12.4 percent ionized. However, a hydrochloric acid solution also at 0.00100 mole/liter is completely ionized.

The difference between hydrochloric and acetic acids, both at the same concentration, is that the stronger acid has more hydronium ions in solution. At 1.000 mole/liter there are, respectively, 1.000 and 0.0042 mole/liter of H_3O^+; these numbers are in the ratio of 240:1. At 0.00100 mole/liter, the two acids have 0.00100 and 0.000124 mole/liter of H_3O^+, a ratio of 8:1. Comparing the two acids at the same overall concentrations, we find the stronger acid attacks metals more vigorously. This property depends on the hydronium-ion concentration.

Because the extent, or degree, of ionization of acetic acid increases as the concentration decreases (in our two examples, from 0.42 to 12.4 percent), a satisfactory description of the ionization process must include the reverse process:

$$H_3O^+ \quad + C_2H_3O_2^- \longrightarrow HC_2H_3O_2 + H_2O$$

Hydronium ion	Acetate ion	Acetic acid	Water
(acid)	(base)		

Here a proton is transferred from the hydronium ion to the acetate ion. The hydronium ion acts as the acid, the acetate ion as the base.

We can combine the two processes into a single equation, with two single-barbed arrows indicating the forward (left-to-right) and reverse (right-to-left) reactions:

$$HC_2H_3O_2 + H_2O \rightleftharpoons H_3O^+ + C_2H_3O_2^-$$

Acid	Base	Acid	Base

ACID-BASE STRENGTHS. The extent of ionization, which governs the hydronium ion concentration, depends on how many protons are associated with water as H_3O^+ and how many are associated with acetate ions as $HC_2H_3O_2$. Ionization is a comparison of the proton-accepting, or basic, properties of the two bases, water and the acetate ion. The negative portion of a weak acid such as acetic acid is a strong base; in this case, the acetate ion is the strong base. In comparison, the chloride ion, formed when hydrochloric acid is ionized, accepts no protons and is a weak base; thus hydrochloric acid is a strong acid.

The *forward* reaction depends on encounters between water and undissociated acid molecules. In these reactions, water is always in excess amounts. A liter of water at 25°C has a mass of 997.0 g and contains 55.34 moles of water. The number of moles per liter does not significantly change on diluting an acetic acid solution that has less than 1 mole of acetic acid present. The ionization rate, which depends on encounters between acid molecules and water, will be less at lower concentrations of acid because there is less undissociated acid in each liter of solution.

The *reverse* reaction depends on encounters between hydronium and acetate ions. Both of these are fewer at lower concentrations. This reverse reaction is more affected by the dilution than the forward process. We should expect the reverse reaction to be relatively less important at lower concentrations. Relatively more of the weak acid is ionized at low concentrations. (Strong acids are always completely ionized in aqueous solution; their ionization reactions do not include a reverse step.)

When ammonia gas NH_3 dissolves in water, the result is an alkaline, or basic, solution. The overall ionization process can be represented by

$$NH_3 \; + \; H_2O \; \rightleftharpoons \; NH_4^+ \; + \; OH^-$$

Ammonia   Water   Ammonium ion   Hydroxide ion
(base)   (acid)   (acid)   (base)

In the forward process, water is the acid and ammonia is the base. The bond joining a proton to the rest of a water molecule is broken, and a bond linking this proton to ammonia is formed. The formed and broken bonds of the forward process are, respectively, undone and remade during the reverse process.

Because ammonia is only partially converted to the ammonium and hydroxide ions in water, it is a weak base. Like weak acids, it becomes more converted as the solution becomes more dilute. Strong bases, e.g., sodium hydroxide, are completely ionized at all concentrations.

It is not the total acid concentration, but rather the hydronium-ion concentration, which is important in most cases. The concentration of a substance is represented by placing the symbol or formula for that substance in brackets. The hydronium-ion concentration is $[H_3O^+]$, and the units usually are moles per liter. (Square brackets generally are used to mean concentration of a component in a reaction mixture.)

Because hydronium-ion concentrations can have any of a wide range of values, scientists often find it convenient to use a logarithmic scale to express these concentrations. The pH is defined by

$$pH = -\log [H_3O^+]$$

Example. What is the pH of a 0.00100 mole/liter solution of hydrochloric acid? of acetic acid? For hydrochloric acid:

$$\begin{aligned} [H_3O^+] &= 0.00100 \text{ mole/liter} \\ [H_3O^+] &= 1.00 \times 10^{-3} \text{ mole/liter} \\ \log [H_3O^+] &= \log [1.00] + \log [10^{-3}] \\ &= 0 + [-3.00] \\ &= -3.00 \\ pH &= 3.00 \end{aligned}$$

For acetic acid:

$$\begin{aligned} [H_3O^+] &= 1.24 \times 10^{-4} \\ \log [H_3O^+] &= \log [1.24] + \log [10^{-4}] \\ &= 0.09 - 4.00 \\ &= -3.91 \\ pH &= 3.91 \end{aligned}$$

Example. A sample of beer has a pH of 4.5. What is the hydronium-ion concentration?

$$\begin{aligned} pH &= 4.5 \\ -\log [H_3O^+] &= 4.5 \\ \log [H_3O^+] &= -4.5 \\ &= 0.5 - 5.0 \\ [H_3O^+] &= 3.2 \times 10^{-5} \text{ moles/liter} \end{aligned}$$

Water serves as a base in the ionization of an acid and as an acid when a base ionizes. Any substance which is either a base or an acid, depending on the circumstances, is said to be *amphiprotic.* Water also ionizes when neither another acid nor another base is present; the reaction, which is referred to as the *autoionization of water,* is

$$H_2O + H_2O \rightleftharpoons H_3O^+ + OH^-$$

Base   Acid   Acid   Base

     In pure water, the hydronium-ion concentration exactly equals the hydroxide-ion concentration since these ions are formed in equal numbers by the same reaction. At 25°C, these concentrations are 1.004×10^{-7} moles/liter; the pH is 6.998. At 37°C, the normal human body temperature, the ionic concentrations in pure water are both 1.547×10^{-7} moles/liter; the pH is 6.810. Since there are more ions at the higher temperature, we conclude that water ionizes to a greater extent when the temperature is increased. Ordinarily it is not practical to measure the pH more precisely than to three significant figures. At 25°C the pH of pure water is 7.00, rounded off to three figures.

     Although we shall reexamine the entire subject of ionization in Chap. 9, one important aspect of hydronium- and hydroxide-ion concentrations will be introduced now. The rate at which water ionizes is the rate at which a proton is transferred from one water molecule to another. The ionization rate depends on how many water molecules are present in a given volume of liquid. The number of water molecules present in 1 liter of dilute solution is very nearly equal to the number of water molecules in 1 liter of pure water. The number of water molecules per liter, therefore, is essentially constant provided the solution concentration is not too high.

     The ionization rate also depends on how much energy is available to break a hydrogen-oxygen bond in water. This is another way of saying that the ionization rate depends on the temperature. If the temperature is kept constant, the ionization rates of water molecules in pure water and in dilute solutions are also constant. In addition, these processes have the same rate.

     The reverse process, the transfer of a proton from a hydronium ion to a hydroxide ion, depends on encounters between a hydronium cation and a hydroxide anion. The greater the concentration of either ion, the more frequent these encounters and the greater the rate of this reverse process. This reverse rate depends on the product of the ionic concentrations $[H_3O^+][OH^-]$.

     Unless some chemical reaction is occurring, the pH of pure water or of some acid or base solution does not change with time. The hydronium-ion concentration, in each case, does not change. If the hydronium-ion concentration is constant, the rate at which it is formed by ionization is exactly balanced by the rate at which it is removed through the reverse process.

The forward ionization process always has a steady value for both pure water and dilute solutions at a given temperature. Therefore the reverse rate and the product $[H_3O^+][OH^-]$ also have a steady value in pure water and dilute solutions at the given temperature.

For the special case of pure water at 25°C, each ion has the concentration of 1.004×10^{-7} moles/liter. The product is

$$[H_3O^+][OH^-] = 1.008 \times 10^{-14} \text{ moles}^2/\text{liter}^2$$

This product has the same value at 25°C for all dilute aqueous solutions. The consequence of this is important: The higher the concentration of one of these ions, the lower must be the other concentration.

Any solution at 25°C which has a pH of 7.00 has an equal number of the two ions. It is neither acidic nor basic and is said to be *neutral*. An acidic solution has more hydronium than hydroxide ions; its pH is less than 7.00. The more acidic a solution is, the lower the pH. With more hydroxide than hydronium ions, the concentration of the latter is less than 1.004×10^{-7} moles/liter and the alkaline solution pH is greater than 7.00.

THE MEASUREMENT OF pH

In an aqueous solution of a strong acid, the transferable protons are all found in the hydronium ions H_3O^+. In a weak acid solution, these protons are distributed between the original undissociated parent acid molecules and the hydronium ions. We distinguish between the *total acidity*, the combined concentration of all the proton-donating molecules in solution, and the hydronium-ion concentration. When the former exceeds the latter, we have a weak acid; when the two concentrations are the same, the acid is strong and completely dissociated.

The standard method of measuring just the hydronium-ion concentration, or pH, of a solution uses a voltaic cell made from special electrodes placed in that solution. The observed voltage depends on the pH; the latter can be calculated directly from the former.

pH INDICATORS. A number of natural plant juices and synthetic dyes have colors which change with different pH values.
Litmus, for example, is a dye extracted from lichens. It is red if the
pH is less than 4.5 but blue if it is greater than 8.3. Berry, pome-
granate, and red-cabbage juice can be made to change color by
the addition of a little weak base, e.g., ammonia water or sodium
carbonate, and change back to the original color with some weak
acid, e.g., lemon juice or vinegar. The synthetic and natural dyes
with this changing-color property are called "indicators." Each in-
dicator has its own characteristic pH color behavior. These are
often used, too, for pH determination.

INTERATOMIC DISTANCES

When discussing the wave nature of light, we mentioned that if the
wavelength of the light, the grating-to-screen distance, and the
diffraction angle were known, the grating slit spacing can be
calculated. A crystalline solid has a regular geometrical appearance; that is, for a given compound, the angles between the crystalline faces are always the same. We presume that this is a consequence of a regular array of atoms within the crystal.

Figure 5.26 is a model of a crystalline solid where the circles
represent the positions of the various atoms. A number of lines are
drawn through the atoms in such a way that along any line the
space between the atoms is regular. Each row of regularly spaced
atoms can serve as a diffraction grating. There are many such rows,
each with its distinct spacing. These spacings are the distances
between parallel planes as in Bragg's diffraction equation. A Bragg
plane is drawn through each regularly spaced atom on any line and
perpendicular to that line. Since the atoms are evenly spaced along
any line, the Bragg planes are parallel and evenly spaced, too. For
the best interference effects, a wavelength comparable in size to
these spacings should be used. X-rays are of this size.

One procedure is to grow a large crystal of the compound of
interest and rotate this crystal through a complete turn of 360°
while exposing it to a beam of monochromatic x-rays. At various
angles of rotation, the several atomic diffraction gratings will, in
turn, be at the proper angles to the x-ray beam for constructive
interference of the waves. From the diffraction angles and the

overall geometry of the diffraction experiment, we can calculate the interatomic distances.

Alternatively, instead of one large crystal, a powder, which is a collection of a great many very small crystals, may be used. With so many small crystals at a very large number of different orientations, it is not necessary to rotate the sample.

The general procedure in determining the interatomic distances, that is, the three-dimensional structure, of a molecule involves these steps. The x-ray crystallographer proposes a hypothetical structure for the molecule. From this proposed model, he calculates the atomic grating spacing for each possible line through the atoms. From these grating spacings, he calculates the angles at which he should observe constructive interference. Then he compares the calculated angles with those experimental angles obtained by means of his diffraction measurements. If the two sets of angles agree, the proposed model corresponds to the actual structure of the molecule; if not, he modifies his model until there is a good agreement. In this way, he eventually determines the three-dimensional geometric structure of the molecule.

From examining a great many molecules with saturated carbon-carbon bonds, the average carbon-carbon distance is 1.54 Å, or 1.54×10^{-8} cm. In diamond, the carbon-carbon distance has been determined quite precisely and is 1.54451 ± 0.00014 Å. For ethane, the same distance is 1.5340 ± 0.0011 Å. The carbon-carbon distance is shorter for the olefins; in ethylene the distance is 1.34 ± 0.02 Å. When there is a triple bond between two carbons, the distance is even smaller; for acetylene, it is 1.205 ± 0.008 Å.

Figure 6.8 is a graph of the interatomic distances R(C—C) as a function of the total bond order. The carbon-carbon interatomic distances for graphite and benzene have been determined by x-ray diffraction; from the graph, one can determine the total bond order. For example, the distance between carbon atoms in benzene lies between the single- and double-bond values and corresponds to a bond order of 1.67. This is interpreted as meaning the carbon-carbon bond in benzene is neither a single bond nor a double bond but some mixture of each.

The molecule butadiene formerly was thought to be H_2C=CH—CH=CH_2, but from x-ray diffraction work, the central carbon-carbon distance corresponds to a bond order of 1.45; each extreme carbon-carbon bond corresponds to a bond order of 1.89. Therefore, the extreme carbon-carbon bonds are somewhat less

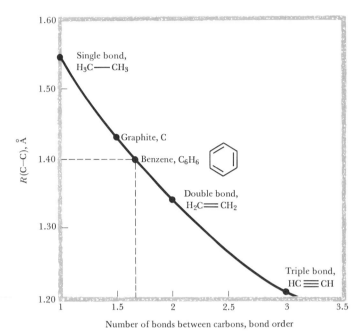

Figure 6.8. Carbon-carbon interatomic distance R(C—C) as a function of bond order.

than a pure double bond, and the central bond is somewhat more than a pure single bond. Chemists can correlate the bond order and the bond distance with the reactivity of these bonds within the molecule.

THE METALLIC BOND

The bonding which holds a collection of metal atoms together is neither covalent nor ionic. A crystalline metal can be considered a regular three-dimensional array of atoms. Each atom donates one or more electrons to be shared by the entire set of atoms. The atoms, stripped of some electrons, are then a regular array of cations bathed in a fluid consisting of electrons.

Since these free electrons are not associated with any particular atoms, they move readily under the influence of an electric field. Metals are good conductors of electricity for this reason. Because the free electrons are irregularly and randomly situated within a metal, their irregular interaction with visible light destroys the regular wave motion of light. Light does not penetrate deeply into metals; metals are opaque.

The arrangement of the atoms within the crystal lattice can be determined by x-ray diffraction (x-rays do penetrate). From this, one can estimate an effective size of the atoms within the crystalline metal. For example, the metal silver is found to form what is called a "face-centered cubic lattice." Each *unit cell* in the large crystal is a cube with a silver atom at each of the eight vertices and a silver atom centered on each of the six faces (Fig. 6.9). The unit cell is that pattern of atoms which is repeated in three dimensions in the regular crystalline lattice, just as wallpaper is the repetition of a plane unit-cell pattern in two dimensions.

From the x-ray diffraction data, the edge of each cube is exactly 4.0776 Å. From the unit cell one can see that the closest approach between neighboring silver atoms involves a vertex atom and a face atom. The diagonal distance across the face is

$$\sqrt{2} \times 4.0776 \text{ Å}$$

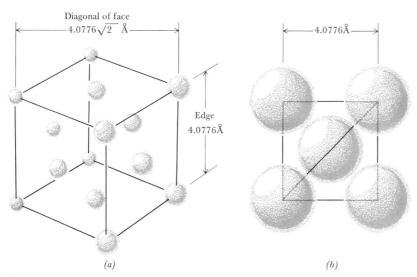

(a)   (b)

Figure 6.9. Silver crystal unit cell and the silver atomic radius. (a) Face-centered cubic lattice arrangement of silver atoms in unit cell of crystal; circles represent centers of the metal atoms. There is one silver atom at each of the eight corners and one at the center of each of the six faces. (b) The face of this cubic unit cell; silver atoms touch along the diagonal of the face. From this information the silver atom radius may be obtained. If the length of the diagonal is determined by x-ray crystallography to be $4.0776 \times \sqrt{2}$ Å, then the radius of one silver atom is one-fourth that number.

and this diagonal is equivalent to four radii *if the silver atoms are assumed to touch.* If R is the *effective atomic radius* of the silver atom in this type of crystal lattice,

$$\text{diagonal} = 4R = 4.0776\sqrt{2} \text{ Å}$$

$$R = \frac{4.0776 \times 1.4142}{4} \text{ Å}$$

$$= 1.4416 \text{ Å}$$

In this way, we can determine quite precisely the effective radii of all the elements, and if we examine ionic crystals, we can determine the effective ionic radii of ions.

Pure silver has an atomic radius of about 1.44 Å. In compounds, the radius is about 1.26 Å. This difference is the effect of an electron leaving the silver atom to form a shared-pair bond. For calcium metal and ion, the atomic radius values are about 1.97 and 0.99 Å. The difference results from the loss of the two $4s$ orbital electrons when calcium becomes a cation. For an atom which takes up one or more electrons to become an anion, there is an expected increase in its effective size. Atomic sulfur has a radius of about 1.28 Å, and that of the divalent anion is about 1.83 Å.

BOND DISSOCIATION

The first evidence that hydrogen is a diatomic molecule H_2 was the observed but unexpected increase in the pressure and number of molecules of a confined sample of this gas at elevated temperatures. This was interpreted as the dissociation reaction

$$H_2 \longrightarrow 2H$$

The higher the temperature, the greater the extent of dissociation.

At any temperature the average energy per molecule can be calculated, but there is also experimental evidence that the molecules do not all have the same energy. A few have high energy; more have less. The distribution of molecules at each energy can be calculated by using a Boltzmann distribution.

If we knew how much energy is required to break the hydrogen-hydrogen bond in H_2, we would be able to state what fraction of diatomic molecules at any temperature have acquired enough energy and are dissociated. From this we would know the total number of molecules—the diatomic H_2 and the monatomic H. The enhanced pressure due to an increase in the number of molecules of gas is directly calculable. Furthermore, we would be able to predict how the pressure varies with temperature because of the dissociation.

Conversely (and in practice), by measuring the total pressure of hydrogen confined to a constant volume over a range of temperatures, the extent of dissociation can be determined at each temperature. The Boltzmann distribution enables us to calculate the minimum energy needed for dissociation.

To completely dissociate 1 mole of diatomic hydrogen confined to a constant volume at 25°C, 103,610 cal is the energy required. We can write

$$H_2(g) \longrightarrow 2H(g)$$
$$Q_v(25°C) = 103,610 \text{ cal/mole}$$

where g refers to the gaseous state of the molecules, Q_v is the energy needed when the volume (subscript v) is constant, the positive sign of the number indicates that the system takes up energy for the dissociation, and 25°C is the temperature at which the number is calculated. (This is the convenient standard temperature at which almost all such data is tabulated, being just slightly above the typical room temperature in the United States.)

If we know the energy needed to dissociate 1 mole of hydrogen-hydrogen bonds, we can calculate the energy necessary to dissociate a single bond. Because chemists work more often with moles than with single molecules, most data is listed in terms of calories per mole.

We have been careful to specify that this energy is for the dissociation at a constant volume. Because there is a net increase in the number of molecules confined to this fixed volume, the gas pressure increases. If the dissociation energy is obtained for the process where the total pressure is maintained at a constant 1 atm, there must be a volume increase to accommodate that molecular-number increase. The energy at constant pressure is 104,200 cal/mole. The equation is

$$H_2(g) \longrightarrow 2H(g)$$
$$Q_p(25°C) = 104,200 \text{ cal/mole}$$

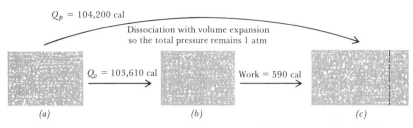

Figure 6.10. *Energy needed to dissociate 1 mole of diatomic hydrogen H_2 into 2 moles of monatomic hydrogen H is slightly greater at constant pressure than at constant volume because of the work needed to push back the ambient atmosphere. The difference between the constant-pressure energy Q_p and the constant-volume energy Q_v required is exactly the work done in the expansion. (a) Large volume containing several moles of diatomic hydrogen H_2 at 1 atm pressure; (b) same volume, but 1 mole of diatomic hydrogen dissociates to form 2 moles of monatomic hydrogen H (with an increase in the total number of moles of gas, the pressure is now greater than 1 atm); (c) the gas is expanded to a larger volume against the ambient pressure until the gas pressure drops to the initial value of 1 atm.*

The subscript p refers to constant pressure. The difference

$$Q_p - Q_v = 590 \text{ cal/mole}$$

is the energy needed to expand the dissociated gas against an ambient pressure of 1 atm from a constant dissociation volume to that larger volume which lets the gas maintain its initial predissociation pressure (Fig. 6.10).

The value Q_p is a measure of both breaking bonds and expanding a gas, and Q_v is a measure of breaking bonds only. Therefore the latter value is the one of interest in comparing bond energies. Most chemical processes occur, not at constant volume, but at constant pressure. For this reason the data ordinarily presented are the constant-pressure values. It is easy to convert from this to constant-volume values or conversely since the work for expansion of the gas is simple to calculate.

THERMOCHEMISTRY

Even though chemists normally carry out chemical reactions under constant-pressure conditions, much of the available data is first obtained by burning substances in constant-volume containers, called "bombs," placed within calorimeters. This experimental method, called "bomb calorimetry," is widely used to measure the heats involved in combustion and in other chemical processes such as neutralization and hydrogenation.

Thermochemistry is the branch of chemistry concerned with the heats transferred during a process. When heat, or energy in another form, is absorbed from the surroundings, as in the dissociation of hydrogen, the process is said to be *endothermic* and the heat value has a positive sign. If heat is given off, as in combustions, the process is *exothermic* and the heat value has a negative sign.

If 1 mole of hydrogen gas and $\frac{1}{2}$ mole of oxygen are pumped into a bomb calorimeter and the mixture is ignited, for example, by a spark plug, an explosion, which is a very rapid chemical reaction, takes place. When the water in the calorimeter is at about 25°C, 67,430 cal are liberated. The reaction is

$$H_2(g) + \tfrac{1}{2}O_2(g) \longrightarrow H_2O(l)$$
$$Q_v(25°C) = -67,430 \text{ cal}$$

where l refers to liquid water. The negative sign indicates the exothermic nature of this combustion.

The same reaction at constant pressure will include a volume contraction because the liquid product takes up less space (18 cm³) than the reactants (36,700 cm³). This contraction occurs with work done by the surroundings on the reaction system. Since this work is a form of energy added to the heat of combustion, we expect more heat to be released in this case than before. The reaction under constant pressure is

$$H_2(g) + \tfrac{1}{2}O_2(g) \longrightarrow H_2O(l)$$
$$Q_p = -68,320 \text{ cal}$$

Table 6.4 lists some thermochemical data. The first two values are the energies, or heats of dissociation, of the two gases H_2 and O_2. The third value is the combustion of hydrogen, and the fourth is the vaporization of 1 mole of liquid water. The next four values are the combustions of the hydrocarbons methane, ethane, ethylene, and acetylene. Ninth and tenth are the combustions of solid carbon and carbon monoxide. Eleventh is the vaporization of solid carbon. (Although this is immeasurable at 25°C, the value can be calculated from higher-temperature data.) Last in the table is the heat of vaporization of hydrogen sulfide at its boiling point. Even though H_2S is about twice as heavy as H_2O, less than half as much energy is needed for vaporization. This difference is attributed to the polar nature of water.

Table 6.4. Heats of Reaction at 25°C at Constant Pressure or Volume

	Reaction	Q_p, cal	Q_v, cal
1	$H_2(g) \longrightarrow 2H(g)$	104,200	103,610
2	$O_2(g) \longrightarrow 2O(g)$	119,120	118,530
3	$H_2(g) + \frac{1}{2}O_2(g) \longrightarrow H_2O(l)$	−68,320	−67,430
4	$H_2O(l) \longrightarrow H_2O(g)$	10,520	9,930
5	$CH_4(g) + 2O_2(g) \longrightarrow CO_2(g) + 2H_2O(l)$	−212,800	−211,620
6	$C_2H_2(g) + \frac{5}{2}O_2(g) \longrightarrow 2CO_2(g) + H_2O(l)$	−310,620	−309,730
7	$C_2H_4(g) + 3O_2(g) \longrightarrow 2CO_2(g) + 2H_2O(l)$	−337,240	−336,060
8	$C_2H_6(g) + \frac{7}{2}O_2(g) \longrightarrow 2CO_2(g) + 3H_2O(l)$	−372,830	−371,350
9	$C(s) + O_2(g) \longrightarrow CO_2(g)$	−94,050	−94,050
10	$CO(g) + \frac{1}{2}O_2(g) \longrightarrow CO_2(g)$	−67,640	−67,340
11	$C(s) \longrightarrow C(g)$	170,890	171,480
12	$H_2S(l) \longrightarrow H_2S(g)\,(\text{at } -61°C)$	4,500	4,500

The first entry in Table 6.4 can be written this way:

$$\tfrac{1}{2}H_2(g) \longrightarrow H(g)$$
$$Q_p = 52,100 \text{ cal}$$

because half the energy is needed to dissociate half as much hydrogen. We can also write

$$2H(g) \longrightarrow H_2(g)$$
$$Q_p = -104,200 \text{ cal}$$

which means that the recombination of 2 moles of monatomic hydrogen to form 1 mole of diatomic hydrogen is an exothermic process which releases 104,200 cal — exactly the energy needed for the dissociation. When the reaction is written in the reverse direction, the sign of the heat of reaction is changed.

There is an important implication to this simple statement. If we add the two reactions

$$H_2(g) \longrightarrow 2H(g) \qquad\qquad Q_p = +104,200 \text{ cal}$$
$$2H(g) \longrightarrow H_2(g) \quad \text{and} \quad Q_p = -104,200 \text{ cal}$$

No net change   No net change

there is no net change in the universe, in the amount of diatomic and monatomic hydrogen, or in the energy.

Suppose the energies did not exactly cancel. This could come about if the exothermic heat released in the recombination is either less or more than the endothermic heat taken up by the dissociation. If more heat is released in the second step than is absorbed in the first, a simple heat engine could be built based on the dissociation of hydrogen followed by the recombination of the atoms. The net effect is the creation of energy. This is contrary to what we should expect from the first law of thermodynamics – the conservation of energy.

It is our experience that such a chemical engine, called a "perpetual motion machine of the first kind" (because it violates the first law), is not possible. If less heat were released in the second (recombination) step than is absorbed in the first (dissociation), the net effect in the universe would be a loss of energy. This, too, is contrary to our experience.

Another example can be devised from reactions 9 and 10 in Table 6.3. We combine them in this way:

$$C(s) + O_2(g) \longrightarrow CO_2(g) \qquad\qquad Q_p = -94,050 \text{ cal}$$
$$\underline{\qquad CO_2(g) \longrightarrow CO(g) + \tfrac{1}{2}O_2(g) \qquad\qquad Q_p = +67,640 \text{ cal}}$$
$$C(s) + \tfrac{1}{2}O_2(g) \longrightarrow CO(g) \qquad\qquad Q_p = -26,410 \text{ cal}$$

We have handled both the molecular symbols and the numerical values in the usual algebraic manner. This, too, must be consistent with the first law of thermodynamics. If it is not, one should be able to design a series of reactions which operate in a cycle so that there is no permanent change in the amounts of carbon, oxygen, carbon monoxide, and carbon dioxide in the universe but a continuous increase or decrease in the total energy.

HESS'S LAW

The rule, consistent with the first law, that we can combine chemical reactions and their heats as we have to obtain a new reaction with its correct heat is called "Hess's law" [Germain Henri Hess (1802–1850)]. It is of considerable use in obtaining the heats of processes not directly observable in the laboratory.

One very practical result of the last set of equations is that we know the production of carbon monoxide from carbon is exothermic and releases 26,410 cal for each mole of carbon monoxide formed. Ordinarily it is not an easy matter to isolate this process; some carbon dioxide may be formed at the same time.

BOND ENERGIES

The *molar bond dissociation energy* is the energy needed to break 1 mole of that bond. The thermochemical data in Table 6.4 can be combined to calculate some of these bond energies. We combine the values for reactions 3, 4, 1, and 2 in this manner:

Reactions	Q_v, **cal**
$H_2O(l) \longrightarrow H_2(g) + \frac{1}{2}O_2(g)$	67,430
$H_2O(g) \longrightarrow H_2O(l)$	−9,930
$H_2(g) \longrightarrow 2H(g)$	103,610
$\frac{1}{2}O_2(g) \longrightarrow O(g)$	59,260
$H_2O(g) \longrightarrow 2H(g) + O(g)$	220,370

The result of these four reactions is the dissociation of 1 mole of gaseous water into 2 moles of monatomic hydrogen and 1 mole of monatomic oxygen. The reactant and products are all gases; there is no energy involved for vaporization. Because two oxygen-hydrogen bonds are broken in this reaction, it follows directly that 110,180 cal are needed to break 1 mole of O—H bonds. This is written as

$$D(O—H) = 110,180 \text{ cal}$$

> **Example.** Calculate the carbon-hydrogen bond energy in methane.
>
> Using the reactions 5, 6, 11, 3, and 1, we can calculate the complete dissociation energy of 1 mole of methane CH_4 into atoms:

Reactions	Q_v, **cal**
$CH_4(g) + 2O_2(g) \longrightarrow CO_2(g) + 2H_2O(l)$	−211,620
$CO_2(g) \longrightarrow C(s) + O_2(g)$	94,050
$C(s) \longrightarrow C(g)$	171,480
$2H_2O(l) \longrightarrow 2H_2(g) + O_2(g)$	134,860
$2H_2(g) \longrightarrow 4H(g)$	207,220
$CH_4(g) \longrightarrow C(g) + 4H(g)$	395,990
$D(C—H) = 99,000$ cal	

This is an average value, the dissociation energy per mole when 4 moles of the C—H bond are broken for each mole of methane.

From the data we can, in a similar way, calculate the total energy necessary to convert 1 mole of ethane into its atoms:

$$CH_3—CH_3(g) \longrightarrow 2C(g) + 6H(g)$$

In this dissociation, six C—H bonds and one C—C bond are broken. We know what is required for the C—H bonds; what is left is the carbon-carbon bond energy. That value for ethane is

$$D(CH_3—CH_3) = 88 \text{ cal}$$

For ethylene and acetylene the double- and triple-bond energies for the carbon-carbon bond are

$$D(CH_2=CH_2) = 163 \text{ cal}$$
$$D(CH\equiv CH) = 230 \text{ cal}$$

In Dalton's atomic theory a chemical reaction is the recombination of the atoms in the reactants to form the products; that is, a chemical reaction is an atomic reorganization. From what we know about the bonds in a molecule we can look at a chemical reaction as the breaking of some bonds and the formation of others. In the combustion of methane

$$CH_4(g) + 2O_2(g) \longrightarrow CO_2(g) + 2H_2O(l)$$

four C—H and two O=O bonds are broken, and two C=O and four O—H bonds are formed.

From what a chemist knows about bond energies he is able to predict the overall thermochemical nature of most reactions—whether it will be exothermic or endothermic. In many cases he can make a good guess about whether or not a reaction is possible.

PROBLEMS

6.1. Using the electronegativities (Table 6.2), which of hydrogen chloride HCl and hydrogen iodide HI do you expect to be less polar?

6.2. Rank the four sodium halides in order from most to least ionic.

6.3. Copper has a face-centered cubic unit cell with an edge of 3.608 Å. What is the radius of the metallic copper atom?

6.4. Lead has a face-centered cubic unit cell with an edge of 4.941 Å. What is its atomic radius?

6.5. Is the constant-pressure hydrogenation of acetylene

$$C_2H_2(g) + H_2(g) \longrightarrow C_2H_4(g)$$

exothermic or endothermic and by how much?

6.6. Does the hydrogenation of ethylene at constant pressure

$$C_2H_4(g) + H_2z(g) \longrightarrow C_2H_6(g)$$

release or absorb heat and how much?

The Geometry of Molecules

At the introduction of Chap. 4 we saw that the molecular formula is not a sufficient description to define a substance. Pentane C_5H_{10}, containing 15 atoms, can be any one of three distinct isomers. Evidently the arrangement of the 5 carbons and 10 hydrogens must be different in each isomer. Thus the atomic *arrangement,* in addition to the mere presence of particular atoms in a molecule, must be an important factor in the behavior of a substance. In this chapter we shall look at the geometry of molecules, that is, at how the atoms are connected and arranged within molecules.

THE STATES OF MATTER

Most compounds have well-defined melting and boiling points. The exceptions are those which decompose or otherwise react, forming new substances, before they melt or evaporate. Sugar, for example, turns into caramel when melted and so does not exist as a gas.

Above their boiling points, substances are gases. Gases completely fill and thus assume the shapes and volumes of their containing vessels. Below the boiling point but above the melting point, a substance is a liquid. Liquids have definite volumes but assume the shapes of their containing vessels. Below the melting point a substance is a solid, with a definite shape and volume. A solid is rigid; in contrast, liquids and gases, because they flow easily, are *fluids*.

At very high temperatures all molecules dissociate into atoms, which further dissociate into ions. A mixture of electrons and ions is called a "plasma." This is the material that stars are composed of. Plasma chemistry has become one of the more exciting areas of research in current chemistry.

Some liquids become increasingly thicker, or more viscous, as they are cooled. If the viscosity is great enough, the liquid actually becomes just as rigid as a solid but without freezing, that is, without forming a well-characterized structure. For example, the mineral quartz (silicon dioxide SiO_2) melts at 1710°C. However, when the liquid is cooled below its melting point, it is so viscous that it does not form the regular solid but forms a rigid liquid called "fused silica." Materials which behave in this way are said to form *glasses*, or *supercooled liquids*.

One way in which we can distinguish between glasses and solids is to repeat the experiments of Max von Laue and the Braggs. A definite and sharp diffraction pattern is obtained by reflecting x-rays from a solid. The pattern from a normal liquid is diffuse rather than sharp, indicating the absence of regular and repeated spacing between atoms. Glasses exhibit the diffraction patterns of liquids. Also, in their formation from the fluid liquid, glasses do not release the heat of fusion which attends the normal solidification of a liquid.

The solid mineral form of silica, quartz, can be obtained from the liquid if cooling occurs on the time scale of geological processes—several hundred thousand years. In the laboratory other techniques are used to synthesize quartz.

Some materials, for example, rubber, are neither solids, glasses, nor liquids. They show, as do liquids, a lack of regular internal structure but are neither rigid like glasses nor fluid like liquids. Such materials are said to be *amorphous;* they have ambiguous physical characteristics. Many synthetic materials are amorphous.

The solid, liquid, gas, plasma, glass, and amorphous forms in which matter occurs are referred to as the *states of matter.*

THE SHAPE OF CRYSTALS

Excluding glasses and amorphous materials, all solids have distinctive shapes. Solid sodium chloride may form different sized aggregates—common table salt is quite fine while halite, or rock salt, is rather coarse. However, no matter what the source or the size, each piece of solid sodium chloride has certain geometric characteristics in common with every other piece of the same substance.

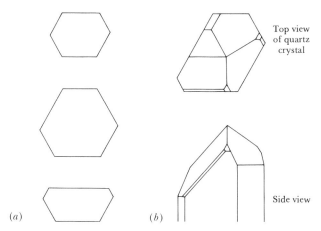

Top view
of quartz
crystal

Side view

(*a*)   (*b*)

Figure 7.1 (a) Three cross-section views of pieces of quartz (as sand, fine dust, or large chunk). Hexagons are not always regular—that is, the six sides are not always equal in length—but the angles are always 120° each and the opposite sides (opposite faces in the prism) are always parallel. (b) A quartz crystal, top and side views.

Each grain or chunk is either a perfect cube, a rectangular solid, or an obvious fragment of these. All the surfaces are flat rather than curved, and almost all the angles are right angles.

Quartz has another set of geometric characteristics peculiar to quartz. Each piece of quartz has three pairs of vertical parallel sides; when viewed from either the top or bottom, the cross section is always a six-sided figure, the hexagon. Unbroken pieces have pyramid-shaped caps and bases (Fig. 7.1).

All pure solids not in the glassy or amorphous state have characteristic geometric shapes called "crystals." Sodium chloride has cubic crystals, and quartz has hexagonal prism crystals. Some crystals are transparent, but others (for example, metal crystals) are opaque. Crystals are seldom found individually. It is not obvious that a piece of metal is a collection of many small crystals oriented in all directions, but if a polished metal sample is gently treated with acid, the exposed crystal faces are attacked at various rates. As a result of this *etching*, the crystal boundaries and profiles are readily seen. The crystals form when the molten metal is cast and allowed to solidify, and the size and shape of these crystals determine the fine properties of the metal.

Natural minerals are commonly collections of assorted crystals. Sandstone is a consolidation of sand grains with other crystalline matter—often calcite $CaCO_3$ or feldspar, a family of aluminum silicates. The sand grains (crystalline quartz) are polished round by continual abrasion with each other. Some sea sand grains are almost spherical rather than angular; nevertheless, each grain can be fractured to reveal the typical quartz geometry, and each grain has the typical x-ray diffraction pattern of quartz.

Sugar and salt are two perfectly clear, colorless crystalline substances which appear opaque because of their fine size. The small particles transmit and reflect light irregularly, but the large crystals are transparent.

When crystals are broken into smaller pieces, the new surfaces formed are always flat, or planar, and meet neighboring surfaces at angles characteristic of the substance. In 1782, René Just Haüy (1743–1822) discovered that angles found in any crystalline substance were not random (see Prob. 7.21 at the end of this chapter). To account for this, Haüy suggested a regularly spaced structure within each crystal; that is, the surfaces bounding each crystal are defined by an internal regularity. Haüy's discovery can be described most simply by the atomic theory of matter. The regular internal structure of crystals proposed by Haüy was confirmed by von Laue and the Braggs, using crystalline materials as diffraction gratings for x-rays.

THE SODIUM CHLORIDE LATTICE

The internal structure of crystalline sodium chloride has been determined by x-ray diffraction studies. The arrangement of the ions can be described as a three-dimensional checkerboard. Each row is an alternation of sodium and chloride ions $Na^+Cl^-Na^+Cl^-Na^+Cl^-$. . . . Rows run in three mutually perpendicular directions, as shown in Fig. 7.2.

Except for the ions at the crystal surface, each sodium ion has as its nearest neighbors six chloride ions, each 2.814 Å away. Each chloride ion has six sodium ions as its nearest neighbors. Because each sodium ion is equidistant from six chloride ions and each chloride ion equidistant from six sodium ions, it is not possible to say that any one chloride ion is bonded to any *given* sodium ion.

Molten sodium chloride is a random, rather than a regular, array of these two ions. Again, because each ion is surrounded by about six ions of the opposite charge, a simple NaCl molecule does not exist. In a gas, however, the molecule is found, together with a

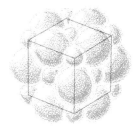

Figure 7.2. Sodium chloride crystal lattice. Each large sphere represents a chloride ion Cl^-; each small sphere, a sodium ion Na^+. Except for the ions at the crystal surface, each ion is surrounded by six ions of the opposite charge—each Na^+ by six Cl^- ions and each Cl^- by six Na^+. The cubic lattice is represented by light lines. The unit cell is defined by a heavy outline, with shaded ions. Each edge of the unit cell is 5.628 Å in length. From the center of an ion to the center of a neighboring oppositely charged ion, the distance is exactly one-half this—2.814 Å—or the sum of the radii of a sodium and a chloride ion.

little of *dimer* (NaCl)$_2$. A dimer has exactly twice as many atoms as a single molecule and hence twice the molecular weight.

METHANE

We can represent methane by its structural formula

$$
\begin{array}{c}
\text{H} \\
| \\
\text{H}-\text{C}-\text{H} \\
| \\
\text{H}
\end{array}
$$

to indicate the presence of four carbon-hydrogen bonds; this is not obvious from the molecular formula CH$_4$ alone. In all reactions of methane, no difference among any of the four hydrogens is detectable; they are all exactly equivalent. From this we deduce that the four carbon-hydrogen distances are the same and so are the four bond dissociation energies D(CH$_3$—H). In addition, the four hydrogens must be arranged symmetrically about the carbon so that each hydrogen is equally accessible to an approaching reactant. As indicated in the structural formula, the hydrogen-carbon-hydrogen angles are all equal.

     Because all five atoms in the structural formula are in one plane, this arrangement is said to be *planar*. However, this is not the only possible arrangement in which the four hydrogens are exactly equivalent. Another possibility is a pyramidal molecule in which the four hydrogens are placed thus: three are at the corners of the triangular base and one is at the apex—the carbon is at the center of the pyramid.

THE TETRAHEDRON

A pyramid with a triangular base is called a "tetrahedron." Figure 7.3a is a regular tetrahedron because its four faces are equilateral triangles. If a hydrogen is placed at each vertex and the carbon at the center (Fig. 7.3b), and if the four carbon-hydrogen bonds are connected, the geometric arrangement is said to be *tetrahedral*. The four hydrogens define and describe a regular tetrahedron.

METHANE. Is methane planar or tetrahedral? Methane reacts with chlorine gas so that two of its hydrogen atoms are replaced by chlorine atoms:

(a)

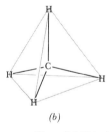

(b)

Figure 7.3. Tetrahedral arrangement of methane CH$_4$. (a) The tetrahedron has four vertices, six edges, and four faces, each an equilateral triangle; (b) if a hydrogen is placed at each vertex and the carbon at the center, the four C—H bonds can be drawn as heavy lines.

$$CH_4 + 2Cl_2 \longrightarrow CH_2Cl_2 + 2HCl$$

If methane were planar, two different dichloromethanes CH_2Cl_2 would be possible

$$
\begin{array}{ccc}
& H & \\
& | & \\
Cl - & C & - Cl \\
& | & \\
& H &
\end{array}
\qquad
\begin{array}{ccc}
& Cl & \\
& | & \\
Cl - & C & - H \\
& | & \\
& H &
\end{array}
$$

depending on whether the two chlorine atoms are near or far from each other. The properties of these two planar molecules would be different, and if the properties are different the molecules are separable.

If methane were tetrahedral, only one dichloromethane would be possible (Fig. 7.4 and Prob. 7.1). There are six ways of putting two chlorines on the four vertices of the tetrahedra; however, these are all congruent because any one can be turned about in space to be made to superimpose exactly on any other.

It is not possible, by any physical or chemical means, to separate dichloromethane into two parts, each with a distinct set of properties. There is only *one* dichloromethane. Therefore the mol-

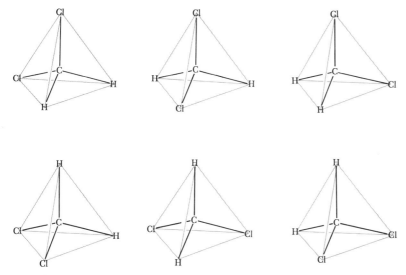

Figure 7.4. Only one dichloromethane can be made from methane if the molecule is tetrahedral. Any one of these six tetrahedra can be made to coincide with all of the others just by rotating it in space. Try to describe the rotation necessary in each case.

ecule, and methane too, must be tetrahedral. Other evidence confirms this structure. The four carbon-hydrogen bonds are directed toward the corners of a tetrahedron.

ETHANE. A similar experiment with ethane gives us information about its structure. Treating ethane with chlorine can give dichloroethane

$$C_2H_6 + 2Cl_2 \longrightarrow C_2H_4Cl_2 + 2HCl$$

This product can be separated into two substances: One has a density of 1.26 g/cm^3, melts at $-35°$C, and boils at 83.6°C; the other has a density of 1.17 g/cm^3, melts at $-97°$C, and boils at 57.3°C. These two distinct substances, each with the same molecular formula and molecular weight of 98.97, are said to be *isomers* of each other. Isomers have properties that are different, but their atomic content is the same; they must be different arrangements of the same atoms. With different properties, the isomers are separable.

The higher-boiling isomer reacts with sodium to form ethylene CH$_2$CH$_2$. Since in ethylene each carbon is bonded to two hydrogens, this is probably so in this dichloro isomer. It must then have one chlorine on each carbon and can be written as CH$_2$ClCH$_2$Cl. The formula states that each carbon is bonded to two hydrogens, one chlorine, and to the other carbon. The reaction with sodium can be written as

$$CH_2Cl—CH_2Cl + 2Na \longrightarrow CH_2{=}CH_2 + 2NaCl$$

The lower-boiling dichloroethane isomer of dichloroethane undergoes a much more complex set of reactions with sodium. However, the evidence indicates that here the two chlorines are bonded to the same carbon. This isomer can be written as CH$_3$CHCl$_2$.

To distinguish between the two isomers, we can number the two carbons 1 and 2. The first isomer CH$_2$ClCH$_2$Cl, with a chlorine on both the first and second carbons, is called "1,2-dichloroethane"; the second CH$_3$CHCl$_2$, with both chlorines on the same carbon, is called "1,1-dichloroethane." We could, just as unambiguously, have called this "2,2-dichloroethane," but convention is to write the lowest numbers possible. The prefix "di-" denotes *two* just as "mono-" refers to one. The fact that there is only one 1,2-dichloroethane tells us a great deal about ethane. If ethane were planar, we would have four 1,2-dichloroethanes.

$$\begin{array}{ccc} & Cl & Cl \\ & | & | \\ H-& C-& C-H \\ & | & | \\ & H & H \end{array} \qquad \begin{array}{ccc} & Cl & H \\ & | & | \\ H-& C-& C-Cl \\ & | & | \\ & H & H \end{array}$$

$$\begin{array}{ccc} & Cl & H \\ & | & | \\ H-& C-& C-H \\ & | & | \\ & H & Cl \end{array} \qquad \begin{array}{ccc} & H & H \\ & | & | \\ Cl-& C-& C-Cl \\ & | & | \\ & H & H \end{array}$$

Because methane and ethane are part of the same paraffin series, and because methane does have a tetrahedral structure, we expect something like this for ethane. Figure 7.5 shows ethane as two interlocking tetrahedra. But even with two tetrahedra we can imagine more than one 1,2-dichloroethane. If we look at this molecule along the carbon-carbon axis with the carbons aligned one behind the other, we can imagine that the molecule can have any one of the orientations shown in Fig. 7.6. In Fig. 7.6 the central circle represents the closer carbon atom; the remote carbon atom is obscured by the closer one. The two hydrogens and one chlorine connected to the closer carbon are represented by bonds which meet at the center of the circle. The three atoms which are connected to the remote carbon are connected by bonds which stop at the circumference of the circle. There are six formations shown. The first three are said to be *eclipsed,* where atoms attached to the remote carbon are obscured by the atoms attached to the closer carbon. The next three are said to be in a staggered conformation. In the *trans* conformation, the chlorines are at maximum separation. There are two *gauche* conformations. If all conformations were possible, then there would be several different types of dichloroethane. Since only one 1,2-dichloroethane is found, the six forms are equivalent and are readily converted from one form to another. There is free rotation about this and about most carbon-carbon single bonds.

The structure of ethane is represented by two carbons linked directly. Each carbon, in addition, is joined to three hydrogens. The four bonds from each carbon, three C—H and one C—C, are directed tetrahedrally. In addition the set of three

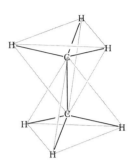

Figure 7.5. Ethane H_3C—CH_3 as two tetrahedra. Each carbon is surrounded tetrahedrally by three hydrogens and one carbon. Each carbon, then, occupies the center of its own tetrahedron and, at the same time, is at a vertex of the tetrahedron about the other carbon. Because the carbon-carbon distance in ethane is 1.5430 Å and the carbon-hydrogen distance is 1.1122 Å, these tetrahedra are not as regular as the methane tetrahedra (Fig. 7.3) in which the four bonds are of equal length.

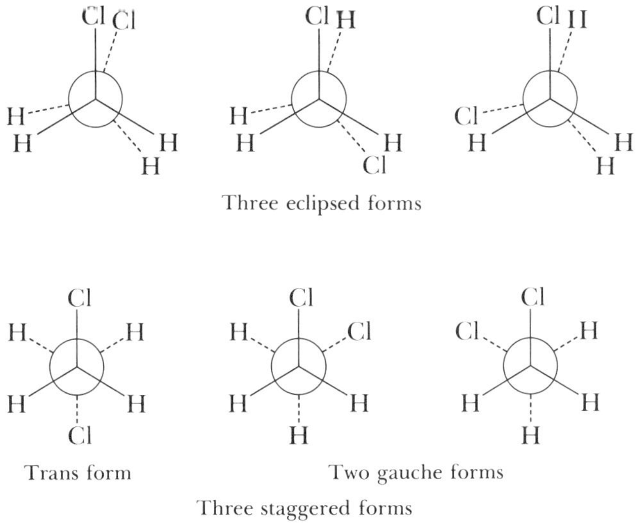

Three eclipsed forms

Trans form   Two gauche forms

Three staggered forms

Figure   *7.6.*

hydrogens on one carbon is free to rotate relative to those hydrogens on the other carbon.

ETHYLENE. When acetylene CH≡CH reacts with chlorine, 1,2-dichloroethylene CHCl=CHCl is formed. This product is separable into two isomers because of differences in their properties (Table 7.1). From chemical and physical evidence the two isomers are represented as follows:

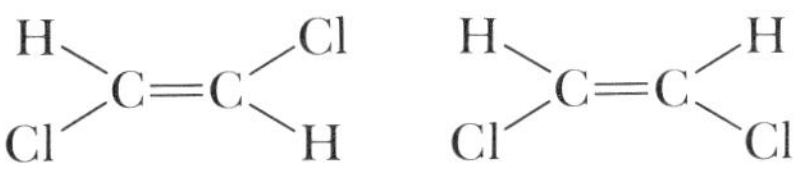

Trans   Cis

In general free rotation does not occur about any double bond. To convert *trans*-1,2-dichloroethylene to *cis*-1,2-dichloroethylene, it is necessary to break one of the double bonds. The amount of energy required for this break is not available at room temperature unless it is supplied by some external source such as a high-energy particle.

    To distinguish between those forms which are readily convertible and those which are not, chemists use the terms *conformation* and *configuration*, respectively. For 1,2-dichloroethane they

Table 7.1. 1,2-Dichloroethylene

Property	Trans	Cis
Molecular weight	96.95	96.95
Density	1.265	1.291
Melting point	$-50°C$	$-80.5°C$
Boiling point	48.4°C	60.1°C

speak of the trans and gauche conformations; for 1,2-di-chloroethylene, the cis and trans configurations.

From x-ray diffraction and other studies, ethylene and sub-stituted ethylenes, in which one or more hydrogens are replaced by other atoms, are planar. All six atoms lie in a single plane.

PARAFFIN ISOMERS

The fourth member of the paraffin series is butane C_4H_{10}. It is possible to separate natural butane into two portions, each with its own set of properties. Except for a common formula and molecular weight, the two substances are different. They are not different conformations because one does not, under ordinary conditions, change even in part to the other. They are isomers of each other. By their chemical properties and such physical means as x-ray diffraction, the structural formulas are known to be

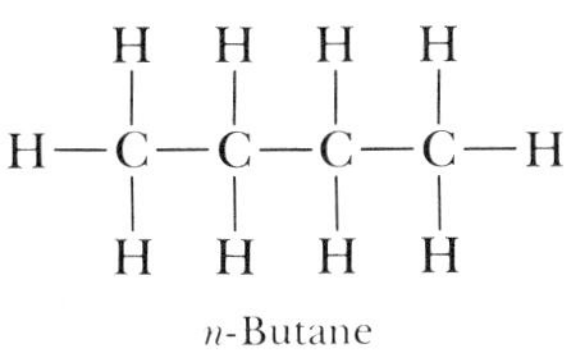

n-Butane

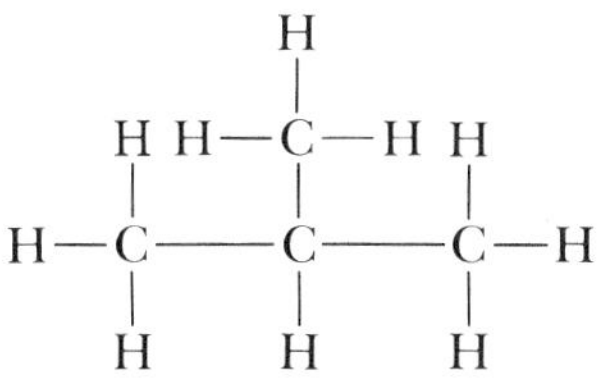

Isobutane, or 2-methylpropane

At the beginning of Chap. 4 we saw that pentane C_5H_{12} has three isomers. Their structural formulas are

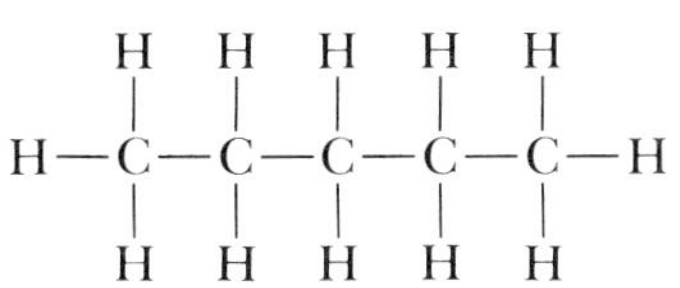

n-Pentane

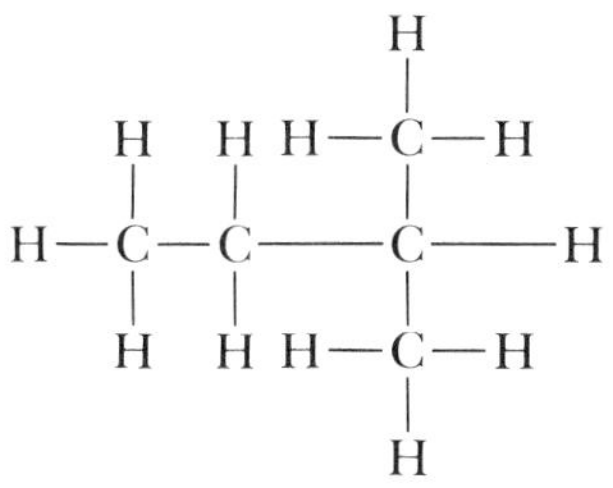

Isopentane, or 2-methylbutane 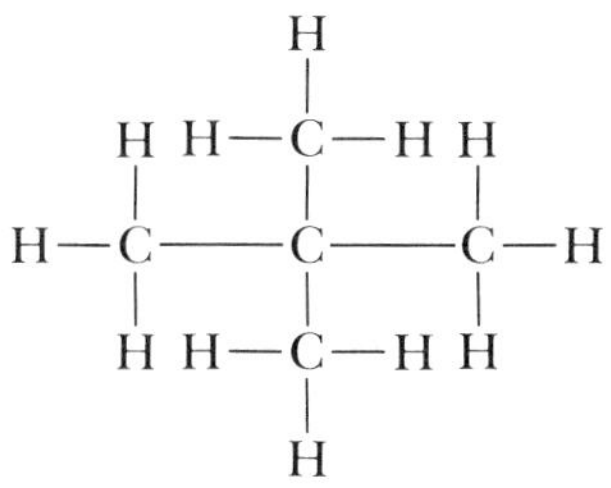   Neopentane, or 2,2-dimethylpropane

There are five isomers of hexane C_6H_{14}; nine heptanes C_7H_{16}; 18 octanes C_8H_{18}; 35 nonanes C_9H_{20}; 75 decanes $C_{10}H_{22}$; and for eicosane $C_{20}H_{42}$ there are 366,319 distinct isomers. The fact that the number of paraffin isomers increases rapidly with the molecular weight is one important reason why there are so many organic compounds (see Prob. 7.2 at the end of this chapter).

It should be kept in mind that the carbon single bonds are directed tetrahedrally; therefore, carbon-carbon-hydrogen angles are not as drawn but 109°, which is the angle formed by lines drawn from two corners to meet at the center of a tetrahedron. There is free rotation about each carbon-carbon single bond. The larger paraffin hydrocarbons are more likely to be kinked and coiled than flat and stretched out. The simplest butane is normal butane, or *n*-butane, and the simplest pentane is normal pentane, or *n*-pentane. By simple, we mean this: In the normal isomers it is possible to start at one end of the molecule and go from carbon to neighboring carbon to the other end, passing each carbon once and only once. These molecules are called "straight-chain" hydrocarbons for this reason even though, in fact, they are twisted in space rather than geometrically truly linear. The other isomers are said to be "branched." The extent of branching, or its absence in the normal straight-chain isomer, governs the chemical and physical properties of the molecule.

When a single hydrogen is removed from a molecule, what is left is a reactive species called a "radical." The names of the radicals derive from the names of the original molecules. From methane CH_4, ethane C_2H_6, and propane C_3H_8, we have the radicals methyl CH_3, ethyl C_2H_5, and either *n*-propyl $CH_3CH_2CH_2$ or isopropyl CH_3CHCH_3. Radicals react quickly because one atom has a valence shell that is incomplete.

There are two names for each branched paraffin. The prefixes iso and neo are commonly used but tell us nothing about the kind of branching in the isomer. Isobutane is more descriptively called "2-methylpropane." This name tells us to replace a hydrogen, on the second carbon in a propane $\overset{1}{C}H_3\overset{2}{C}H_2\overset{3}{C}H_3$, with a methyl radical. Neopentane is made by replacing both second-carbon hydrogens with methyl radicals—hence the name 2,2-dimethylpropane. The name 2-methylbutane tells us to take a normal butane molecule and replace a hydrogen on the second carbon with a methyl (see Prob. 7.3).

One of the most important branched hydrocarbons is iso-octane:

$$
\begin{array}{c}
\quad\ \ \text{H} \qquad\qquad\ \ \text{H} \\
\quad\ \ | \qquad\qquad\ \ | \\
\text{H}\ \ \text{H}\!-\!\text{C}\!-\!\text{H}\ \ \text{H}\ \ \text{H}\!-\!\text{C}\!-\!\text{H}\ \ \text{H} \\
|\quad\ \ |\qquad\ \ |\quad\ \ |\qquad\quad | \\
\text{H}\!-\!\text{C}\!-\!\!-\!\!-\!\text{C}\!-\!\!-\!\!-\!\text{C}\!-\!\!-\!\!-\!\text{C}\!-\!\!-\!\!-\!\text{C}\!-\!\text{H} \\
|\qquad\ \ |\qquad\ \ |\qquad\ \ |\qquad\quad | \\
\text{H}\ \ \text{H}\!-\!\text{C}\!-\!\text{H}\ \ \text{H}\qquad \text{H}\qquad\ \ \text{H} \\
\qquad\quad | \\
\qquad\quad \text{H}
\end{array}
$$

2,2,4-trimethylpentane

The longest straight-chain sequence has five carbons, thus pentane. The three methyls are placed two on the second and one on the fourth carbon of pentane.

If the fuel combustion within a gasoline engine does not go smoothly, a knocking sound develops. Isooctane was once thought to be the best fuel for this engine. In contrast, *n*-heptane knocks badly. Isooctane is given an octane rating of 100; *n*-heptane, 0. The octane rating of any hydrocarbon blend is determined by selecting the *n*-heptane–isooctane mixture which knocks as much as the hydrocarbon blend to be tested. The octane rating is the percent isooctane in the comparison standard.

Other hydrocarbons (some olefins, for example) perform even better than isooctane. By using these, the octane rating of a gasoline can exceed 100. Another common technique used to suppress knocking is the addition of a compound, usually tetraethyl lead $Pb(C_2H_5)_4$. Because the lead can accumulate within the engine cylinders, 1,2-dibromoethane CH_2BrCH_2Br is added to form the volatile lead bromide $PbBr_2$. In this form the metal lead is exhausted into the atmosphere, then becoming a major source of pollution. The possibilities for precipitating the lead before it is exhausted into the atmosphere are being explored by many automobile manufacturers.

CYCLIC HYDROCARBONS

In addition to the straight- and branched-chain unsaturated paraffin hydrocarbons, all with the general formula C_nH_{2n+2}, there is yet another molecular arrangement possible. These are the *cyclic* saturated hydrocarbons, some of which are shown in Fig. 7.7.

Like the ordinary saturated paraffins, the bonds are all single; the carbon-carbon distances are all about 1.54 Å. To maintain the tetrahedral bonding of the noncyclic paraffins, the rings, except for cyclopropane, are puckered rather than flat. Three dif-

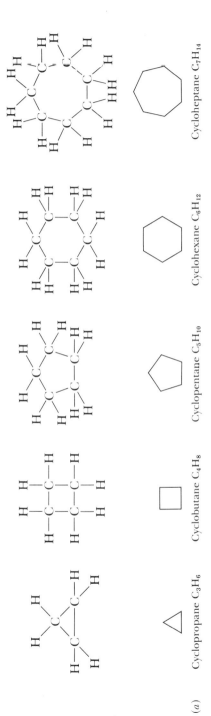

(a)

Cyclopropane C_3H_6 Cyclobutane C_4H_8 Cyclopentane C_5H_{10} Cyclohexane C_6H_{12} Cycloheptane C_7H_{14}

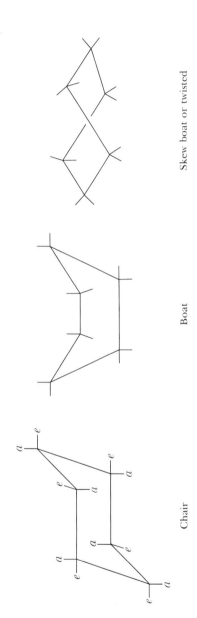

Chair   Boat   Skew boat or twisted

(b)

Figure 7.7. (a) Some cyclic saturated hydrocarbons. (b) Three nonplanar conformations of cyclohexane. Hydrogens are said to be axial or equatorial (a or e), if their C—H bonds are perpendicular or parallel to the plane of the four carbons not at the tips of the boat or chair.

ferent conformations of cyclohexane are shown. You can visualize forming these rings from the straight-chain paraffin by removing a hydrogen from each end and joining the extreme carbons. These cyclic compounds have two fewer hydrogens than the noncyclic paraffins; their general formula is C_nH_{2n}.

BENZENE. In 1825 Michael Faraday discovered a liquid hydrocarbon in illuminating gas. By analysis its molecular formula was found to be C_6H_6; it was called "benzene." Nearly a half-century of investigation was required before the structure was known.

We are confident that the structural formula for propane $CH_3CH_2CH_3$ is

$$
\begin{array}{c}
\text{H} \quad \text{H} \quad \text{H} \\
| \quad\;\; | \quad\;\; | \\
\text{H}-\text{C}-\text{C}-\text{C}-\text{H} \\
| \quad\;\; | \quad\;\; | \\
\text{H} \quad \text{H} \quad \text{H}
\end{array}
$$

because the number of chlorine-substituted isomers found exactly equals the number of isomers predicted. In Fig. 7.8 just a few of the 30 isomers found and predicted are shown (see Prob. 7.4). The number of isomers is large because hydrogens on an end carbon are not equivalent to those on the interior carbon.

In benzene, the situation is much simpler. It is possible to replace, one by one, the six hydrogens in benzene C_6H_6. There is only one monochlorobenzene C_6H_5Cl; there are three dichlorobenzenes $C_6H_4Cl_2$, three trichlorobenzenes $C_6H_3Cl_3$, three tetrachlorobenzenes $C_6H_2Cl_4$, one pentachlorobenzene C_6HCl_5, and one hexachlorobenzene C_6Cl_6. The initial hydrocarbon and all its substituted chloroderivatives are 13 in number. The obvious conclusion is that, unlike the unequal hydrogens in propane, the hydrogens in benzene are equivalent.

In 1865 Friedrich August Kekule (1829–1896) proposed a cyclic, or ring, structure for benzene. To maintain the usual carbon valence of 4, the structural formulas

Propane C_3H_8, one isomer

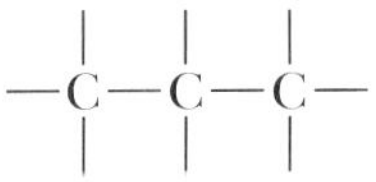

CH₃CH₂CH₃

$CH_3CH_2CH_3$

Monochloropropane C_3H_7Cl, two isomers

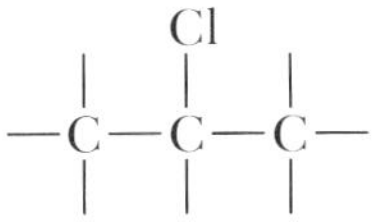

$CH_3CHClCH_3$
2-Chloropropane

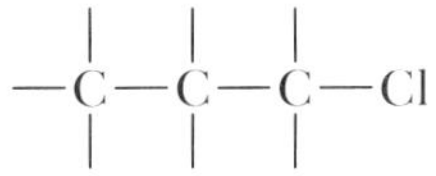

$CH_3CH_2CH_2Cl$
1-Chloropropane

Dichloropropane $C_3H_6Cl_2$, four isomers

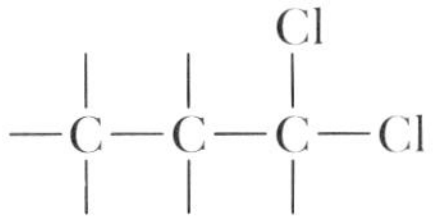

$CH_3CH_2CHCl_2$
1,1-Dichloropropane

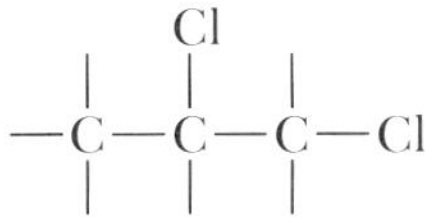

$CH_3CHClCH_2Cl$
1,2-Dichloropropane

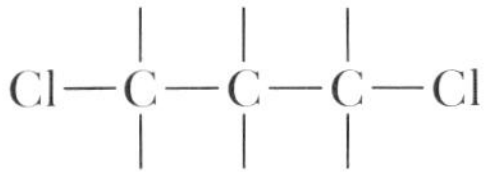

$CH_2ClCH_2CH_2Cl$
1,3-Dichloropropane

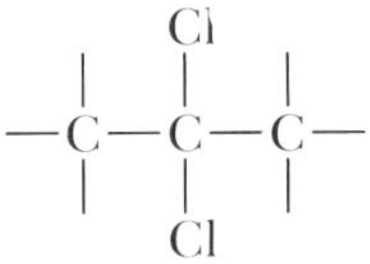

$CH_3CCl_2CH_3$
2,2-Dichloropropane

Trichloropropane $C_3H_5Cl_3$, five isomers

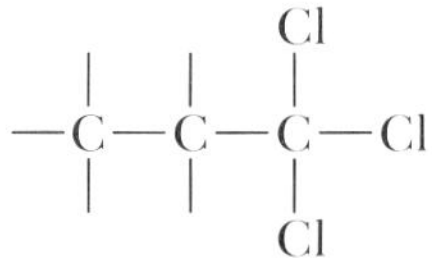

$CH_3CH_2CCl_3$
1,1,1-Trichloropropane

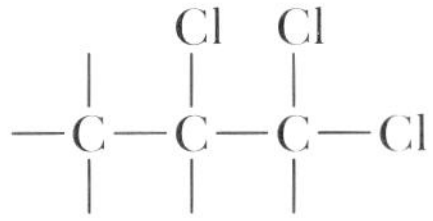

$CH_3CHClCHCl_2$
1,1,2-Trichloropropane

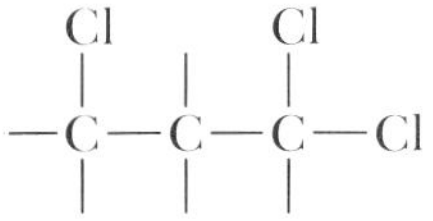

$CH_2ClCH_2CHCl_2$
1,1,3-Trichloropropane

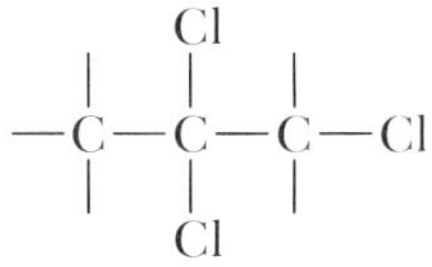

$CH_3CCl_2CH_2Cl$
1,2,2-Trichloropropane

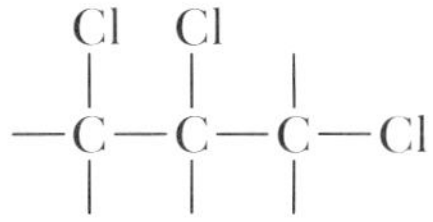

$CH_2ClCHClCH_2Cl$
1,2,3-Trichloropropane

Figure 7.8. Propane and some substituted chloropropanes. There are a total of 30 distinct mole-cules. Each carbon is bonded to another carbon, to a hydrogen, or to a chlorine atom. The hydrogens are not shown. Each of the 30 isomers has its own properties and can be isolated from the others. Because only these 30 isomers — no more and no less — are known to exist, we are con-fident that the structural formula of propane is as written.

were often used. There are two equivalent forms which differ only in the positions of the double bonds. The ↔ bond indicates this equivalence.

However, benzene is unlike other unsaturated molecules in that it cannot be hydrogenated as easily as the olefins. When hydrogenated, it forms cyclohexane C_6H_{12}, which is further evidence for the ring structure of benzene. Benzene, we know from x-ray diffraction, is a perfect regular hexagon with equal carbon-carbon bond distances of 1.40 Å, which is between 1.54 Å of a single bond and 1.34 Å of a double bond. All the carbon-carbon bonds in benzene are of the order 1.67 (Fig. 6.8). They are neither double nor single bonds but somewhere in between.

Benzene is considered to have the special structure represented as

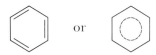

Benzene and several of its twelve chloro derivatives are shown in Fig. 7.9. To distinguish between positions, the carbons are num-

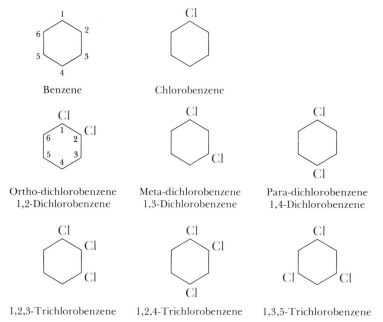

Figure 7.9. Benzene and some substituted chloro derivatives. There are 13 in all: this smaller number, compared to the 30 propane compounds in Fig. 7.8, is owing to the symmetry of the benzene ring and the equivalence of all the positions on that ring.

bered clockwise from one to six, beginning at the top carbon. The prefixes ortho, meta, and para are often used instead of the 1,2, 1,3, and 1,4 designations (Prob. 7.5).

AROMATIC HYDROCARBONS

Benzene and some (but not all) of its derivatives have pleasant odors compared to those molecules without benzene rings. For this reason any molecule with one or more such rings is said to be *aromatic*. The others we have already looked at—the straight, branched, or cyclic hydrocarbons, saturated or unsaturated, substituted or not—are called "aliphatic" compounds.

When soft, or bituminous, coal is heated at high temperatures in the absence of air, a very large number of gases and liquids are driven off, leaving coke, a solid porous residue. In addition to benzene, there are a number of high-boiling, polyring aromatics in *coal tar* (Fig. 7.10). Remember that although the hydrogens are not shown in the shorthand drawing there are hydrogens in the structure.

The aromatic hydrocarbons, in comparison with other organic compounds, have rather high boiling points. The carbon-carbon bond distances are all between the single- and double-bond values. The bonding is typical of aromatic compounds. The only carbons which are bonded to hydrogen are those belonging to a single ring. In naphthalene, for example, carbons labeled 9 and 10 have no hydrogens; of the other eight carbons, each has one hydrogen.

GRAPHITE. From Fig. 7.10 we can see that the larger the aromatic hydrocarbon, the relatively fewer hydrogens there are. In benzene, carbon and hydrogen atoms are present in equal numbers. The larger the molecule, the more likely any carbon is to be part of two or more rings. These carbons have no hydrogens.

The ultimate limit to building ring systems is a flat surface entirely covered by aromatic carbon rings in the way a floor can be covered with hexagonal tiles (Fig. 7.11). This is the structure of graphite, one form of pure carbon. A collection of these planes spaced about 3.5 Å apart—too far for strong bonding—is the structure of solid graphite. The planes can slip past each other easily, and this property makes graphite a convenient lubricant for locks and other metal parts.

Each plane of hexagonal rings can be considered a single

Benzene   C$_6$H$_6$ (bp 78°C)

Naphthalene   C$_{10}$H$_8$ (bp 218°C)

Anthracene   C$_{14}$H$_{10}$ (bp 354°C)

Phenanthrene   C$_{14}$H$_{10}$ (bp 340°C)

Naphthacene   C$_{18}$H$_{12}$ (sublimes)

Pyrene   C$_{16}$H$_{10}$ (bp 365°C)

Chrysene   C$_{18}$H$_{12}$ (bp 450°C)

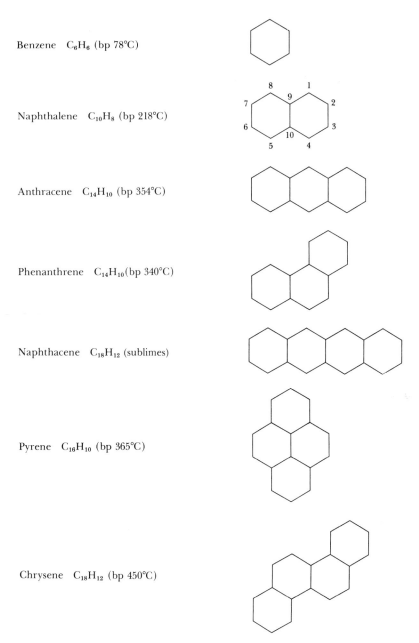

Figure 7.10. Some aromatic, polycyclic compounds found in coal tar (the thick liquid obtained from soft coal). Their boiling points are among the highest of the organic compounds. Naphthacene sublimes; that is, it goes directly from the solid to the vapor state. Benzo(a)pyrene boils only under reduced pressures, and is a carcinogen (a cancer-producing substance).

Benzo(a)pyrene $C_{20}H_{12}$
 (bp 310 at 10 torr pressure)

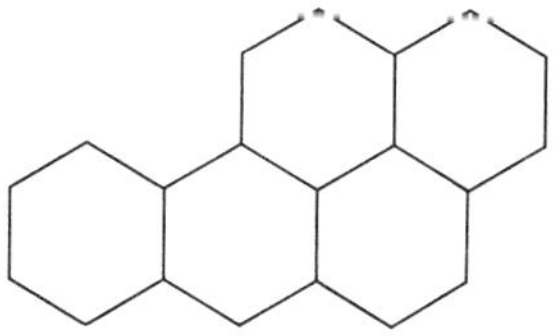

Picene $C_{22}H_{14}$ (bp 520°C)

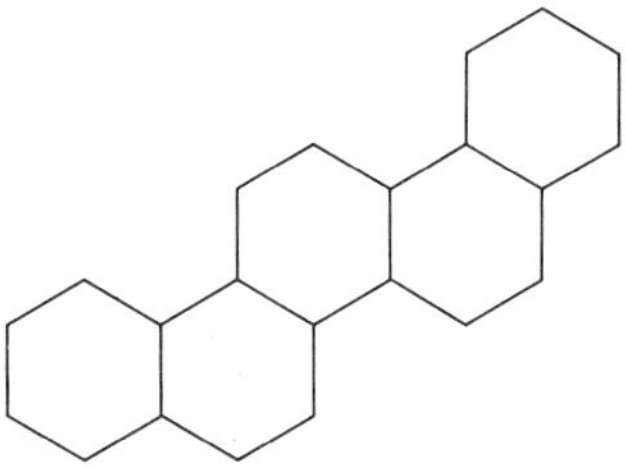

Figure 7.10. Continued.

molecule. The molecular weight is then enormous. Since graphite particles vary in size, the molecular weight also varies from particle to particle. As in the case of sodium chloride, it is convenient to say that 1 mole of carbon atoms, 12.0111 g of graphite, is the molecular weight of this substance. Graphite sublimes at about 3500°C and boils at 4200°C.

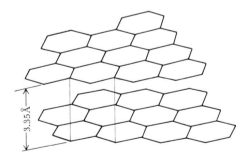

Figure 7.11. Sheet structure of graphite. A plane is completely covered by aromatic rings. The more rings a compound has, the relatively fewer hydrogen atoms contained within the compound. Graphite is the limiting case, with a very large number of carbon atoms and rings and no hydrogens (except possibly at the surface boundaries). Graphite is soft and fragments readily.

DIAMOND. Graphite is the limit of the large planar aromatic hydrocarbons, and diamond is the limit of tetrahedrally bonded pure carbon. One unit cell of the diamond lattice is shown in Fig. 7.12. Each carbon is at the center of a tetrahedron and bonded to four other carbons. One individual diamond is a single molecule, but here, too, 1 mole of diamond is 12.0111 g.

In both diamond and graphite the bonding is regular throughout the entire particle of these substances, but at the surface the lattice abruptly stops. A surface carbon cannot be surrounded by three other carbons, in the case of graphite, or four, in diamond. Surface atoms, for this reason, are not equivalent to those in the interior of the graphite or diamond particle. But for almost all except the very finest particles, the fraction of the total atoms located at the surface is small.

Diamond is the hardest natural substance known. We measure *hardness* by the ability of one substance to scratch another.

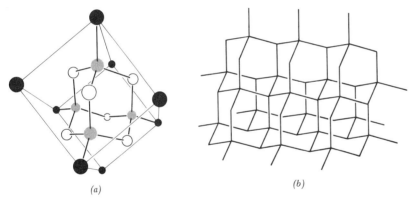

(a)   (b)

Figure 7.12. (a) Diamond unit cell. Diamond, like graphite, is composed only of carbon. Each large black circle represents a carbon nucleus at the eight vertices of the cubic unit cell. The open circles represent carbon nucleii at the center of each of the six cube faces. The gray circles are the four carbon atoms within the cube. All the carbons are exactly equivalent; only by defining a particular unit cell can we place certain carbon atoms at the unit-cell vertices. Another arbitrarily chosen unit cell could put these same atoms within the new cell or at the centers of the cell faces. The four interior carbon atoms are bonded tetrahedrally to four other carbons. (b) Diamond crystal is built up, joining very large numbers of these unit cells together, the front face of one touching the back of another, the top of one touching the bottom of another, and the right side face of one to the left face of another. In this way each carbon atom is bonded tetrahedrally to four others. Notice the long zigzag chains running through the lattice. The straight segments represent the carbon-carbon bonds of length 1.54 Å. The nuclei of the carbon atoms are at the bends of the zigzag chains. If the carbon atoms were drawn to scale, they should touch each other, obscuring the bonds and the general lattice features. The angles of bonding along the zigzag chains are uniformly 109°, the tetrahedral angle. Two chains run through each carbon atom; each carbon is bonded tetrahedrally to four other carbons.

Diamond scratches everything and nothing (except a very few synthetic materials) scratches it. Diamond is also quite *brittle*; that is, it breaks when struck. Hardness is not an indication of resistance to impact but to scratching only. Diamond boils at 4200°C, like graphite.

SILICA AND SILICATES. Silicon appears just below carbon in the periodic table; the electronic structures are $1s^2 2s^2 2p^6 3s^2 3p^2$ for silicon and $1s^2 2s^2 2p^2$ for carbon. Both elements require four more electrons to complete the outer s and p subshells; both have a valence of 4.

Silicon, the second most abundant element in the earth's crust, is not found as a pure metal but as an oxide, silica, which has two oxygens for each silicon atom. In quartz, the common form of silica, each silicon atom is tetrahedrally surrounded by and bonded to four oxygens (Fig. 7.13). Like diamond, the silica tetrahedra are connected to each other (by sharing oxygens), forming a continuous and regular network throughout the crystal—a single enormous molecule. For convenience, we assume the hypothetical single molecule to be silicon dioxide SiO_2, with a molecular weight of 60.085.

The tetrahedron can be represented by $[SiO_4]$. Tetrahedra are joined by sharing oxygens in common. Even though each silicon is bonded to four oxygens, each oxygen is bonded to and shared by two silicons. In effect, there are two oxygens for each silicon, thus the formula SiO_2. By comparing the unit cell top view in Fig. 7.13 with the prism top view in Fig. 7.1, we can see a relationship. The unit cell is a parallelepiped with an obtuse angle of 120°; this is the hexagon angle in the prism. The prism is simply a series of these unit cells repeated regularly in three dimensions.

The silica tetrahedron is almost ubiquitous in the earth's crust. Amethyst is quartz with impurities which give this gem mineral a purple color; jasper is quartz colored red-brown by iron oxides. Opal is an amorphous form of silica, containing water molecules; the tetrahedra are packed irregularly rather than in a repeated lattice.

Silicates are compounds of silica with other metal oxides. Feldspars are composed of silica; and aluminum, sodium and potassium oxides. Asbestos is a fibrous crystalline form of magnesium, silicon, and oxygen; talc has the same elements but a different composition and crystalline structure. A typical garnet contains silica

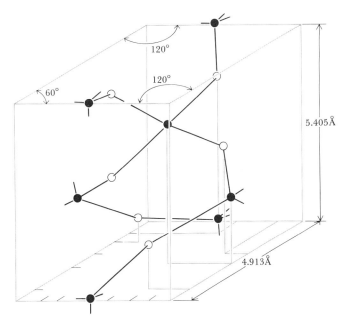

Figure 7.13. Quartz unit cell. Each black sphere represents a silicon nucleus; each white sphere is an oxygen nucleus. This unit cell is not cubic: the base is a parallelogram with four equal sides and an acute angle of 60°; the vertical sides are perpendicular to the base and are 5.405 Å high. There is one silicon atom on the top and bottom edges of the front and back faces or sides, one silicon atom on the left and right sides, and one silicon entirely within the unit cell prism. The oxygens are all within the unit cell. Each silicon atom is bonded tetrahedrally to four oxygens as seen with the interior silicon atom; each oxygen atom is bonded to two silicon atoms. The silicon atoms on the edges are bonded to oxygens from four different unit cells; the silicon atoms on the faces are bonded to oxygens from two unit cells.

and calcium and aluminum oxides. Topaz is an aluminum fluorosilicate, beryl is an aluminum silicate with some beryllium oxide, and mica is a crystalline mineral found in sheets or thin plates, containing SiO_2, Al_2O_3, K_2O, and H_2O. Ordinary clay is a hydrous mixture of aluminum silicates with fine quartz and feldspar particles. Granite, a rock made by the solidification of molten materials, is a mixture of quartz and feldspar.

In all of these materials, the silicon atoms are always bonded tetrahedrally to four oxygens. The oxygens in silicates are not all bonded each to two silicon atoms; some are bonded to metal ions. The fraction of oxygens bonded to silicon and to metal ions determines the continuity of the silica tetrahedron network and, usually, the ease of melting of the mineral.

If the silica tetrahedra are arranged irregularly, the substance is not crystalline; rather, it is amorphous, as is opal. Quartz which is melted and then cooled retains the random structure of the liquid when it becomes a rigid glass. Commercial glasses are made from sand (quartz), lime (calcium oxide CaO), and sodium oxide Na_2O. The metal oxides are added to lower the melting point and to make the manufacture of glass products easier.

Vitreous silica, a common name for silica in the glassy state, has the unusual property of not expanding very much when heated. Red-hot vitreous silica can be dropped into cold water without shattering. Ordinary commercial glass cannot withstand this drastic thermal treatment because it does significantly expand and contract with an increase and decrease in temperature. When abruptly chilled, the outside of a commercial soda-lime-silica glass cools and contracts more rapidly than the inside, producing a stress large enough to break the glass. *Borosilicate* glasses are made from silica and boron oxide B_2O_3 in a continuous network. These glasses are almost as good as vitreous silica in being resistant to thermal shock, and the borosilicates are widely used as cooking vessels.

Pure silica melts at about 1710°C. When then cooled below this temperature, vitreous silica is formed because the very high viscosity of the melt does not permit the tetrahedra to line up in a regular lattice; the random liquid structure is frozen in. When the composition is about 22 percent Na_2O and 78 percent SiO_2, it melts at 1000°C. It is cheaper to melt such a mixture. However, the more sodium oxide present, the less viscous is the melt, that is, the more likely it is to crystallize rather than form a rigid glass. There is then a limit to how much metal oxide can be added to silica in glass manufacture. The less viscous the melt, the more rapidly it must be cooled to form a glass. Above a metal oxide limit for glasses, no matter how fast the melt can be cooled, it is too slow for glass formation. Near this limiting composition, glass can be made which is barely stable.

Special glass compositions have been made where gentle heat treatment converts part of the material into small crystals imbedded in a glassy matrix. This matrix can have impressive durability to both thermal and mechanical shock and is also used in ovenware. In Fig. 7.14 several silica and silicate structures are shown.

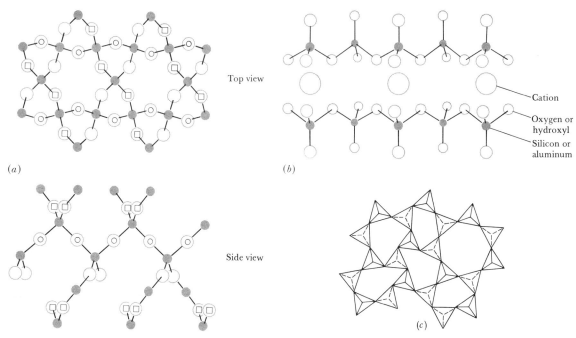

Top view

Cation

Oxygen or
hydroxyl

Silicon or
aluminum

Side view

(*a*)

(*b*)

(*c*)

Figure 7.14. (a) Structure of quartz. The small dark spheres represent the silicon atoms, each of which is bonded tetrahedrally to four oxygen atoms (represented by the larger circles). The hexagonal nature of quartz can be seen from the top view; the helical pattern can be seen from the side view and Fig. 7.13. All silicas and silicates have a tetrahedral structure. A silicon atom is at the center, and one oxygen is at each of the four vertices. Since quartz is crystalline, the tetrahedra must be arranged in a regular lattice. The oxygen atoms are labeled according to distance from viewer. (b) Part of a silicate mineral. Here each silicon or aluminum atom is surrounded tetrahedrally by four oxygens or hydroxyl groups —OH. In contrast to (a), this mineral has more oxygen per silicon atom. Only three out of four oxygens are shared by two tetrahedra. As a consequence, the silicate network is not continuous and has a net negative charge. To maintain electric neutrality in the mineral, the cations (from constituent metal oxides) are located regularly within the crystal lattice. (c) A view of vitreous silica, pure amorphous SiO_2, in which the tetrahedra are not arranged regularly. If a glass were made with silica and sodium oxide Na_2O then not all oxygens would be joined to two silicon atoms; the continuous network would be interrupted.

FUNCTIONAL GROUPS

We have already looked at two large classes of organic compounds, the aromatic and aliphatic hydrocarbons, and some of their chlorine derivatives. An aliphatic or aromatic hydrogen can be replaced by any of a large number of substituents. Because a particular substituent converts the parent molecule into a new molecule which functions like or has properties characteristic of the substituent, the added atom or group of atoms is called a "functional group."

Alcohols are those compounds in which a hydrogen (not bonded to an aromatic-ring carbon) is replaced by a hydroxyl group OH. If the hydroxyl group replaces a hydrogen right on an

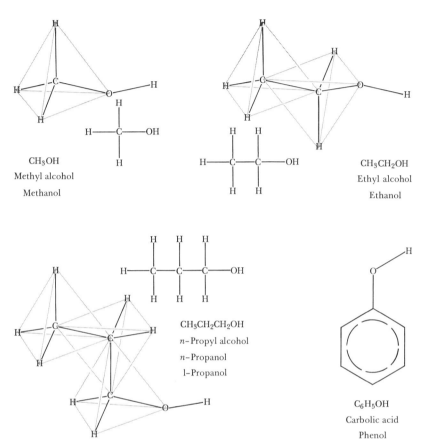

CH₃OH

Methyl alcohol

Methanol

CH₃CH₂OH

Ethyl alcohol

Ethanol

CH₃CH₂CH₂OH

n–Propyl alcohol

n–Propanol

1–Propanol

C₆H₅OH

Carbolic acid

Phenol

Figure 7.15. Some alcohols and phenol. An alcohol is an organic molecule containing the hydroxyl group —OH attached to a carbon atom. If the hydroxyl group is attached to an aliphatic group, the compound is represented by the general formula ROH. If the hydroxyl is attached to an aromatic ring, the compound is called a "phenol" ArOH. Because there is free rotation about the single bonds, each molecule has a continually changing conformation.

aromatic ring, the compound is a *phenol*. Figure 7.15 gives the structural formulas for a few common alcohols and phenols; their names derive from the parent hydrocarbon. A benzene ring with one hydrogen removed is the phenyl radical C_6H_5. For convenience, the alcohols are collectively represented by the formula ROH and the phenols by ArOH, where R and Ar are the parent molecule radicals (see Prob. 7.7).

Although these compounds look like the inorganic hydroxides, they are not alkaline. In phenol, 103 kcal/mole are required to remove the OH and only 85 kcal/mole are required to remove the H. Phenol is a weak acid and is sometimes called "carbolic acid." Depending on the circumstances of a reaction, either the hydroxyl or the hydrogen is removed from an alcohol or phenol.

Another important group of organic compounds has the *carboxylic acid* functional group

$$-C\underset{O-H}{\overset{\displaystyle O}{\Big\langle}}$$

Some carboxylic acids RCOOH are shown in Fig. 7.16. Most of these are weak acids, such as acetic acid CH_3COOH (the principal constituent in vinegar).

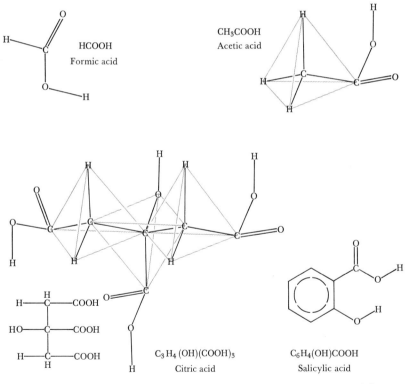

Figure 7.16. Some carboxylic acids. Carboxylic acids contain the group —COOH and donate a proton upon ionization. If the carboxylic group is attached to an aliphatic group, the acid is represented as RCOOH; if it is attached to an aromatic group, then the acid is represented by Ar-COOH. Formic acid was first obtained from the bodies of ants. Acetic acid is the principal substance in vinegar. Citric acid is present in citrus fruits. Salicylic acid and its derivatives are used as analgesics or anodynes (e.g., aspirin).

Ethers are molecules in which two radicals are connected by an oxygen. They have the general form ROR′, ArOR, and ArOAr′, where the primes indicate that the two radicals need not be the same. Some ether structures are given in Fig. 7.17. Diethyl-ether $CH_3CH_2OCH_2CH_3$ is the common anesthetic ether. Ethers, in general, are water-insoluble and good solvents for many non-

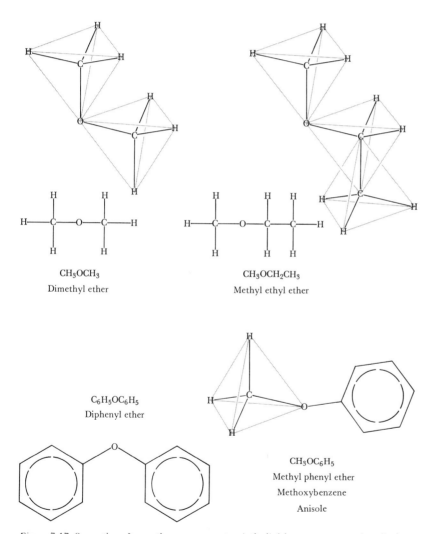

CH_3OCH_3

Dimethyl ether

$CH_3OCH_2CH_3$

Methyl ethyl ether

$C_6H_5OC_6H_5$

Diphenyl ether

$CH_3OC_6H_5$

Methyl phenyl ether

Methoxybenzene

Anisole

Figure 7.17. Some ethers. In an ether an oxygen atom is the link between two organic radicals or groups. The two linked groups can be identical aliphatic radicals (represented by ROR), two different aliphatic groups (ROR′), two aromatic groups (ArOAr or ArOAr′), or mixed aliphatic and aromatic (ROAr).

polar and nonionic molecules. Because they are themselves non-polar, there is not much association among ether molecules in the liquid state and their boiling points are low. After a substance is extracted from a mixture by selective solution in ether, it is easy to remove the solvent by evaporation and obtain the pure solid or liquid as a residue.

When two hydrogens on a common carbon are replaced by a single oxygen, the compound formed is either an *aldehyde*

$$\overset{\displaystyle H}{\underset{\displaystyle }{R-C}}=O$$

or a *ketone*

$$\underset{\displaystyle \underset{\displaystyle O}{\|}}{\overset{\displaystyle R \qquad R'}{\underset{\displaystyle C}{\diagdown\diagup}}}$$

depending on the oxygen being placed on an end or interior carbon. A few representative aldehydes and ketones are given in Fig. 7.18. Ketones are often used as solvents for lacquers, while aldehydes react with many other compounds and are widely used in building new molecules.

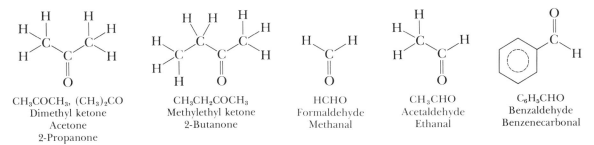

CH₃COCH₃, (CH₃)₂CO — Dimethyl ketone — Acetone — 2-Propanone

CH₃CH₂COCH₃ — Methylethyl ketone — 2-Butanone

HCHO — Formaldehyde — Methanal

CH₃CHO — Acetaldehyde — Ethanal

C₆H₅CHO — Benzaldehyde — Benzenecarbonal

Figure 7.18. Some ketones and aldehydes. A ketone contains the carbonyl group C=O, *the carbon of which is bonded to two other groups. The general formula for a ketone is RR'CO. The two groups can be identical or different, aliphatic or aromatic. An aldehyde RCHO is a special case of a ketone, where one of the groups is a hydrogen atom. In formaldehyde H₂CO both groups are hydrogen.*

If one or more hydrogens in ammonia NH_3 are replaced by radicals, the substituted ammonia is called an "amine." These compounds have unpleasant, decomposed, fishy odors. Figure 7.19 shows the structures of ammonia and a few simple amines.

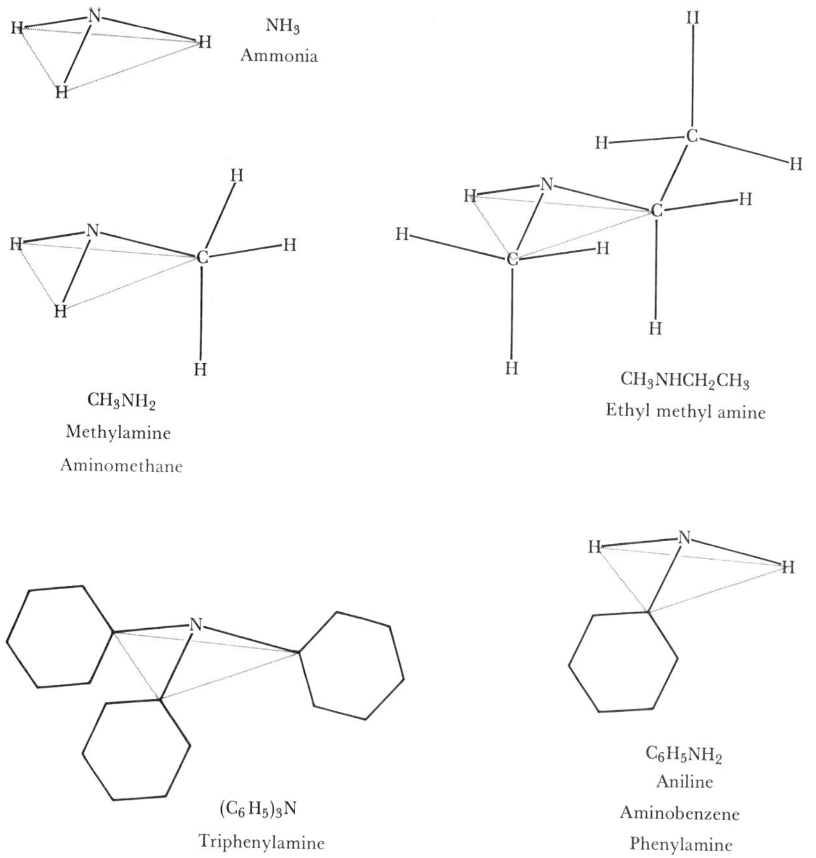

Figure 7.19. Ammonia and some amines. The ammonia molecule is pyramidal. No atom is within the tetrahedron. When one hydrogen of the parent ammonia molecule is replaced by a group (as in methylamine), the resulting compound is a "primary amine." "Secondary" and "tertiary" amines are made by replacing two and three hydrogens; the pyramid base spreads out to accommodate the increased bulk.

MULTIFUNCTIONAL MOLECULES

A wider range of properties is possible when combinations of two or more functional groups are present in a single molecule. The kind and number of functional groups and their geometric arrangement on the parent molecule all contribute to the behavior of the multifunctional molecule.

There are two compounds with the formula $C_4H_4O_4$, each of which has a carbon-carbon double bond and two carboxylic acid groups not attached to the same carbon:

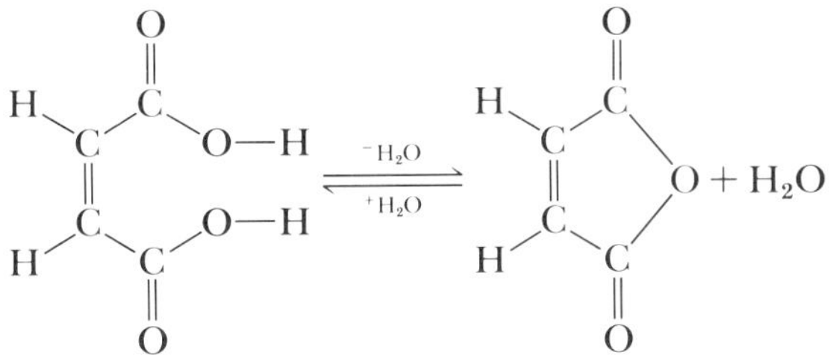

Maleic acid (cis isomer) 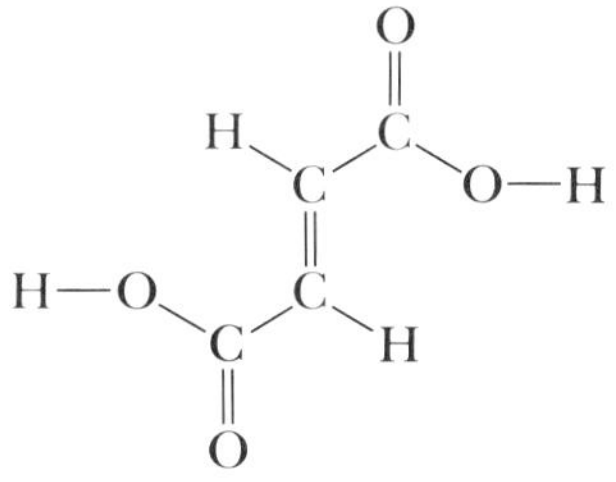   Fumaric acid (trans isomer)

These isomers, fumaric and maleic acids, have quite different properties, which are listed in Table 7.2. Because water is more easily removed from it, we know maleic acid is the cis isomer.

Table 7.2

Property	Fumaric acid	Maleic acid
Density, g/cm³	1.635	1.590
Melting point, °C	287	130
Boiling point, °C	290	Decomposes at 135°C
Solubility at 25°C, g		
in 100 cm³ water	0.70	79
in 100 cm³ alcohol	5.8	70
in 100 cm³ ether	0.72	8

MULTIPLE HYDROXYL GROUPS

If two or more hydroxyl groups are present, the molecule is called a "polyhydric alcohol." Figure 7.20 shows three simple polyhydric alcohols.

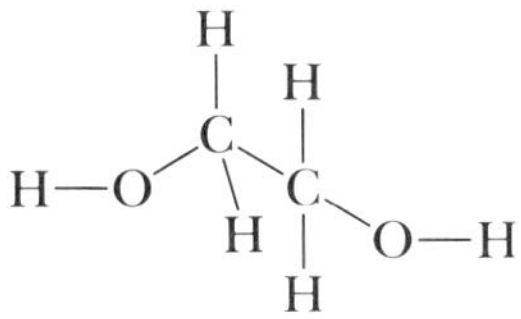

CH_2OHCH_2OH
Ethylene Glycol
1,2-Ethanediol

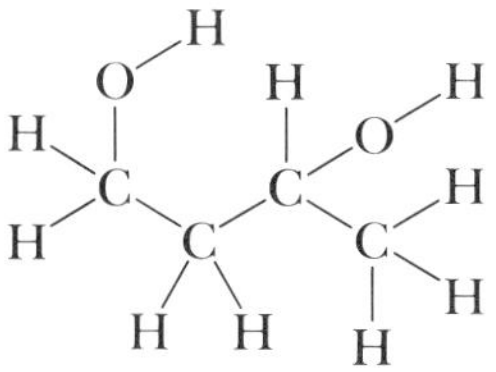

$CH_2OHCH_2CHOHCH_3$
1,3-Butanediol

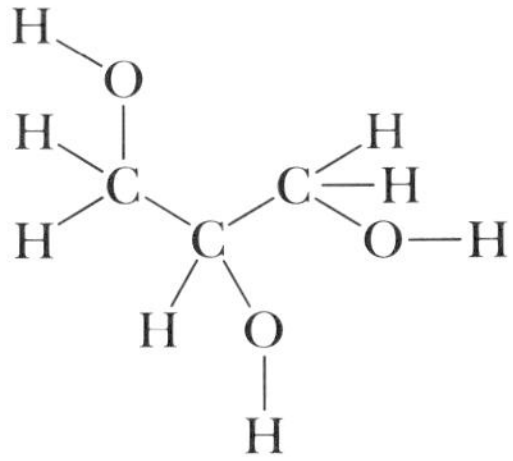

$CH_2OHCHOHCH_2OH$
Glycerol
Glycerin
1,2,3-Propanetriol

Figure 7.20. Compounds having more than one hydroxyl group.

Glucose $C_6H_{12}O_6$ is a sugar found in grapes In human beings starches and sugars are *digested*, or broken down chemically, into glucose and other molecules to be used as the principal source of energy. One mole, 180.16 g, of glucose is oxidized to carbon dioxide and water in an exothermic series of reactions, releasing 673,000 cal of energy. Originally glucose was thought to be a six-carbon straight-chain molecule, but certain reactions and x-ray diffraction-structure studies indicate the molecules are mostly in one of two ring configurations and only a small fraction of the total molecules are in the chain form at any one time. The three structures are shown in Fig. 7.21.

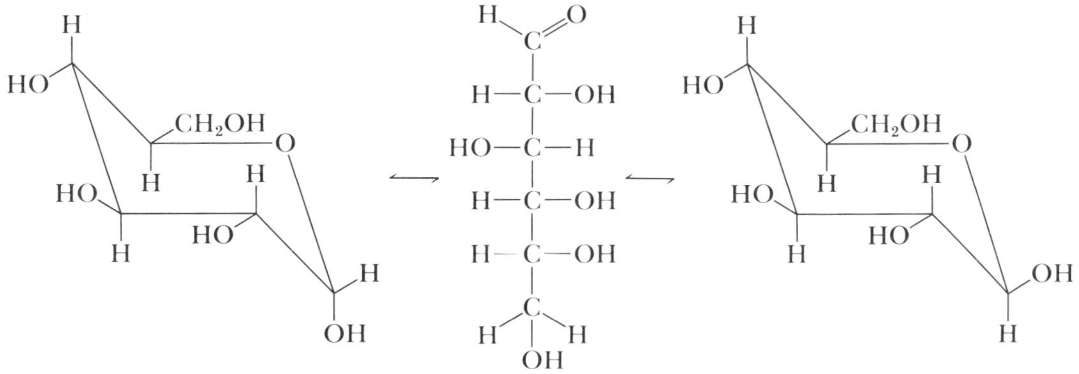

α-D-Glucose (chair form)   Open-chain aldehyde form   β-D-Glucose (chair form)

Figure 7.21. The forms of glucose, a hexose, or six-carbon, sugar. Almost all the glucose molecules are in the form of a puckered ring or chair form of cyclohexane. There are five carbons (not explicitly shown) and one oxygen in each ring. The sixth carbon is attached to the ring as a side group. The alpha and beta forms of the ring differ only in the direction of a single hydroxyl group. In the α form this OH group is axial and in the β form equatorial (see Fig. 7.7b). Each carbon is tetrahedrally bonded to four other atoms, except the end carbon in the open-chain form, which has the carbonyl group C=O. The open chain is actually a zigzag with free rotation about all single bonds. A ring opens up to the chain and then recloses either to the same or to the alternate chair form. The designation D describes a structural property of these sugars to be discussed in the text.

Another sugar with the same molecular formula—$C_6H_{12}O_6$—is fructose. It has three forms, a five-membered ring, a six-membered ring, and a straight chain (Fig. 7.22). Sugars with six carbons, such as glucose and fructose, are called "hexoses." Almost all simple sugars found in nature are either hexoses or *pentoses* (sugars having five carbon atoms).

Common table sugar, which is sucrose, is one glucose and one fructose (Fig. 7.23) joined through an oxygen by combining a hydroxyl from each and removing a water molecule. For this

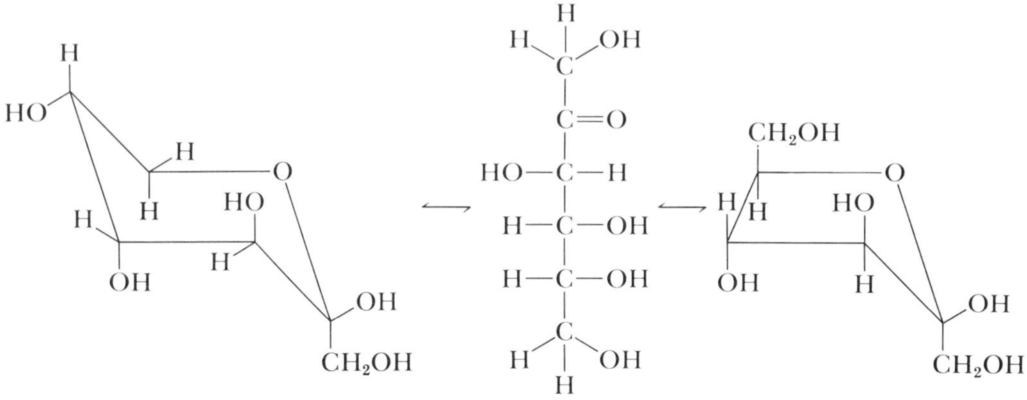

β-D-Fructose　　Open-chain keto form　　β-D-Fructofuranose

Figure 7.22. The forms of D*-fructose, another hexose.* D*-fructose can form either a six- or five-atom ring and the open-chain form. The open chain contains the carbonyl group of a ketone and for this reason is called the "keto form." The five-atom ring is called "fructofuranose" because it can be derived from the molecule furan, also a five-atom ring. The furanose ring is bent rather than planar.*

reason the sucrose formula is not exactly the sum of glucose and fructose but has one water less: $C_{12}H_{22}O_{11}$.

Sucrose is called a "disaccharide" because it is made from two simpler sugars. Hexoses and pentoses are called "monosaccharides." Starches and cellulose are very large molecules made up of many hexoses and pentoses and for this reason are called

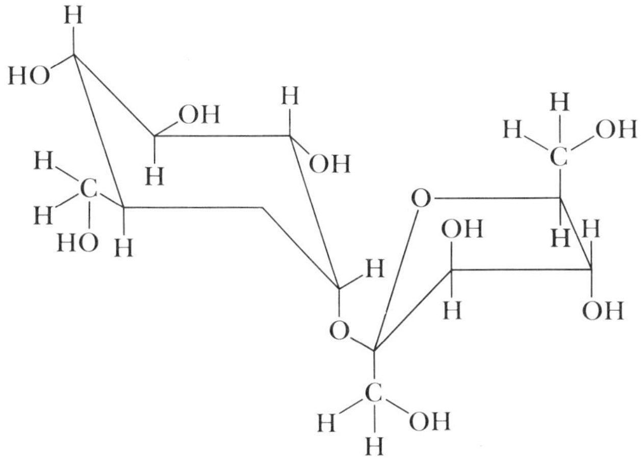

Sucrose

Figure 7.23. Common table sugar, sucrose, is a disaccharide because the molecule is made from two simpler hexoses: the six-atom ring glucose and the five-atom ring fructose.

"polysaccharides." In almost all sugars, starches, and celluloses, there are twice as many hydrogens as oxygen atoms, just like water. Collectively, these compounds are called "carbohydrates."

> **Example.** In the field of nutrition the food-energy-value unit is the *large* calorie, actually 1,000 cal, or 1 kcal. For the comparison of different foods, it is usual to state the large calorie value of 100 g of each food. From the heat of combustion of glucose, 673,000 cal/mole, what is its energy value? (Molecular weight = 180.16.)

$$100 \text{ g glucose} \times \frac{673,000 \text{ cal}}{180.16 \text{ g glucose}} \frac{1 \text{ kcal}}{1,000 \text{ cal}} = 374 \text{ kcal}$$

AMINO ACIDS

An *amino acid* is a molecule containing both an amine ($-NH_2$) and a carboxylic acid ($-\overset{\overset{\text{O}}{\|}}{C}-OH$ or $-COOH$) group. If the two functional groups are attached to the same carbon, the substance is an α-amino acid (Fig. 7.24). Amino acids are important compounds in all living organisms. Proteins are made by joining together these bifunctional amino acids. The interesting and important aspects of amino acids arise from the fact that each molecule contains both the basic amino group and the acidic carboxylic group.

ESTERS

The reaction between an alcohol ROH and a carboxylic acid $R'\overset{\overset{\text{O}}{\|}}{C}OH$ resembles the neutralization reaction between a base MOH and an acid HX:

$$ROH + HO-\overset{\overset{\text{O}}{\|}}{C}R' \longrightarrow RO-\overset{\overset{\text{O}}{\|}}{C}R' + H_2O$$
$$MOH + HX \longrightarrow M^+ + X^- + H_2O$$

One obvious difference is that the product M^+X^- is an ionic salt while the product $ROOCR'$ is not. The general molecules $RO\overset{\overset{\text{O}}{\|}}{C}R'$

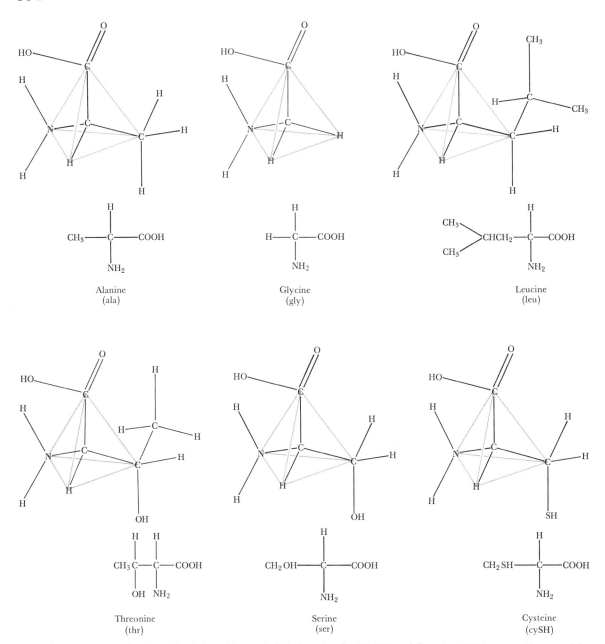

Figure 7.24. Some natural amino acids. Amino acids contain both the carboxylic (COOH) and the amino (NH₂) groups. These are alpha amino acids because the two groups are attached to the same carbon atom. Below each tetrahedral figure is the plane formula, the name, and the common abbreviation for these six common amino acids.

or ROOCR' are called "esters." A reaction in which an ester is the product is an *esterification*.

The volatile esters have pleasant odors and are responsible for many fruit and leaf smells. Propyl acetate, made from propyl alcohol and acetic acid

$$CH_3CH_2CH_2OH + HO\overset{\overset{\displaystyle O}{\|}}{C}CH_3 \longrightarrow CH_3CH_2CH_2O\overset{\overset{\displaystyle O}{\|}}{C}CH_3 + H_2O$$

has the odor of pears. Methyl salicylate, from methyl alcohol and salicylic acid, is the ester in wintergreen. Isoamyl acetate

$$(CH_3)_2CHCH_2CH_2O\overset{\overset{\displaystyle O}{\|}}{C}CH_3$$ called "pear oil," or "banana oil," contributes to the odor of these fruits. These esters can be artificially made. In fact, proper blends of esters can approximate the natural smells of most fruits and berries.

There is another difference between ester and salt formation. In making a salt, the acid contributes a hydrogen and the base a hydroxide; in making an ester, it is possible to determine from which molecules the hydrogen and hydroxyl groups come.

Oxygen has three stable isotopes O^{16}(99.76 percent), O^{17}(0.04 percent), and O^{18}(0.20 percent). Water containing the last isotope is heavier than water containing either of the first two. Repeated evaporation of water leaves a residue rich in O^{18}. This isotope is stable and can be detected in a mass spectrometer.

Although acetic acid donates a proton to a base in a chemical reaction, it is possible to imagine the esterification

$$CH_3\overset{\overset{\displaystyle O}{\|}}{C}{-}OH + HOCH_2CH_3 \longrightarrow CH_3\overset{\overset{\displaystyle O}{\|}}{C}{-}O{-}CH_2CH_3 + H_2O$$

Acetic acid   Ethyl alcohol   Ethyl acetate

to occur by either of two mechanisms:

1. Water is formed from the acid proton and the alcohol hydroxyl.
2. Water is formed from the acid hydroxyl and the alcohol proton.

This problem can be resolved by preparing an ethyl alcohol rich in O^{18}. The two mechanisms

$$\text{CH}_3\overset{\overset{\displaystyle O}{\|}}{\text{C}}-\text{O}\boxed{\text{H}+\text{HO}^{18}}\text{CH}_2\text{CH}_3 \xrightarrow{\ 1\ } \text{CH}_3\overset{\overset{\displaystyle O}{\|}}{\text{C}}\text{OCH}_2\text{CH}_3 + \text{H}_2\text{O}^{18}$$

$$\text{CH}_3\overset{\overset{\displaystyle O}{\|}}{\text{C}}\boxed{\text{OH}+\text{H}}\text{O}^{18}\text{CH}_2\text{CH}_3 \xrightarrow{\ 2\ } \text{CH}_3\overset{\overset{\displaystyle O}{\|}}{\text{C}}\text{O}^{18}\text{CH}_2\text{CH}_3 + \text{H}_2\text{O}$$

will lead to (1) water rich in O^{18} or (2) ethyl acetate rich in O^{18}. Evidence from the mass spectrograph indicates ethyl acetate rich in O^{18} is found. The second mechanism is the correct one, contrary to what one may have predicted (see Prob. 7.8).

FATS

In addition to carbohydrates and proteins, a third important class of compounds is found in living systems. These are the *lipids,* constituents of plants and animals which are insoluble in water and soluble in ether or hexane. Among the lipids are the *fats,* esters of glycerin (the trihydric alcohol propane-1,2,3-triol), and long-chain carboxylic acids, the *fatty acids.*

Since glycerin has three hydroxyl groups, it can be esterified with three acids. If the three acids are the same, the fat is a *simple glyceride;* if not, it is a *mixed glyceride.* The acids can be saturated or unsaturated. The great majority of naturally occurring fatty acids have an even number of carbons. A typical fat, shown in Fig. 7.25, is made from two saturated fatty acid molecules, stearic acid $C_{17}H_{35}COOH$, and from one unsaturated fatty acid, oleic acid $C_{17}H_{33}COOH$.

Because there is free rotation about all the single bonds, the fat assumes many shapes and converts readily from one to another. The carbon-hydrogen bonds are nonpolar and the keto groups $C{=}O$ polar. Because there are many more of the former than the latter, fats typically are not water-soluble.

Fats can be broken up into glycerin and the alkali salts of the fatty acids by the action of strongly alkaline agents. Since a hydroxyl group —OH is added to each fatty acid part and hydrogen to the glycerin part, in effect, water is added to a fat to form glycerin and fatty acids. The process, for this reason, is called a "hydrolysis." This term is also used for the conversion of any ester

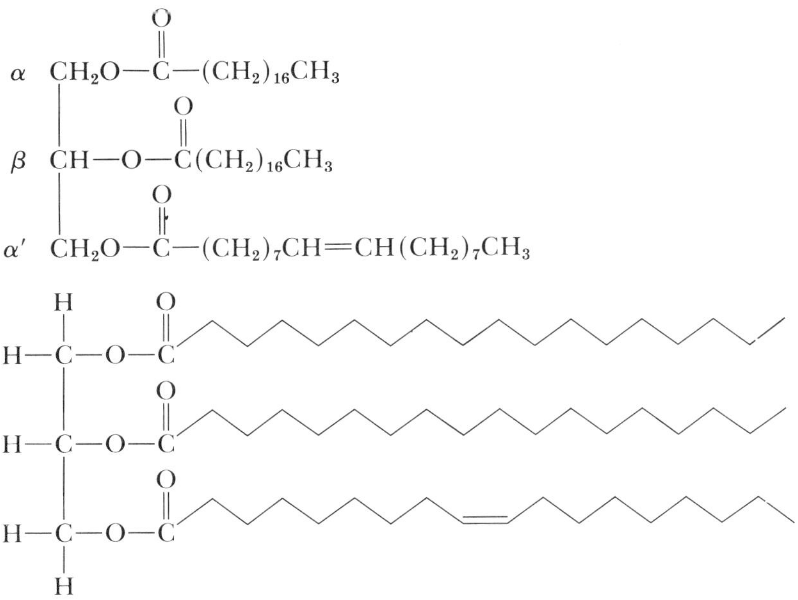

$$\alpha \quad \underset{|}{CH_2}O-\overset{\overset{O}{\|}}{C}-(CH_2)_{16}CH_3$$

$$\beta \quad \underset{|}{CH}-O-\overset{\overset{O}{\|}}{C}(CH_2)_{16}CH_3$$

$$\alpha' \quad CH_2O-\overset{\overset{O}{\|}}{C}-(CH_2)_7CH{=}CH(CH_2)_7CH_3$$

Figure 7.25. A fat is the ester of glycerin and three long-chain fatty acids. In this example the fat is composed of two stearic acid molecules and one oleic acid molecule. The three carbons of the glycerin moiety are labeled α, β, *and* α'. *This particular fat has the name* α, β-stearo-α'-olein, *or* α-oleo-α', β-stearin. *Because there is free rotation about all the single bonds, the fat molecule continually changes its conformation.*

into its alcohol and acid and for the breaking up of a polysaccharide into smaller sugars.

The hydrolysis of fat by alkali such as sodium hydroxide is called "saponification" because the products formed, the salts of the fatty acids, are soaps.

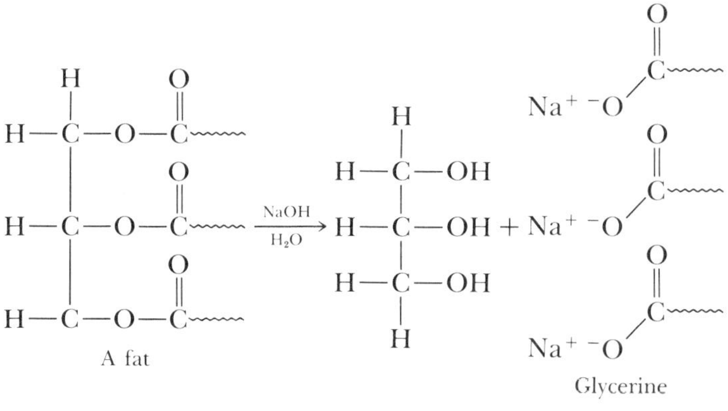

A fat $\xrightarrow[\text{H}_2\text{O}]{\text{NaOH}}$ Glycerine

A *soap* is any metal salt of an acid with seven or more carbon atoms. The carboxylic end of a soap molecule is ionic because a strong base has removed the proton, leaving an anion. This negatively charged end is water-soluble because of the strong forces between it and the polar water molecules. The rest of the soap molecule is nonpolar and water-insoluble and mixes well with other nonpolar substances, e.g., greases, oils, and fats. For this reason, a soap is a good cleansing agent; one end mixes well with water-insoluble dirt, and the other mixes well with water. In this way dirt can be washed away by water (Fig. 7.26).

AMIDES

When water is removed from a carboxylic acid, the product is an *acid anhydride*. The addition of water will convert the anhydride back to carboxylic acid. We have already looked at this pair of reactions in maleic acid, a dicarboxylic acid. However, if the molecule is a monocarboxylic acid, then two molecules must combine to form the anhydride:

Carboxylic acid   Acid anhydride

The addition of water to the anhydride is another example of *hydrolysis*.

When ammonia NH_3 is used in place of water, the process is called "ammonolysis":

Acetic anhydride   Acetamide

Acetic acid

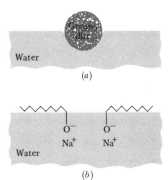

(a)

(b)

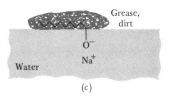

(c)

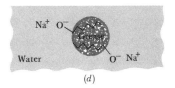

(d)

Figure 7.26. Soap action. (a) Ordinarily a grease- or oil-containing dirt will not mix with water. (b) Soap molecules have one end which can dissolve in water and another end which cannot. Because of this property, soap molecules are concentrated at the surface of water rather than being distributed uniformly throughout. (c) The water-insoluble end of a soap molecule can dissolve in a grease or oil, a common ingredient of dirt. (d) The dirt particle can then mix with water through the agency of soap molecules; in effect, the grease drop has an ionic coat which dissolves in water, and water has a hydrocarbon surface which mixes with grease.

The new molecule formed CH_3CONH_2 is acetamide. In general, the substance formed from a carboxylic acid RCOOH by replacing

$$\overset{\text{O}}{\overset{\|}{}}$$

the hydroxyl with the amine group is an *amide* $R\overset{\text{O}}{\overset{\|}{C}}NH_2$ or $RCONH_2$.

If an amine RNH_2 is used in the ammonolysis of an acid anhydride, the amide formed also contains R. In some cases, the amine can react directly with the carboxylic acid:

$$CH_3\overset{\text{O}}{\overset{\|}{C}}OH + H_2NC_6H_5 \longrightarrow CH_3\overset{\text{O}}{\overset{\|}{C}}NHC_6H_5 + H_2O$$

The important feature of all amides is the presence of the four atoms in this arrangement:

$$-\underset{\underset{\text{O}}{\|}}{C}-\underset{\underset{\text{H}}{|}}{N}-$$

ABSORPTION SPECTROSCOPY

The molecular geometry of a substance can be determined from its chemical properties or by diffraction techniques. These are often lengthy investigations, but they can be speeded up if we know which functional groups are present in the molecule.

Esters have characteristic fruity smells, and amines have rotten-fish odors. Chemists can identify these and other groups by a number of tests. But a convenient and rapid physical method is also available, based on the way in which a substance absorbs light. Figure 7.27 is a representation of the instrument used in this method, a *spectrophotometer.*

COLOR

A substance appears white if the light from it reaching our eyes is a uniform blend of all the wavelengths found in the visible spectrum, i.e., from 380 to 800 nm (Fig. 4.37). This light can be either transmitted or reflected by the substance. If no light is transmitted or reflected, the substance is black. A gray object sends a uniform blend of low-intensity light to our eyes.

If the visible light we see is not a uniform blend, then one

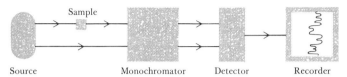

Figure 7.27. An absorption spectrophotometer. Light from the source is split into two beams; one passes through a sample, the other through air. Both beams then pass through a monochromator, which passes both beams, in narrow-wavelength segments, to the detector where the beams are compared, wavelength by wavelength. If a particular wavelength is not absorbed by the sample, the intensities of the two beams at that particular wavelength are equal; but if there is some absorption, the intensity of the beam through the sample will be less than the other beam at the wavelength absorbed by the sample. The recorder writes a record of these comparisons, called the "absorption spectrum" of the sample.

region of the visible spectrum is more intense than another and we sense the color associated with that dominant region. For example, a solution of copper sulfate absorbs strongly in the region from 700 to 1000 nm. This means it selectively removes red and near infrared light but not purples, blues, greens, or yellows. What is transmitted is light that is deeply blue.

Dyes are molecules made especially to selectively absorb certain wavelengths and transmit others associated with the colors desired. Dyes must have other properties, too, depending on their use. In coloring fabrics the dye molecule is bonded to the fabric substance. It should be stable under the conditions of use.

When the wavelengths absorbed are in the visible or in the ultraviolet region, the absorption is due to an electronic transition (Fig. 5.24) from a lower to a higher permitted energy level. If no part of the visible spectrum is absorbed in such a transition, the visible spectrum is unaffected and no color results.

INFRARED ABSORPTION SPECTRA

Molecules move about with a kinetic energy that depends on the temperature. They have various electronic energies described by quantum mechanics. In addition, there are twisting and rotation about chemical bonds, rotations of the entire molecule in space, and vibrations between atoms joined by bonds. These vibrational and rotational energies cannot have any arbitrary values but can assume only certain levels, as described by quantum mechanics; that is, vibrational and rotational energies are quantized.

Transitions from one permitted vibrational or rotational energy to another require the absorption or emission of energy $\Delta\epsilon$

in the form of electromagnetic radiation. The frequency of this radiation ν is related to the energy by the Planck equation $\Delta\epsilon = h\nu$. Because the vibrational and rotational energies are limited in number, so also are the possible energy differences $\Delta\epsilon$ between these levels and the frequencies ν and wavelengths λ of the radiation absorbed or emitted. The frequencies and wavelengths absorbed in vibrational and rotational transistions are all in the infrared region of the spectrum. The record from an absorption spectrophotometer in this region, usually from 2.5 to 15μ, is the *infrared absorption spectrum.*

The permitted energies and the differences between these permitted levels depend on the atomic constituents and geometry of the molecule, the molecular mass, and the interatomic bonds. The stiffness of a bond is analogous to the stiffness of a spring in that it determines the frequency of vibration between masses connected by the bond or spring. The free rotation about a single bond can be hindered if the atoms rotating are large enough to interact with each other.

No two different substances have molecules with exactly the same bonds and masses. For this reason, the vibrational and rotational transitions and the infrared spectra are unique for each substance. The infrared (ir) spectra for three compounds are shown in Fig. 7.28. Just as human beings have individual fingerprints, molecules have individual ir spectra.

An unknown substance can be quickly identified by its ir spectrum. For example, if it has a spectrum which is identical in detail to that shown in Fig. 7.28*a*, it can only be acetophenone and nothing else. Also, when a compound is made, its purity can be determined in this way. A sample of synthesized pure allyl alcohol must have uniquely the spectrum of Fig. 7.28*b*. Any additional absorptions indicate the presence of impurities. Thus ir helps the scientist to determine the purity as well as the identity of a compound.

FUNCTIONAL GROUPS AND ir SPECTRA

Acetophenone (Fig. 7.28*a*) absorbs strongly at 13 and 14.35 μ. An examination of many spectra indicates that aromatic molecules absorb strongly in the region between 11.5 and 14.5 μ. All ketones absorb strongly at wavelengths in the regions 5.5 to 6.0 and 7.2 to 9.2 μ. Acetone (Fig. 7.28*c*) and acetophenone (Fig. 7.28*a*) are both ketones and have these characteristic absorptions.

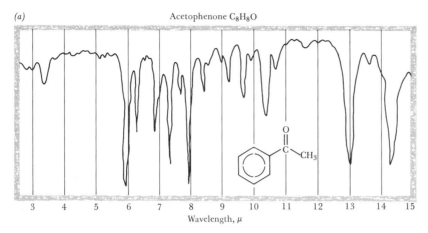

(a) Acetophenone C_8H_8O

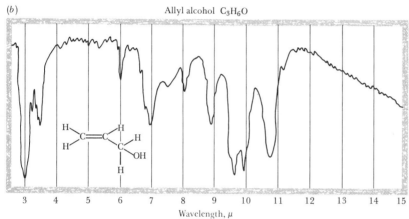

(b) Allyl alcohol C_3H_6O

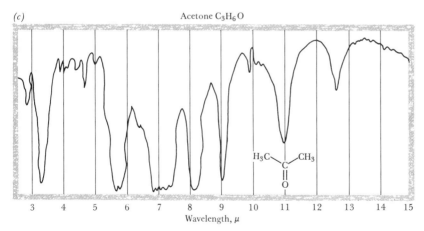

(c) Acetone C_3H_6O

Figure 7.28. The infrared absorption spectrum of three compounds. The wavelength is plotted along the horizontal axis in micron units; the absorption is plotted along the vertical axis. The low points, or troughs, in these curves occur at wavelengths of absorption; the deeper the trough, the more strongly is light absorbed at that particular wavelength. High points, or crests, occur at weak or no absorption. The pattern of crests and troughs is characteristic of a substance.

By looking at structural formulas and ir spectra, it is possible to assign certain absorption wavelengths to particular functional groups and parent molecules. An experienced chemist can learn much about the structure of a molecule from its ir spectrum.

THE IDENTIFICATION OF A COMPOUND

The simple sugar glucose is converted to ethyl alcohol by the action of microorganisms. This fermentation is important in the production of alcoholic beverages. If the process is allowed to continue longer than a certain time, the alcohol will be converted to acetic acid, the principal constituent of vinegar. A number of by-products are also made, and these give wines and distilled liquors their characteristic tastes and smells.

One such by-product can be isolated from a typical wine and identified. An analysis indicates that it is 54.53 percent carbon, 9.15 percent hydrogen, and 36.32 percent oxygen by weight. The atomic composition is determined in Table 7.3. Therefore, the molecule may be C_2H_4O, $C_4H_8O_2$, $C_6H_{12}O_3$, or so on. The basic molecule has a molecular weight

$$(2 \times 12.01) + (4 \times 1.008) + (1 \times 16.00) = 44.05$$

Then the true molecular weight may be 44.05, 88.10, 132.15, or some other whole-number multiple. A molecular-weight determination indicates the value is about 90; therefore, the proper formula is $C_4H_8O_2$, its molecular weight being 88.10.

Table 7.3

Element	Sample, g/100 g	Atomic weight	Relative no. of atoms	
C	54.53	12.01	4.540	2.000
H	9.15	1.008	9.077	3.999
O	36.32	16.00	2.270	1.000

With this information a chemist will probably consult a table listing compounds according to formulas. In the case of $C_4H_8O_2$ he will find 13 common isomers (Fig. 7.29a). The ir spectra of two of these isomers are shown in Fig. 7.29b and c. It turns out that the by-product spectrum is identical to Fig. 7.29b; the by-product is the ester ethyl acetate.

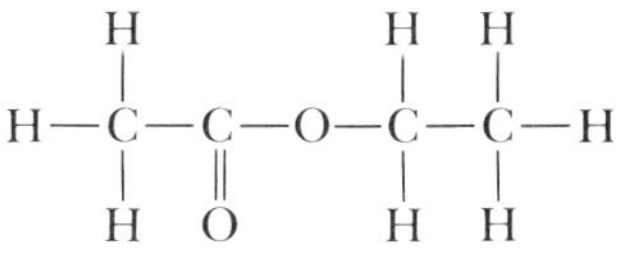

Ethyl acetate

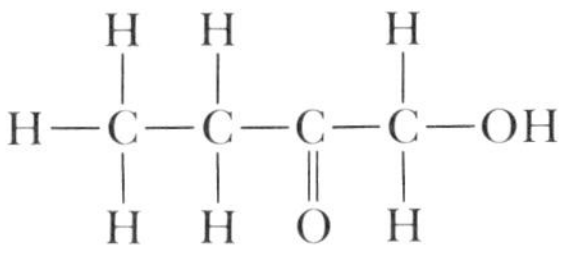

1-Hydroxy-2-butanone

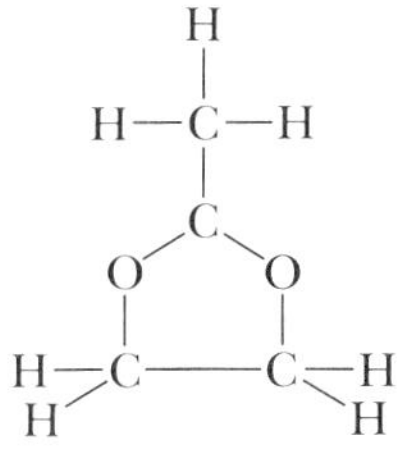

2-Methyl-1,3 dioxolane

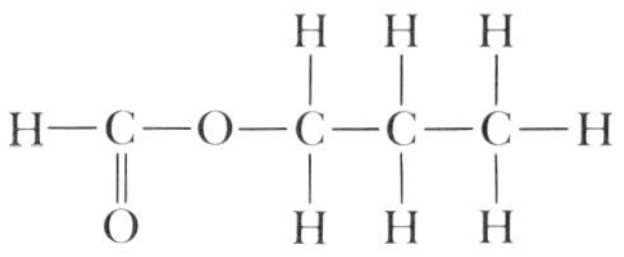

n-Propyl formate

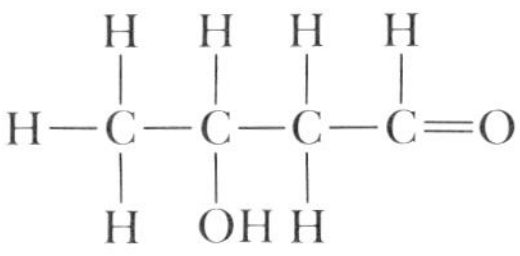

Aldol

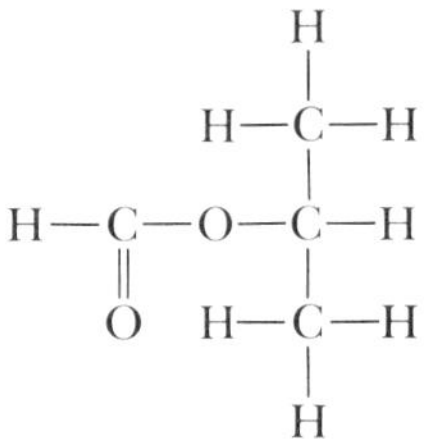

Isopropyl formate

Figure 7.29(a) Thirteen isomers with the formula $C_4H_8O_2$. (Continued on next page.)

(a) *Continued*

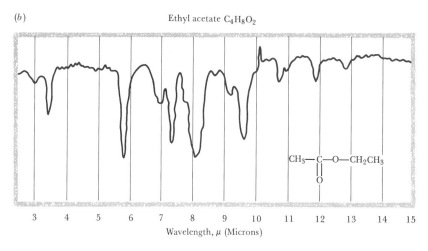

(b) Ethyl acetate C₄H₈O₂

CH₃—C—O—CH₂CH₃
  ‖
  O

(c) Paradioxane *p*—Dioxane C₄H₈O₂

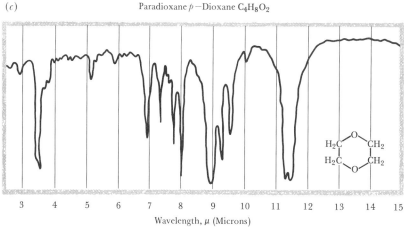

Figure 7.29. Continued. (b,c) Infrared spectra of isomers are quite different even though they have exactly the same atoms. Thus the ir spectrum of a substance depends upon its molecular structure.

Even if the 13 isomer spectra were not available, certain of the isomers could be eliminated by inspecting the by-product spectrum. Ethers absorb strongly at about 9.0 μ. Since this compound does not, the ether isomers—*m*-dioxane, *p*-dioxane, and 2-methyl-1,3-dioxolane—are removed from consideration. The absorptions of a carboxylic acid are absent, and so the three carboxylic acid isomers can be eliminated. Also from the spectrum, erythrol can be shown not to be the by-product because there is no olefin double-bond absorption. Eventually, there will be only the four esters left.

Esters absorb at 5.8 μ. Experienced spectroscopists can recognize the ethyl and acetate group absorptions in the spectrum. The conclusion is that the by-product is ethyl acetate.

Before ir spectrophotometers were available, the identification involved much more work. One procedure involved the determination of the molecular weight and formula. Some of the physical properties, such as the boiling point (77.2°C), density (0.90 g/cm³), and *miscibility* (mixing ability, or solubility with other liquids), were measured and recorded. Since ethyl acetate is a well-known compound, all these data would be available in tables and handbooks. Only ethyl acetate would have exactly all the properties of the by-product. From this, the by-product was identified.

At one time, of course, ethyl acetate was not known. Then chemists did even more work in establishing the identity of this by-product. First, typically, the molecular weight, formula, and physical properties were determined. From the fruity smell, it seemed probable that it was an ester. One reaction, with strong alkali, hydrolyzed the by-product, forming two separable compounds. These were identified from their physical and chemical properties to be ethyl alcohol and acetic acid. Since an alcohol and an acid react to form an ester, the by-product was very likely the ethyl ester of acetic acid, ethyl acetate.

To be even more certain, the chemist could go to his collection of pure compounds, select ethyl alcohol and acetic acid, and form the ester. This could be purified and its properties determined. If all the physical and chemical properties of the prepared ester were equal to the properties of the original by-product, the conclusion, again, would be that ethyl acetate was the by-product.

Although there are many physical and chemical properties that can be determined, the chemist does not measure all of them. He selects the minimum number which allows him to eliminate all but one possible substance. The newer and less well known a substance, the more detailed must the analyses and property determinations be. The ir spectrum remains one of the more sensitive and selective tests for a compound.

OPTICALLY ACTIVE CRYSTALS AND MOLECULES

The top view of the hexagonal quartz prism in Fig. 7.1 is typical of about one-half the quartz crystals found in nature. Figure 7.30 shows both this view and its mirror image. The two crystals cannot

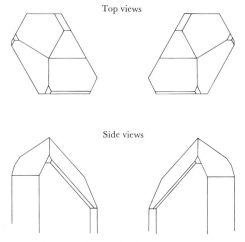

Top views

Side views

Figure 7.30. Right- and left-hand quartz hexagonal prisms, top and side views.

be made to coincide but are related as are the right and left hands. They are called "right-handed" and "left-handed" quartz, respectively. Both are found in nature.

If the two forms were not distinguishable in any other way, it would be arbitrary which one we called "right-handed." However, if a cross-sectional slice is cut from each crystal and examined by polarized light (Fig. 4.36), a new physical property is observed (Fig. 7.31).

All quartz crystals found in nature can be classified as either right-handed or left-handed by examining their prism tops. In general, those in the former group rotate polarized light to the right and those in the latter to the left. The optical activity of quartz and certain other minerals showing handedness vanishes when the crystal is melted. The rotation of polarized light, here, is a property of the crystalline structure.

Jean Baptiste Biot (1774–1862) discovered that many naturally occurring organic compounds are optically active. Sugars, camphor, turpentine, and tartaric acid (from the sediment in wine-making), for example, show this property either in the pure state or when dissolved. For these compounds, optical activity does not depend on a particular crystal form but is a property of the molecules themselves.

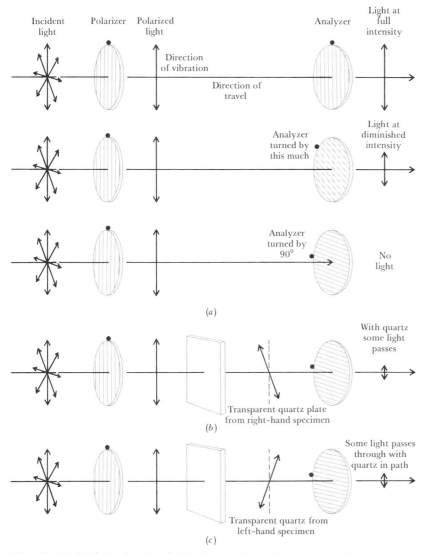

Figure 7.31. (a) Polarizer permits only light vibrating in one direction to pass. When analyzer is aligned parallel to polarizer, this light passes through the analyzer, too. When analyzer is rotated the light intensity is diminished, and at a 90° rotation no light gets through. The small black dot on the analyzer serves to measure the rotation angle. (b) When the analyzer is rotated by 90° but a plate cut from the crystal to the right in Fig. 7.30, some light gets through the analyzer which must be turned a few degrees beyond 90° to stop all the light. Evidently, the quartz plate rotated the polarized light exactly those few degrees. (c) Using a plate cut from the other crystal in Fig. 7.30, the analyzer must be rotated a few degrees less than 90° to stop all the light. This plate also rotates the polarized light but in the opposite direction. If the two plates (b and c) are of the same thickness the effects are identical in magnitude. The additional rotation away from 90°, in either direction, to block out all the light is the measure of the extent of light rotation by the plates (and indicates the handedness of the plates).

THE CONFIGURATION OF
OPTICALLY ACTIVE MOLECULES

There are two crystalline substances found in wine barrels that are salts of the isomeric dicarboxylic acids, tartaric acid and racemic acid. Each has the structural formula

```
        H
        |
HO — C — COOH
        |
HO — C — COOH
        |
        H
```

Tartaric acid rotates polarized light to the right, but racemic acid is optically inactive.

Sodium ammonium tartrate and sodium ammonium racemate are the salts obtained from replacing one carboxylic acid hydrogen by a sodium ion Na^+ and the other by an ammonium ion NH_4^+. Louis Pasteur (1822–1895) noticed that the tartrate salt crystals were all of a single-handedness, or *chirality*, while the racemate salt crystals were mixed, or half right- and half left-handed. It was possible to separate by hand the racemate crystals into two groups. One group contained those crystals of the same handedness as the tartrates, and the other group contained crystals of the opposite handedness.

Pasteur showed that the first group, when dissolved, rotated polarized light in the same direction (to the right) as pure tartaric acid; the second group, in solution, rotated the light by the same amount but to the left. Pasteur demonstrated, in 1848, that racemic acid is really an equal mixture of naturally occurring tartaric acid and its crystalline mirror image.

Since the natural isomer rotates polarized light to the right, it is said to be *dextrorotatory* and is called "*d*-tartaric acid," or "(+)-tartaric acid." The mirror-image isomer, rotating light to the left, is *levorotatory* and is called "*l*-tartaric acid," or "(−)-tartaric acid." The pair are said to be *optical isomers*. The equal molecular mixture of these two isomers is optically inactive because of the exact canceling of dextro- and levorotations. This mixture, racemic acid, is sometimes called "*d,l*-tartaric acid." Any optically inactive equal mixture of optically active molecules and their mirror images is called a "racemic" mixture.

Since racemic acid and *d*- and *l*-tartaric acids are isomers with the same two-dimensional structural formulas the only way in

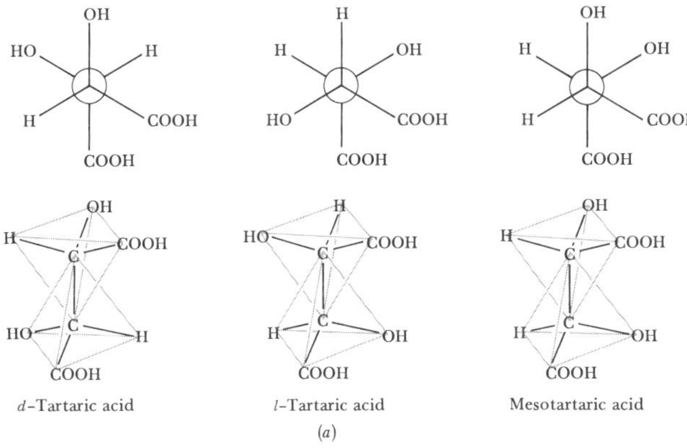

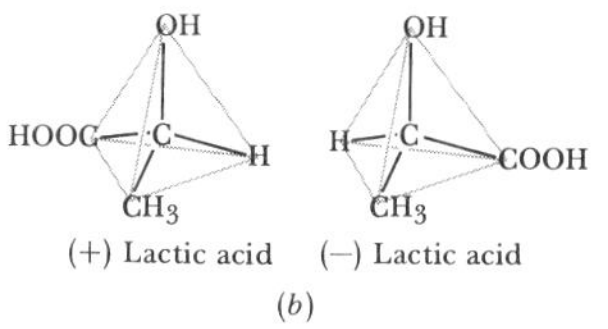

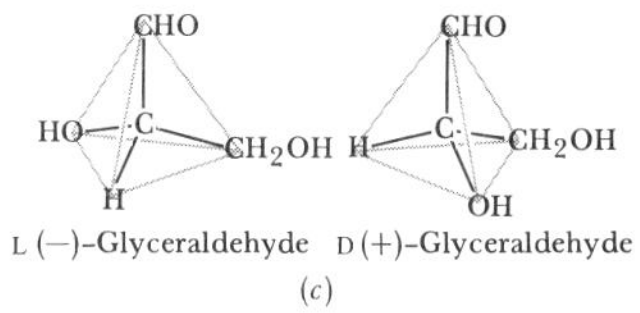

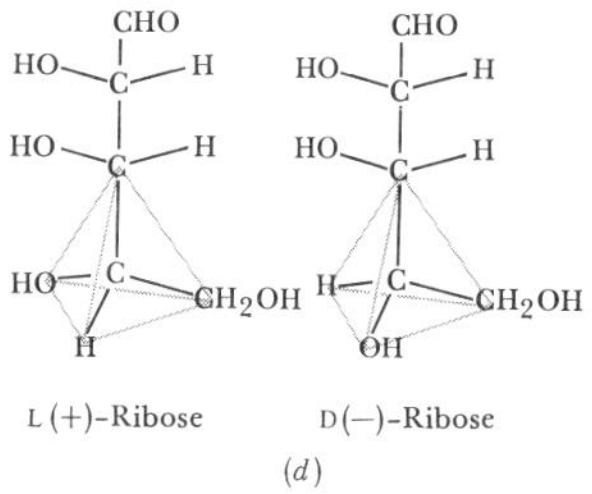

Figure 7.32. (a) d-tartaric acid, l-tartaric acid, and mesotartaric acid; (b) optical isomers of lactic acid; (c) the glyceraldehyde optical isomers; (d) the ribose optical isomers.

which their different optical properties can arise is through different three-dimensional configurations. Once the tetrahedral nature of carbon bonds was established, such optical isomers were easily pictured. The assignment as to which isomer is dextrorotatory was made from an x-ray diffraction-structure determination. In Fig. 7.32a the two isomers are shown.

Later a third isomer, mesotartaric acid, was identified. It is optically inactive. One-half of each molecule is dextrorotatory; the other half is levorotatory. Mesotartaric acid is said to be internally compensated. Unlike racemic acid, it cannot be resolved into two optically active parts.

Lactic acid found in sour milk is optically inactive. The same substance in the fluid of muscles is dextrorotatory. The former is a racemic mixture of two optical isomers; the latter is the dextrorotatory form alone (Fig. 7.32b).

Both the lactic and tartaric acids have something in common. Each contains one or two *asymmetric carbons*, i.e., carbons attached to four different groups. Mirror-image isomers are found when a substituted methane has four different atoms or groups. If two of these are not different, the molecules are identical (Probs. 7.9 to 7.11 at the end of this chapter). Some molecules contain no asymmetric carbons yet are optically active. Others, like mesotartaric acid, contain asymmetric carbons but are inactive. Nevertheless it is a good but not general rule that optically active molecules contain asymmetric carbons.

NATURALLY OCCURRING OPTICALLY ACTIVE MOLECULES

The α-amino acids (Fig. 7.24) each contain a carbon attached to a carboxylic acid group —COOH, an amine group —NH_2, a hydrogen H, and (except for glycine) a radical R different from all these. Except for glycine, all natural amino acids are optically active.

Monosaccharides have one or more asymmetric carbons, too, and are optically active. Glucose, with several (Fig. 7.21), for example, has a carbon atom fifth in line from the aldehyde —CHO in the chain structure. It is bonded to a hydrogen H, a hydroxyl OH, the —CH_2OH, and the rest of the molecule. The same carbon is also asymmetric in the glucose ring structure.

Although the lactic acid from muscle tissue is dextrorotatory, when the carboxylic acid is neutralized, the lactate salt is levorotatory. There is no change in the handedness of the molecule. The nature of the four groups attached to the asymmetric carbon must determine the sign of rotation of polarized light. For this reason it is necessary to describe an optically active molecule in two ways: the configuration of the four groups on the asymmetric carbon and the direction of rotation of light.

The basic molecule of reference for this is the simplest carbohydrate, the triose glyceraldehyde (Fig. 7.32c). The L and D, as in L(−)-glyceraldehyde, refer to the spatial arrangement of the four different groups on the central asymmetric carbon. The negative sign (−) indicates a levorotatory molecule.

In the combustion or oxidation of sugar to, eventually, carbon dioxide and water, each carbon atom can be thought to go through a series of oxidation steps, from the alcohol —CHOH to the aldehyde —CHO and then to the carboxylic acid —COOH. In a monosaccharide the L or D characterization is determined by that asymmetric carbon most remote from the carbon of highest oxidation. For ribose (Fig. 7-32d), it is the fourth carbon. L(+)-ribose has the same arrangement of the —OH, —H, and —CH_2OH groups as does L(−)-glyceraldehyde. The same is true for D(−)-ribose and D(+)-glyceraldehyde.

The amino acid designations follow a similar rule (see Fig. 7.33). All the amino acids found in proteins have the L structure (but may be dextrorotatory or levorotatory). The D structure is found occasionally in the walls of bacteria and spores and in some antibiotics. The monosaccharides in sugars, starches, and cellulose are almost all in the D configuration.

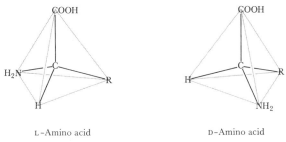

L–Amino acid   D–Amino acid

Figure 7.33. The optical isomers of α-amino acids.

When molecules with asymmetric carbons are made in the laboratory, the optically inactive racemic mixture is obtained. In living systems, in contrast, there is a consistent preference for one or the other of a pair of optical isomers.

Pasteur's technique of handpicking crystals is usually the most tedious method available for separating optical isomers. A second method involves the reaction between optically active acids and bases to form several separable insoluble salts. Adrenalin, a powerful hormone, which, among other things, raises one's blood pressure, is produced in the human body as the levorotatory isomer exclusively. Containing an amine group, it can act as a base and combine with an acid.

Synthetic adrenalin, made from a coal-tar chemical, is an equal mixture of the dextrorotatory and levorotatory isomers. If this racemic mixture is treated with pure L(+)-tartaric acid, two salts are formed. Each salt contains one L(+)-tartaric acid and either the dextro or the levo adrenalin isomer. The properties of these two salts, e.g., their solubilities, are not the same; thus they are separable. A simple analog, using right and left hands, illustrates the differences in physical properties between these two adrenalin–tartaric acid salts. When two persons hold hands, it is ordinarily one's right and the other's left hand which are coupled. The two persons linked in this way form a certain two-body package. If the same two persons are linked by their right hands, as when shaking hands, an entirely different two-body package is formed.

There is an important class of organic bases derived from vegetable matter, the *alkaloids*. These usually have powerful physiological effects on animal organisms. Examples are caffeine, mescaline, morphine, nicotine, quinine, and strychnine (Fig. 7.34). The natural alkaloids are optically active. Nicotine from tobacco leaves

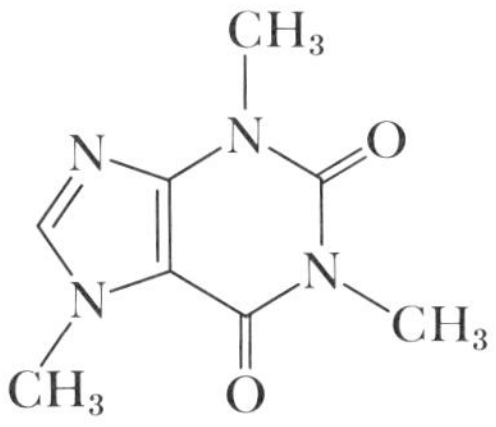

Caffeine

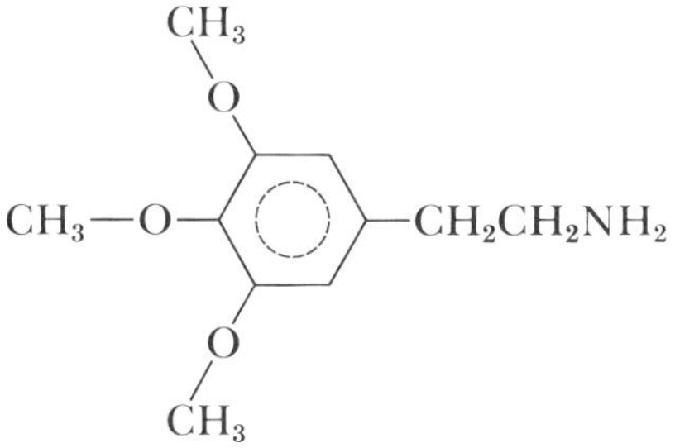

Mescaline

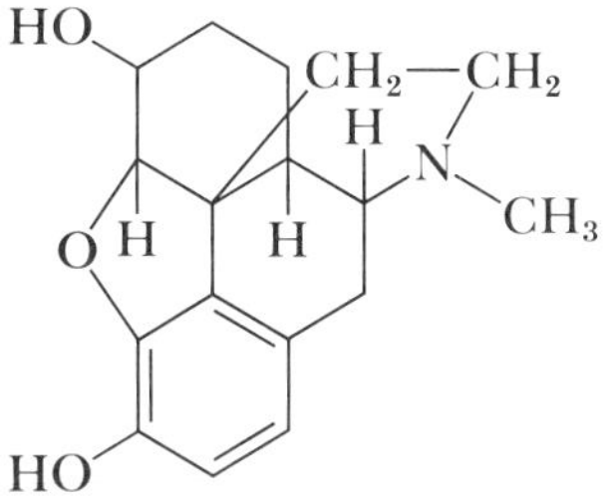

Morphine

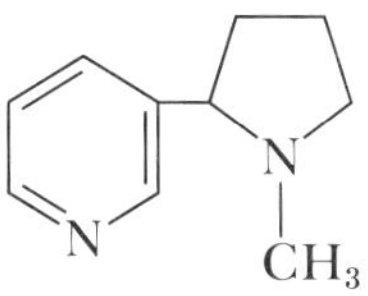

Nicotine

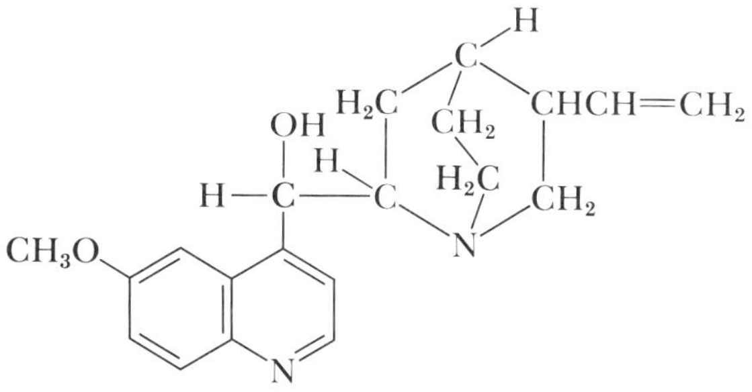

Quinine

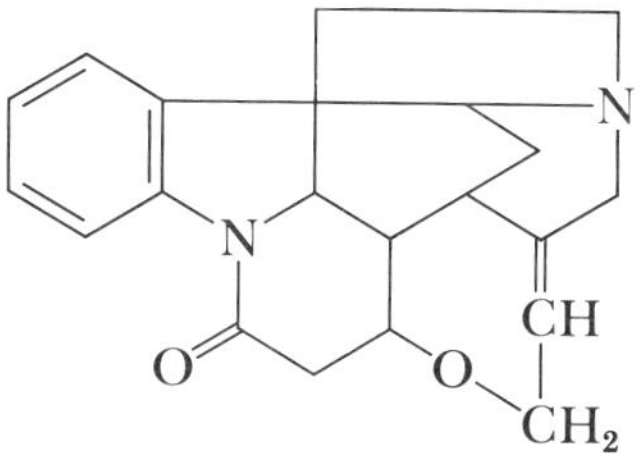

Strychnine

Figure 7.34. Some natural alkaloids.

is always levorotatory. These bases can be used to separate racemic mixtures of acidic optical isomers.

The physiological properties of optical isomers are frequently remarkably different. Levorotatory nicotine is much more poisonous than the dextrorotatory form. The natural (−)-adrenaline is about ten times more active in its effects than the synthetic dextrorotatory isomer. Natural L(+)-ascorbic acid, vitamin C, prevents scurvy; D(−)-ascorbic acid does not.

Sometimes it is possible to utilize biological differences to obtain separate optical isomers. D(−)-Fructose is the natural configuration of this hexose. The synthetic fructose, made from glycerin, is a mixture of the D(−) and L(+) isomers. Certain yeasts will ferment the former but not the latter, leaving the pure synthetic L(+)-fructose.

A green mold found on cheese, *Penicillium glaucum*, will preferentially attack L(+)-tartaric acid first and then D(−)-tartaric acid. If the process is stopped in time, the levorotatory D(−)-isomer can be obtained uncontaminated with its optical isomer.

Malic acid is a dicarboxylic acid found in green apples. A synthetic mixture of the dextro and levo forms can be converted to their salts and fed to a rabbit. Only the former is found in the urine because only the levo isomer is utilized by the rabbit.

Since Pasteur's pioneering work, the relationship between the chemical behavior of a molecule and its three-dimensional geometry or *stereochemistry*, has been a principal area of investigation for chemists.

MOLECULES, TASTE, AND ODOR

Having looked at the several ways in which the structures of molecules are determined, we can now examine the relation between structure and the physiological functions of taste and odor.

The mechanisms of taste and smell are still not completely understood. But any proposed model of these functions must account for the relation between molecular geometry and taste or smell. The collection of experimental observations in Fig. 7.35 suggests a number of models.

NATURAL MACROMOLECULES

A *macromolecule* is a very large molecule. Those found in living organisms are made by joining together many smaller molecules, for example, monosaccharides and amino acids. Sucrose (Fig. 7.23) is a disaccharide made by joining the two monosaccharides D(+)-glucose and D(−)-fructose. The bond connecting these, called a

(a)

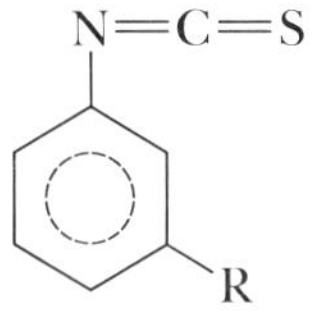

R may be CH$_3$, OCH$_3$, Br, or Cl

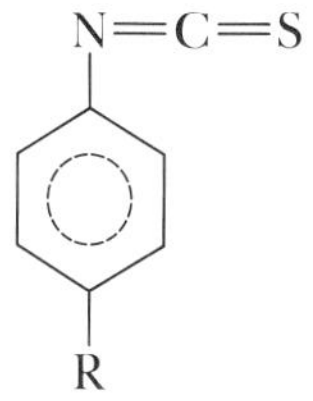

Pungent   Sweet, aniselike

H—C≡N

Hydrogen cyanide (almond smell, poison)   Benzaldehyde (almond smell, found in almond)

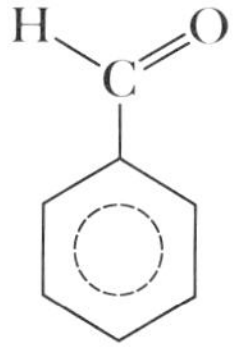

(b)

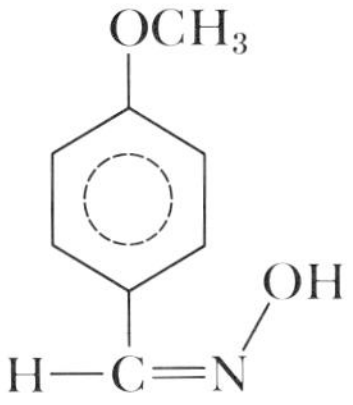

Anti-alpha-anisaldoxime (very sweet) 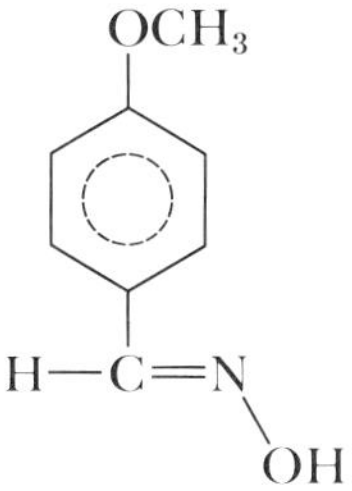   Syn-alpha-anisaldoxime (tasteless)

(c)

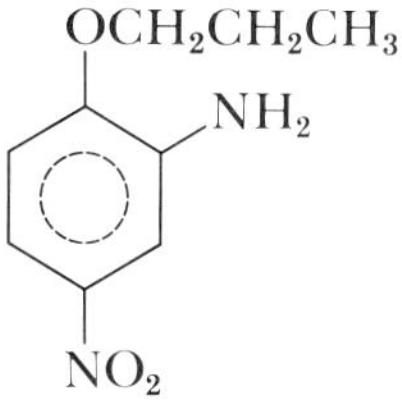

2-Amino-4-nitropropoxybenzene (very sweet) 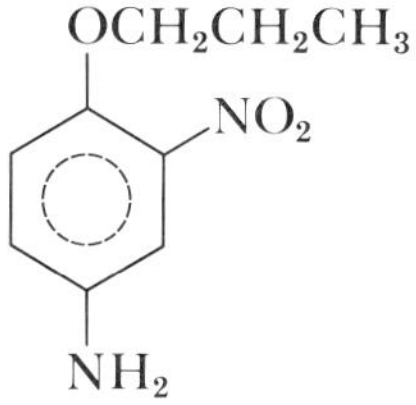   4-Amino-2-nitropropoxybenzene (tasteless)

(d)

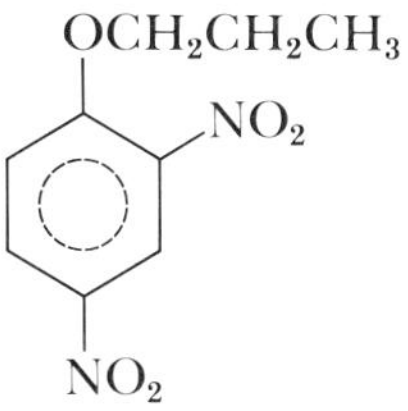

2,4-Dinitropropoxybenzene (bitter)

Figure 7.35. (a) The R group may be CH$_3$, OCH$_3$, Cl, Br. In the meta position the substance has a pungent odor; and in the para position, sweet and aniselike. (b) Hydrogen cyanide is the posion gas used in gas chamber executions. Benzaldehyde is found in bitter almond kernels along with a little HCN. Both compounds have a bitter-almond odor. (c) The hydroxyl group direction governs the taste of the molecule. (d) The taste depends on the way a few atoms are placed in the basic molecule.

"glycoside bond," is made from two hydroxyls (one from each hexose) by removing a water molecule.

There are three important polysaccharides made from D-glucose and other hexoses. Because glucose is quite water-soluble, it cannot be stored in either plants or animals. But large chains of D-glucose joined by glycoside linkages are relatively insoluble. In plants these are in the form of *starches;* in animals, they are found as *glycogens.* Through hydrolysis both starches and glycogens can be completely broken down into D(+)-glucose. Neither starches nor glycogens are molecules of a single molecular weight. The glycogens are larger; by osmotic-pressure determinations, they seem to have molecular weights considerably in excess of 5 million.

Cellulose, also made from glucose units, is a branched-chain network serving as the structural support in most plants. Cotton fibers, for example, are about nine-tenths cellulose.

Polysaccharides which incorporate some amino acids are found in animal cartilage, bone, and corneas. These are usually bound to proteins. The repeated units in starch, glycogen, and cellulose are not D-glucose $C_6H_{12}O_6$ but this hexose from which water is removed, $C_6H_{10}O_5$, in the formation of glycoside linkages. We refer to this glucose-less-one-water as a glucose *residue.* Thus these polysaccharides are chains of repeated D-glucose residues.

Because an amino acid is bifunctional, its amino group can react with one molecule and its carboxylic acid group with another. In Fig. 7.36, the three amino acids alanine, glycine, and leucine join into one large molecule at the formation of two amide $(-\overset{\overset{\textstyle O}{\textstyle \|}}{C}NH-)$ linkages. The formula for the large molecule is $C_{11}H_{21}O_4N_3$. (Much more information is conveyed by using the amino acid symbols in the correct sequence: ala-gly-leu.) Because the full amino acids are not present in this molecule, water having been removed, we say it contains three *amino acid residues.* In living systems the amide linkages joining the residues are called "peptide linkages." The molecule in Fig. 7.36, having three amino acid residues is called a "tripeptide." Any long chain of amino acid residues in living organisms is called a "polypeptide"; a shorter chain, with fewer residues, is called an "oligopeptide."

Proteins are very large polypeptides. Often the term polypeptide is arbitrarily applied to sequences of less than about seventy-five residues and longer chains are called proteins. The special properties of proteins depend on which and how many

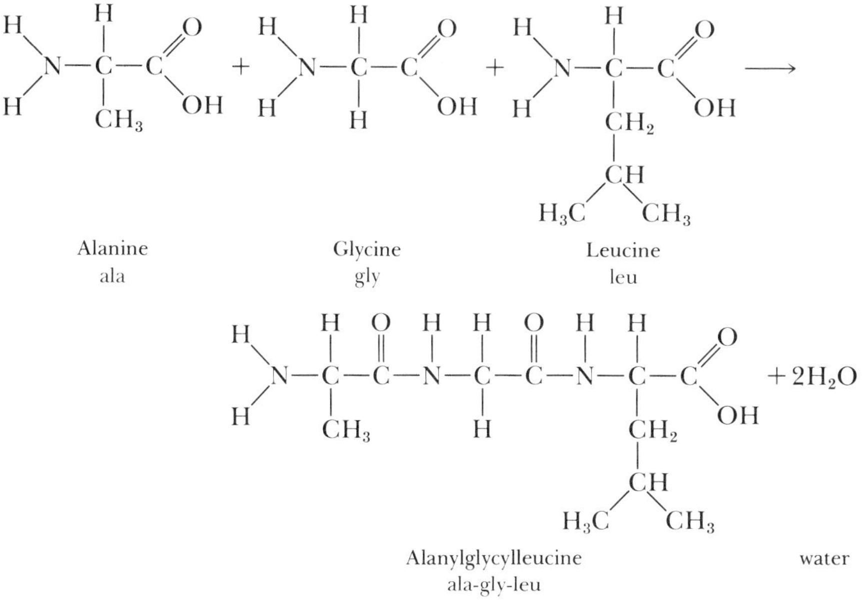

Figure 7.36. Formation of a tripeptide.

amino acid residues are in each chain. Proteins, like fats and carbohydrates, can be used for energy. In muscle fibers, protein interactions result in contracting and performing all the mechanical work done in an organism, for example, pumping and moving.

An important class of proteins is the *enzymes*. These make possible all the chemical processes necessary for life. Without enzymes these vital reactions would function either too slowly or not at all. Some enzymes are purely protein, while others have a nonprotein part. A *coenzyme* is an easily removed nonprotein, while a *prosthetic group* is more firmly attached to the main enzyme protein.

Enzymes are highly specific *catalysts;* that is, they speed up a reaction without undergoing a permanent change themselves. An enzyme usually has an influence on only a very small number of specific reactions. The effectiveness of an enzyme depends in a markedly sensitive way on the temperature of the organism and on the acidity of the medium in which it acts. Both these conditions determine the geometric structure of the protein. The function of an enzyme depends on the three-dimensional configuration of the amino acid residue sequence.

An indication of the critical role of enzymes in maintaining life comes from the observation that the most potent poisons are those which inhibit or interfere with enzymes. Cytochrome oxidase is one enzyme involved in the utilization of oxygen. The cyanide ion CN^- acts as a poison by combining with the copper and iron atoms in this enzyme, destroying the enzyme function. Nerve gas, a chemical warfare agent, reacts with an enzyme necessary for most nervous system functions. If as little as one-fifth of a drop (about 10 mg) is present and dispersed in 1 m^3 of air, a person exposed to this will die in 1 min from asphyxiation and, occasionally, heart failure. Penicillin is believed to interfere, through enzyme destruction, with cell-wall formation in pneumococci, gonococci, and streptococci, thus controlling diseases caused by these microbes.

Other natural macromolecules in common use are silk, wool, and rubber. Silk is made from mostly glycine and alanine in long polypeptide chains called "silk fibroin." The amino (—NH) and carbonyl groups (C=O) are both polar. There is a strong interaction between these electric dipoles, keeping the polypeptide chains close together and in sheets (Fig. 7.37). This type of bonding through electric dipoles is especially strong because the small size of the hydrogen atom permits close approach of the two dipoles. It is called "hydrogen bonding."

Wool and hair fibers, as well as feathers, nails, and horn, are made from the protein *keratin*. These fibers can be stretched to about twice their original lengths, and, when released, they return to their initial size. The x-ray diffraction pattern of stretched keratin fibers resembles that of silk fibroin. The unstretched diffraction pattern indicates a folding of the protein chain into a helix

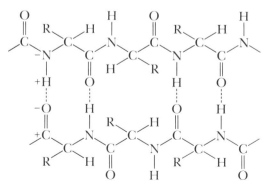

Figure 7.37. Silk fibroin.

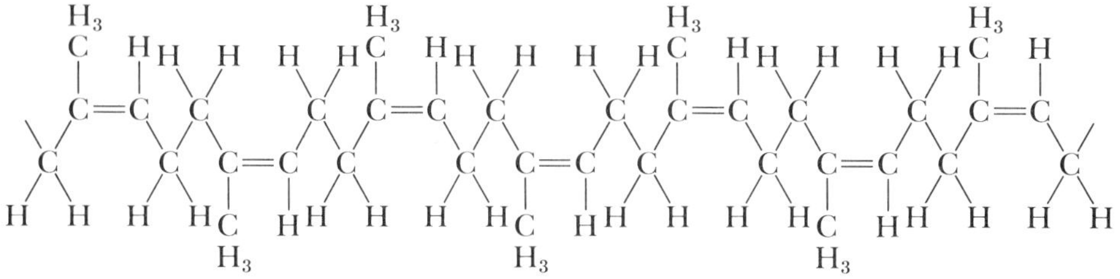

Natural rubber, poly(*cis*-isoprene)

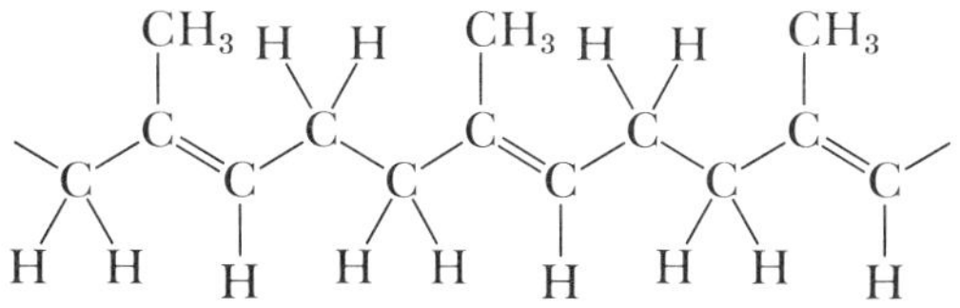

Gutta percha, poly(*trans*-isoprene)

Figure 7.38. Natural rubber is a chain of cis isoprene, since the CH$_3$ and H attached at each end of the double bond are in the cis configuration. When they are trans, the substance is gutta percha.

configuration. This α-helix is maintained through hydrogen bonding.

Natural rubber, obtained from rubber trees, is made from isoprene units joined in long chains (see Fig. 7.38).

THE IONIZATION OF AMINO ACIDS

To understand how a protein functions it is necessary to know which amino acids are present, in what order they are linked, and the three-dimensional shape of the polypeptide chain. The analysis of a protein begins with hydrolysis of the chain, breaking it up into individual amino acids, which are then separated and identified. One principal separation method depends on the ionizations of the carboxylic and amine groups.

Any amino acid can donate a proton

$$R-\overset{\overset{\displaystyle H}{|}}{\underset{\underset{\displaystyle NH_2}{|}}{C}}-COOH + H_2O \rightleftharpoons R-\overset{\overset{\displaystyle H}{|}}{\underset{\underset{\displaystyle NH_2}{|}}{C}}-COO^- + H_3O^+$$

(Anion)

or accept a proton

$$\begin{array}{ccc} \overset{\displaystyle H}{\underset{\displaystyle NH_2}{R-C-COOH}} + H_3O^+ & \rightleftharpoons & \overset{\displaystyle H}{\underset{\displaystyle NH_3{}^+}{R-C-COOH}} + H_2O \\ & & \text{(Cation)} \end{array}$$

It can also transfer the proton from its carboxylic acid to its amine group

$$\begin{array}{ccc} \overset{\displaystyle H}{\underset{\displaystyle NH_2}{R-C-COOH}} & \rightleftharpoons & \overset{\displaystyle H}{\underset{\displaystyle NH_3{}^+}{R-C-COO^-}} \\ & \text{(Zwitterion, or dipolar ion)} & \end{array}$$

Each of these three proton-transfer processes is *reversible;* i.e., both it and its reverse process can happen. The hydronium-ion concentration, or pH, determines which of these is important. When the pH of the solution is low, a high hydronium-ion concentration promotes the reverse of the first reaction and the forward of the second reaction. In highly acidic media the amino acid will act as a base, accept a proton, and have a positive charge.

If the solution has a high pH, the hydroxide ions present will accept protons from the amino acid carboxylic acid group. The amino acid will act as an acid, donate its proton, and have a negative charge. At some intermediate pH the amino acid will neither donate a proton to some other molecule nor accept a proton from another acid. It will be a doubly charged ion formed in the third proton-transfer reaction. Such an ion, with an equal number of negatively and positively charged groups, has no net electric charge; it is electrically neutral. This kind of ion has a special name—a hybrid, or *zwitterion*. The pH at which a particular amino acid is found mostly as zwitterions is called the "isoelectric point" and is often abbreviated as IpH.

ELECTROPHORESIS

Each amino acid has its own characteristic isoelectric point. By adjusting the hydronium-ion concentration of its solution, the amino acid can be made cationic, anionic, or zwitterionic. At any particular pH, some amino acids will be present in solution as cations and some as anions; if the pH coincides with an amino acid isoelectric point, the amino acid will be found mostly as zwitterions. Table 7.4 lists the isoelectric points for four amino acids and the ionic forms of these at several hydronium-ion concentrations. As the hydronium-ion concentration is reduced and the pH of the solution is increased, each amino acid progressively is found as a cation, then as a zwitterion, and finally as an anion.

The difference in isoelectric points is the basis for one important technique of separating amino acids into pure fractions. When a positive and a negative electrode are placed in a solution of several amino acids, the cations migrate to the negative electrode, the anions to the positive electrode, and the zwitterions (if present) to neither electrode. This selectivity in ionic migration and the separation technique using ionic migration are both called "electrophoresis."

Using the information of Table 7.4, Fig. 7.39 illustrates how the four amino acids are isolated from each other. Progress in the electrophoresis can be followed by optical means, such as examining how the solution absorbs or refracts (bends) light, which depends on the nature and concentration of amino acid in the solution.

Since ionic migration can be measured, this same technique is also used to determine the isoelectric point of a pure amino acid or of a larger polypeptide or protein. The pH can be varied and the direction of ionic migration observed. The pH at which migration does not occur is the isoelectric point.

CHROMATOGRAPHY

The attraction to and retention of a substance at the surface of a solid or liquid is called "adsorption." A cigarette filter works by collecting some of the smoke components on the cellulose acetate or charcoal-powder filter surface. When a detergent is dissolved in water, analysis reveals that more detergent is present at the water-and-air surface than throughout the bulk of the liquid. Both these are examples of adsorption.

A technique for the fractionation of mixtures of amino acids (and other substances as well) depends on the selective adsorption of the several amino acids at a solid surface. In Fig. 7.40, a glass column is packed with a powder, called the "adsorbent," on which surface the substances to be separated, called *adsorbate*, can be ad-

Table 7.1. Ionic Form and Net Electric Charge of Four Amino Acids at Several pH Values†

	Amino acids			
	Cysteine	Serine	Leucine	Alanine
Isoelectric point	5.07	5.69	6.04	6.11
At pH = 5.00 $(H_3O^+) = 1.00 \times 10^{-5}$ moles/liter	+	+	+	+
At pH = 5.07 $(H_3O^+) = 8.51 \times 10^{-6}$ moles/liter	±	+	+	+
At pH = 5.38 $(H_3O^+) = 4.17 \times 10^{-6}$ moles/liter	−	+	+	+
At pH = 5.69 $(H_3O^+) = 2.04 \times 10^{-6}$ moles/liter	−	±	+	+
At pH = 5.86 $(H_3O^+) = 1.38 \times 10^{-6}$ moles/liter	−	−	+	+
At pH = 6.04 $(H_3O^+) = 9.12 \times 10^{-7}$ moles/liter	−	−	±	+
At pH = 6.08 $(H_3O^+) = 8.32 \times 10^{-7}$ moles/liter	−	−	−	+
At pH = 6.11 $(H_3O^+) = 7.76 \times 10^{-7}$ moles/liter	−	−	−	±
At pH = 6.50 $(H_3O^+) = 3.16 \times 10^{-7}$ moles/liter	−	−	−	−

† + Cation
 − Anion
 ± Zwitterion

Amino acid	Symbol	IpH, isoelectric point	If pH < IpH, the amino acid is a cation	If pH = IpH, the amino acid is a zwitterion	If pH > IpH, the amino acid is an anion
Cysteine	☐	5.07	+	− +	−
Serine	◯	5.69	+	− +	−
Leucine	⬠	6.04	+	− +	−
Alanine	⬡	6.11	+	− +	−

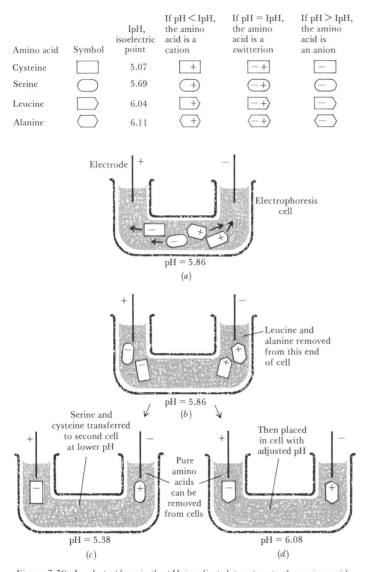

Figure 7.39. *In electrophoresis the pH is adjusted to separate the amino acids.*

sorbed. A solution of the mixed substances is poured in at the top and then washed down with a solvent.

Adsorption can be a chemical or a physical process. An adsorbate joined very firmly to the adsorbent will move down the column very slowly if at all; a weakly held adsorbate will travel down the column rapidly with the solvent. Each substance has its own

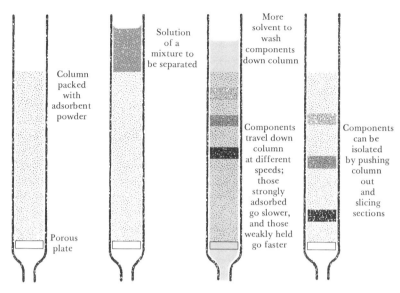

Column
packed
with
adsorbent
powder

Porous
plate

Solution
of a
mixture to
be separated

More
solvent to
wash
components
down column

Components
travel down
column
at different
speeds;
those
strongly
adsorbed
go slower,
and those
weakly held
go faster

Components
can be
isolated
by pushing
column
out
and
slicing
sections

Figure 7.40. Column chromatography.

rate of travel. Therefore, after a certain time the components of the mixture will be at different distances along the column. Some of the first substances to be thus separated were colored pigments. At separation, different colored bands were seen along the column. For this reason, the general process is called "chromatography."

After a chromatographic separation, the column material can be carefully pushed out and sliced into sections, each of which contains one component of the original mixture. These components can be washed off the adsorbent with strong solvents and then dried. [A very simple variation of column chromatography is *paper chromatography* (Prob. 7.12).]

THE SEQUENCE OF AMINO ACIDS

There is a useful reaction between an amine group and 2,4-dinitrofluorobenzene:

In a polypeptide (see Fig. 7.36), one end of the sequence has a free, or unreacted, amine group and the other end a carboxylic acid group.

When any polypeptide is treated with 2,4-dinitrofluoroben-zene, only the first amino acid in the sequence will combine with this reagent. After hydrolysis of the entire polypeptide into individual amino acids, that amino acid joined to dinitrobenzene is the first amino acid in the chain. The hydrolysis, if not complete, will break up the polypeptide into shorter chains, and these can be pieced together to determine the sequence in the original polypeptide.

Example. A hexapeptide (containing six amino acids) is treated with 2,4-dinitrofluorobenzene and then completely hydrolyzed. The individual amino acids are separated by electrophoresis and chromatography. There is one each of the six amino acids shown in Fig. 7.24. Glycine is the first amino acid in the hexapeptide, and is represented by *gly. (The asterisk indicates it is the first amino acid.)

A less drastic hydrolysis of the original hexapeptide produces fragments which can be separated. These, in turn, can be treated with reagent and further hydrolyzed into fragments. Five fragments obtained in this less severe hydrolysis are

*gly-ala
ser-cySH
leu-ser-cySH
cySH-thr
ala-leu

What is the correct sequence in the hexapeptide?
Since glycine begins the chain, we can write

*gly-ala

There is only one alanine present, and the fifth fragment indicates it is followed by leucine. Thus,

*gly-ala-leu

Using information from the third and then the fourth fragments, we reconstruct the chain

*gly-ala-leu-ser-cySH-thr

This is the only possible sequence consistent with the fragments. The second fragment is consistent with the third but gives no more information (Prob. 7.13).

Example. How many different hexapeptides can be made from this same set of six different amino acids?

There are six possible beginnings to the chain. After the first amino acid is selected, five remain from among which the second position can be filled. This is a total of $6 \times 5 = 30$ possible combinations for the first and second positions in the hexapeptide. Similarly, there are four possibilities for the third position, three for the fourth, two for the fifth, and only one for the last position. This is a total of

$$6 \times 5 \times 4 \times 3 \times 2 \times 1 = 720$$

arrangements. The product of these six numbers, in order from 6 to 1, is represented by 6! and is read as "six factorial."

Example. With 10 different amino acids, each used once, how many distinct polypeptides can be made?

There are 10!, or 3,628,800, different ways in which to order the 10 different amino acids in a decapeptide. This is close to the total number of all compounds known today. Obviously not all these possible decapeptides have been made. It is evident that the possible number of all polypeptides is enormous, and it is also evident that the total number of all possible compounds far exceeds the number actually known today.

SYNTHETIC MACROMOLECULES

The polysaccharide cellulose is a macromolecule comprising a single repeating unit, the D-glucose residue. We have already used the word "dimer" to indicate a molecule which is twice as large as another, called the "monomer." The dimer of NaCl is $(NaCl)_2$, or Na_2Cl_2, and is found in the vapor of sodium chloride. We call molecules which are three and four times as large as a basic monomer molecule "trimers" and "tetramers," respectively.

A macromolecule built from a single unit molecule, or residue, repeated a large number of times is called a "polymer." We can call cellulose poly(D-glucose); it is a polymer of D-glucose residues. If a protein were made from just one amino acid residue, it would be a polymer of that amino acid, for example poly(alanine). Since polypeptides and proteins ordinarily contain many different amino acid residues, they are called "macromolecules" but not polymers. However, silk (Fig. 7.37), is made up almost entirely of alternating alanine and glycine residues. To a good approximation, it can be called a "poly(L-alanylglycine)." Natural rubber (Fig. 7.38) is poly(*cis*-isoprene).

When cellulose is treated with concentrated nitric and sulfuric acids, some or all of the polysaccharide hydroxyl groups —OH are replaced by nitrate groups —ONO_2. The product, nitrocellulose, is more easily dissolved than the parent cellulose. With camphor, it becomes celluloid and can be fabricated into sheets, films, etc. Guncotton, made by nitrating cotton, is widely used in smokeless powder for ammunition. Cellulose acetate, made by treating cellulose with the anhydride of acetic acid (acetic anhydride), is soluble. It can be converted into an insoluble product and, at the same time, spun into rayon fibers.

Celluloid, guncotton, and rayon, made by modifying a natural macromolecule or polymer, are called "derived macromolecules," or "derived polymers." Although the production of derived polymers is still important industrially, it has been exceeded in variety and in amount by the production of *man-made*, or *synthetic*, *polymers*.

The synthetic polymers are made by joining together one or more kinds of monomer into macromolecules (Fig. 7.41). Formaldehyde reacts with phenol (Fig. 7.41*d*) to form one of the family of phenolaldehyde polymers called "Bakelite." Others can be made by starting with substituted phenols. The synthetic macromolecules in Fig. 7.41 are called "condensation polymers." When monomers are joined to form polymers of this class, a molecule, such as water, is eliminated.

The second large class of polymers is made by joining unsaturated monomers. Ethylene $CH_2{=}CH_2$ can be converted to a radical $\cdot CH_2{-}CH_2 \cdot$ by breaking one bond. To this reactive species other ethylene molecules can be added at both ends, making a larger radical. The addition continues until the monomer supply is exhausted or the radical somehow is converted to a stable form. The process is called "addition polymerization."

In contrast to condensation polymerization, no molecules are eliminated in joining on monomers. The product is polyethylene. Because the repeating unit can be thought of as either —(CH_2CH_2)— or —(CH_2)—, it is also called "polythene." Because ethylene is an olefin, polymers made by joining ethylene or substituted ethylenes are called "polyolefins."

One substituted ethylene is vinyl chloride CH_2—$CHCl$. Using this monomer, polyvinyl chloride is made (Fig. 7.42). Since the vinyl group CH_2=CH—is involved in making polyolefins, the general process is also called "vinyl polymerization." Other examples of vinyl polymerization and polyolefins are given in Fig. 7.42.

If two monomers, each of which can form a polymer, are both present during the polymerization, the polymer chain will be a mixture of both—a *copolymer* (Fig. 7.43).

Using such techniques as osmotic pressure, the molecular weights of these synthetic polymers can be shown to be as high as several million.

> **Example.** If the average molecular weight of a polyethylene sample is 2.0×10^6, how many carbons are in the polymer chain?
>
> Except for the ends of the chain, the polymer is simply a concatenation of CH_2 groups. Each has a "molecular weight" of 14. To have a molecular weight of 2 million, there must be an average of
>
> $$\frac{2.0 \times 10^6}{14} = 1.4 \times 10^5$$
>
> of these CH_2 groups and thus 140,000 carbons. The two carbons at the ends of the chain may or may not be in the form of methyl CH_3 groups. In either case, we are correct in neglecting their influence among the other carbons. The average polyethylene molecule of this size was made from 70,000 ethylene monomers CH_2=CH_2.

NONLINEAR POLYMERS

The synthetic molecules discussed in the last section are all *linear polymers* in the sense that normal hydrocarbons are linear. This linearity is the consequence of using only bifunctional monomers, i.e., monomers which can join to just two other monomers.

When a trifunctional monomer is introduced into the polymerizing reaction, that molecule can join to three others. Rather than having the polymer chain extending in two directions from it, as in a linear polymer, this trifunctional monomer can have three chains. In Fig. 7.44 the polyester made from succinic acid and glycol is linear. With just a little of the trifunctional glycerin, the chain becomes *branched*. With more glycerin replacing glycol in the reaction, two long chains can be linked, or bridged, by a branching chain; this is called a "cross-linked" polymer.

Monomer or monomers used

(a)

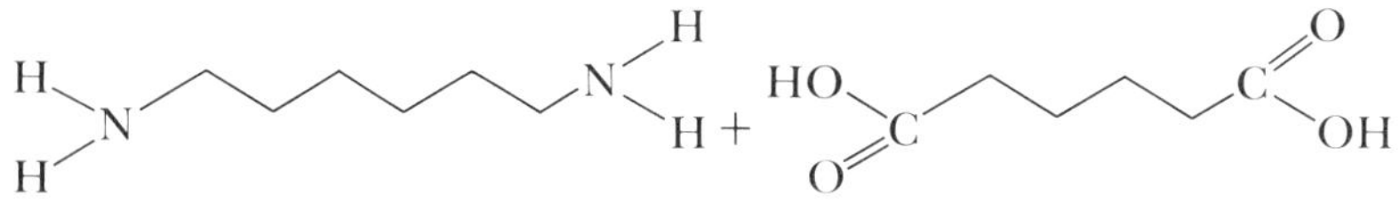

$NH_2(CH_2)_6NH_2$
Hexamethylene diamine

$HOOC(CH_2)_4COOH$
Adipic acid

(b)

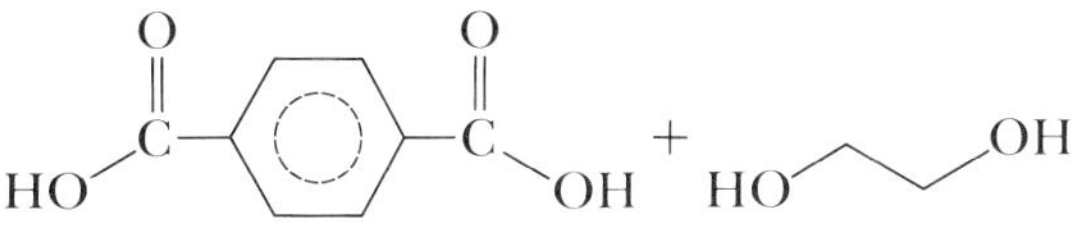

$HOOCC_6H_4COOH$
Terephthalic
acid

$HO(CH_2)_2OH$
Ethylene glycol

(c) HO

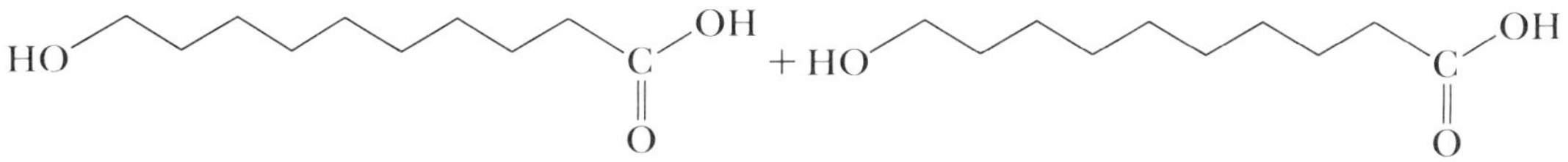

$HO(CH_2)_9COOH$
10-Hydroxydecanoic acid

(d)

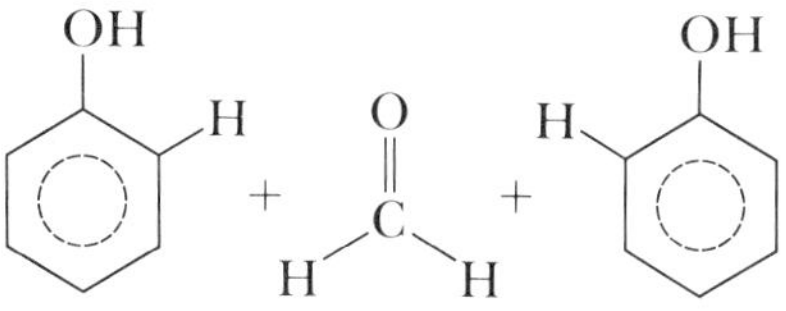

C_6H_5OH
Phenol

CH_2O
Formaldehyde

(e)

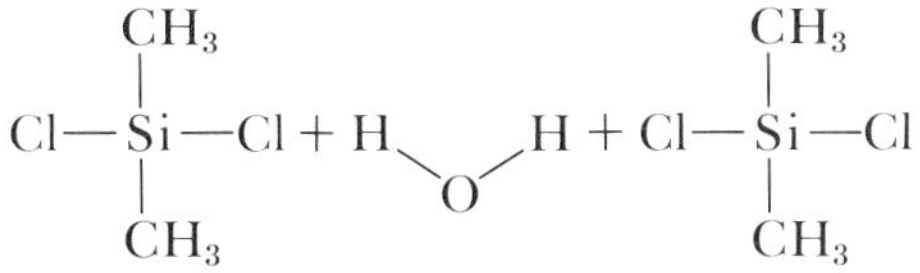

$SiCl_2(CH_3)_2$
$Si(CH_3)_2Cl_2$
Dimethyldichlorosilane

Figure 7.41. Common condensation polymers.

Repeating unit

Molecule eliminated during condensation

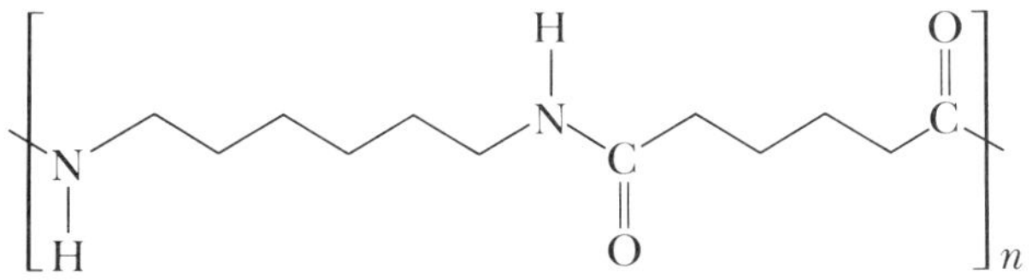

Poly(hexamethylene adipamide)
Nylon 66
A polyamide

H_2O

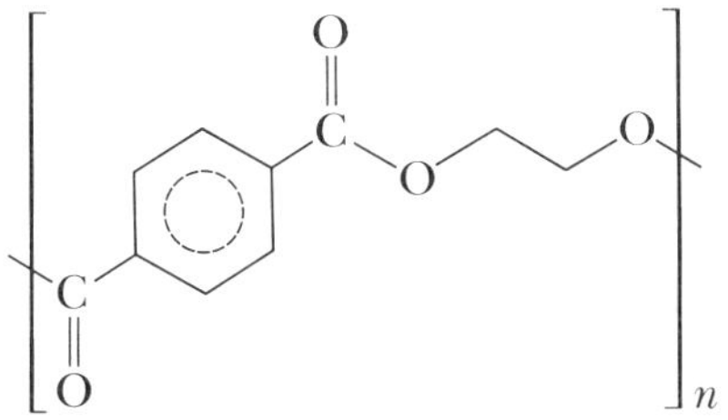

Poly(ethylene terephthalate)
Dacron, terylene
A polyester

H_2O

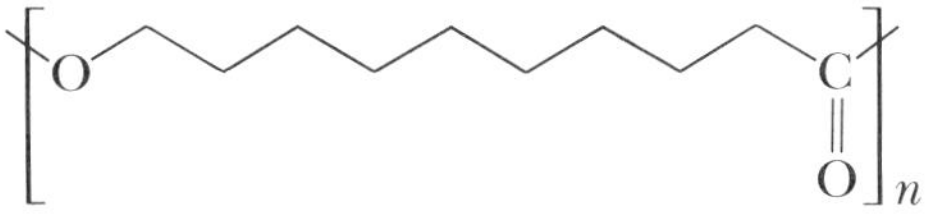

A polyester

H_2O

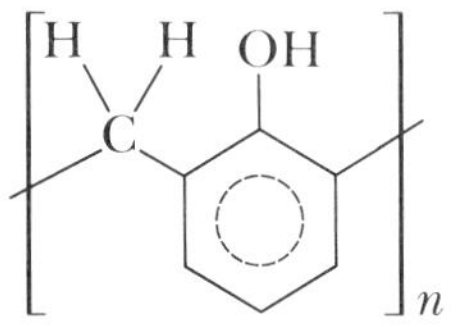

Phenol-formaldehyde
A Bakelite

H_2O

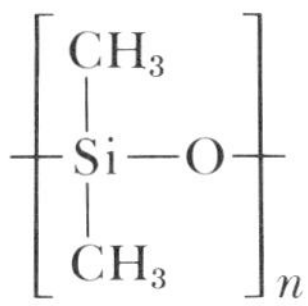

Poly(dimethyl silicone)
A silicone

HCl

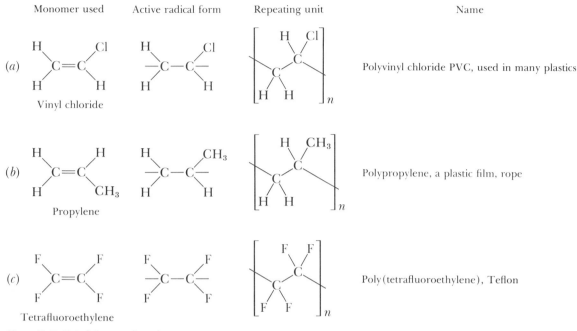

Monomer used	Active radical form	Repeating unit	Name

(a) Vinyl chloride — Polyvinyl chloride PVC, used in many plastics

(b) Propylene — Polypropylene, a plastic film, rope

(c) Tetrafluoroethylene — Poly(tetrafluoroethylene), Teflon

Figure 7.42. Polyolefins or polyvinyls.

Natural rubber (Fig. 7.38) poly(*cis*-isoprene) contains double bonds along its polymer chain. Sulfur-containing compounds can be added to the olefinic bonds of two separate polyisoprene chains, thus producing cross-linked, or *vulcanized*, rubber. This process was discovered by Charles Goodyear (1800–1860).

With sufficient cross-linking, all the chains are joined. The entire sample of polymer is a single enormous molecule. We have already seen that specimens of crystalline and vitreous quartz and diamond can also be considered to be single molecules.

Vinyl chloride   Vinylidene chloride — A copolymer of vinyl and vinylidene chlorides, e.g., saran

Figure 7.43. Formation of a copolymer, saran.

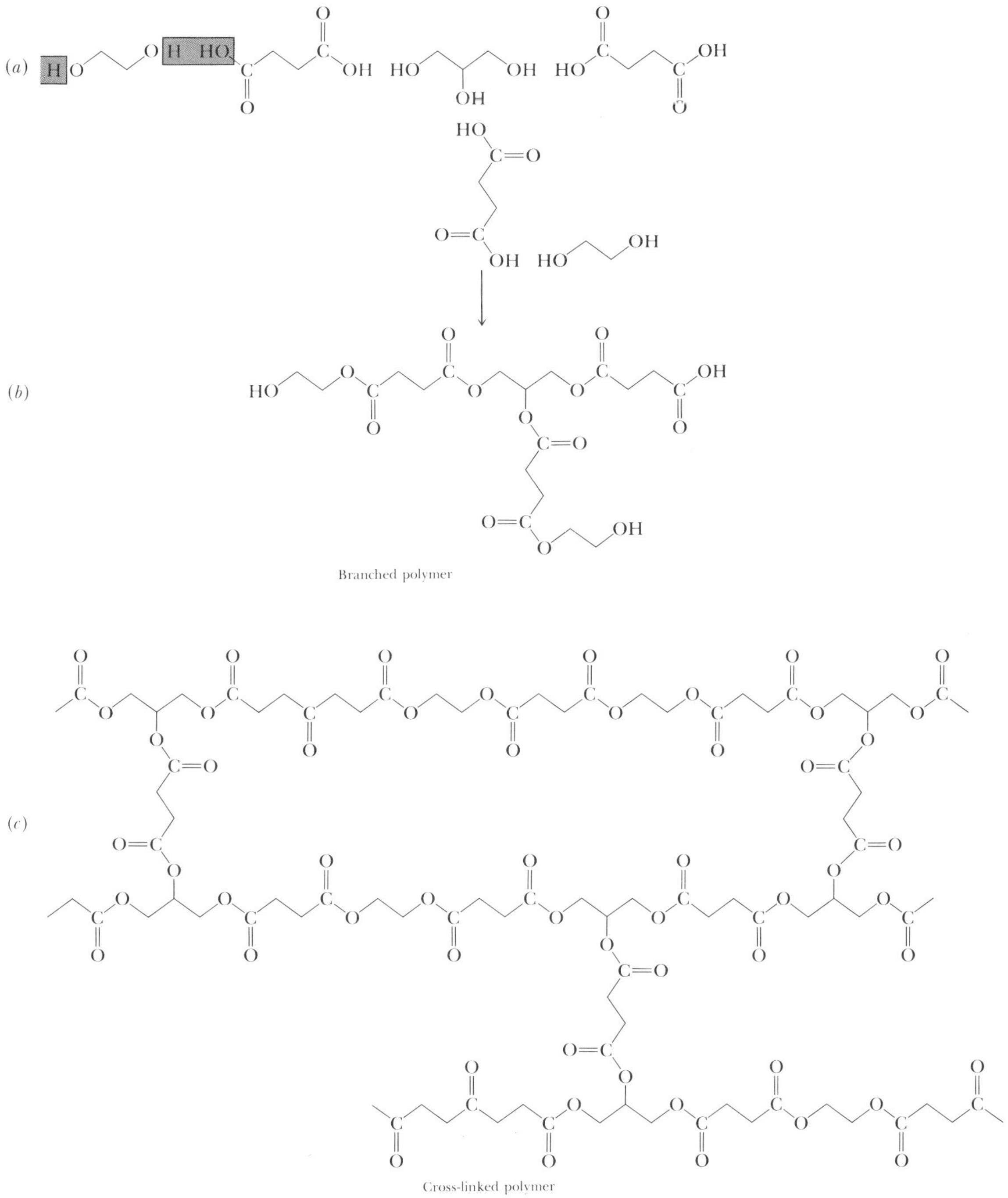

Figure 7.44. (a) A linear polyester. (b) With some trifunctional glycerin, branching occurs. (c) With more glycerin, cross-linking is possible. Natural rubber is vulcanized by addition of sulfur, or a sulfur-containing compound to pairs of double bonds on neighboring chains of polyisoprene (Fig. 7.38).

THE CONFORMATION OF MACROMOLECULES

The x-ray diffraction pattern of silk fibroin indicates that protein to be crystalline. The regular arrangement of the polypeptide chains can be described by the strong hydrogen bonding between neighboring chains (Fig. 7.37). From electronegativity values (Table 6.2), we can see that the positive end of the—NH—group is close to the negative end of the carbonyl—$\overset{\overset{\textstyle O}{\textstyle \|}}{C}$— group.

The synthetic polymer polyethylene has the crystalline structure shown in Fig. 7.45. Poly(tetrafluoroethylene) is polyethylene with all hydrogens replaced by fluorines. This has not a planar, but rather a helical conformation, as in Fig. 7.46. A planar zigzag conformation is a special, flattened-out helix. In poly(tetrafluoroethylene), repulsion between polar carbon-fluorine bonds of neighboring chains twists this polymer out of the purely planar conformation into the regular helix shown.

Even nonpolar and nonionic substances can be liquefied and solidified. Methane melts at $-184°C$ and boils at $-162°C$, oxygen at -218 and $-183°C$, and argon at -189 and $-186°C$. From these low temperatures we know little energy is needed to break up the regular order of the solids during melting and little is needed to separate the molecules in evaporation. Confirming this, the heats of vaporization, about 2,210 cal/mole for methane, 1,600 cal/mole for oxygen, and 1,500 cal/mole for argon, are smaller than the values for water (9,700) and benzene (7,360).

The same weak forces which keep methane solid or liquid also act between neighboring polyethylene planar zigzag chains and neighboring poly(tetrafluoroethylene) helices. If only a single C—H from one polyethylene chain is close to a C—H from a second chain (or from a folded-back segment of the same chain), the attractive force is too small to maintain this proximity; but if many such weak attractive interactions occur at once, the result is a strong attraction. The stability of regular crystalline arrays in these polymers depends, then, on the simultaneous closeness at a great many points along the chains or chain segments.

Because polyethylene and poly(tetrafluoroethylene) are regular linear repetitions of —CH_2— and —CF_2—, respectively, this contact along extended chain segments is possible; i.e., two chain segments pack together nicely, as shown in Fig. 7.45. If a polymer chain lacks the regularity found in polyethylene, there is less

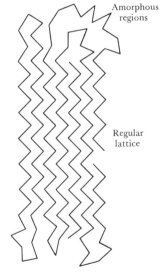

Figure 7.45. The planar zigzag conformation permits the packing of chain segments in a regular lattice. Amorphous regions occur where a long polyethylene chain turns around before returning to the crystalline lattice region.

chance of good packing of neighboring chains or helices, fewer points of weak attraction, and not as much crystallinity. A branched, or cross-linked, polymer is largely or completely amorphous, especially if the branches are long and irregularly placed on the main chain.

RANDOM-COIL CONFORMATION OF POLYMERS

The planar zigzag structure of crystalline polyethylene is related to the consistently trans conformation at every carbon-carbon bond. At room temperature the thermal energy available is not enough to overcome the attractive forces between neighboring chain segments in a crystalline region.

Above 137°C the crystalline structure of polyethylene is destroyed, and there is sufficient energy to separate chain segments and cause rotation away from the trans conformation at some of the carbon-carbon bonds. A rotation about any bond through one-third of a full turn, 120°, moves the chain segments on each side of that bond out of a common plane (Fig. 7.47). If at each bond there is a random choice of three conformations—trans, (+)-gauche, and (−)-gauche—the polymer conformation is a three-dimensional coil. A projection of this coil in two dimensions is shown in Fig. 7.48.

Because the transition at 137°C from crystalline to amorphous polyethylene resembles an ordinary solid-to-liquid melting, we can say polyethylene has a melting point of 137°C. But polyethylene does not become a liquid at this temperature. Objects made from this polymer retain their shapes. As the temperature is increased above 137°C the objects become progressively more limp. At a high enough temperature, polyethylene flows and resembles molten paraffin wax.

The random-coil conformation is also found below 137°C when polyethylene is dissolved. The attraction between solvent molecules and chain segments is greater than chain-segment–segment attraction. The former prevails; the chain is completely surrounded by solvent molecules, i.e., solvated and isolated from other chains. Free rotation occurs at carbon-carbon bonds within the chain, and a random coil results.

With simple reasoning, we can estimate the probability of either a coiled or an extended polymer chain. Since three points always lie in a single plane (are *coplanar*), the three carbons in propane C_3H_8 are always coplanar. For butane C_4H_{10} the last

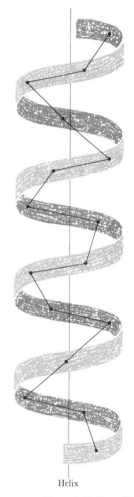

Helix

Figure 7.46. Crystalline polytetrafluoroethylene has a regular helical conformation. The carbon atoms are represented by the black dots and are bonded tetrahedrally to two neighboring carbons and to two fluorines (not shown). The carbon-carbon bonds are indicated by the straight lines, and the helical conformation of the carbon is indicated by the helical ribbon.

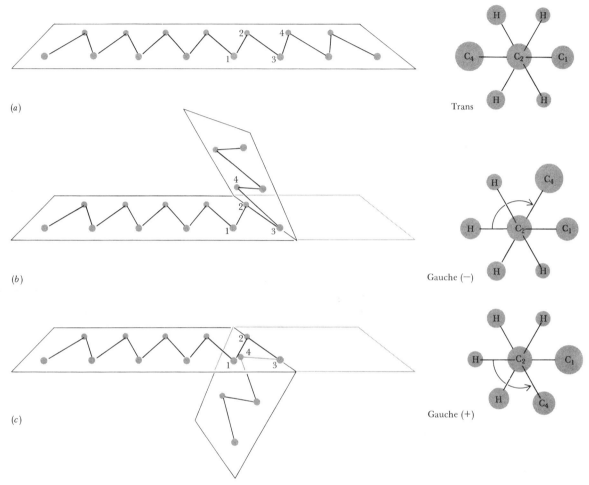

Figure 7.47. (a) In the planar zigzag structure, the trans conformation is at every C—C bond. (b,c) Rotation about a single C—C bond to a gauche conformation moves part of the chain out of this common plane.

carbon lies in the plane of the first three carbons only if the bond from the second to the third carbon is trans. Rotation to either of the two gauche conformations pushes the last carbon out of this plane. If each conformation about this bond is equally likely, only one-third of the butane has four coplanar carbons. Rotation about the last carbon-carbon bond cannot affect the coplanarity since only hydrogens, not carbons, are moved.

Figure 7.48. Random-coil, three-dimensional conformation (projected on a plane).

With pentane C_5H_{12} the five carbons are coplanar only if the bonds from the second to the third and from the third to the fourth carbons are both trans. With nine equally possible combinations of trans, (+)-gauche, and (−)-gauche in these two bonds, only one-ninth of the pentane is planar.

Rotations about these bonds occur rapidly, about several billion times per second. The conformation of any bond is continually changing, as is the conformation of any entire molecule. Thus, and more correctly, we can say that one-ninth of the pentane molecules will have the planar conformation at any time.

Now we can generalize for any size straight-chain hydrocarbon molecule. All rotations, except at the first and last bonds, affect the chain conformation. If we know how many bonds are present, we can calculate the number of combinations of bond conformations and the chance that all these are trans.

Example. What fraction of n-decane molecules in a liquid are planar?

With ten carbons in $C_{10}H_{22}$ there are nine carbon-carbon bonds. Excluding the first and last, the seven inner bonds affect the planarity. Since each of these has three conformations, there is a total of

$$3 \times 3 \times 3 \times 3 \times 3 \times 3 \times 3 = 3^7 = 2{,}187$$

different combinations for the seven bond conformations, for example $TT + GT - GT + G$. Only one of these, the all-trans combination $TTTTTTT$, will keep all ten carbon atoms in one plane. The chance that any n-decane is planar zigzag is 1/2,187, or 4.572×10^{-4}. This is also the fraction of the total number that are planar zigzag.

Example. One g of a polyethylene sample of molecular weight 14,000 is dissolved in benzene. How many molecules have an entirely planar zigzag conformation?

Since each repeating unit CH_2 has a "molecular weight" of 14, there are 1,000 of these and 1,000 carbon atoms. Of the 999 carbon-carbon bonds, the first and last are excluded and the 997 others must all be in the trans conformation. Because each bond can have any of three conformations, there are 3^{997} distinct conformation combinations for the 997 bonds. This number is calculated by using logarithms:

$$N = 3^{997}$$
$$\log N = 997 \log 3$$
$$= 997(0.4771)$$
$$= 475.7$$
$$= 0.7 + 475$$
$$N = 5.012 \times 10^{475}$$

This number is so large as to defy comprehension. For comparison the number of seconds since the solar system began is about 1.4×10^{17}, and the number of atoms in the earth is about 10^{50}.

If only one combination of conformations, the purely trans set, keeps all 1,000 carbon atoms coplanar, the fraction of all the polyethylene molecules that are planar zigzag is

$$\frac{1}{N} = 2 \times 10^{-476}$$

In 1 g of this polyethylene there are

$$1 \text{ g} \times \frac{1 \text{ mole}}{1.4 \times 10^4 \text{ g}} \frac{6.02 \times 10^{23} \text{ molecules}}{\text{mole}} = 4.3 \times 10^{19} \text{ molecules}$$

Using the fraction of these that are planar, there are

$$2 \times 10^{-476} \times 4.3 \times 10^{19} \text{ molecules} = 9 \times 10^{-457} \text{ planar molecules}$$

This number is so small that the difference between it and zero also defies comprehension.

We conclude that no molecules in this sample are entirely planar. There can be shorter segments in any molecule, however, which are transiently planar.

From the previous examples we can see that the larger the molecule, the more carbon-carbon bonds, the more bonds that must have the trans conformation for planarity, and the less chance of this. Above the melting point or in solution, the straight-chain polymer molecules have a completely random-coil conformation.

THE GLASSY POLYMER STATE

Above its melting point a polymer has a random-coil conformation. However, at temperatures below the melting point there is insufficient thermal energy to undo the partial or complete regularity found in crystalline polymers.

If the rate of cooling from above to below the melting point is slow compared to the rate at which the polymer bonds can assume different conformations, the chances for high crystallinity are good (Fig. 7.49). But if the cooling is very rapid or if rotation about the polymer chain bonds is hindered by bulky attached groups, the random structure is frozen in below the melting point. This is called the "glassy polymer state." It differs from the silicate glassy state, which is a three-dimensionally linked network. It also differs from a cross-linked amorphous polymer, which cannot be softened much without breaking chemical bonds. A glassy polymer can be melted and resolidified many times without damage.

(*a*)

(*b*)

Figure 7.49. (a) Random-coil conformation found above melting point, can be frozen in by rapid cooling to glassy polymer state. (b) With slower cooling, crystalline regularity is more likely.

POLYMER PROPERTIES

The various synthetic polymers, with a variety of properties, have been an increasingly important set of raw materials. The production of polymers and the fabrication of these into useful objects is one of the major chemical industries today.

A *plastic* material is readily deformed when placed under force or stress and maintains its new shape after the stress is removed. Plastic materials can be shaped or molded. Since many, but not all, polymers are moldable at room temperature or when warmed, the name *plastics* is frequently used for these materials.

A *rigid* object keeps its shape and size under stress, provided the stress does not exceed a limit characteristic of the material. Rigidity in polymeric substances is attained by crystallinity, cross-linking, or the glassy state. Partially crystalline polyamides, e.g., nylon, are sufficiently rigid to be made into machine screws and nuts for special applications where metals cannot be used.

Natural rubber can be molded and then further treated with heat and vulcanizing agents to be made into a more rigid vulcanized material. Similarly, polymerization can be interrupted for shaping and then continued to a rigid cross-linked synthetic material. If the cross-linked substance cannot be resoftened by warming and then reshaped, it is no longer plastic. These materials are said to be *thermoset*. The phenol-aldehyde polymers (Fig. 7.41*b*), common in electrical applications, are typical rigid thermoset substances.

If enough stress is applied to a rigid body, it will break, in contrast to a plastic body which is deformed. The brittleness, or ease of breaking, of a rigid body is lessened if the energy introduced by the stress can be used to rotate rather than break bonds. The amorphous regions in a partially crystalline polymer serve this purpose.

An *elastic* body is deformed under stress but returns to its original shape and size when the stress is removed. Some steels, e.g., those used in springs, are highly elastic. Rubber, as in a rubber band, is also elastic. But natural *rubber elasticity* differs from metal elasticity in several ways. A simple experiment illustrates one of these differences. (Fig. 7.50 and Prob. 7.14).

When a metal is heated, it expands. Polymers which are at least partially crystalline contract with an increase in temperature, which can be described by the greater number of opportunities for random bond rotation at higher temperatures. We have seen that random rotations lead to a coiled conformation. Coils are shorter than the extended polymer chain; thus, the more coiled and the fewer straight crystalline segments, the shorter the polymer specimen.

The selection of monomer (or monomers) and control of polymerization and fabrication conditions permit predictable manipulation of the polymer structure and its physical and chemi-

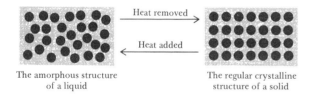

The amorphous structure
of a liquid

The regular crystalline
structure of a solid

Heat removed →
← Heat added

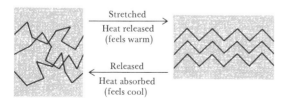

Stretched
Heat released
(feels warm) →

← Released
Heat absorbed
(feels cool)

Figure 7.50. Stretching and releasing a rubber band releases and absorbs heat, comparable to heat removed and added in freezing and melting.

cal properties. Polymers have certain advantages over metals and wood. They are less dense than metals, they soften at lower temperatures, and many are relatively inexpensive. In addition, objects made from polymers are light, easy to form, and often cheap. Polymers are also less susceptible to the corrosion and rot which affect metal and wood. Provided it is not used at too high a temperature or subjected to too great a stress, an object made from polymers is often preferred over one made from other materials.

Resistance to corrosion and rot are not unqualified advantages. Objects made from natural materials, such as wood and cotton, are *biodegradable* (consumed by living things such as microorganisms) when discarded. The accumulation of non-biodegradable synthetic polymer containers and wrappings is an important factor in pollution. Also, the incineration of halogen-containing polymers — polyvinyl chloride and teflon, for example — can introduce new contaminants into our atmosphere.

DEOXYRIBONUCLEIC ACID

Determination of the final structures in this chapter has been called the most important biological discovery of the twentieth century. Living organisms have structural units called "cells." Each cell, with a few exceptions, has a subunit, the "nucleus." In 1868 Johann F. Miescher (1844–1895) discovered a substance, isolated from salmon sperm cells and also from pus cells, which he called "nuclein."

Nuclein was found to have a relatively high phosphorus and nitrogen content and to be more acidic than protein. Later, it was renamed "nucleic acid" even though some nucleic acids are found outside the nucleus. Nucleic acid is composed of large molecules, with molecular weights of 100 million not uncommon.

By treatment with acids, alkalis, or special enzymes, a nucleic acid can be hydrolyzed into units called "nucleotides." Each nucleotide is made from an organic base, phosphoric acid H_3PO_4, and a pentose, either ribose or deoxyribose. Nucleic acids containing ribose are called "ribonucleic acid" or RNA for short. Those with deoxyribose are *deoxyribonucleic acid*, or DNA. The organic bases found in RNA are adenine A, guanine G, cytosine C, and uracil U. The first three and thymine T are in DNA (Fig. 7.51).

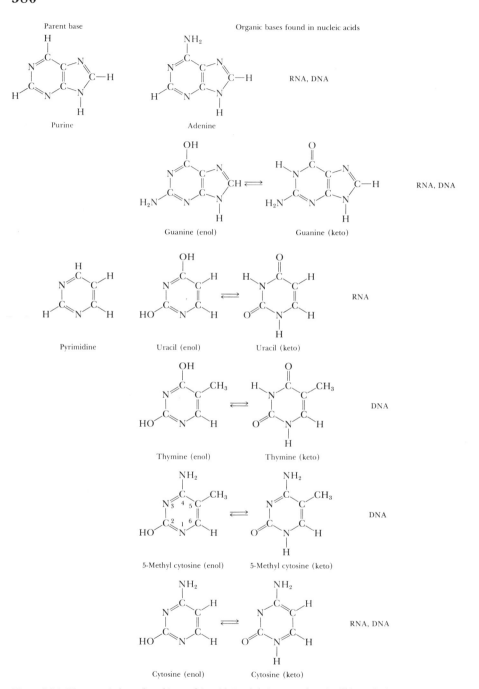

Figure 7.51. The organic bases found in nucleic acids (and their parent bases). All but adenine can exist in either of two tautomeric forms, the enol (with an OH) and the keto (with a C=O). There is a rapid conversion back and forth between the enol and keto forms.

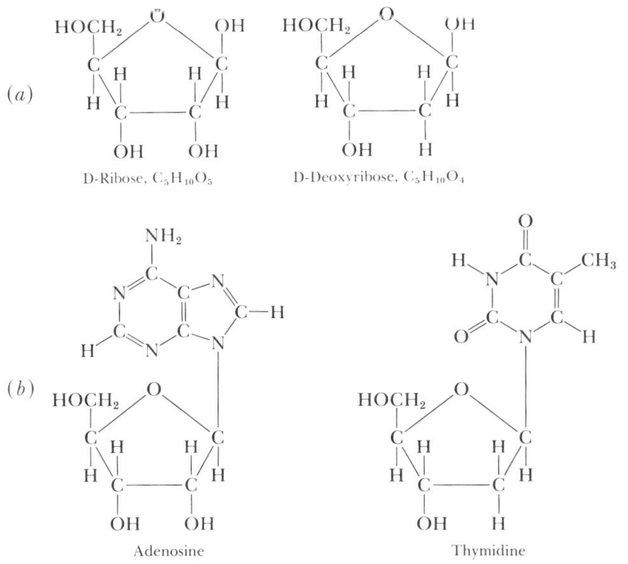

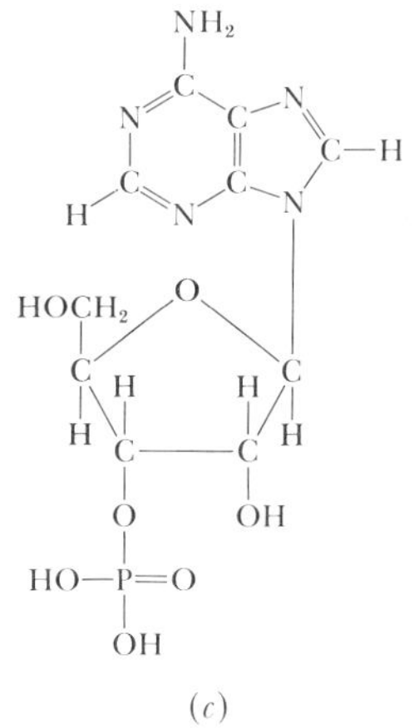

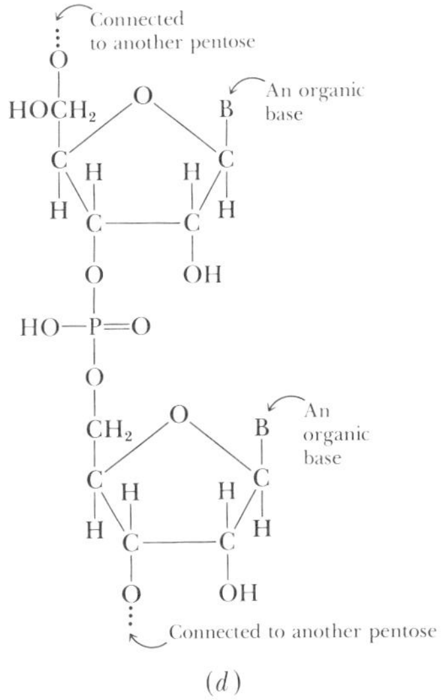

Figure 7.52. (a) The pentoses found in RNA and DNA. (b) Two nucleosides; adenosine (purine-pentose) from ribose and adenine; thymidine from deoxyribose and thymine (pyrimidine-pentose); (c) a nucleotide, adenosine monophosphate by adding phosphoric acid $PO(OH)_3$ to the nucleoside adenosine; (d) a polynucleotide, where phosphates connect the pentoses.

The sugars D-ribose and D-deoxyribose have a ring structure and differ from each other only in the fact that a single oxygen is absent in the latter. The organic bases are joined to the pentoses to form *nucleosides*. A phosphoric acid added to this makes a *nucleotide*. Both RNA and DNA are macromolecules of nucleotides — poly(nucleotides). The sugars are connected by phosphoric acid (Fig. 7.52). The polymer main chain is a regular alternation of a pentose and phosphoric acid residue. DNA chains are longer than RNA. The variety found in RNA and DNA comes from their sizes and the combinations of organic bases.

Among the interesting facts discovered about the nucleic acids over the course of many decades of research one concerns the organic-base composition of DNA. For most DNA, the number of adenines is equal to the number of thymines. Also, guanine and cytosine are found in equal numbers. This suggests a possible pairing of bases, adenine with thymine AT and guanine with cytosine GC.

Based on the x-ray diffraction work of Maurice H. F. Wilkins (b. 1916), an ingenious suggestion by Francis H. C. Crick (b. 1916)

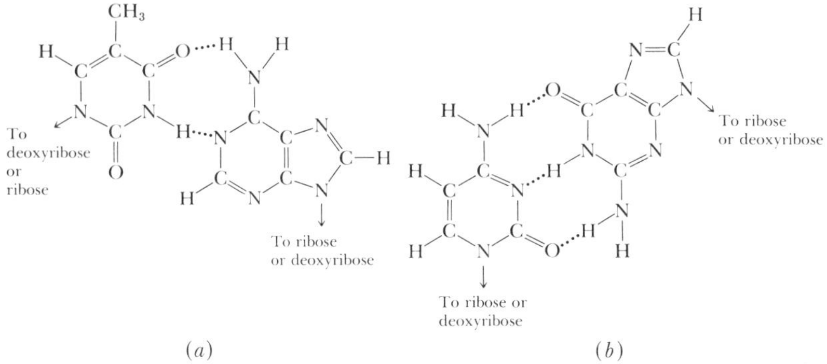

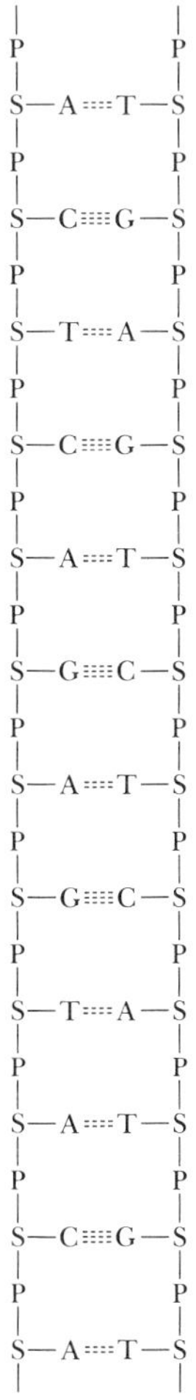

Figure 7.53. (a) The smaller pyrimidine (thymine) and the larger purine (adenine) held together by polar-group attractions; (b) a pyrimidine (cytosine) with a purine (guanine); (c) the ladder structure for two polynucleotide chains.

and James D. Watson (b. 1928) in 1953 established the structure of DNA and elucidated its biological activity. When cells divide during growth, DNA is replicated and each new daughter cell receives one complete set of DNA.

A ladder structure (Fig. 7.53) consists of two uprights, the pentose-phosphate chain, connected by rungs of organic-base pairs. If the two polynucleotide chains are separated, as when two chromosomes of a pair go to separate daughter cells, each single polynucleotide chain will accept only the proper companion organic base opposite each base on the intact chain. Thus only a thymine can be accommodated opposite an adenine and only a guanine opposite a cytosine.

This strict pairing is understandable for two reasons. First, adenine and thymine each have two polar groups placed in exactly the right way for strong attraction, and cytosine and guanine, similarly, have three polar groups properly directed for attraction. Second, one pyrimidine and one purine base in a pair is just the right distance between the pentoses. Two pyrimidines make too short a rung, and two purines are too long. A regular ladder using all four bases must have the bases paired in the manner described.

The individual polynucleotide thus serves as a pattern for the formation of a new companion polynucleotide. This new polynucleotide is an exact replica of the one removed. In this way DNA maintains its identity during cell division and, in reproduction, from generation to generation. The patterns of organic bases on RNA are governed by those on DNA. These, in turn, serve as guides, or *codons*, for the proper amino acid sequence in enzymes.

The Crick-Watson-Wilkins structure has one more feature. The ladder structure (Fig. 7.53) was presented only for the convenience of using a two-dimensional figure. The correct structure has the ladder uprights each twisted into a helix. The DNA molecule is a *double helix* (Fig. 7.54), with the deoxyribose-phosphate chains

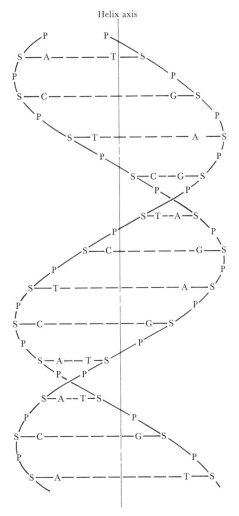

Figure 7.54. The twisted ladder or double-helix structure of DNA.

outside and the organic-base pairs holding these together. For their momentous work, Crick, Watson, and Wilkins were awarded the Nobel Prize in Medicine and Physiology in 1962.

PROBLEMS

7.1. Construct a regular tetrahedron by cementing or gluing together six sticks of equal length, as in Fig. 7.3*a*, or construct one from heavy paper using compass, straight edge, scissors, and tape:

a. Draw a large circle and mark the radius along the circumference, thus dividing the circle into six equal arcs with six points.

b. Connect an alternate set of three circumference points, forming a large equilateral triangle.

c. Draw a light line from each triangle vertex to the unused opposite circumference point, bisecting the side opposite the vertex.

d. Connect the midpoints of the three triangles sides, forming four smaller equilateral triangles.

e. Cut out the large triangle.

f. Crease the sides of the inner small triangle and fold up the three large triangle vertices so that they come together at the tetrahedron apex.

g. Tape the tetrahedron sides together for rigidity.

Now with soft pencil, chalk, or colored tape, label two of the vertices "H" and two "cl." Show that all six tetrahedra in Fig. 7.4 can be made simply by moving your one paper tetrahedron about in space; i.e., show that there is only one dichloromethane. (Save the tetrahedron for other problems in this chapter.)

7.2. Draw the nine heptane isomers.

7.3. Draw the structural formula and write the general compositional formula of 2,2,3,3,4,4-hexamethylpentane.

7.4. Draw structural formulas, write the composition formulas, and write the names for propane and the 29 substituted chloropropanes, a few of which are shown in Fig. 7.8.

Hint: There is one propane and one octachloropropane; two each of monochloro- and heptachloropropanes; four each of the dichloro and hexachloro isomers; five each of the trichloro- and pentachloropropanes; and six tetrachloropropanes. Can you suggest why these numbers of isomers (1,2,4,5,6,5,4,2,1) come in pairs?

7.5. Draw the structures of all 12 chlorobenzenes and give the name of each compound (see Fig. 7.9).

7.6. How many isomers can be made by replacing two hydrogens in naphthalene (Fig. 7.10) by two chlorines?

7.7. As in Fig. 7.15 draw the geometric structures and structural formulas for 2-propanol (also called isopropyl alcohol, isopropanol, or rubbing alcohol), meta-cresol (or *m*-cresol), and para-cresol (*p*-cresol). The cresols, found in coal tar, are better germicides but less toxic than phenol.

7.8. Match these structures with the proper names:

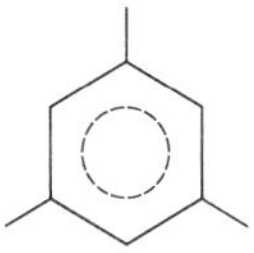

a. 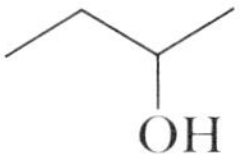 *b.* 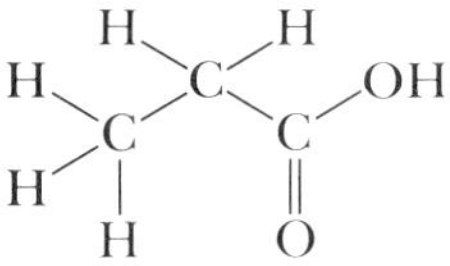 *c.* 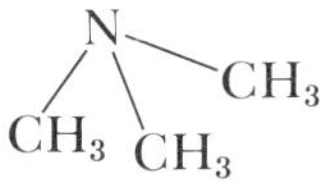 *d.*

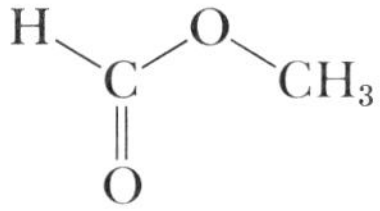

e. 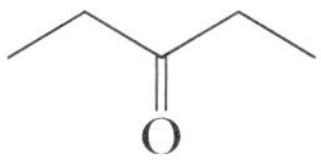 *f.* 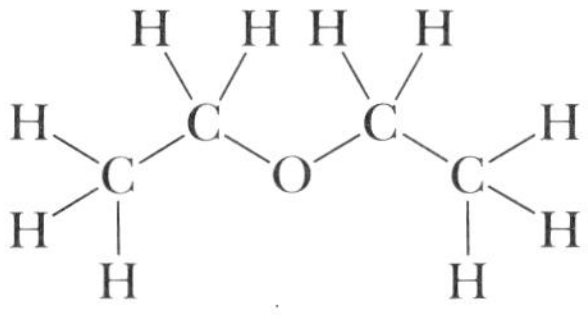 *g.*

    i. Methyl formate   v. Diethyl ether
    ii. Trimethyl amine   vi. Propionic acid
    iii. 3-pentanone, or diethyl ketone   vii. 2-butanol
    iv. 1,2,5-tribromobenzene

Which is a carboxylic acid? an alcohol? an ester? an aromatic? Which probably has an agreeable odor?

7.9. Show that CH_2ClF has only one isomer, using the tetrahedron made for Prob. 7.1. You can do this by marking one vertex Cl, another F, and the other two either H or unmarked. In an unsubstituted methane there are four choices for placing the chlorine atom; once this choice is made, there are three positions for the fluorine — 12 combinations in all. Show that you can produce all 12 combinations with your single CH_2ClF tetrahedron simply by holding it in different orientations.

7.10. Show that lactic acid has two distinct isomers. Do this with two tetrahedra, labeling the four vertices as in Fig. 7.32*b*. Can you move these around so that they both have the same substituents in the same orientation or arrangement?

7.11. Which of these molecules have optical isomers?

Bromoethane
1-bromo-1-chloroethane
1-bromo-2-chloroethane

Draw the pair of optical isomer structures.

7.12. From a piece of absorbent white paper towel, cut a rectangular strip about 3 by 25 cm. About 4 cm from one narrow end, mark a small dot with a water-base ink (not ball-point pen ink). Stand the strip in a tall glass into which you have put water to a depth of about 3 cm so that the end of the paper strip with the dot just touches the bottom of the glass, wetting the paper below but not as high as the dot. Support the paper strip by folding it over the lip of the glass and securing it, if necessary, with a rubber band or tape. As the water travels up the absorbent paper, it carries some of the ink with it. Does your ink appear to have one or more than one colored ingredient? If it has only one, you can repeat this experiment mixing a little red, green, and yellow food coloring with the ink.

7.13. A decapeptide, with 10 amino acid residues, has these fragments when partially hydrolyzed:

leu-thr
cySH-gly-ala
ala-leu-leu
gly-cySH-gly
*ser-cySH
cySH-ser-gly-cySH

The asterisk indicates the amino acid beginning the sequence. When completely hydrolyzed, the polypeptide is found to be made from two each of serine, cysteine, glycine, and leucine and one each of alanine and threonine. From this information reconstruct the correct amino acid sequence.

7.14. Do the experiment with a rubber band (Fig. 7.50). Abruptly stretch the rubber band to its maximum length (without breaking it) and *immediately* touch your upper lip to the middle of the stretched rubber. Does it feel warm or cool? Hold it extended and away from your lip for about 10 s. Abruptly allow the rubber band to return to its normal length and again *immediately* touch it to your upper lip. Now does it feel warm or cool?

7.15. A DNA strand contains 10,000 deoxyriboses and 10,000 organic bases. If the bases A, T, C, and G are equally available to be attached to the pentoses, how many different combinations of bases can be used in the strand?

7.16. A string of beads can be used as a model for a long polymer chain since it is flexible and can be either coiled or fully extended. If each bead represents a single carbon atom, how many are needed for a polyethylene molecule whose molecular weight is 6 million? If each bead is $\frac{1}{4}$ in. in diameter, how long will the fully extended string-of-beads model be?

7.17. Some large proteins and polypeptides have acidic and basic functional groups along their main chains. Such molecules are called "polyions," or "polyelectrolytes." Each polyion has its characteristic isoelectric point, at which pH the total negative charge of the several acidic groups exactly balances the total positive charge of the several basic groups. For example, IpH = 4.82 for gelatin. Blood has a pH of about 7.4 and varies by no more than 0.1 of a pH unit from this value. What do you think happens to the conformation of a polyion below, at, and above its isoelectric point? What is the net charge (i.e., positive, negative, or neutral) and probable conformation of gelatin in blood?

7.18. Examine the fiber-content label on clothing, carpets, and upholstery. In addition to the natural fibers (cotton, wool and fur, silk, and linen), how many synthetic materials do you find?

7.19. Look at the containers on your bathroom and kitchen shelves and in your refrigerator. How many of these are biodegradable, and how many are not? If you can look at a collection of children's toys, how many are made of biodegradable paper and wood, how many of iron (which rust when discarded), and how many of synthetic non-degradable materials?

7.20. Para-aminobenzoic acid (PABA) is made from benzoic acid (a carboxylic acid group on a benzene ring C_6H_5COOH) by putting an amine group on the para-ring position. Draw its structure. PABA is a bacterial growth vitamin. A synthetic substance para-aminobenzenesulfonamide, one of the first sulfa drugs, is effective against streptococcal diseases, pneumonococcal pneumonia, and gonococcal infections. The sulfonamide group is $-SO_2NH_2$. Draw the structure of this sulfa drug and suggest why it might be effective.

7.21. Consider these crystal profiles for a hypothetical substance:

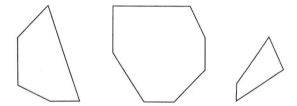

As in Haüy's analysis, select the largest and most nearly perfect of these, carefully trace its outline on a piece of paper, using a ruler to draw the straight-line segments of its profile. Next, extend four of the sides until they meet, forming a perfect-crystal's square outline. Show that the lengths of the extended lines are simply related; i.e., when a crystal is broken, the fragment split off is not just a random shape. Assume the least of these extended line segments is a unit, and then mark this unit all along each edge of the reconstructed perfect crystal. Next connect these marks and show that you can generate all the shapes of the other crystal profiles. Finally, draw a checkerboard on this large reconstructed square, with light pencil lines connecting the unit marks on opposite sides of the reconstructed square. At the intersection of these light lines make heavy dots. Do the sides of all the original imperfect crystals pass through these dots in a regular manner?

Nuclear Phenomena

Dalton's concept of indivisible, immutable atoms has been modified in two important ways. First, it has been shown that an atom *is not* a fundamental particle; it can be broken up into more basic units, the fundamental particles of Table 5.5: electrons, protons, and neutrons.

Second, Dalton considered that in all reactions the identity of an atom remains unchanged. From this, physical and chemical changes involve some rearrangement of atoms and some redistribution of electrons, but neither of these affects the nature of the nuclei involved. However, with the discovery of natural radioactivity came evidence that the nuclei themselves *were* altered during the emission of alpha and beta particles. A change in the nucleus changes one element to another.

In this chapter we shall look at a variety of processes in which nuclei are affected. These have given us more information about the nature of matter, provided important tools for scientific investigation, and inaugurated a major technological revolution. An important aspect of nuclear processes is the redefinition of the laws of conservation of mass and energy. This redefinition arises from the theory of relativity which, with quantum mechanics, brought about two major changes in twentieth-century scientific thinking.

THE ETHER—SPACE

The density of air decreases with elevation of altitude: the higher up we go, the less air there is. Our atmosphere is confined to a thin layer, estimated to be at most a few hundred kilometers thick, above the surface of the earth. There is an extremely low density of matter in the space between us and the sun or stars. Yet light travels from these hot radiating bodies to the earth. How does light exist in empty space? Because waves require a medium in which the oscillations can occur, the idea of a light-bearing, or *luminiferous, ether* was advanced—a weightless, frictionless, invisible, undetectable medium, pervading all space, for light to travel in.

As the earth rotates on its axis, one full turn a day, a point on the equator moves at a speed of 464 m/s (see Prob. 8.1 at the end of this chapter). Making a complete orbit about the sun each year, the earth moves at an average speed of 29.77 km/s (see Prob. 8.2). In addition to these motions, the earth, with all the other planets, is carried along with the sun as this star moves in the universe.

If the ether fills all space, then the earth must move through this medium. By measuring how fast the earth (or an observer on the earth) travels through the ether, we shall know the absolute speed of the earth. The attempt to measure this absolute motion is one of the most famous investigations in science, the Michelson-Morley experiment.

THE MICHELSON–MORLEY EXPERIMENT

More time is required for a boat to travel upstream on a river than downstream. If the boat is free to drift, it will move downstream at the approximate speed of the water. Going upstream with its engine operating, the effective speed (which is its speed measured against the land) is the boat speed *minus* the water speed; going downstream the effective speed is the boat speed *plus* the water speed. If the distance along the bank between the two stations is known, together with the boat speed and the upstream and downstream travel times, the speed of the water can be calculated.

Albert A. Michelson with Edward W. Morley (1838–1923) used a variation of this technique to measure the motion of the earth through the ether. Their apparatus (Michelson's interferometer, Fig. 4.6) enabled them to detect very small differences in the travel time of light beams. In Fig. 8.1 this experiment is shown. As seen in Fig. 8.1, the interference pattern resulting from the experiment can be seen by the observer at the telescope. A shift in this pattern should be observed as the interferometer is turned in space. In 1887 Michelson and Morley reported the remarkable results of many careful experiments. They observed no shift. They were not able to detect the motion of the earth relative to the ether.

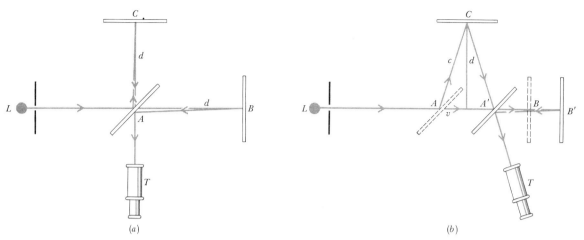

(a) (b)

Figure 8.1. The Michelson-Morley Experiment.

      (a) Light from a source L is split into two beams at a semimirror A. Half is reflected to mirror C, back through A at telescope T. The other half beam passes through A, to B, back to A, then to T. The beams recombine to form an interference pattern at T. The distances AC and AB can be made equal to d. The transit times for the paths ACA and ABA are equal:

$$t_1 = \frac{ACA}{c} = \frac{2d \ cm}{c \ cm/s} = \frac{2d}{c} \ s$$

$$t_2 = \frac{ABA}{c} = \frac{2d \ cm}{c \ cm/s} = \frac{2d}{c} \ s$$

Then there is no difference in path length, nor in transit time for the two complete paths LACAT and LABAT. The interference pattern seen at T does not change as the entire interferometer (L,A,B,C and T) is rotated in space.

      (b) With the earth moving in the ether the effective speed of light should depend on whether the light is moving with, against, or across the ether motion. Suppose the earth and interferometer move in the ether with a velocity v in the direction LAB. As light goes from A to C (with speed c) the mirrors also move (with speed v). After reflection at C light goes back to the semimirror, now at position A'. The two right triangles defined by ACA' have a common side d. The three sides of each right triangle are in the ratio $c : v : \sqrt{c^2 - v^2}$ by the Pythagorean theorem. Then the path ACA' can be calculated from

$$\frac{AC}{d} = \frac{CA'}{d} = \frac{c}{\sqrt{c^2 - v^2}}$$

$$\text{and} \quad ACA' = AC + CA' = \frac{2dc}{\sqrt{c^2 - v^2}}$$

The transit time for ACA' is this distance divided by the speed of light; $t_1 = 2d/\sqrt{c^2 - v^2}$.

      The other beam has the path AB'A'. Light moves to mirror B' with a relative speed of $c - v$, and back to A' with a relative speed $c + v$. The transit time for the path AB'A' is

$$t_2 = \frac{d}{c - v} + \frac{d}{c + v} = \frac{d(c + v) + d(c - v)}{(c - v)(c + v)} = \frac{2dc}{c^2 - v^2}$$

The ratio of the transit times is

$$\frac{t_1}{t_2} = \frac{2d}{\sqrt{c^2 - v^2}} \frac{c^2 - v^2}{2dc} = \sqrt{1 - v^2/c^2} < 1$$

or $t_1 < t_2$; the first transit time is shorter. If the entire interferometer is rotated through a quarter turn, the second path will then have the shorter transit time. From this, the interference pattern seen at T should shift with rotation. In the Michelson-Morley experiment no shift was seen upon rotation. Therefore, this simple picture of motion through an ether must be modified.

      FitzGerald suggested that a body undergoes contraction in the direction of its motion, so that a distance d contracts, only along the direction of motion, to $d\sqrt{1 - v^2/c^2}$. In the Michelson-Morley experiment path AB'A', but not ACA', is shortened. The second transit time

$$t_2 = \frac{2d(\sqrt{1 - v^2/c^2})c}{c^2 - v^2} = \frac{2d}{\sqrt{c^2 - v^2}} = t_1$$

With the FitzGerald contraction the two transit times are equal. Turning the interferometer in space cannot change the interference pattern seen at T.

We know the earth moves through space with a speed of at least 30 km/s, or $\frac{1}{10,000}$ the velocity of light. The interferometer was sensitive enough to detect this small a change in the observed speed of light. Something was fundamentally wrong with the assumptions of the experiment.

THE LORENTZ–FITZGERALD CONTRACTION

In 1893, George Francis FitzGerald (1851–1901) suggested that a body traveling through the ether undergoes a change in shape. If it moves with a velocity v relative to the ether, then the dimension in the direction of this travel contracts from a length d to

$$\left(\sqrt{1 - \frac{v^2}{c^2}}\right) d$$

The dimensions perpendicular to the motion in the ether are not affected. Thus a sphere moving in the ether becomes flattened. The FitzGerald factor $(1 - v^2/c^2)^{\frac{1}{2}}$ is ordinarily indistinguishable from unity unless v is very large.

This contraction of the interferometer in the direction of the motion in the ether (Fig. 8.1*b*) exactly compensates for the difference in travel time of the two beams. The travel times are equal and thus not dependent on the direction in which the interferometer is turned. No shift in interference pattern is seen as the instrument is turned.

Another approach, resulting in the same conclusion, was made by Hendrik Antoon Lorentz (1853–1928). Maxwell's equations describe the behavior of electric and magnetic fields. A competent observer of electromagnetic phenomena will discover these laws. However, making no special assumptions about the nature of space and time, another competent observer who is moving with respect to the same phenomena will find laws different from Maxwell's equations to describe what he observes and measures.

Lorentz guessed that this ought not to be the case. Maxwell's equations should be correct descriptions of electromagnetic phenomena whether the observer is at rest or moving relative to those phenomena. For this to be true, he showed that spatial measurements in the direction of this motion and durations of time must both be affected by the speed of the observer relative to the observed phenomena. The exact ways in which space and time are changed by this relative speed are called the "Lorentz transformations." They predict the contraction suggested by FitzGerald, now called the "Lorentz-FitzGerald contraction."

SPECIAL RELATIVITY

The failure of the Michelson-Morley experiment to detect the absolute motion of the earth through space or through the ether can be accounted for by the Lorentz-FitzGerald contraction. Other investigators devised different experiments to measure this absolute motion; still others devised experiments to detect the spatial contraction. Consistently, they failed. A separate compensating phenomenon was invented to account for each lack of success.

Albert Einstein looked at the problem of absolute motion in space from a fundamentally different point of view. He proposed that absolute motion has no meaning at all. All motion must be measured relative to something, a *frame of reference.* An observer at rest with respect to the earth ordinarily uses the earth (which is moving on its axis and in its orbit) as his frame of reference. Distances and velocities can be measured by using fixed points on the earth. The speed of an automobile is determined from the time needed to travel between two landmarks. The same observer can also determine the speed of an automobile while riding on a train or flying in an airplane. If he uses marks that are fixed with respect to the train or plane, he can obtain the automobile speed *relative* to the train or plane. The train and plane are possible frames of reference.

For objects moving within the solar system it is often more convenient to use the sun rather than the earth as the frame of reference. The same information can be represented more simply by choosing a particular frame of reference. Certain remote stars also constitute a convenient frame of reference.

In Einstein's *special theory of relativity*, the adjective "special" refers to a particular set of frames of reference—those which move with respect to each other at constant velocities. These are called "inertial systems." If a physical law is found to be valid in one inertial system, it must be equally valid in any other inertial system; that is, the physical law must have exactly the same form no matter which inertial system the observer uses to make his measurements and formulate his laws.

Because absolute motion, i.e., motion *not* measured with respect to some inertial system, has no meaning, all attempts to detect it, as was done in the Michelson-Morley experiment, must fail. Differences in the speed of light in space, as measured by observers in different inertial systems, can be used to determine absolute motion. But, consistent with the meaninglessness of absolute motion, the measured speed of light in space will have the same numerical value for observers in all inertial systems. The speed of light is independent of the speed of the light source.

From these basic premises of special relativity, Einstein showed that the Lorentz transformations describe how space and time appear to be affected by the velocity between two inertial systems. The Lorentz transformations apply to all physical measurements and are not restricted to electromagnetic phenomena.

THE EINSTEIN–LORENTZ TRANSFORMATIONS

Suppose an object that is at rest in one inertial system is measured by an observer in a second inertial system, moving at a constant velocity v relative to the first inertial system. The dimension of the object in the direction of this relative motion appears to be shorter, according to the Lorentz-FitzGerald contraction. To ask whether the object really contracts has no meaning because the measured length is the only length we can know.

If the mass of the object in the first inertial system is m_0, its *rest mass* (since it is at rest in this system), its mass measured by the observer in the second system is

$$m = \frac{m_0}{\sqrt{1 - v^2/c^2}}$$

Since the denominator is less than 1, the mass m is greater than the rest mass m_0.

The faster an object moves, the greater its mass appears to an observer not moving with the object. Thus a greater force is needed to accelerate the faster, more massive object. Before Einstein developed his special relativity theory, it had been observed that the charge-to-mass ratio was smaller for faster-moving high-energy beta particles. In 1910, highly accelerated cathode rays were produced, and both the electron velocities and charge-to-mass ratios were measured. The relativistic mass increase, calculated from the known electron velocity, resulted in a calculated lower charge-to-mass ratio, which agreed (within the precision of the experiment) with the observed charge-to-mass ratio. The correctness of the mass equation has been confirmed many times in subsequent experiments.

The time interval T_0 of an event in the first inertial system appears to lengthen, to an observer in the second inertial system, to

$$T = \frac{T_0}{\sqrt{1 - v^2/c^2}}$$

Thus a clock that is at rest in the first system seems to run slow to the observer in the second system. This phenomenon is called the "time dilation effect."

One interesting aspect of time transformation concerns *simultaneity*. Two events, not at the same place, which appear to be simultaneous to an observer in one inertial system will not appear simultaneous to an observer in another inertial system. The concept of simultaneity has meaning only when we specify the inertial system from which the observation is made.

GENERAL RELATIVITY

In 1915 Einstein published his *general theory*. In this more complicated theory, measurements are considered where the observed and the observer are not moving at a constant relative velocity; that is, the velocity may change continuously during the measurements. The general theory of relativity is a more complete theory of gravitation than the simpler Newtonian law.

Three crucial tests were proposed for this new theory. The first test concerned the orbit of the planet Mercury about the sun. Each of the nine planets has an elliptical orbit (Fig. 8.2a). Mercury is both the smallest planet and the closest to the sun. Its orbit is not a perfect closed ellipse but a spiral; that is, its major axis rotates, or precesses, by a small but measurable amount each turn (Fig. 8.2c). The presence of the other eight planets affects the orbit of Mercury and can account for almost all of its nonelliptical orbit; however, general relativity accounts for the rest.

General relativity describes the effect of a gravitational field on electromagnetic radiation. In particular, two other crucial tests were suggested, predicting phenomena unknown in 1915. The second test depended on the bending, or refracting, of light by gravity

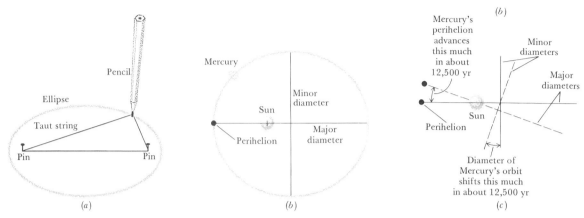

Figure 8.2. *General relativity and the orbit of the planet Mercury. (a) To draw an ellipse, a string is looped about two pins and a pencil. The pencil, holding the string taut, traces a closed curved figure, the ellipse. If the two pins coincide, the ellipse is a circle. (b) The ellipse of Mercury's orbit can be drawn if the pins are 2.382×10^7 km (12,850,000 nautical or 14,800,000 land miles) apart and the string loop is 13.966×10^7 km long. This ellipse has a major diameter 11.5×10^7 km across and a minor diameter 11.336×10^7 km across. The sun, whose diameter is 1.392×10^6 km, is placed at one pin position, called a "focus," and the planet, with a diameter of 5000 km, moves about the elliptical orbit. The perihelion is the point on the orbit of closest approach to the sun. The geometric form of planetary orbits was discovered by Johann Kepler (1571–1630) from astronomical observatory data, especially those collected by Tycho Brahe (1546–1601). Kepler offered three laws of motion to describe these orbits: (1) planets revolve about the sun, and their orbits are elliptical; (2) the line joining the sun to a planet sweeps out equal areas in equal times (that is, Mercury must move fastest at its perihelion and slowest when it is at the far end of its major diameter); and (3) the period of an orbit, i.e., how long it takes a planet to make one complete trip about the sun, is proportional to the major diameter raised to the $\frac{3}{2}$ power. Sir Isaac Newton showed that Kepler's three laws are logical consequences of the inverse-square law of gravitation and classical (that is, Newtonian) mechanics. (c) If Mercury were the only planet in the solar system, its orbit would be a perfect closed ellipse. After each orbit of 87.97 days, the new perihelion would coincide with the previous one. The perihelion, however, is seen to advance by a small amount each orbit. About 225,800 yr are required for the perihelion itself to make a complete turn. There are 360° in a full turn; 60 min (60′) in a degree; and 60 seconds (60″) in 1 min. In 1 yr the perihelion advances about 5.74″. The calculated influence of the other planets can account for 5.31″ of this annual advance. General relativity predicts an additional advance of 0.43″/yr, which is just the discrepancy between the observed and prerelativity calculated advance.*

as it passes a massive body. During a solar eclipse, the moon obscures the sun, and stars, otherwise invisible in daylight, are visible. In 1919 the pattern, or arrangement, was photographed for a cluster of stars almost in the line of sight of the eclipsed sun; the light from this cluster of stars passed near the hidden sun. Six months earlier, a photograph had been obtained for the same cluster at night when its light did not pass near the sun. As Einstein predicted, there was a distortion in this cluster pattern during the eclipse. Furthermore, the magnitude of the observed distortion (1.98 ± 0.12 and 1.61 ± 0.30 minutes in two different observations) agreed with Einstein's predicted value (1.75 minutes). Additional confirmation was obtained during two other eclipses: 1.72 ± 0.11 minutes in 1923, and 1.82 ± 0.15 minutes in 1928.

The third test, the *gravitational red shift*, was based on the prediction that an electromagnetic oscillation should slow down when leaving a gravitational field. For example, the frequency of visible light from a star should shift toward the red end of the visible spectrum. However, the observed wavelengths will increase at the same time, and so the speed of light will remain unchanged. The line spectrum (see Fig. 5.20) for each element is well known; in the spectrum obtained from a massive star, each wavelength shifted by the predicted amount toward the red.

THE MASS–ENERGY EQUATION

By doing work on a body, its motion can be increased. The body has a higher velocity and a higher kinetic energy. The increase in kinetic energy is equal to the work done. At the same time, the higher velocity is attended by an increase in mass. Einstein showed that no matter what kind of energy increase, there will be a concomitant mass increase; also with an energy decrease, there will be a mass decrease. His equation relating the change in energy E and change in mass m is quite simple:

$$m = \frac{E}{c^2} \quad \text{or} \quad E = mc^2$$

The constant c^2, converting the mass to an equivalent energy or vice versa, is simply the square of the speed of light.

We have used energy in several units: ergs, joules, and calories. Mass has been expressed in grams and kilograms. For processes on the atomic scale it is also convenient to use atomic mass units, based on the single C^{12} isotope with a mass of exactly 12 atomic mass units, or 12 amu. To examine the magnitude of the mass-energy effects we shall have to see, for example, how many grams are gained or lost with a gain or loss of 1 cal.

> **Example.** What is the mass equivalent of 1 erg, 1 joule, and 1 cal?
> Since
>
> $$1 \text{ erg} = 1 \text{ g-cm}^2/\text{s}^2$$
> $$m = \frac{1 \text{ erg}}{c^2}$$
> $$= \frac{1 \text{ g-cm}^2/\text{s}^2}{8.987554 \times 10^{20} \text{cm}^2/\text{s}^2}$$
> $$= 1.1126498 \times 10^{-21} \text{ g}$$

Because 1 joule $= 10^7$ ergs, the corresponding mass will be 10^7 times larger:

$$m = 1.1126498 \times 10^{-14} \text{ g}$$

One cal is equivalent to 4.184 joules; thus

$$m = 4.184 \times 1.1126498 \times 10^{-14} \text{ g}$$
$$= 4.655 \times 10^{-14} \text{ g}$$

Example. What are the energy equivalents in ergs, joules, and calories of 1 g?
    Since

$$1.1126 \times 10^{-21} \text{ g} = 1 \text{ erg}$$

$$1 \text{ g} = \frac{1}{1.1126 \times 10^{-21}} \text{ erg}$$

$$= 8.9876 \times 10^{20} \text{ ergs}$$
$$= 8.9876 \times 10^{13} \text{ joules}$$

$$= 8.9876 \times 10^{13} \text{ joules} \times \frac{\text{cal}}{4.184 \text{ joules}}$$

$$= 2.148 \times 10^{13} \text{ cal}$$

Example. What are the energy equivalents of 1 amu?
    The mass of a single C^{12} atom is its atomic weight (exactly 12) divided by Avogadro's number

$$\frac{12 \text{ g/mole}}{6.0225 \times 10^{23} \text{ atoms/mole}}$$

and 1 amu is just one-twelfth this, or

$$\frac{1}{6.02252 \times 10^{23} \text{ g}} = 1.66043 \times 10^{-24} \text{ g}$$

Then 1 amu is

$$1.66043 \times 10^{-24} \text{ g} \times \frac{8.9876 \times 10^{20} \text{ ergs}}{\text{g}}$$

$$= 1.4923 \times 10^{-3} \text{ ergs}$$
$$= 1.4923 \times 10^{-10} \text{ joules}$$

$$= 1.4923 \times 10^{-10} \text{ joules} \times \frac{1 \text{ cal}}{4.184 \text{ joules}}$$

$$= 3.567 \times 10^{-11} \text{ cal}$$

These important relations are given in Table 8.1 for future reference. Next we shall see why this energy-mass relationship was not observed before the discovery of nuclear phenomena.

> **Example.** To melt 1 mole of ice, 1,436 cal of energy are required. What is the mass increase when 1 mole (18.015 g) of ice is converted to liquid water?

$$1.436 \times 10^3 \text{ cal} \times 4.655 \times \frac{10^{-14} \text{ g}}{\text{cal}} = 6.685 \times 10^{-11} \text{ g}$$

This mass increase is too small to be detectable.

> **Example.** When 1 mole of TNT (2,4,6-trinitrotoluene) is oxidized, 820,700 cal are released in the exothermic reaction

$$4C_7H_5N_3O_6 + 21O_2 \longrightarrow 28CO_2 + 10H_2O + 6N_2$$

If that much heat is lost to the surroundings, what is the difference in mass between the reactants and products?

For 1 mole of TNT (227.13 g) and $\frac{21}{4}$ moles of O_2 (167.99 g), the energy lost is

$$8.207 \times 10^5 \text{ cal} \times \frac{4.655 \times 10^{-14} \text{ g}}{\text{cal}} = 3.820 \times 10^{-8} \text{ g}$$

Even for this highly exothermic reaction, the mass loss is undetectable.

Thus far, we have seen no experimental evidence for the Einstein mass-energy equation. Confirmation will come when we look at nuclear reactions. Since large energy changes in our examples are associated with very small mass differences, it follows that measurable mass differences will be associated with extremely large energies. Nuclear reactions involve this magnitude of energy.

There is no violation of the conservation laws of mass or energy. When 1 mole of ice was melted, it absorbed heat from its surroundings. With more energy, the mass of the water increased; with less energy, the mass of the surroundings decreased. The net effect was no mass change in the universe. Similarly in the TNT reaction, the energy released as heat or light is absorbed by the rest of the universe. Where the TNT and oxygen lost some mass in being converted to water, carbon dioxide, and nitrogen, the rest of

Table 8.1. Energy-Mass Conversion Units

Mass	Equivalent energy
1 g	8.9876×10^{20} ergs
	8.9876×10^{13} joules
	2.148×10^{13} cal
1 amu	1.4923×10^{-3} ergs
	1.4923×10^{-10} joules
	3.567×10^{-11} cal

Energy	Equivalent mass
1 erg	1.1126×10^{-21} g
	6.7011×10^{2} amu
1 joule	1.1126×10^{-14} g
	6.7011×10^{9} amu
1 cal	4.655×10^{-14} g
	2.803×10^{10} amu

the universe gained exactly this lost mass. A photon has an effective mass (see Chap. 5, the Compton effect, and Prob. 5.16). Thus, even when light is emitted from a source such as exploding TNT, the mass of the photons of the emitted light, together with the mass equivalent of the heat released, equals the mass loss of the explosion reaction system.

THE DETECTION AND MEASUREMENT OF RADIOACTIVITY

Becquerel discovered natural radioactivity when a photographic film was darkened by emanations from uranium (see Fig. 5.11). The amount of darkening depends on the intensity of the particular ray. Since alpha and beta rays are discrete particles and gamma rays are high-energy photons, we can measure how many of these are emitted from a specimen in a given time. These rays are separately measurable since they can be isolated by an electric field. Alternatively, filters, paper, or metal sheets can be used to remove alpha particles or both alpha and beta particles, so that only the more penetrating rays reach the film. Persons who work with radioactive material wear *film badges* to determine whether or not their exposure to radioactivity has exceeded safe levels.

A simple charged electroscope (Fig. 4.8) will discharge, i.e., its leaves will move closer together, if ions are present in the air. Emissions from a radioactive element moving through air will knock off electrons from or add electrons to the molecules in air. These produced ions neutralize the electric charge on the leaves, lessening the electrostatic repulsion between them. The rate at which the leaves approach each other is a measure of the number of ions in the air and thus a measure of the activity of the radioactive element.

A modification of the electroscope is the *Geiger counter* (Fig. 8.3a). *Ionizing radiation* from a radioactive element will ionize the tube gas, making it a conductor and enabling a current to flow between the wire and cylinder. The Geiger counter differs from the film and electroscope devices in that discrete particles are individually counted.

A *scintillation counter* is a sensitive screen which sparkles each

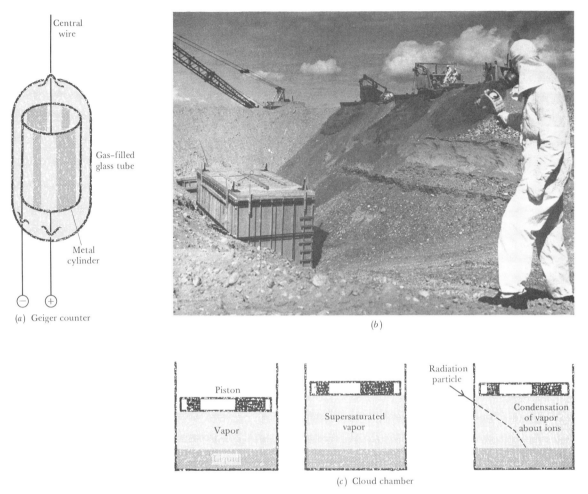

Central
wire

Gas–filled
glass tube

Metal
cylinder

⊖ ⊕

(*a*) Geiger counter

(*b*)

Piston

Vapor

Supersaturated
vapor

Radiation
particle

Condensation
of vapor
about ions

Liquid

(*c*) Cloud chamber

Figure 8.3. Two devices to detect and measure ionizing radiation. (a) In a Geiger counter an electric potential not quite large enough to cause a current flow is placed across the gas between the central wire and the metal cylinder. Each time an energetic particle, from radioactive emission, for example, passes through the counter, it knocks electrons from some gas molecules, creating ions. The partially ionized counter gas is a better conductor than before. Therefore, a current can flow between the central wire and the metal cylinder. Each time a particle enters the counter, another brief current pulse is counted. (b) This man, in a protective suit, is checking to see that no radiation is escaping from a lead container of radioactive wastes that are being buried. The problem of radioactive waste disposal is of increasing seriousness. (Battelle-Northwest Unit of AEC.) (c) Saturated vapor is in contact with a liquid in a Wilson cloud chamber (Charles T. R. Wilson, 1869–1959). When the piston is abruptly moved up, the vapor cools and momentarily contains more molecules than it ordinarily would at that lower temperature. It is supersaturated, and some molecules will condense if dust or other nuclei are present. An energetic particle entering the cloud chamber will knock electrons from molecules in the vapor, creating ions along its path. Since water is a polar molecule, it will be attracted to and condense on these ions. The condensed vapor or cloud indicates the path taken by the energetic particle. From the way the path changes on collision and how it is influenced by electric and magnetic fields, the nature of the particle is deduced. In a bubble chamber, ionizing radiation produces boiling along a particle's path.

time it is hit by a high-energy (fast-moving) particle (Fig. 5.13*a*). It measures radioactivity by counting individual particles.

Dust-free air can be made, momentarily, to contain more water vapor than it should under equilibrium conditions. Ionizing radiation passing through will be detected by the condensation, or *cloud track*, formed along the path of the moving particle causing the ionization. A dust-free liquid can be heated above its normal boiling point. Ionization radiation can provide ionic nuclei for vaporization. The boiling first appears as a track of small bubbles along the moving-particle path. This device is called a "bubble chamber" (Fig. 8.3*c*).

The upper limit of speed, for light and for matter, is the speed of light in a vacuum. Light moves more slowly in other media than it does in a vacuum. The ratio of the speed of light in a vacuum to the speed of light in a particular medium is the *index of refraction* of light in that medium. This index is a measure of how the path of light is bent, or refracted, in crossing the boundary between a vacuum and the medium. The index of refraction of pure water at 20°C for light of wavelength 589.3 nm is 1.33335. Thus, the speed of light of this wavelength in water is

$$\frac{2.997925 \times 10^{10} \text{ cm/s}}{1.33335} = 2.24842 \times 10^{10} \text{ cm/s}$$

In 1934 Pavel Cerenkov (b. 1904) observed a glowing in a liquid exposed to high-energy particles. Ilya Frank (b.1908) and Igor Tamm (1895–1971) described this *Cerenkov radiation* as a result of a particle moving into the liquid faster than the speed of light in that liquid. An electron moving in water faster than 2.25×10^{10} cm/s will emit Cerenkov radiation when it enters water. A *Cerenkov counter* is used to detect particles moving above the speed of light in the liquid of the counter.

GROUP–DISPLACEMENT LAW

Natural uranium is found always with other elements—some are stable but most are radioactive. The Curies isolated polonium and radium from a uranium ore. The radioactive elements can be separated from each other by chemical means. Radium, for example, has an atomic number of 88 and lies directly below barium in the periodic table; its chemical properties are those of the alkaline earths. Whatever chemical reactions are used to isolate barium or calcium, such as forming the insoluble sulfate $BaSO_4$, can be used to isolate radium.

Once a pure radioactive isotope is obtained, it continues to emit particles. Depending on how fast it emits these particles, gradually other isotopes are found in the sample, and at the same time there is less of the original pure isotope. By analysis one can determine exactly which isotopes are formed from a given pure isotope; that is, one can determine the *daughter* isotopes formed from a particular *parent* isotope.

Table 8.2 lists the sequence in *radioactive decay* when $_{92}U^{238}$ emits an alpha particle $_2He^4$ and becomes $_{90}Th^{234}$. The $_{90}Th^{234}$ isotope, in turn, decays by emitting a beta particle and becomes $_{91}Pa^{234}$. In each case notice that the sum of the subscripts of the two products equals the subscript of the parent isotope to the left of the arrow; in the first decay, $92 = 2 + 90$. These subscripts are the electric charges on the three involved species. The same is true of the superscripts; in the first decay, $238 = 4 + 234$. The superscripts are the nucleon numbers, or *mass numbers*, of the involved species.

The overall decay process, from uranium to lead, is

$$_{92}U^{238} \longrightarrow {}_{82}Pb^{206}$$

Since each alpha particle $_2He^4$ carries off four nucleons, and since there is an overall loss of $238 - 206 = 32$ nucleons, we know a total of eight alpha particles were emitted in the several decay steps from uranium to lead. We can write

$$_{92}U^{238} \longrightarrow 8{}_2He^4 + {}_{82}Pb^{206}$$

The superscripts balance: $238 = (8 \times 4) + 206$. However, the subscripts do not:

$$92 \neq (8 \times 2) + 82 = 98$$

The charge of the products, eight alpha particles and one lead, exceeds that of the parent uranium by six. If six beta particles are also emitted, this will reduce the charge of the products by six and not affect the nucleon number:

$$_{92}U^{238} \longrightarrow 8{}_2He^4 + 6{}_{-1}e^0 + {}_{82}Pb^{206}$$

Table 8.2. The Radioactive Sequence Beginning with U^{238}

$$_{92}U^{238} \xrightarrow{4.5 \times 10^9 \text{ yr}} {}_2He^4 + {}_{90}Th^{234}$$

$$_{90}Th^{234} \xrightarrow{24.1 \text{ days}} {}_{-1}e^0 + {}_{91}Pa^{234}$$

$$_{91}Pa^{234} \xrightarrow{1.18 \text{ min}} {}_{-1}e^0 + {}_{92}U^{234}$$

$$_{92}U^{234} \xrightarrow{2.48 \times 10^5 \text{ yr}} {}_2He^4 + {}_{90}Th^{230}$$

$$_{90}Th^{230} \xrightarrow{7.52 \times 10^4 \text{ yr}} {}_2He^4 + {}_{88}Ra^{226}$$

$$_{88}Ra^{226} \xrightarrow{1.622 \times 10^3 \text{ yr}} {}_2He^4 + {}_{86}Rn^{222}$$

$$_{86}Rn^{222} \xrightarrow{3.825 \text{ days}} {}_2He^4 + {}_{84}Po^{218}$$

$$_{84}Po^{218} \xrightarrow{3.05 \text{ min}} {}_2He^4 + {}_{82}Pb^{214}$$

$$_{82}Pb^{214} \xrightarrow{26.8 \text{ min}} {}_{-1}e^0 + {}_{83}Bi^{214}$$

$$_{83}Bi^{214} \xrightarrow{19.7 \text{ min}} {}_{-1}e^0 + {}_{84}Po^{214}$$

$$_{84}Po^{214} \xrightarrow{1.6 \times 10^{-4} \text{ s}} {}_2He^4 + {}_{82}Pb^{210}$$

$$_{82}Pb^{210} \xrightarrow{22 \text{ yr}} {}_{-1}e^0 + {}_{83}Bi^{210}$$

$$_{83}Bi^{210} \xrightarrow{5.01 \text{ days}} {}_{-1}e^0 + {}_{84}Po^{210}$$

$$_{84}Po^{210} \xrightarrow{138.4 \text{ days}} {}_2He^4 + {}_{82}Pb^{206}$$

$_{82}Pb^{206}$ Stable

Now both the nucleon or mass number and electric charges are balanced. (However, this simple exercise in arithmetic does not tell us the order in which alpha and beta particles are emitted.)

The simple displacement laws which describe the decay products in the table were discovered in 1902 by Rutherford and Soddy, as well as by several other independent workers. When an alpha particle is emitted, the mass, or nucleon number, is reduced by four and the atomic number, or nuclear charge, is reduced by two. When a beta particle is given off, there is no change in the mass number but an increase of one in the atomic number. This comes about by the conversion of a neutron in the nucleus into a proton and the emitted electron. With an increase of one in the number of protons, the atomic number increases by one.

There are three stable isotopes of lead: $_{82}Pb^{206}$, $_{82}Pb^{207}$, and $_{82}Pb^{208}$. And there is one isotope which decays extremely slowly: $_{82}Pb^{204}$. Since the mass number can change only by alpha emission, it can change only in units of four nucleons. U^{238} forms Pb^{206} by losing eight alpha particles. There is no way in which U^{238} can naturally decay to any of the other stable lead isotopes. To form $_{82}Pb^{207}$ it must lose $238 - 207 = 31$ nucleons or $7\frac{3}{4}$ alpha particles, an impossible condition.

In addition to the decay sequence of Table 8.2, there are other important series. The disintegration of $_{90}Th^{232}$ ends with the stable lead isotope $_{82}Pb^{208}$. That starting with $_{92}U^{235}$ terminates at $_{82}Pb^{207}$. The lead found in uranium ore has an atomic weight of 206.06, and that in thorium ore an atomic weight of 207.8. [This definite difference between two chemically indistinguishable leads suggested the concept of isotopes to Soddy (1913).]

In each decay reaction of Table 8.2, a radioactive isotope is converted to an isotope of another element. The changing of one element into another is called "transmutation." The transmutations in Table 8.2 are natural, not man-made, processes.

THE HALF–LIFE OF DECAY

Once a pure radioactive isotope is isolated, several important aspects of the decay process become evident. One ng $(1.000 \times 10^{-9} \text{ g})$ of the pure isotope $_{88}Ra^{226}$ emits 37 alpha particles/s. This much radium can be converted into radium salts: 1.707 ng of the bromide $RaBr_2$, 1.314 ng of the chloride $RaCl_2$, or

1.425 ng of the sulfate $RaSO_4$. In each case the activity is the same: 37 emissions/s. The activity does not depend on the chemical form of the radium. Nor is the activity affected by extremes in temperature, very high pressures, or electric or magnetic fields. Because the nucleus is a very small part of the whole atom and is shielded by the approximately eighty-eight electrons about it, radioactivity must be a property of the nucleus alone.

If 1 ng of radium, in any compound or pure, emits 37 alpha particles/s; 1 μg (10^{-6} g) emits exactly 1,000 times as many, or 3.7×10^4/s; and 1 g, 3.7×10^{10}/s. The activity is exactly proportional to the mass of radium present, and to the number of radium nuclei involved in the decay process.

The activity of a sample of any radioactive isotope depends only on the number of nuclei of that isotope present. With the repeated emission of either alpha or beta particles, the number of parent nuclei continually decreases. Thus the rate of emission also decreases.

In the case of radium, in 1 ng or 10^{-9} g, there are 2.66×10^{12} atoms. A loss each second of 37 radium nuclei reduces the total number of parent radium nuclei only by an insignificant amount. Even in a year less than 0.05 percent of the radium nuclei are used up. The emission rate of radium decreases slowly on a laboratory time scale but drops significantly over geological periods of time.

The immediate daughter nucleus of U^{238} (Table 8.2) is Th^{234}, which has a much greater activity than either uranium or radium. One ng of $_{90}Th^{234}$ (2.57×10^{12} atoms) emits 8.56×10^5 beta particles each second. This is nearly 25,000 times the rate from an equal mass of Ra^{226}. We should expect, therefore, the Th^{234} nuclei to be used up more rapidly and the emission rate to decline significantly in a shorter time.

The beta emission rate depends on the number of Th^{234} nuclei present in a sample. Thus counting this rate is a measure of how much of this thorium isotope is left. As the counted emission rate declines, so must the number of Th^{234} nuclei.

The decay of Th^{234} has an interesting and important feature. Table 8.3a is a record of how much is left of this isotope over a period of time. During each time interval of 24.1 days, exactly one-half of the remaining Th^{234} nuclei are converted to daughter nuclei. (This daughter Pa^{234} is also radioactive and a beta emitter, and so care must be taken to continuously separate the formed

Table 8.3. Radioactive Decay of Th^{234} and Ra^{226}

a. **Th^{234}**

Time from beginning of counting, days	Time interval, days	Th^{234} left	Th^{234} used up
0	0	$1 = 100\%$	0
24.1	24.1	$\frac{1}{2}$	$\frac{1}{2}$
48.2	24.1	$\frac{1}{4}$	$\frac{3}{4}$
72.3	24.1	$\frac{1}{8}$	$\frac{7}{8}$
96.4	24.1	$\frac{1}{16}$	$\frac{15}{16}$
120.5	24.1	$\frac{1}{32}$	$\frac{31}{32}$

b. **Ra^{226}**

Time from beginning of counting, yr	Time interval, yr	Ra^{226} left	Ra^{226} used up
0	0	$1 = 100\%$	0
1,622	1,622	$\frac{1}{2}$	$\frac{1}{2}$
3,244	1,622	$\frac{1}{4}$	$\frac{3}{4}$
4,866	1,622	$\frac{1}{8}$	$\frac{7}{8}$
6,488	1,622	$\frac{1}{16}$	$\frac{15}{16}$
8,110	1,622	$\frac{1}{32}$	$\frac{31}{32}$

daughter nuclei.) We say that Th^{234} has a *half-life* of 24.1 days (that is, 24 days, 2 hr, and 24 min).

The radioactive decay of Ra^{226}, presented in Table 8.3*b*, has a half-life of 1,622 yr. Obviously we cannot wait this long in the laboratory for 50 percent of a radium sample to be used up, but the rate of emission enables us to calculate in how long a time one-half of it will decay. Observe that, except for the time and time-interval columns in Table 8.3*a* and *b*, the numbers are identical. In Fig. 8.4, the decay curve is plotted for each of these radioactive isotopes. The two curves are identical if we adjust the time scale. For all decaying isotopes, the decay curve is represented by Fig. 8.4 provided we have the time axis marked in half-life units.

That the half-life is a constant follows from what we know about the decay process. The more nuclei, the faster the decay; with fewer nuclei, the slower the decay. With half the nuclei, the emission rate is one-half. Only half as many nuclei will disintegrate in a given time, but only half as many need disintegrate for a 50 percent reduction in that number. The half-life of its disintegration is a basic property of a radioactive isotope. If the half-life is sufficiently short, it is possible to identify a radioactive isotope just by how fast its activity declines in time.

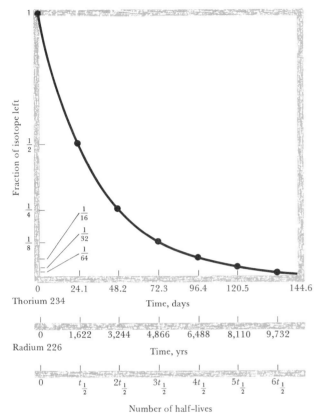

Figure 8.4. The number of radioactive nuclei, measured by the emission activity of these nuclei, decreases with an increase in time. The unit $t_{\frac{1}{2}}$, called the half-life, is the time required for one-half the nuclei to decay.

RADIOACTIVE DATING

Since the activity of a radioactive isotope cannot be altered by physical or chemical means, the half-life measured today is the half-life that isotope has had since its formation. Using Fig. 8.4, if we know what fraction of the initial amount of isotope remains, we can tell how long the disintegration process has been going on. This is the main technique used today for determining the age of mineral deposits, the earth's crust, and so on.

The time scale, in half-lives, years, or seconds, has evenly spaced intervals (Fig. 8.4). The fraction remaining during these evenly spaced time intervals does not decrease by a constant difference but does decrease by a constant ratio (namely, by one-half during each half-life interval). The differences between the successive fractions $1, \frac{1}{2}, \frac{1}{4}, \frac{1}{8}$, and so on, are not constant. The ratios of two successive fractions are consistently $\frac{1}{2}$. We recall that the logarithm of a ratio

$$\log (N/M) = \log N - \log M$$

A constant ratio means a constant difference in logarithms. Figure 8.5 permits us to estimate ages by radioactive decay.

Isotopes used for geological dating must have been present when the geological structure was formed and, in part, must be converted to daughter nuclei. If an isotope has too short a half-life, it will be completely converted soon after it is formed. On the other hand, a half-life that is too long means its emission rate is too slow for convenient detection. The conversion of uranium to lead goes at a rate suitable for these geological periods. The U^{238} decay, in 14 steps, ends with the stable isotope Pb^{206} (Table 8.2). The slowest decay step is the first one; all the others go quite fast relative to the first decay. In effect, the uranium-to-lead process has a half-life of 4.5×10^9 yr.

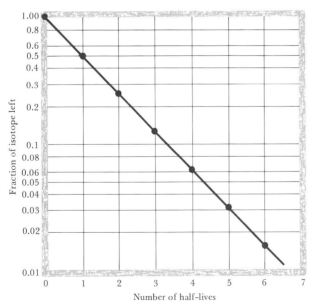

Figure 8.5. Radioactive-decay curve using a logarithmic scale. In most cases it is simpler to work with a straight-line graph.

Example. A piece of granite is found to contain 3.4 ng $_{49}U^{238}$ and 1.3 ng $_{82}Pb^{206}$, among other isotopes of these two metals. If we assume that all of the Pb^{206} present is *radiogenic*, which means it is formed from the radioactive decay of U^{238}, how long ago was the granite formed?

The masses of the parent and daughter isotopes are first converted to moles:

$$3.4 \times 10^{-9} \text{ g U} \times \frac{\text{mole}}{238 \text{ g}} = 1.4 \times 10^{-11} \text{ moles U}$$

$$1.3 \times 10^{-9} \text{ g Pb} \times \frac{\text{mole}}{206 \text{ g}} = 6.3 \times 10^{-12} \text{ moles Pb}$$

Since one decayed U^{238} nucleus ends up as a Pb^{206} nucleus, 6.3×10^{-12} moles of Pb^{206} is the stable end product of 6.3×10^{-12} moles of decayed U^{238}. Since we know how much uranium has decayed and how much there is now, we know how much there was initially:

Now	1.4×10^{-11}
Used up	0.63×10^{-11}
Originally	2.0×10^{-11}

We know what fraction of the initial uranium isotope is present:

$$\frac{1.4 \times 10^{-11}}{2.0 \times 10^{-11}} = 0.70$$

From the master curve (Fig. 8.5) this fraction corresponds to 0.51 half-life, or

$$0.51 \text{ half-life} \times \frac{4.5 \times 10^9 \text{ yr}}{\text{half-life}} = 2.3 \times 10^9 \text{ yr}$$

Today the accepted age of the earth is about 4.8 billion yr. This is based upon the terrestrial amounts of radiogenic Pb^{206}, Pb^{207}, and Pb^{208} from U^{238}, U^{235}, and Th^{232}. The assumption is made that some of these lead isotopes were also present at the beginning of the earth and their amounts were increased by the decay of the uraniums and thorium. To distinguish between radiogenic and original lead, analyses of meteorites have been made. Their lead content is identified with the original lead; the remainder in the terrestrial samples is assumed to be radiogenic. The analysis of isotopes in moon samples provides further information about the ages of the moon, earth, and solar system.

> **Example.** Assume the entire earth (about 10^{50} atoms) was once pure Ra^{226} (with a half-life of 1,622 yr). Then calculate how long a time would be required for all this to decay so that not one single atom of radium remains.
>
> If initially there are 10^{50} atoms, then at the end of one half-life period, only half of this, or $\frac{1}{2} \times 10^{50}$, remains. At the end of two half-lives, $\frac{1}{2} \times \frac{1}{2} \times 10^{50}$, or $(\frac{1}{2})^2 \times 10^{50}$, is left; at the end of three half-lives, $(\frac{1}{2})^3 \times 10^{50}$ atoms of Ra^{226} are left. In general, at the end of n half-lives, there will be $(\frac{1}{2})^n \times 10^{50}$ atoms. We want this number to be less than 1; less than one atom is no atom. In symbols

$$(\tfrac{1}{2})^n \times 10^{50} < 1$$

and if we multiply both sides of this inequality by 2^n,

$$10^{50} < 2^n$$
$$50 \log 10 < n \log 2$$
$$50 < n(0.3010)$$
$$\frac{50}{0.3010} < n$$
$$166.1 < n$$

Thus there must be 166.1 half-lives, or

$$166.1 \text{ half-lives} \times \frac{1{,}622 \text{ yr}}{\text{half-life}} = 269{,}000 \text{ yr}$$

This is much less than the age of the earth. Any Ra^{226} present at the formation of the earth has surely decayed. The Ra^{226} found today was formed more recently, in the decay series shown in Table 8.2.

Seven isotopes of carbon are known to exist, with mass numbers 10 to 16. Only two are stable: $_6C^{12}$ (98.89 percent) and $_6C^{13}$ (1.11 percent). The others are radioactive and, except for $_6C^{14}$, have relatively short half-lives. C^{14} has a half-life of 5,770 yr and is found mixed uniformly with the two stable isotopes in atmospheric carbon dioxide, (dissolved) in water, and in living organisms.

Comparing the half-life of C^{14} with the age of the earth, any

$_6C^{14}$ present at the beginning of the earth should have decayed within 800,000 yr. Even if the entire earth were initially pure C^{14}, there would be none present today. Therefore, C^{14} must be created continuously and decay continuously to account for the present level of this isotope.

The creation process occurs in the upper atmosphere when high-energy particles strike $_7N^{14}$. The formed $_6C^{14}$ is oxidized to carbon dioxide and mixes with air of the lower atmosphere. Carbon dioxide, including the radioactive molecules, is used by plants to make carbohydrates and proteins, ingested by animals, and returned to the atmosphere as respiration and combustion end products. All carbon dioxide, including the radioactive molecules, is involved in this *carbon cycle* (Fig. 11.13). Because the half-life of C^{14} is much longer than the turnover rate of carbon within an organism, the concentration of this isotope is the same in air, (dissolved) in water, and in living matter.

Once an organism dies, it no longer takes up carbon dioxide or carbon-containing foods. Through radioactive decay the concentration of the C^{14} isotope relative to the stable carbon isotopes decreases in time. Therefore dead matter should have less $_6C^{14}$ than living matter relative to its total carbon content.

Willard F. Libby (b. 1908) and coworkers examined the total carbon and C^{14} content of wood samples obtained from a variety of places and collected around the beginning of this century; they also examined the content in seal meat and oil from Antarctica. The total carbon content is obtained by burning part of the sample and collecting and weighing the carbon dioxide formed. The amount of C^{14} is determined by counting the beta emissions, according to the decay process

$$_6C^{14} \xrightarrow[\text{years}]{5,770} {}_{-1}e^0 + {}_7N^{14}$$

Because the half-life is long and the C^{14} content low, the rate of emission is slow. Special techniques were devised by Libby and coworkers to exclude extraneous counts from other sources. For example, all potassium contains the beta-emitting isotope $_{19}K^{40}$ ($t_{\frac{1}{2}} = 1.3 \times 10^9$ yr), and potassium is almost ubiquitous, being found in minerals, stone, glass, and so on. Heavy shielding was used against outside sources of beta particles. In addition, the basic Geiger counter was surrounded with auxiliary counters. When any of the latter were triggered first by an outside electron, the basic counter was shut off. Thus it counted only emissions from the samples of interest.

For each gram of carbon in each of 18 samples, a range from 14.53 to 16.31 counts/min was found; the average of these is 15.3 counts/min-g C. This value was taken to be the radioactivity in living organisms. Older samples have fewer counts per gram of carbon. From this, the time that sample stopped taking up radioactive carbon can be estimated. C^{14} dating has become a major archeological tool for ages up to 15,000 yr. For this work Libby received a Nobel Prize in Chemistry in 1960.

> **Example.** In 1968 a prehistoric campfire was unearthed in Chile, along with tools, a mastodon bone, and some pieces of charcoal. From the analysis of the last, there were 3.90 counts/min-g C. How long ago was the wood for the charcoal separated from a living tree?
>
> We assume the atmospheric C^{14} content was the same then as today. If the initial activity were 15.3 and the present activity 3.90 counts/min-g C, only 3.90/15.3, or 0.255, of the initial radioactive isotope is left. From the master curve (Fig. 8.5), this fraction corresponds to 1.97 half-lives. Thus, the age is

$$1.97 \text{ half-lives} \times \frac{5{,}770 \text{ yr}}{\text{half-life}} = 11{,}400 \text{ yr}$$

ARTIFICIAL TRANSMUTATION

An emitted alpha particle will lose energy and slow down by colliding with particles in its path. A further and anomalous effect was noticed by Rutherford's coworkers when alpha particles travel through air. To investigate this phenomenon, Rutherford directed emitted alpha particles through a gas-filled tube to a scintillation screen. This screen was shielded by a thin metal foil capable of stopping alpha particles but not high-energy protons. He experimented with several different gases in his tube, including air, moisture-free air, nitrogen, oxygen, and carbon dioxide. The number of protons hitting the screen depended only on the number of nitrogen atoms in the alpha-particle path.

Rutherford's conclusion was that an alpha particle caused the nitrogen nucleus to emit a proton. By considering the energies and momenta of the alpha particle and the emitted proton, it was possible to show that the recoil particle from the alpha-nitrogen collision has 17 nucleons. The reaction is written as

$$_7N^{14} + {}_2He^4 \longrightarrow {}_1H^1 + {}_8O^{17}$$

An abbreviated way of writing the same process is

$$N^{14}(\alpha,p)O^{17}$$

The first symbol in the parentheses is the particle taken up; the second, the emitted particle. The first isotope is the initial nucleus; the second, the final nucleus. Although the subscripts are redundant (since oxygen always has a nuclear charge of 8, and so on), we often write them out explicitly for convenience. This discovery, announced by Rutherford in 1919, is the first man-made, *artificial*, or *induced, transmutation*. It is also good evidence that protons exist in nuclei.

In 1932 James Chadwick discovered the neutron by the induced transmutation of beryllium:

$$_4Be^9 + {}_2He^4 \longrightarrow {}_6C^{12} + {}_0n^1 \qquad \text{or} \qquad Be^9(\alpha,n)C^{12}$$

From its properties, Chadwick showed the emitted particle to have no charge and the mass of a proton; therefore, he considered it to be the neutron predicted by Rutherford in 1920.

THE POSITRON

A charged electroscope (Fig. 4.8) will slowly discharge even if no ionizing radiation is brought near it. In part, this can be caused by the natural radioactivity in the earth's crust, a source of experimental difficulty in Libby's C^{14} method. However, if ionization- and radiation-detection instruments are taken to high altitudes or sent up in balloons, the amount of ionization detected increases with the distance from the earth's crust. C. T. R. Wilson proposed the source of this ionization to be radiation coming from outside the earth—*cosmic rays.*

Studying cosmic rays with a Wilson cloud chamber in 1932, Carl D. Anderson (b. 1905) observed, in addition to tracks made by electrons, other tracks by particles with mass equal to the electron mass but positively charged. Two years earlier, P. A. M. Dirac (b. 1902) used theoretical arguments to show that such a particle is plausible and that when it combines with an ordinary electron, both particle and electron vanish with a release of energy.

This new particle is called a "positron," and its symbol is $_1e^0$ or β^+. The combination of a positron with an electron and the conversion of their masses to radiant energy is called the "annihilation of matter." Positrons were not observed or recognized before Anderson's discovery because of this annihilation and their consequent short life, electrons being plentiful in ordinary matter. The matter we ordinarily observe is not subject to this annihilation. The positron is the first example of what we call "antimatter," characterized by annihilation with matter. The positron can also be called the "antielectron."

> **Example.** How much energy is produced when a positron-electron annihilation occurs, and what is the radiant-energy wavelength if one, and if two gamma-ray photons are made at the annihilation?
>
> The combined mass of the positron-electron pair is twice the electron mass, or 1.822×10^{-27} g. From Table 8.2 this is equivalent to
>
> $$\epsilon = 1.822 \times 10^{-27} \text{ g} \times \frac{8.988 \times 10^{20} \text{ ergs}}{1.000 \text{ g}}$$
> $$= 1.638 \times 10^{-6} \text{ ergs}$$
>
> If this energy appears as a single photon of frequency ν, it is also
>
> $$\epsilon = h\nu$$
>
> The frequency is
>
> $$\nu = \frac{\epsilon}{h}$$
> $$= \frac{1.638 \times 10^{-6} \text{ ergs}}{6.626 \times 10^{-27} \text{ erg-s}}$$
> $$= 2.472 \times 10^{20}/\text{s}$$
>
> The photon wavelength is
>
> $$\lambda = \frac{c}{\nu}$$

$$= \frac{2.998 \times 10^{10} \text{ cm/s}}{2.472 \times 10^{20}/\text{s}}$$

$$= 1.213 \times 10^{-10} \text{ cm}$$

$$= 0.01213 \text{ Å}$$

This is a very short, high-energy gamma ray.

If two photons are produced and they share equally the annihilation energy, each will have one-half the above energy, one-half the frequency, and twice the wavelength, or 0.02426 Å. Both wavelengths have been observed at a positron-electron annihilation.

The inverse process of annihilation has also occurred. When a short-wavelength gamma ray, not longer than 0.01213 Å, is taken up by a heavy nucleus, both an electron and a positron have been observed to be emitted.

ARTIFICIAL RADIOACTIVITY

The alpha particles emitted by natural radioactive isotopes have well-defined energies characteristic of the parent isotope. Thus the emitted alpha particles are good investigative tools: one variable, an uncertainty in speed or energy, is eliminated if all the alpha particles used come from the same parent.

Following the work of Rutherford, Chadwick, and co-workers, one of the more important discoveries was made in 1934 by Irene Curie (1897–1956, daughter of Marie and Pierre Curie) and her husband Frederic Joliot (1900–1958). An aluminum foil was exposed to alpha particles emitted by polonium. During the bombardment with alpha particles, neutrons, protons, and positrons were detected being emitted by the foil. When the alpha source was removed, the neutrons and protons stopped coming off the foil, but the positrons continued. The emission rate of the positrons showed the typical decay form of Fig. 8.4; the half-life was about 2.6 min.

If an alpha particle is captured by an aluminum nucleus, the atomic number increases by two and the nucleus is converted into phosphorus:

$$_{13}\text{Al}^{27} + _{2}\text{He}^{4} \longrightarrow _{15}\text{P}^{30} + _{0}n^{1} \qquad \text{or} \qquad \text{Al}^{27}(\alpha,n)\text{P}^{30}$$

The emission of the uncharged neutron does not affect the atomic number, and a free neutron is known to decay spontaneously into a proton and an electron, with a half-life of 12 min:

$$_0n^1 \longrightarrow {_1}H^1 + {_{-1}}e^0$$

The isotope $_{15}P^{30}$ is not found in nature and was suspected to be radioactive.

The treated, positron-emitting aluminum foil was dissolved in acid and treated with reagents to separate-out all phosphorus compounds—in one case, as a gas and, in another, as a precipitate. In both cases the positron emitter was found with the suspected phosphorus and not in the dissolved aluminum. The Joliot–Curies wrote the decay as

$$_{15}P^{30} \longrightarrow {_1}e^0 + {_{14}}Si^{30}$$

Similar results were found by exposing boron and magnesium to alpha particles:

$$_5B^{10}(\alpha,n){_7}N^{13} \xrightarrow{\;10.0\ min\;} {_1}e^0 + {_6}C^{13}$$

$$_{12}Mg^{24}(\alpha,n){_{14}}Si^{27} \xrightarrow{\;42\ s\;} {_1}e^0 + {_{13}}Al^{27}$$

(Both $_6C^{13}$ and $_{13}Al^{27}$ are stable.) These were the first instances of man-made radioactive isotopes. Today, by various means, about one-thousand artificial radioactive isotopes have been made.

DEUTERIUM

Chemical properties of an element are determined almost entirely by the nuclear charge and by the electrons about the nucleus. The several isotopes of an element have essentially identical chemical properties. Physical properties, such as average speed, can depend on the mass differences between isotopes of an element. But for $_{92}U^{235}$ and $_{92}U^{238}$, this difference is about 1 percent, and for Cl^{35} and Cl^{37}, less than 6 percent.

For appreciable mass differences, isotopes of light elements must be compared. The lightest element is hydrogen. Although only one isotope, $_1H^1$ or the proton, was known, a search was begun for a possible second one $_1H^2$. With both a proton and neutron in its nucleus, this isotope is twice as massive as the usual $_1H^1$.

In 1931 Ferdinand Brickwedde (b. 1903) slowly evaporated 4 liters of liquid hydrogen until about 1 ml remained. Because the heavier isotope, if it existed, should evaporate more slowly, it should be more concentrated in the residual liquid. Then Harold C. Urey (b. 1893) and George M. Murphy (1903–1968) examined the line spectrum of this sample. The wavelengths of the discrete lines (Fig. 5.20) depend just a little on the nuclear mass. Urey and Murphy observed faint lines, in addition to the normal spectrum, at exactly the wavelengths predicted for a doubly massive hydrogen nucleus. The heavier isotope is called "deuterium" ($_1H^2$, $_1D^2$, or simply D). Its nucleus is a *deuteron*.

Hydrogen and oxygen gases are made by the electrolysis of dilute aqueous salt solutions. The process is continuous: the gases are collected, and water is added. Therefore, an electrolytic cell is another source of the concentrated isotope deuterium because protons, being lighter, move more rapidly to the electrodes than do the deuterons. Thus more protons than deuterons are reduced to gas. Then proportionally more deuterium is in the cell water after long electrolysis than before. By repeated electrolysis, rather pure *heavy water* can be made. This is D_2O; both hydrogens are deuterium. First isolated by Gilbert N. Lewis (1875–1946), heavy water was found to have a density of 1.105 and a molecular weight of 20.03, to freeze at 3.82°C and to boil at 101.42°C.

Deuterium is a stable isotope. It can be accelerated to collide with other deuterons. In 1934 Rutherford and coworkers observed this reaction:

$$_1H^2 + {_1H^2} \longrightarrow {_1H^3} + {_1H^1}$$

also

$$_1H^2 + {_1H^2} \longrightarrow {_0n^1} + {_2He^3}$$

The new isotope $_1H^3$ is called *tritium*; its nucleus is a *triton*, one proton, and two neutrons. It is radioactive, emits a beta particle, and has a half-life of 12.26 yr.

PARTICLE ACCELERATORS

The early investigators of nuclear behavior used the emissions from naturally radioactive isotopes. There is a coulombic repulsion between a positive target nucleus and an alpha particle used as a probe. The coulombic repulsion depends on the two positive charges, $+2$ for the alpha and $+Z$ for the nucleus of atomic number Z, and varies as the product of these: $+2Z$. An incoming alpha particle must have enough speed or kinetic energy to over-come this repulsion. The higher the atomic number of the target nucleus, the greater this speed must be.

Alpha particles from radioactive isotopes have definite kinetic energies. For this reason an alpha particle from a given iso-tope will have enough energy to collide with some nuclei but not with those with greater than a certain atomic number (nuclear charge). We have already seen an example of each of these cases. In Rutherford's discovery of transmutation, an alpha particle was taken up by a nitrogen nucleus, with $Z = +7$. In contrast, in his scattering experiment (Fig. 5.13), a gold nucleus, with $Z = +79$, repulsed alpha particles traveling toward it. Therefore, in order to see what happens between alpha particles and heavier nuclei, it is necessary to accelerate the alpha particle to greater speeds. And other positive particles, such as the proton and deuteron, are not emitted from radioactive isotopes. These, too, must be accelerated artificially to overcome this repulsion between positively charged bodies.

An electric charge in an electric field has a force exerted on it by that field. The stronger the field, the greater the force; and the greater the force, the more the acceleration of the body bearing the charge. By using electric fields, charged particles (alpha, beta, positrons, and deuterons) can be accelerated to high speeds and high kinetic energies.

The unit of energy or work is the joule; the unit of charge is the coulomb. A difference of one volt was defined to exist (page 160) between two positions when 1 joule of work is needed to move 1 coul of charge from one position to another. The volt is the work per unit charge, 1 joule/coul. A body bearing a charge of 1 coul and moved by an electric field through 1 V has work of 1 joule done on it. If all this work appears as kinetic energy, the kinetic energy of the body is increased by 1 joule.

An electron has a charge of 1.60210×10^{-19} coul. When it is moved by an electric field from one position to a position 1 V lower, its kinetic energy is increased by 1.60210×10^{-19} joules, or 1.60210×10^{-12} ergs. This much energy has a special name, the *electron-volt* (eV). This is a convenient energy unit for nuclear processes because the particles have charges that are whole-number multiples of the electron charge. Knowing the charge and the voltage used in the acceleration, the energy of the accelerated particle can be found simply by multiplication. One thousand eV is abbreviated as keV; 1 million, as MeV; and a billion, as beV or GeV (for gigaelectron-volts).

John D. Cockcroft (1897–1967) and Ernest T. S. Walton (b. 1903) produced ionized hydrogen atoms (protons) in a discharge tube. In 1932 they built a *voltage multiplier* to produce about 700,000 V, used to accelerate the protons to an energy of 700 keV. These high-energy protons bombarded a lithium target, and alpha particles were formed. The reaction

$$_3\text{Li}^7 + {}_1\text{H}^1 \longrightarrow {}_2\text{He}^4 + {}_2\text{He}^4$$

or

$$\text{Li}^7(p,\alpha)\text{He}^4$$

was the first disintegration of a stable nucleus by a particle, the proton, not emitted from a radioactive isotope.

In 1931 Ernest O. Lawrence (1901–1958) and M. Stanley Livingston (b. 1905) built the first *cyclotron*. A beam of positively charged particles, such as protons, is accelerated toward a negatively charged electrode and, at the same time, forced into a curved path by a magnetic field (Fig. 8.6). The curvature is enough to put the beam into a spiral path. The two electrodes, or *dees*, are alternatingly made positive and negative to give the proton beam repeated accelerations. When the beam emerges, it has a high speed and can be directed toward a target.

The kinetic energy of a mass m moving at a velocity v is

$$\tfrac{1}{2}mv^2$$

The energy of an accelerated particle depends only on the electric charge and the voltage used. Thus an electron and a proton will be accelerated to the same energy (but in opposite directions) by the same voltage. Since its mass is only 1/1,836 the mass of a proton, the electron speed must be greater by a factor of

$$\sqrt{1,836} = 42.8$$

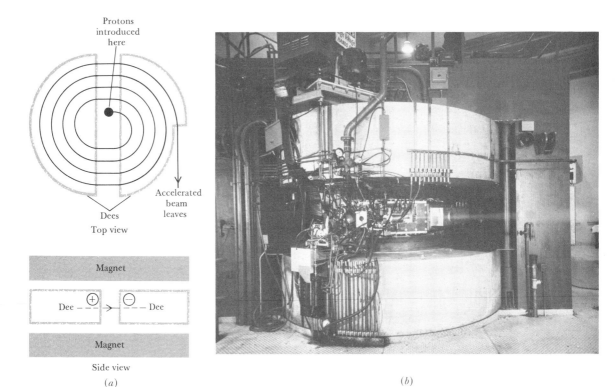

Figure 8.6. The cyclotron. (a) Spiral path of the electron. The path is curved under the influence of magnets; each time the electrodes are reversed, the proton is accelerated by being forced from one to the other dee. (b) The 60 cyclotron at the Argonne National Laboratory was built to study nuclear reactions and also to produce radioisotopes. The beam is a stream of deuterons traveling at a speed of 28,000 mi/s. (Argonne National Laboratory.)

The electron speed in a cyclotron rapidly approaches the speed of light. Its mass, consequently, increases. This makes the precise timing of the cyclotron charge alternation much more difficult.

Lawrence and coworkers also built a *linear accelerator* in 1931 (Fig. 8.7). In the past four decades these accelerators have been greatly improved. The energies obtained have been increased from Cockcroft and Walton's initial 700 keV to about 33 GeV. Today accelerators with a 300- to a 500-GeV capacity are being built.

HIGH–ENERGY PARTICLES

In the first accelerated-particle experiment (the Cockcroft–Walton disintegration of lithium by proton bombardment), the resulting

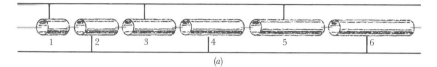

Figure 8.7. Linear accelerator. (a) Cylindrical cavity through which the electron beam passes is positively charged. As the pulse leaves one cylinder, the charge is reversed, forcing the electron into the next, oppositely charged cylinder. The electrostatic forces accelerate the electron. The rate of alternation of charge is several hundred million changes per second. Since the electron is progressively speeded up, the cylinders are made progressively longer. (b) The linear accelerator at Stanford, California, stretches over the countryside. (Stanford University and the U.S. Atomic Energy Commission.)

two alpha particles were seen to have quite high energies. An inspection of the masses involved will indicate why this is so:

$$_3\mathrm{Li}^7 \; + \; _1\mathrm{H}^1 \longrightarrow \, _2\mathrm{He}^4 \, + \, _2\mathrm{He}^4$$

$$\quad 7.01601 \quad 1.007825 \qquad 4.00260 \quad 4.00260$$

The combined lithium and hydrogen mass is 8.02384 amu; the two heliums have a total of 8.00520 amu. There is a discrepancy of 0.01864 amu, which is equivalent to

$$1.864 \times 10^{-2} \text{ amu} \times \frac{1.4923 \times 10^{-3} \text{ erg}}{\text{amu}} \times \frac{1 \text{ eV}}{1.60210 \times 10^{-12} \text{ erg}}$$

$$= 1.736 \times 10^7 \text{ eV}$$
$$= 17.36 \text{ MeV}$$

Cockcroft and Walton found each alpha particle to have an energy of about 8.6 MeV; combined, they had an energy of about 17.2 MeV, in good agreement with the Einstein mass-energy prediction. In comparison, the alpha particles from natural uranium have energies less than 5 MeV.

We can imagine the proton and target nucleus combining and then disintegrating:

$$_3\text{Li}^7 + {_1}\text{H}^1 \longrightarrow [_4\text{Be}^8] \longrightarrow {_2}\text{He}^4 + {_2}\text{He}^4$$

The Be^8 isotope is not found in nature and when formed is very unstable. Its half-life has been estimated to be about 10^{-16} s. Such short-lived structures are called "compound nuclei" and are distinct from the more stable radioactive nuclei.

This reaction is attended by a mass decrease and an energy release. Not all nuclear reactions do this. In Rutherford's transmutation

$$_2\text{He}^4 + {_7}\text{N}^{14} \longrightarrow [_9\text{F}^{18}] \longrightarrow {_1}\text{H}^1 + {_8}\text{O}^{17}$$
$$\phantom{_2\text{He}^4}4.00260 \quad 14.00307 1.007825 \quad 16.99914$$

there is a mass increase of 0.00130 amu.

That a compound nucleus is not the same as a radioactive isotope can be seen from the usual decay of F^{18}:

$$_9\text{F}^{18} \longrightarrow {_1}e^0 + {_8}\text{O}^{18} \quad \text{and} \quad {_9}\text{F}^{18} + {_{-1}}e^0 \longrightarrow {_8}\text{O}^{18}$$

This isotope can either emit a positron, or its nucleus can capture an electron, to form O^{18}. The half-life for either process is about 1.87 hr. *Electron capture* is sometimes an alternative to positron emission.

Over a thousand radioactive isotopes have been produced artificially by the action of high-speed particles on target nuclei. Protons, deuterons, alpha particles, electrons, and positrons have been used and have been emitted from either the transient compound nucleus or the radioactive isotope formed. Thus, there are available more than a thousand isotopes with a selection of emitted particles, half-lives, and particle energies.

THE ANTIPROTON

The first antimatter particle discovered was the antielectron, or positron. In general, if a particle has a companion antiparticle, the properties of the latter are defined. Both must have the same mass; if unstable, the same half-life; and if charged, then oppositely. They are formed together, as a pair, and they are annihilated as a pair when they come together.

The energy released at annihilation of a proton and its antiparticle is equivalent to twice the mass of a proton. Conversely, this much energy, at least, is needed to form a proton-antiproton pair. We can estimate this minimum energy. Since each is approximately 1 amu,

$$1 \text{ amu} \times \left(\frac{1.4923 \times 10^{-3} \text{ ergs}}{1 \text{ amu}}\right) \left(\frac{1 \text{ eV}}{1.60210 \times 10^{-12} \text{ ergs}}\right) = 931.46 \text{ MeV}$$

Then the pair, 2 amu, is equivalent to twice this: 1.863 GeV. Actually more than three times this, or 6.2 GeV, is required because the produced pair has kinetic energy too.

In 1955 Emilio Segrè (b. 1905), Owen Chamberlain (b. 1920) and coworkers accelerated a proton beam to this high energy in a new type of cyclotron, called a "bevatron." The beam was directed to a copper target. Although many other fragments were produced in addition to the expected proton-antiproton pair, they did detect a particle whose mass is equal to that of the proton, whose charge is equal to that of an electron, and which is annihilated at contact with a nucleon (either a proton or a neutron). If isolated from other matter, the antiproton is stable like the proton.

In addition to the antiproton, the antineutron has been detected. For each of the three fundamental particles there is a complementary antiparticle. An antihydrogen atom can be made from an antiproton nucleus and an orbiting positron. Except for the negative nucleus and positive electron, the antihydrogen has properties identical to those of the usual hydrogen atom. The two atoms annihilate each other, proton-antiproton and electron-positron, on contact.

In 1965 an antideuteron was observed. This is a nucleus with two nucleons: an antiproton and an antineutron.

THE STABLE NUCLEI

The number of protons in a nucleus is equal to the atomic number Z of the element. Except for hydrogen ($Z = 1$), all nuclei contain two or more protons. Because all protons are positively charged, and because those in a nucleus are packed close together, we expect strong repulsive coulombic forces to act to push these nucleons apart. Evidently, other forces must act to keep protons together. If we consider the gravitational force of attraction between nucleons, we find this much too small to compensate for the repulsive electric force (see Prob. 8.5 at the end of this chapter).

Of the approximately one thousand six hundred known isotopes, both those found in nature and those artificially made, only about two hundred sixty-eight are stable. Stability, then, although important for the continuity of chemical species, is not a general property of all nuclei. To learn more about stability we can begin with a census—a record of the atomic number Z, the mass (or nucleon) number A, and the neutron number $N = A - Z$ for each stable isotope.

As with any census we can expect some inaccuracy in our count. For example, at one time Pb^{204} was thought to be stable. Today we believe it has a remarkably long half-life of about 1.4×10^{19} yr. This means that a 1-g sample of lead, which is about 1.37 percent Pb^{204}, will emit two alpha particles a year. It is not surprising that the radioactivity of this isotope was unobserved until recently. Nor should it surprise us if some in the current list of 268 stable isotopes turn out to be unstable. It is not unreasonable to consider stability to be a relative property. We call an isotope stable if it has never been observed to be radioactive, that is, if its half-life is long enough to make the emission of particles unobservably infrequent.

Another source of inaccuracy in our census is contamination. In a few cases the emission from a supposed radioactive isotope has been traced to the presence of a radioactive impurity. When the impurity is removed, the isotope no longer emits. In spite of these uncertainties, the census in Appendix B, a Table of the Stable Isotopes, may give us some ideas about the stability of nuclei.

The census gives us the following patterns in stability:

1. Only 81 of the naturally occurring elements have stable iso-

topes. Elements of atomic number 84 or higher, element 43 (technetium), and element 61 (promethium) do not have a single stable isotope.

2. Some elements are *simple;* i.e., they each have only one stable isotope. Others are more complex; i.e., they have more than one stable isotope. Tin ($Z = 50$) has the most: 10 stable isotopes. For all elements with odd atomic numbers, there are, at most, two stable isotopes. Almost all the simple elements have odd-numbered atomic numbers.

3. Of the 268 stable isotopes, 212 have an even number of protons and only 56 have an odd number.

4. Of the stable isotopes, 207 have an even number of neutrons and 61 have an odd number.

5. The combination of both even proton number Z and even neutron number N is the most common among the stable isotopes (Table 8.4) and the combination of odd Z and odd N is the least common.

Table 8.4. Even-Odd Proton-Neutron Distribution of the Stable Isotopes

Protons, Z	Neutrons, N	Number of isotopes
Even	Even	156
Even	Odd	56
Odd	Even	51
Odd	Odd	5
Total		268

Some stable isotopes are more common than others. Oxygen is the most common element in the earth's crust; therefore, we expect to find more of O^{16} than of Pb^{208}. A careful analysis (Prob. 8.6 at the end of this chapter) indicates that 80 percent of the earth's crust is made up of just four isotopes:

$$_8O^{16} \qquad _{14}Si^{28} \qquad _{20}Ca^{40} \qquad _{26}Fe^{56}$$

In all four, both Z and N are even.

We interpret the greater occurrence of even proton and even neutron numbers to be a measure of the greater stability of the isotopes of oxygen, silicon, calcium, and iron. Evenness in these numbers suggests pairing. We recall that stability in chemical bonds is associated with electron pairing.

MAGIC NUMBERS

The pairing of electrons does not by itself lead to a complete description of chemical stability. The observed chemical inertness of the gases helium, neon, argon, krypton, and xenon is related to the fact that their electron subshells contain a complete set of electrons. The valence of atoms in a stable compound is also related to the formation of closed electron subshells.

Suppose nucleons are placed in shells within their nucleus. If these *nuclear shells* have definite capacities (by analogy with electron shells), we expect extra stability for those isotopes whose nuclear shells are completely filled. We have chosen to measure stability by abundance. An examination of Appendix B and other criteria of stability gives us information about the capacities of these shells.

When either the proton number Z or the neutron number N, or both, have one of the values 2, 8, 20, 28, 50, 82, or 126, the isotope or isotopes are more stable. We can cite some of the evidence.

1. Tin ($Z = 50$) has the greatest number of stable isotopes, 10.
2. The heaviest elements, all radioactive, decompose by alpha and beta emission until an isotope of lead ($Z = 82$) is formed.
3. The two heaviest stable nuclei are $_{82}Pb^{208}$ ($Z = 82$, $N = 126$) and $_{83}Bi^{209}$ ($N = 126$).
4. The chance of a neutron being captured by a target nucleus is smaller if Z or N is a magic number.
5. The alpha particle $_2He^4$ is very stable, but with exactly twice the number of nucleons, $_4Be^8$ is very unstable ($t_{\frac{1}{2}} = 3 \times 10^{-16}$ s). With just one more neutron, $_2He^5$ is even less stable ($t_{\frac{1}{2}} = 2 \times 10^{-21}$ s). The first nuclear shells are filled with two protons and two neutrons.

For their development of the nuclear shell theory, J. Hans D. Jensen (b. 1907), Marie Goeppert-Mayer (b. 1906), and Eugene P. Wigner (b. 1902) were awarded the Nobel Prize in Physics in 1963.

BINDING ENERGIES AND MASS DEFECTS

The stability of a nucleus can be described by how much energy is needed to make it decay: The more energy required, the more

Table 8.5. Energy Requirements for Nuclear Decay

Parent nucleus and mass, amu	Daughter nuclei and mass, amu	Combined daughter nuclei mass, amu	Mass gain or loss, amu	Energy, MeV
$_2He^4$	$\longrightarrow$ $_2{}_1H^1 + 2{}_0n^1$		Gain	Required
4.00260	$2(1.007825) + 2(1.08665)$	4.032980	0.03038	28.30
$_{88}Ra^{226}$	$\longrightarrow$ $_{86}Rn^{222} + {}_2\alpha^4$		Loss	Released
226.0254	$222.0175 + 4.00260$	226.0201	0.0053	4.9
$_{26}Fe^{56}$	$\longrightarrow$ $_{24}Cr^{52} + {}_2\alpha^4$		Gain	Required
55.9349	$51.9405 + 4.00260$	55.9431	0.0082	7.6
$_4Be^9$	$\longrightarrow$ $_5B^9 + {}_{-1}\beta^0$		Gain	Required
9.01219	$9.01333 + 0.0005486$	9.01388	0.00169	1.57
$_{87}Fr^{223}$	$\longrightarrow$ $_{88}Ra^{223} + {}_{-1}\beta^0$		Loss	Released
223.0198	$223.0186 + 0.0005486$	223.0191	0.0007	0.7

stable it is; and conversely, the less energy, the less stable it is. In Table 8.5 five possible nuclear-decay reactions are examined. The difference in mass between the parent and all the daughter nuclei is calculated, and then this difference is converted to an equivalent energy.

Energy is required to fragment helium into two protons and two neutrons, for alpha emission by Fe^{56} and for beta emission by Be^9. These three nuclei are stable. One isotope of polonium $_{84}Po^{212}$ emits alpha particles with nearly 8.8 MeV in energy; other radioactive emissions have less energy. Thus, under natural circumstances, there is not enough energy available for the fragmentation of a helium nucleus. In contrast, Ra^{226} and Fr^{223} do not need additional energy to decay. The first spontaneously emits an alpha, and the second a beta particle.

Several other criteria are used for nuclear stability against complete fragmentation into protons and neutrons. These are applied to the eight nuclei listed in Table 8.6. The *mass decrement* δ is the difference in mass between all the fragmentation daughter protons and neutrons, W, and the original parent nucleus, M. The *binding energy* E_B is the energy equivalent (in megaelectron volts). The *binding energy per nucleon* E_b has a maximum value for Fe^{56}, among all the stable nuclei. The *mass defect* Δ is the difference between the isotopic mass M and the nucleon number A. Since C^{12} is the standard for atomic masses, M is exactly 12 and the defect is 0 for C^{12}. The *packing fraction* f is the mass defect per nucleon, multiplied by 10,000 for convenience. Fe^{56} is seen, again, to be the most

Table 8.6. Binding Energy and Mass Defects for Several Stable and Unstable Isotopes

Isotope	Isotopic mass, amu	W†	$W - M = \delta$	E_B‡	E_b§	Mass defect, $M - A$	f¶
$_2\text{He}^4$	4.00260	4.032980	0.03038	28.30	7.075	+0.00260	+6.50
$_6\text{C}^{12}$	12.00000	12.098094	0.098094	91.37	7.614	0	0
$_{26}\text{Fe}^{56}$	55.9349	56.463400	0.5285	492.3	8.791	−0.0651	−16.2
$_{37}\text{Rb}^{85}$	84.9117	85.705445	0.7937	739.3	8.698	−0.0883	−10.4
$_{50}\text{Sn}^{122}$	121.9034	123.015130	1.1117	1035.5	8.488	−0.0966	−7.92
$_{82}\text{Pb}^{208}$	207.9766	209.733440	1.7568	1636.4	7.867	−0.0234	−1.12
$_{88}\text{Ra}^{226}$	226.0254	227.884370	1.8590	1731.6	7.662	+0.0254	+1.12
$_{92}\text{U}^{238}$	238.0508	239.984990	1.9342	1801.6	7.570	+0.0508	+2.13

Hydrogen atomic mass = 1.007825 amu
Neutron mass = 1.008665 amu

† W = Constituent mass = $1.008665N + 1.007825Z$
‡ E_B = Binding energy = $931.46(W - M)$ MeV
§ E_b = Binding energy per nucleon = $\dfrac{E_B}{A}$
¶ f = Packing fraction = $\dfrac{(M - A)}{A} \times 10^4$

stable nucleus, with the least (greatest negative) mass defect and packing fraction.

In Table 8.7, stability and magic numbers are compared. In part *a* of the table the most stable isotope Sr^{88} has a neutron number 50 and is the most abundant. In part *b*, among three *isobars* (nuclei with the same nucleon number A), the most stable is S^{36}, with a neutron number 20. And in part *c*, among five *isotones* (with the same neutron number N), Pb^{208} is the most stable.

MESONS AND OTHER PARTICLES

Outside the nucleus a free neutron is unstable and undergoes *beta decay* in about 17 min, on the average, after it becomes free.

$$_0n^1 \longrightarrow _{-1}e^0 + _1\text{H}^1$$

Beta decay does not occur in the nuclei of the 268 stable isotopes. The packing of protons and neutrons within these nuclei makes the neutrons stable. Protons repel each other outside but not inside a nucleus. Obviously, both protons and neutrons behave differently within and outside nuclei.

To compensate for the strong electrostatic repulsion between nuclear protons, we postulate a special attractive *nuclear*

Table 8.7.

a. Relative Abundance, Binding Energy, and Packing Fraction Are Greatest for That Strontium Isotope Whose Neutron Number Is Magic ($N = 50$)

Strontium isotope	Neutron number, N	Natural abundance, %	Atomic mass, amu	Binding energy		Packing fraction, f
				Total E_B, MeV	Per nucleon E_b, MeV	
$_{38}Sr^{84}$	46	0.56	83.9134	728.9	8.677	−10.3
$_{38}Sr^{86}$	48	9.86	85.9094	748.8	8.707	−10.5
$_{38}Sr^{87}$	49	7.02	86.9089	757.3	8.705	−10.5
$_{38}Sr^{88}$	50	82.56	87.9056	786.5	8.733	−10.7
Total		100.00				

b. Among These Isobars (the Same Nucleon Number), the Greatest Binding Energy and Packing Fraction Are Found in S^{36}, Whose Neutron Number Is Magic ($N = 20$)

Isotope	Neutron number N	Atomic mass M, amu	E_B, MeV	E_b, MeV	f
$_{16}S^{36}$	20	35.96709	308.70	8.575	−9.142
$_{17}Cl^{36}$	19	35.9673	307.8	8.550	−9.08
$_{18}Ar^{36}$	18	35.96755	306.70	8.519	−9.014

c. Among Isotones with the Same Neutron Number $N = 126$, the Greatest Binding Energy and Packing Fraction Are Found in the Nucleus That Is Doubly Magic ($Z = 82, N = 126$)

Isotope	Proton number Z	Neutron number N	E_b, MeV	f
$_{82}Pb^{208}$	82	126	7.8673	−1.12
$_{83}Bi^{209}$	83	126	7.8478	−0.938
$_{84}Po^{210}$	84	126	7.8338	−0.814
$_{85}At^{211}$	85	126	7.8109	−0.592
$_{86}Rn^{212}$	86	126	7.7943	−0.44

force which keeps protons from flying away from each other. Since this force is not observed between extranuclear particles, it must have a very short range of effectiveness, about 10^{-13} cm, the size of a nucleus.

Mirror nuclei are isobars such that the neutron number of one is the proton number of the other. Examples, with masses, are

$_1H^3(3.01605)$ and $_2He^3(3.01603)$

$_4Be^9(9.01219)$ and $_5B^9(9.01333)$

For the first pair the binding energies per nucleon are 2.83 and 2.57 MeV; for the second pair, 6.46 and 6.26 MeV. The corresponding mass defects are: 0.01605 and 0.01603; 0.01219 and 0.01333. The closeness of these values indicates that the nuclear force is approximately the same for proton-proton, proton-neutron, and neutron-neutron interactions.

The last column of Appendix B lists the neutron-to-proton ratio N/Z for the most abundant isotope of each of the 81 stable elements. Except for H^1, which has no neutrons, this ratio increases from 1, for the lighter elements, to $1\frac{1}{2}$, for the heavier ones. There are approximately equal numbers of neutrons and protons in the nuclei, with an excess of neutrons especially in the heavier stable nuclei. Since the heavier nuclei have more protons, with the attendant stronger coulombic repulsion, it is believed a greater excess of neutrons is needed to maintain stability.

The approximately equal numbers of protons and neutrons suggest pairing off of these nucleons. If beta decay of neutrons, and positron decay of protons, and their reverse are all imagined to occur within a nucleus

$$_0n^1 \rightleftharpoons\ _{-1}e^0 + \ _1p^1$$
$$_1p^1 \rightleftharpoons\ _1e^0 + \ _0n^1$$

then protons and neutrons are formed from each other by the exchange of electrons or positrons. In radioactive nuclei an electron or positron can escape; in stable nuclei it cannot.

If the exchange is rapid enough, protons and neutrons have the same averaged properties—neither proton nor neutron. They are essentially equivalent indistinguishable particles. A proton-neutron pair is more stable with than without this exchange. An analogous situation obtains in benzene (page 316). The carbon-carbon bonds are neither single nor double but some hybrid of these. This benzene molecule is more stable than a six-carbon ring with three single and three double bonds.

Hideki Yukawa (b. 1907) showed in 1935, by theoretical arguments, that, rather than an electron or positron, a particle 140 times as massive is needed for this intranuclear exchange. He also predicted that this particle will decay outside a nucleus in about 1 μs into an electron or positron. The next year Carl D. Anderson and Seth H. Neddermeyer (b. 1907) observed particles with these properties in cosmic rays. A variety of these, called "mesons" because they are intermediate in mass to electrons and nucleons, have been identified.

Emitted alpha particles have well-defined energies. About 95 percent of those from Ra^{226} have 4.78 MeV in energy; the rest have 4.59 MeV. But for beta emission, particle energies may be found over a wide range of values. Even with gamma rays considered, mass-energy did not seem to be conserved.

In 1930 Wolfgang Pauli suggested that very small uncharged particles could be emitted also to maintain the mass-energy balance in beta decay. Enrico Fermi (1901–1954) further developed this idea and named the particle the "neutrino." Very small and uncharged, it is extremely difficult to detect. Not until 1956 were means available to prove its existence. Other particles, called "baryons," or "hyperons," have been discovered. These are a little heavier than the two nucleons and decay into protons, neutrons, and mesons.

As the collection of known fundamental particles has grown in number (electrons, protons, neutrons, mesons, baryons, and their antimatter companion particles) to well over one hundred, so has the idea that we do not really understand what we mean by a *fundamental particle*. The fundamental particles obviously are involved in the structure of matter, but not all are necessarily ultimate particles.

NEUTRON BOMBARDMENT

With Chadwick's discovery of the neutron

$$Be^9(\alpha,n)C^{12} \qquad B^{11}(\alpha,n)N^{14}$$

this particle began to be used as a probe with nuclei. Because the neutron is not charged, it is not subject to the same coulombic repulsions met with in proton and alpha-particle bombardments; therefore, neutrons of low energy can strike a nucleus. This is fortunate because the accelerators mentioned cannot accelerate a neutral particle to high speeds and energies.

In a typical means of producing neutrons, alpha particles from Ra^{226} are used to enter nuclei of Be^9. As in Chadwick's first reaction, a neutron is emitted. Although neutrons are not accelerated by electric fields, it is convenient to use the same energy units for all particles. A 5-MeV neutron, for example, from a radium-beryllium source, has the same mass, speed, and energy as does a

5-MeV proton. Neutron speeds can be determined by a variation of the rotating-sector device (Fig. 3.6). Alternatively, the neutron speed can be obtained by diffraction and the neutron's deBroglie wavelength. B^{10} *captures* neutrons and emits alpha particles: B^{10} $(n,\alpha)Li^6$. From counting the alpha particles emitted by the otherwise stable B^{10}, the presence and number of neutrons can be determined.

A neutron moving through a substance will make occasional collisions with the atoms of that substance. Each time it strikes an atom, part of its energy can be transferred to the atom. Lighter atoms are more effective than heavier atoms in reducing the energy and speed of a neutron. Eventually the neutron will have a speed and an energy comparable to those of atoms and molecules at room temperature. Such neutrons are said to be *slow*, or *thermal*, *neutrons*. Their energies are about 0.03 eV (a considerable reduction from 5 million eV), and their speeds are about 2×10^5 cm/s. Substances, such as paraffin, heavy water, and graphite, used to slow down neutrons are called "moderators." In spite of being a light element, boron is not a good moderator because 20 percent of natural boron is B^{10} and this isotope captures neutrons.

The chance of neutron capture by a target nucleus can be measured by the effective area this target presents to neutrons. A heavy atom nucleus has a diameter of about 10^{-12} cm and a cross-sectional area of about 10^{-24} cm^2. A convenient unit for effective area is the *barn*, defined to be equal to 10^{-24} cm^2. *Neutron-capture cross sections* vary considerably from isotope to isotope and also depend on the neutron speed.

FISSION

In 1934 Fermi and Segrè bombarded uranium with slow neutrons and discovered, among the products, several distinct beta emitters. The usual procedure in such work is to determine the half-life for each emission process and then search through the list of known radioactive isotopes and their half-lives. Each beta emitter can be identified by its half-life. Fermi and Segrè could find no agreement between their products and the then-known beta emitters in the atomic number range $Z = 86$ to 92. For this reason they believed a uranium nucleus $Z = 92$ captured a neutron and then emitted an electron, increasing the atomic number by 1 to $Z = 93$:

$$_{92}U^{238} + _{0}n^{1} \longrightarrow _{92}U^{239}$$
$$_{92}U^{239} \longrightarrow _{-1}e^{0} + _{93}X^{239}$$

This would be a completely new *transuranium* element.

Fermi's discovery that uranium has an appreciable thermal neutron-capture cross section, that is, that the chance of uranium capturing a neutron is high, stimulated much work in this area. In 1938 Otto Hahn (1879–1968) and Fritz Strassmann (b. 1902), repeating the experiment, found one of the formed beta emitters to be inseparable and thus indistinguishable from barium; it was a new barium isotope.

The correct interpretation of what had happened was given early in 1939 by Lise Meitner (1878–1968) and her nephew, Otto Frisch (b. 1904). They suggested that neutron capture made the uranium nucleus sufficiently unstable to break up into two large fragments. This is in contrast to the serial emission of alpha or beta particles in a radioactive-decay sequence. Meitner and Frisch called this new process "fission."

A simple liquid-drop model was used to describe fission (Fig. 8.8). If the uranium-neutron compound nucleus is sufficiently deformed from its normal spherical shape, the short-range attractive nuclear and weak forces will not be effective over the long dimension of the ellipsoidal nucleus. Coulombic repulsion will dominate, and the nucleus will break up into two fragments which fly apart. The fragments have considerable energy, a total of about 200 MeV – much more than was then available by other means.

If one fragment is barium ($Z = 56$), then when uranium ($Z = 92$) breaks in two, the other fragment must be krypton ($Z = 36$) to conserve the number of protons in the parent uranium nucleus.

Bohr and John Wheeler (b. 1911) considered the stabilities of U^{235} and U^{238} after slow neutron capture and argued that it was the lighter isotope, rather than the more common U^{238}, which underwent fission. Small relatively pure samples of the two isotopes were prepared by mass spectrometry. Walter Zinn (b. 1906) and Leo Szilard (1898–1964) found the thermal neutron-capture cross sections to be 3 barns for the heavier isotope and about 100 barns for the lighter isotope. Furthermore, U^{235} is fissionable with slow neutrons (0.03 eV) while U^{238} requires a fast neutron (about 1 MeV) for fission. Since fission of U^{235} releases 200 MeV and requires only 0.03 eV, the energy released is greater than required by a factor of almost 7 billion.

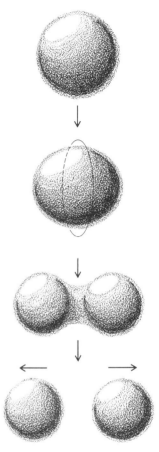

Figure 8.8. Liquid-drop model of nuclear fission.

Meitner and Frisch predicted a net production of neutrons after fission (Fig. 8.8). The heavier parent has a higher neutron-to-proton ratio than is found in the lighter fragments (Appendix B). Each fragment can either expel neutrons or undergo beta decay to readjust the neutron-to-proton ratio to a lower, more characteristic value. Szilard and Zinn discovered for each uranium nucleus split, between two and three neutrons are released.

The fission fragments are not equal, one being about 50 percent heavier than the other. The heavier fragment mass is usually in the range 130 to 149, and the lighter from 85 to 104. The quantity of energy released does not vary too much with the particular fragments formed. In Table 8.8 one typical fission reaction is analyzed. The impressive energy release is not from the annihilation of discrete fundamental particles but rather a change in mass defect and in binding energy. The energy released has been measured by calorimetry and found to agree with the value in the table.

Table 8.8. One Typical U^{235} Fission by Thermal Neutron Capture

$$_{92}U^{235} + _{0}n^{1} \longrightarrow [_{92}U^{236}] \longrightarrow [_{54}Xe^{141}] + [_{38}Sr^{95}]$$

$$[_{54}Xe^{141}] \longrightarrow _{0}n^{1} + _{54}Xe^{140}$$

$$_{54}Xe^{140} \xrightarrow{16\text{ s}} _{-1}e^{0} + _{55}Cs^{140}$$

$$_{55}Cs^{140} \xrightarrow{66\text{ s}} _{-1}e^{0} + _{56}Ba^{140}$$

$$_{56}Ba^{140} \xrightarrow{12.8\text{ days}} _{-1}e^{0} + _{57}La^{140}$$

$$_{57}La^{140} \xrightarrow{40.2\text{ hr}} _{-1}e^{0} + _{58}Ce^{140} \quad \text{(stable)}$$

$$[_{38}Sr^{95}] \longrightarrow _{0}n^{1} + _{38}Sr^{94}$$

$$_{38}Sr^{94} \xrightarrow{1.3\text{ min}} _{-1}e^{0} + _{39}Y^{94}$$

$$_{39}Y^{94} \xrightarrow{29\text{ min}} _{-1}e^{0} + _{40}Zr^{94} \quad \text{(stable)}$$

Overall reaction

$_{92}U^{235}$	$+ _{0}n^{1}$	$\longrightarrow$	$_{40}Zr^{94}$	$+ _{58}Ce^{140}$	$+6_{-1}e^{0}$	$+2_{0}n^{1}$
235.0439	1.008665		93.9061	139.9053	0.0033	2.017330

Initial mass	236.0526	
Final mass	235.8320	
Mass loss	0.2206 amu	

Energy equivalent   0.2206 amu × 931.46 MeV/amu = 205.5 MeV

Note: Most of this (all but about 10 MeV in energy) is released promptly.

NUCLEAR CHAIN REACTIONS

The fission process has two very important features: a great amount of energy is released, and more neutrons are produced than used up. If a produced neutron is captured, causing the fission of the capturing nucleus and the release of other neutrons, a series, or chain, of neutron capture, fission, neutron release, neutron capture, and so on, is begun. This is called a "chain reaction."

Because two or more neutrons are released for each nuclear fission, the number of free neutrons increases. Then the number of neutrons captured and the number of nuclei split must also increase. From this, the rate of energy release, or the power, will increase quite spectacularly. We can calculate the expected magnitudes involved in the fission process.

> **Example.** Assume that a chain reaction starts with a single neutron and 1 mole of U^{235}, that each released neutron is captured by a fissionable nucleus, that two neutrons are produced at each fission, and that 10^{-8} s is an average generation time between one step and the next in the fission chain reaction. How long a time is required for all the nuclei to fission?
>
> There are 6×10^{23} nuclei, and so this many neutrons are required. Starting with a single neutron, the number in successive generations is
>
> 1, 2, 4, 8, 16, 32, . . .
>
> The cumulative neutron sum, which is also the number of split nuclei, is
>
> 1, 1 + 2, 1 + 2 + 4, 1 + 2 + 4 + 8, . . .
>
> or
>
> 1, 3, 7, 15, 31, 63, . . .
>
> These numbers are
>
> 2 − 1, 4 − 1, 8 − 1, 16 − 1, 32 − 1, 64 − 1, . . .
>
> or
>
> $2 - 1, 2^2 - 1, 2^3 - 1, 2^4 - 1, 2^5 - 1, \ldots$

and eventually the general number

$$2^n - 1 \approx 2^n$$

In n generations a cumulative total of 2^n neutrons will be produced and 2^n nuclei split.

We want this number to be

$$2^n = 6 \times 10^{23}$$

Then

$$\log 2^n = \log 6 + \log 10^{23}$$
$$n \log 2 = \log 6 + 23 \log 10$$
$$n(0.3010) = 0.7781 + 23$$
$$n = \frac{23.7781}{0.3010}$$
$$= 79.00 \text{ generations}$$

Each generation has a time duration of 10^{-8} s, and so 79 generations require 7.9×10^{-7} s.

Example. How much mass disappears and how much energy is released during the fission of 1 mole (235 g) of U^{235}?

In a single fission the mass loss is about 0.22 amu (Table 8.8). For 1 mole of fissioned nuclei, this amounts to 0.22 g. The energy equivalent of this is (from Table 8.2)

$$0.22 \text{ g} \times \frac{8.99 \times 10^{13} \text{ joule}}{1 \text{ g}} = 2.0 \times 10^{13} \text{ joules}$$

$$0.22 \text{ g} \times \frac{2.15 \times 10^{13} \text{ cal}}{1 \text{ g}} = 4.7 \times 10^{12} \text{ cal}$$

Example. How many grams of ice at its freezing point can be converted into steam at the boiling point by this much heat?

To melt 1 g of ice, about 80 cal are needed; to raise the temperature from 0 to 100°C, 100 cal; and to convert the liquid to a vapor, 540 cal. A total of 720 cal/g are required.

Then

$$4.7 \times 10^{12} \text{ cal} \times \frac{1 \text{ g}}{720 \text{ cal}} = 6.5 \times 10^9 \text{ g}$$

The density of ice is 0.92 g/cm³; thus

$$6.5 \times 10^9 \text{ g} \times \frac{1 \text{ cm}^3}{0.92 \text{ g}} = 7.1 \times 10^9 \text{ cm}^3$$

A cubic block of ice with this volume has an edge

$$s = (7.1 \times 10^9)^{\frac{1}{3}}$$
$$= 1.9 \times 10^3 \text{ cm}$$
$$= 19 \text{ m}$$
$$= 62 \text{ ft}$$

This block of ice has about thirty times the volume of a six-room house.

Example. What is the volume of 1 mole of U^{235}? (Uranium metal has a density of 19.0 g/cm³.)

$$1 \text{ mole U} \times \frac{235 \text{ g}}{\text{mole}} \frac{\text{cm}^3}{19.0 \text{ g}} = 12.4 \text{ cm}^3$$

This volume is less than $\frac{1}{2}$ oz.

Example. What are the mass and volume of TNT needed to give off the same energy?

The heat of slow combustion of TNT in oxygen is 821 kcal/mole; the energy of rapid explosion in the absence of oxygen is 285 kcal/mole. The second process will be used for comparison. The molecular weight of TNT is 227 g/mole, and its density is 1.65 g/cm³. The mass

$$4.7 \times 10^{12} \text{ cal} \times \frac{\text{mole}}{2.85 \times 10^5 \text{ cal}} \frac{2.27 \times 10^2 \text{ g}}{\text{mole}} = 3.7 \times 10^9 \text{ g TNT}$$
$$= 3.7 \times 10^6 \text{ kg TNT}$$

This is about 4.0×10^3 tons, or 4.0 kt, TNT—a useful unit in military explosives.

The volume is

$$3.7 \times 10^9 \text{ g} \times \frac{\text{cm}^3}{1.65 \text{ g}} = 2.2 \times 10^9 \text{ cm}^3$$

A cube with this volume has an edge

$$s = (2.2 \times 10^9)^{\frac{1}{3}} \text{ cm}$$
$$= 1.3 \times 10^3 \text{ cm}$$
$$= 13 \text{ m}$$

> This is about 43 ft an edge, or ten times the volume of a six-room house.

A small mountain of ice is converted to steam in a flash, using an energy source which can be held in a tablespoon. About fifteen million times that quantity of a conventional chemical explosive is required to do the same thing. Simple calculations like these were done and understood by many scientists in 1939, but performing theoretical calculations was no guarantee that so colossal an energy release was attainable. The importance of this knowledge, its possible military application, and the international political conditions of that year made the exploitation of nuclear energy one of the major technological enterprises of all time.

THE FIRST SELF–SUSTAINING CHAIN REACTION

In the summer of 1939 the probability of a second world war appeared to be quite high. Because of its racial policies, many of Germany's scientists (and most of its best) left for countries such as Great Britain and the United States.

On August 2, 1939, the most respected scientist of his time, Albert Einstein, sent a letter to President Franklin Roosevelt advising him of the work of Fermi and Szilard and of the possibility of a chain reaction and its implications, and urging him to establish a liaison between his administration and a group of nuclear physicists. With this letter, Szilard sent an outline of the support needed to pursue the investigation and development of a nuclear chain reaction. World War II began with the German invasion of Poland on September 1, 1939, and the consequent declaration of war by Great Britain and France on Germany on September 3, 1939. By April 1940, scientists working on fission chains imposed a voluntary censorship on their work.

The energy of fission of one U^{235} nucleus, about 200 MeV, is distributed among a few beta and gamma particles and two or three neutrons (Table 8.8). The neutrons, carrying off most of this, have each about 100 MeV. These high-energy neutrons are likely to be taken up by the most common isotope U^{238}. It was necessary, therefore, to rapidly reduce the fission-produced neutron energy to the desired thermal value, about 0.03 eV, before the neutron collided with U^{238}. A substantial amount of moderator was required. Fermi (who headed this project) and his coworkers chose graphite as the moderator.

The thermal neutrons so prepared can be captured by U^{235} with consequent fission and production of more neutrons, captured by U^{238} without a net increase of neutrons, taken up by an impurity, or can escape from the uranium-moderator system. To enhance the chance for success in a fission chain process, the rate of neutron escape must be reduced. Escape is more likely from fissioning nuclei located near the surface of the uranium-moderator material and less likely from those nuclei buried deeper within. For a given mass of fissionable material, the spherical shape has the least surface and promotes the chance for neutron capture over neutron escape (Fig. 8.9). Also, the larger the sphere of fissionable material, the smaller the fraction of neutrons escaping.

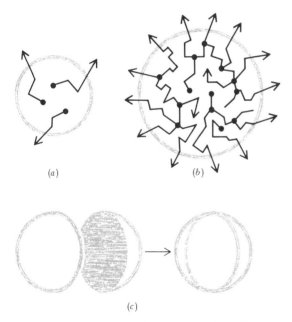

Figure 8.9. (a) In a subcritical mass or size of fissionable material too many neutrons originate near the surface and escape before capture. There is no build-up in the number of neutrons. (b) When the fissionable material is larger than a critical mass or critical size, a greater fraction of the neutrons originate deep within the material. There are more neutron-nuclei encounters, more neutron capture, more fission, and an increase in the number of neutrons. The fission process is accelerated. (c) The assembled sphere of at least critical size has less surface than the two hemispheres. Fewer neutrons can escape before capture from the assembled sphere.

Because a sphere is not the most convenient shape with which to work, a compromise was reached. Fermi's team at the University of Chicago alternated layers of uranium and uranium oxide bricks with graphite bricks in a rounded stack, or pile, containing in all 12,400 lb of the metal and metal oxide. The entire system, first called a "pile," is more commonly termed a "nuclear reactor."

The criterion for maintenance or buildup of neutrons is the *multiplication,* or *reproduction, factor k* defined by

$$k = \frac{\text{number of neutrons within the system produced in one generation of fissions}}{\text{number of neutrons produced in the immediately preceding generation}}$$

If k exceeds unity, there is a net increase in neutrons and the fission rate also increases. If k is exactly 1, the number of useful neutrons and the fission rate remain steady. And when k is less than 1, the number of neutrons and the fission rate both decline. In Table 8.9, the growth or decline in the number of neutrons with time and generation is shown.

The neutron level within the pile was monitored with a boron-10 counter. A too-rapid fission rate and energy release could damage the pile and be hazardous to the members of the project. To be able to promptly start or stop the chain reaction without tearing down the pile, cadmium control rods were used. Cd^{113} has

Table 8.9. Fraction of Thermal Neutrons Remaining in a Fission Chain Nuclear Reactor as a Function of the Multiplication Factor and Number of Fission Generations

Generation in chain	Time, s	Multiplication factor, k				
		0.95	0.999	1.000	1.001	1.05
0 (first)	0 (start)	1.000	1.000	1.000	1.000	1.000
1	10^{-8}	0.950	0.999	1.000	1.001	1.05
10	10^{-7}	0.5999	0.990	1.000	1.01	1.63
100	10^{-6}	5.92×10^{-3}	0.906	1.000	1.10	1.32×10^2
1000	10^{-5}	5.25×10^{-23}	0.372	1.000	2.69	1.55×10^{21}

If $k = 1.05$, then in 10 generations after the start the neutron enhancement is

$$1.05 \times 1.05 \times 1.05 \cdots \times 1.05 = (1.05)^{10}$$

$$\begin{aligned} N &= (1.05)^{10} \\ \log N &= 10 \log 1.05 \\ &= 10(0.02119) \\ &= 0.2119 \\ N &= 1.63 \end{aligned}$$

If the chain begins with 100 neutrons, in 10 generations, or 10^{-7} s, there will be 163 neutrons.

the unusually high thermal neutron capture cross section of 27,000 barns. Pushing in or withdrawing these rods captures more or fewer neutrons, thus making the multiplication factor less or more than unity. The initial neutrons used to start the chain came from a beryllium-radium source.

On December 2, 1942, Fermi and his team (which included Zinn, Szilard, Compton, and Wigner) established the first self-sustaining fission chain reaction. The multiplication factor was estimated to be 1.0006. About $\frac{1}{2}$ W of power was produced from about 10 billion fissions/s. This was the beginning of the age of nuclear energy, an occasion ranking in importance with the discovery of fire and the inventions of the wheel and engine.

NUCLEAR–POWER REACTORS

The nuclear reactor is an attractive alternative to fossil fuel as a source of electric energy. The burning of coal and oil by electrical utilities is the main source of the air pollutant sulfur dioxide. Also, there is a limit to our supply of these fuels—estimates of no more than 200 yr for coal are frequently cited. Coal and petroleum are found highly concentrated in deposits. Uranium, in contrast, is distributed widely in rocks but in minute amounts. Granites typically contain about 5 ppm U (5 parts uranium in 1 million parts of granite). The chemistry of uranium is similar to that of zirconium, uranium replacing zirconium in the crystalline lattice of zirconium minerals. But not much uranium is needed for producing energy. From the information of the first pile, to produce 1 kw of electric energy continuously for 1 yr only 0.38 g of U^{235} or 53 g of natural uranium are consumed (Prob. 8.8). There is enough uranium in the crust of the earth to produce electric energy for several millenia.

The first commercial nuclear-power reactor was put into operation at Shippingport, Pennsylvania, in December 1957 and supplied 60,000 kW to the greater Pittsburgh area. The heat produced in a reactor is used to generate steam and rotate a turbine and then an electric-power generator (Fig. 8.10). The Tennessee Valley Authority plans to have two reactors at Brown's Ferry, Alabama, each with 1,065,000 kW. Three more of this size are being built, two in Pennsylvania and one in California. As a rough estimate about 1 percent of the electric power now used in this country is made from nuclear energy. Since more of these reactors are being built than fossil-fuel-burning steam generators and hydroelectric plants, it is very likely that reactors will supply the majority of this energy by the end of the twentieth century.

Although nuclear reactors produce no sulfur dioxide, other technological problems and possible hazards must be considered. The material of a reactor is subject to damage from both the high temperatures and the energetic particles formed within the reactor. For this reason a sacrifice in efficiency must be made for safety.

For a given power output, more heat is wasted at a lower efficiency; that is, more heat is dumped into the air or into a body of water (see Chap. 5). Thus there is a real danger to organisms living in a lake used as the cold reservoir for a power station (Figs. 8.10 and 5.2). *Thermal pollution* results from the artificial increase in temperature from this heat dumping. With an increase in temperature, several things happen to living organisms. First, as with almost all chemical reactions, processes within an organism speed up, and the organism may die of exhaustion. Second, the proper functioning of enzymes and other proteins depends on their special geometric structures. At elevated temperatures, these proteins are *denatured;* i.e., their structures are altered, and their functions are destroyed. Third, because oxygen is less soluble in warmer water, fish and other living organisms may suffocate.

Every precaution is necessary to prevent the personnel and surrounding community from being contaminated by radioactivity. This could come about by an uncontrolled overheating and explosion in the reactor. The proper disposal of radioactive waste by-products will continue to grow as a serious problem as more nuclear reactors are used.

THE TRANSURANIUM ELEMENTS

Both U^{235} and U^{238} can capture thermal neutrons. Their capture cross sections are 100 and 2.7 barns. When uranium is exposed to slow neutrons and after all the fission products are accounted for, there is a product, not separable from uranium, whose half-life is 23.5 min. In 1936 Hahn, Meitner, and Strassmann suggested that this is another uranium isotope

$$_{92}U^{238} + _{0}n^{1} \longrightarrow _{92}U^{239}$$

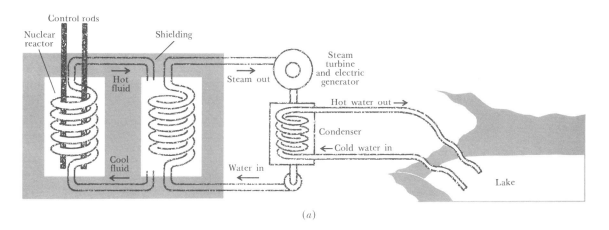

(a)

(b)

Figure 8.10. (*a*) *Production of electric energy by a nuclear reactor.* (*b*) *Nuclear-power plant in Massachusetts* (*Yankee Atomic Electric Company*).

U^{239} is a beta emitter. In 1940 Edwin M. McMillan (b. 1907) and Philip H. Abelson (b. 1913) discovered, with this new uranium isotope, a second beta emitter whose half-life is 2.35 days. The amount of this second substance corresponded exactly to the amount of U^{239} used up in its beta decay. Since beta decay is attended by an increase in atomic number, a new element with atomic number $Z = 93$ must be formed. McMillan called it "neptunium"; its symbol is Np, and it is formed by the reaction

$$_{92}U^{239} \xrightarrow{\text{23.5 min}} {}_{93}Np^{239} + {}_{-1}e^0$$

Minute amounts of neptunium have been found in uranium ores. Since this element follows uranium in the periodic table, neptunium and other elements to be described are called "transuranium elements."

Also in 1940, Glenn T. Seaborg (b. 1912), McMillan, and coworkers bombarded uranium with accelerated deuterons to make Np^{238}:

$$_{92}U^{238} + {}_{1}H^2 \longrightarrow {}_{93}Np^{238} + 2{}_{0}n^1$$

This product decays by beta emission to form an element which they called "plutonium":

$$_{93}Np^{238} \xrightarrow{\text{2.1 days}} {}_{94}Pu^{238} + {}_{-1}e^0$$

This plutonium isotope has a half-life of 89 yr and decays by alpha emission.

In 1941 Seaborg, Segrè, and others identified the decay product of the first neptunium isotope (Np^{239}) as Pu^{239}:

$$_{93}Np^{239} \xrightarrow{\text{2.35 days}} {}_{94}Pu^{239} + {}_{-1}e^0$$

This isotope is quite stable, with a half-life of 24,360 yr. It has a thermal neutron-capture cross section of 270 barns—even greater than that of U^{235} (100 barns) and, even more important, it is more fissionable than U^{235} with slow neutrons. Because it has a long half-life, and because its parent isotopes are relatively unstable, Pu^{239} will accumulate in reactors. It can be separated from uranium by chemical means. Nuclear reactors specially designed for the production of artificial nuclei are called "breeders."

Elements 43 and 61, with no stable isotopes (Appendix B) are not found on earth. In 1937 Segrè and C. Perrier discovered the first isotope in a deuteron-bombarded molybdenum and called it "technetium $_{43}$Tc." It is the first man-made element. The second man-made element, promethium $_{61}$Pm, has been found among fission products.

Pu^{239} has been bombarded with alpha particles and neutrons to form two more transuranium elements: curium $_{96}$Cm^{242} and americium $_{95}$Am^{241}. These and their products have been bombarded with accelerated alpha particles, boron, carbon, nitrogen, and even oxygen nuclei to build nuclei of elements up to 104 and 105. Most of this work has been done by Seaborg and Albert Ghiorso (b. 1915). With about five atoms in each experiment, Ghiorso, using special techniques, was able to learn something about the chemistry of lawrencium $_{103}$Lr.

The other transuranium elements are named in honor of places (berkelium $_{97}$Bk and californium $_{98}$Cf) or of men (einsteinium $_{99}$Es, fermium $_{100}$Fm, mendelevium $_{101}$Md, and nobelium $_{102}$No). Element 104, unofficially, is called either "rutherfordium" or "kurchatovium"; and 105, also unofficially, is called "hahnium."

THE FIRST FISSION WEAPONS

A simple nuclear reactor allowed to run uncontrolled will generate enough heat to disrupt the reactor, breaking it up into subcritical pieces, and the chain reaction will soon stop. For military purposes a much more rapid process is needed. The fission of a maximum number of nuclei must occur before disruption can interfere with the chain reaction.

U^{238} can capture too many neutrons without itself undergoing fission. Because the number of neutrons will not increase rapidly enough, preliminary calculations indicated a bomb could not be made from this isotope. The two logical ingredients for a bomb were U^{235} and Pu^{239}, each with a large capture and fission cross section for both slow and fast neutrons. Pu^{239} can be made in a breeder. The moderator used is heavy water D_2O, which absorbs virtually no neutrons. Plutonium is separable by chemical means from its parent uranium.

Because the two uranium isotopes U^{235} and U^{238} have identical chemical properties, other means were needed to *enrich* the U^{235} content, i.e., to raise its abundance in the bomb material. The large-scale technique used to separate U^{238} (99.27 percent abundant) and U^{235} (0.72 percent abundant) was based on the very slight mass difference between the isotopes.

Uranium hexafluoride UF_6 boils at 56°C. Fortunately fluorine has only one stable isotope $_9F^{19}$ (18.99840 amu). Then (ignoring the very small amounts of U^{234}, which is 0.0057 percent abundant) uranium hexafluoride can have one of two possible molecular weights. With U^{238} (238.0508 amu), it is

$(U^{238}F_6)$ 352.0412

and with U^{235} (235.0439 amu) it is

$(U^{235}F_6)$ 349.0343

A mixture of these isotopes at the same temperature will have the same average kinetic energy. Since kinetic energy is one-half the mass times the speed squared $mv^2/2$, the heavier isotope will have the slower average speed.

> **Example.** What is the ratio of the speeds of the two isotopic forms of UF_6?
>
> Designate the lighter isotope uranium hexafluoride by the subscript 1 and the heavier by subscript 2. Then
>
> $$\tfrac{1}{2}m_1v_1^2 = \tfrac{1}{2}m_2v_2^2$$
>
> $$\frac{v_1^2}{v_2^2} = \frac{m_2}{m_1}$$
>
> $$= \frac{352.0412}{349.0343}$$
>
> $$= 1.008615$$
>
> $$\frac{v_1}{v_2} = 1.0043$$

There are considerably fewer U^{235} atoms than U^{238} atoms in the UF_6 gas mixture, in the ratio $0.72 : 99.27$. But since it moves slightly faster, an individual $U^{235}F_6$ molecule will hit a wall more frequently than an individual $U^{238}F_6$ molecule. If the wall has a

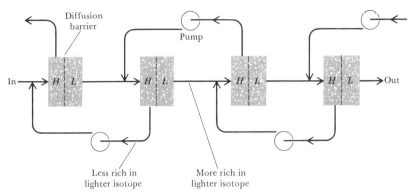

Figure 8.11. Isotope separation by gas diffusion. When some UF$_6$ *gas has passed through a barrier (dashed line) it has more of the lighter isotope (L) than the gas which has not passed (H).*

hole through which UF$_6$ can pass, the lighter molecules will pass through at a rate slightly greater than that predicted from their number alone (Fig. 8.11).

The ratio of passage of light to heavy uranium hexafluoride through a hole will be

$$\frac{0.72}{99.27} \times \frac{1.0043}{1}$$

which includes both their relative number (0.72:99.27) and their relative speed (1.0043:1). The uranium hexafluoride collected on the other side of the wall will be slightly richer in the lighter molecule compared to the initial mixture. If this mixture is allowed to pass through a second wall, the gas collected after this second passage will have the ratio

$$\frac{0.72}{99.27} \times \frac{1.0043}{1} \times \frac{1.0043}{1} \qquad \text{or} \qquad \frac{0.72}{99.27}\,(1.0043)^2$$

After three passages, the collected gas will have the ratio

$$\frac{0.72}{99.27}\,(1.0043)^3 \qquad \text{and so on}$$

> **Example.** If 99 percent pure U^{235} is to be made by this gas-diffusion process, how many passages are needed?
>
> The final product ratio of light to heavy uranium hexafluoride is to be 99:1, which must equal

$$\frac{0.72}{99.27} \; (1.0043)^{\,n} = \frac{99.00}{1.00}$$

Then

$$(1.0043)^{\,n} = \frac{99.00}{1.00} \times \frac{99.27}{0.72}$$

$$= 13,600$$

$$n \log (1.0043) = \log 13,600$$

$$n = \frac{\log 13,600}{\log 1.0043}$$

$$= \frac{4.1335}{0.00186}$$

$$= 2,220 \text{ successive passages}$$

Ideally, the gas-diffusion process must be repeated over two thousand times to produce 99 percent U^{235} (Fig. 8.11). In actual practice about five thousand barriers were required, and the separation factor was found to be smaller than 1.0043 — closer to 1.002. The barriers, or walls, were made from thin sheets of a zinc-silver alloy etched with acid to form the very small holes. Each barrier had about a billion holes, none of which was bigger than $0.01\ \mu$ (or $\frac{4}{10,000,000}$ in.) in diameter. Figure 8.11 is a schematic sketch of a few of the barriers in the cascade series. A gaseous-diffusion plant was begun in 1943 at Clinton, Tennessee, and began operation 2 yr later. About 1 kg of purified U^{235} was made each day. This separation process was one of the major engineering feats of our time.

To destroy the military and industrial capacities of Germany in 1941, an estimated 500,000 tons of TNT were thought to be required. If all the atoms in 1 kg of U^{235} undergo fission, the energy is equivalent to that of 20,000 tons of TNT. Therefore, 25 kg of U^{235} could do the job. But if the fission is only 1 or 5 percent efficient, 100 or 20 times that much — 2,500 or 500 kg of U^{235} — are needed. About 3 or 4 yr would be needed to produce that much. Balancing this was the fear that Axis scientists might accomplish this before the Allies (Prob. 8.10).

It was evident that when bombs could be made, they would not be small. The critical size for a chain process placed a lower limit on the amount of fissionable material in each bomb. The minimum estimate using pure U^{235} was 1 kg, and using plutonium about 300 g. For practical purposes the bomb had to be kept in two subcritical masses before use. To activate it the two parts must be joined together, constituting a critical mass, and this assembly must be done in an extremely short time. One method used was to shoot one part into or against the other, as a gun shoots a bullet. Another method compressed the fissionable material into a smaller, more compact volume, which also increased the multiplication rate above one.

Germany surrendered to the Allied forces on May 7, 1945, before an Allied bomb was prepared. On July 16 a nuclear device was tested in Alamogordo, New Mexico, under the supervision of J. Robert Oppenheimer (1904–1967). Observers were stationed 5.7 and 9.7 miles from the bomb, which was mounted on a steel tower. When the bomb was exploded at 5:30 A.M., with a flash of light brighter than daylight, the temperature was estimated to be 10,000,000°—about as hot as the interior of the sun. Forty seconds later a pressure pulse knocked down two men standing at the closer observation post. A mushroom-shaped cloud traveled up 40,000 ft (8 miles). The tower was vaporized; a crater was left. All the effort of nearly 6 yr proved to be successful.

The war continued against Japan in the Pacific. On February 19, 1945, Iwo Jima was invaded, and it was conquered by March 16. There were 4,590 Americans and over 20,000 Japanese dead. In the battle for Okinawa (April 1 to June 30) the respective losses were 12,520 and 110,071. It seemed evident then that an invasion of Japan itself would involve a large number of casualties on each side.

The decision was made to use either a U^{235} or a Pu^{239} bomb to end the war sooner and without an invasion. On August 6, 1945, an estimated 20-kt uranium bomb (equivalent to 20,000 tons of TNT) was dropped over Hiroshima and exploded in midair, about 1,850 ft above ground level; 68,000 civilians were killed and 76,000 were injured in a total population of 256,300 (Fig. 8.12). Three days later a plutonium bomb was exploded over Nagasaki; 38,000 died and 21,000 were injured in a city of 173,800. Japan surrendered on August 14, 1945, and World War II came to an end after 2,172 days.

To understand the significance of the new weapon system, we must first look at some civilian casualties in World Wars I and II (Table 8.10). Advances in airplane technology enabled nations to more effectively bomb enemy civilian populations. Most of the London casualties in World War II were inflicted during 57

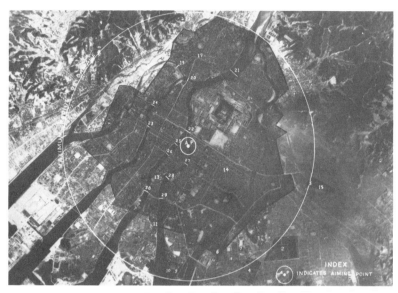

Figure 8.12. Total area of devastation by the atomic bomb dropped on Hiroshima on August 6, 1945. Radius of large circle, 1.2 mi; area, 4.7 mi². (U.S. Air Force, Washington, D.C.)

successive nights of air raids, September 7 to November 2, 1940. Although the fatalities in target cities in the table are comparable, *a single bomb* did the damage in Hiroshima and in Nagasaki.

Before the Alamogordo test a 10-ton TNT bomb was the largest single deliverable explosive; the largest bomber, a B-29, was

Table 8.10. A Comparison of Civilian Casualties by Aerial Bombing

City	Time	Killed	Injured	Remarks
London	April 1915	670	1,962	567 explosive bombs
London	1939–1945	30,000	50,000	50,000 explosive bombs
Hamburg	July 24– August 3, 1943	50,000		Fire storm—800,000 homeless; 53% of homes destroyed
Dresden	February 1945	130,000		Fire storm
Tokyo	March 9, 1945	83,000	102,000	1,667 bombs dropped from 279 planes; 5,300 killed/mi² for 15.8 mi²
Hiroshima	August 6, 1945	68,000	76,000	1 uranium bomb; fire storm; 15,000 killed/mi² for 4.7 mi²
Nagasaki	August 9, 1945	38,000	21,000	1 plutonium bomb; 20,000 killed/mi² for 1.8 mi²

able to carry one of these. The new 20-kt *nuclear bombs* (first called "atomic bombs") had 2,000 times the blast power of the conventional TNT type. To deliver the same destruction, an armada of 2,000 B-29 bombers would have been needed.

Technology was defined previously as the means we have to control or manipulate nature. Advanced technology can be tersely described: fewer men can do more in less time. This is the significance of a nuclear bomb. It has changed the very character of warfare.

THERMONUCLEAR REACTIONS

The enigma of how the sun's energy is made was elucidated in 1938 by Hans Bethe (b. 1906), who proposed a *carbon-nitrogen cycle* (Table 8.11a). The sum of the six steps is simply the union, or *fusion*, of four protons to form a helium nucleus and two positrons. The mass loss for this process is 25.71 MeV in the form of neutrinos and photons. A second cycle, with the same overall effect, is given in part *b* of the table.

Table 8.11. Fusion of Nuclei as the Source of Solar and Stellar Energy

a. Bethe's carbon-nitrogen cycle

$$_6C^{12} + _1H^1 \longrightarrow _7N^{13} + \gamma$$
$$_7N^{13} \longrightarrow _6C^{13} + _1e^0 + \nu$$
$$_6C^{13} + _1H^1 \longrightarrow _7N^{14} + \gamma$$
$$_7N^{14} + _1H^1 \longrightarrow _8O^{15} + \gamma$$
$$_8O^{15} \longrightarrow _7N^{15} + _1e^0 + \nu$$
$$_7N^{15} + _1H^1 \longrightarrow _6C^{12} + _2He^4 + \gamma$$

Total $\quad 4_1H^1 \quad \longrightarrow \quad _2He^4 + \quad 2_1e^0 \quad + 4\gamma + 2\nu$
Mass $\quad 4(1.007825) \quad\quad 4.00260 \quad 2(0.00055)$

$4.00260 + 2(0.00055) - 4(1.007825) = -0.02760$ amu $\quad$ mass lost
$0.02760 \times 931.46 = 25.71$ MeV (released as photons and neutrinos for each helium nucleus formed)

b. Proton-proton cycle

$$2(_1H^1 + _1H^1 \longrightarrow _1H^2 + _1e^0)$$
$$2(_1H^2 + _1H^1 \longrightarrow _2He^3)$$
$$_2He^3 + _2He^3 \longrightarrow _2He^4 + _1H^1 + _1H^1$$

Total $\quad 4_1H^1 \longrightarrow _2He^4 + 2_1e^0$

Same energy-mass as in carbon-nitrogen cycle; neutrinos and gamma rays are produced here, too.

Both of these nuclear reactions require minimum temperatures of several ten-million degrees. For this reason they are called "thermonuclear" processes. It is believed that about 95 percent of the solar energy comes from the proton-proton cycle and the rest from the carbon-nitrogen cycle. In more massive stars the Bethe cycle predominates.

Example. The total power output of the sun in all directions is 3.86×10^{33} ergs/s. How many fusions/s must occur to supply this power?

One fusion releases

$$25.71 \text{ MeV} \times \frac{1.60210 \times 10^{-6} \text{ erg}}{1 \text{ MeV}} = 4.119 \times 10^{-5} \text{ erg}$$

Then

$$\frac{3.86 \times 10^{33} \text{ ergs}}{\text{s}} \frac{1 \text{ fusion}}{4.119 \times 10^{-5} \text{ ergs}} = 9.37 \times 10^{37} \text{ fusions/s}$$

Example. What is the mass loss/s in the sun? If the sun has a total mass of 1.99×10^{33} g, what fraction is lost each second?

Since we know the fusions per second and the mass loss per fusion, the mass loss per second is

$$\frac{9.37 \times 10^{37} \text{ fusions}}{\text{s}} \frac{0.02760 \text{ amu lost}}{\text{fusion}} \frac{1 \text{ g}}{6.0225 \times 10^{23} \text{ amu}}$$

$$= \frac{9.37 \times 2.760 \times 10^{35} \text{ g}}{6.0225 \times 10^{23} \text{ s}}$$

$$= 4.29 \times 10^{12} \text{ g/s}$$

As a fraction of the total, the mass loss is

$$\frac{4.29 \times 10^{12} \text{ g/s}}{1.99 \times 10^{33} \text{ g}} = 2.16 \times 10^{-21}$$

In 10 billion yr the loss has only been

$$10^{10} \text{ yr} \times \frac{3.14 \times 10^{7} \text{ s}}{1 \text{ yr}} \frac{4.29 \times 10^{12} \text{ g}}{\text{s}} = 1.35 \times 10^{30} \text{ g}$$

This is

$$\frac{1.35 \times 10^{30}}{1.99 \times 10^{33}} = 6.78 \times 10^{-4} \quad \text{or} \quad 0.0678 \text{ percent of the solar mass}$$

Example. Estimate an upper limit to the life expectancy of our hot and radiating sun.

If we assume fusion will continue to the end, that the sun is now essentially all hydrogen, and at the end it will be all helium (none of which is reasonable), we can proceed.

When four protons undergo fusion, 4.03130 amu of hydrogen combine and 0.02760 amu vanish. From this, when 4.03130 g H fuse, 0.02760 g is destroyed. A mass of pure hydrogen equal to the sun mass will undergo fusion until this much mass is destroyed:

$$1.99 \times 10^{33} \text{ g H} \times \frac{0.02760 \text{ g destroyed}}{4.03130 \text{ g H}} = 1.36 \times 10^{31} \text{ g destroyed}$$

If the present destruction rate continues to the end (another doubtful assumption),

$$1.36 \times 10^{31} \text{ g} \times \frac{1 \text{ s}}{4.29 \times 10^{12} \text{ g}} \times \frac{1 \text{ yr}}{3.14 \times 10^{7} \text{ s}} = 1.01 \times 10^{11} \text{ yr}$$

over 100 billion yr. Since only the interior of the sun is hot enough for fusion, the age will be less than this.

Temperatures of the magnitude of those at the sun's interior are needed for fusion. We have already seen that these temperatures can be created on earth, namely, in a fission bomb. The fusion of four protons (about 4 amu) yields 25.71 MeV; the fission of one U^{235} (235 amu) yields about 200 MeV. Fission, using $\frac{235}{4}$, or about sixty times as much material, produces $\frac{200}{25.71}$, or only eight times as much power as does fusion. Using equal masses, more energy can be obtained from fusion.

In the design of fusion weapons, reactions such as these are of interest:

$$_1\text{H}^2 \; + \; _1\text{H}^3 \longrightarrow \; _2\text{He}^4 \; + \; _0n^1$$
$$(2.01410) \quad (3.01601) \quad (4.00260) \quad (1.008665)$$

$$_1\text{H}^3 \; + \; _1\text{H}^3 \longrightarrow \; _2\text{He}^4 \; + \; 2_0n^1$$
$$(3.01601) \quad (3.01601) \quad (4.00260) \quad 2(1.008665)$$

From the indicated masses, the losses in mass are 0.01884 and 0.01209 amu. The released energies 17.55 and 11.26 MeV are carried off mostly by the produced neutrons. These fast neutrons can be used to fission U^{238}. Since fusion is not a chain reaction, there is no minimum or critical size.

Thermonuclear bombs are set off by a fission explosion, whose high temperatures make fusion possible. The liberated fast neutrons from fusion produce a third nuclear explosion, the fission of natural uranium. The total energy is about half from each fusion and fission. On November 1, 1952, the United States set off a device which vaporized the Pacific Island of Elugelab. Its power was estimated to be 10.4 Mt, or 502 times as powerful as those fission bombs dropped on Japan. On August 23, 1953, the Soviet Union tested an even more effective fusion bomb. Then, on February 28, 1954, the United States detonated its second, improved experimental thermonuclear device: 15 Mt. By May 20, 1956, this country was able to drop such a bomb from an airplane. The increase in blast power from fission to thermonuclear bombs is as great as the increase from TNT to fission.

THE FIREBALL

A nuclear bomb is not merely larger than a conventional military explosive; it has features of its own that make it much more lethal than an equivalent amount of TNT. To see how these effects come about we shall first look at the device for a few seconds following detonation.

Gamma rays and high-speed neutrons are immediately released; this is called the "prompt nuclear radiation." With a temperature of at least 10,000,000°, all the material of the bomb is vaporized. The atoms are ionized and moving at very high speeds. The pressure exceeds 1 million atm. If the explosion happens in the lower atmosphere, called an *air burst*, then this energy is transferred to the immediately surrounding air. This, in turn, is raised to a high temperature and becomes luminous enough to radiate considerable amounts of energy. The surface of this hot gas, the *fireball*, is at about 7500 K. At this temperature, from Wien's displacement law, the wavelength of maximum blackbody radiation is 3850 Å, at the ultraviolet-visible part of the spectrum. *Thermal radiation* from the fireball is adequately described by the blackbody-radiation law and is mostly in the ultraviolet, visible, and infrared regions of the spectrum.

For a 1-Mt burst, the fireball has a diameter of 440 ft in 0.7 ms and 7,200 ft in 10 s. Expanding this fast, it compresses and imparts a motion to the nonluminous air touching the fireball. This high-pressure air moves out from the fireball as a *blast wave*. The fireball expands and cools. In less than 1 min thermal radiation ceases and it looks like a white spherical cloud. Moving up to higher altitudes, at speeds of about 200 mph, the sphere becomes doughnut-shaped and draws up with it a column of cooler air and debris from the earth's surface. Eventually it may assume a typical mushroom shape.

The fission products of the bomb and the induced radioactivity by neutron capture in the debris and air constitute still another danger—the *delayed*, or *residual, nuclear radioactivity*. If the bomb is exploded at or near the surface of the earth (a *surface burst*), a crater is formed (Fig. 8.13*a*). Neutron capture by the atoms in the soil adds considerably to this residual radioactivity. The heavier particles—visible and up to 3 mm in size—fall back to the earth within 24 hr. This is called the "early fallout"; because these particles are not carried very far from the explosion area, it is also called "local fallout." Finer particles, carried by winds, are found worldwide and are called "delayed," or "worldwide fallout."

The total energy of a fission bomb is roughly partitioned as shown to the right. A thermonuclear bomb using both fission and fusion will have relatively less residual radiation.

Blast and shock	50%
Thermal radiation	35%
Initial nuclear radiation	5%
Residual nuclear radiation	10%

THE DESTRUCTIVENESS OF A NUCLEAR BOMB

The fast neutrons in prompt radiation collide with and transfer energy most effectively to the lighter atoms used as moderators. But carbon, oxygen, hydrogen, and nitrogen are the elements of proteins. Neutron radiation, therefore, harms living organisms by altering the structure and function of proteins.

The unit for measuring gamma or x-rays is the *roentgen* (R), the amount of these rays which can produce 2.08×10^9 ion pairs in 1 cm^3 of dry air at 0°C and 1 atm (STP). A lethal dose is a whole-body exposure of 1,000 R. Approximate intensities are given in Table 8.12 for six weapons of the kiloton (kt) and megaton (Mt)

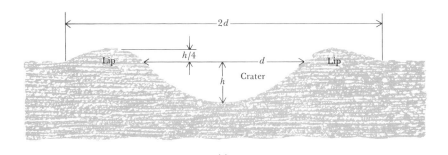

(a)

(b)

Figure 8.13. (a) Crater produced by a surface blast at or over dry soil. For each bomb size are listed the maximum altitude or height of burst above which a significant crater will not be formed, the inner diameter of the crater d, and the depth of the crater h. The height of the lip about the crater rim is about one-fourth the crater depth. The crater diameter, including the lip, is about twice the inner diameter. (b) Formation of the "plume" characteristic of 20 kt fission bomb in shallow water; plume is more than one mile high. (Atomic Energy Commission.) All dimensions are given in meters.

Size of bomb	Maximum height above ground of burst for crater, m	Diameter of crater d, m	Depth of crater h, m
1 kt	1.5	40	9
10 kt	3.0	85	20
20 kt	3.8	98	24
1 Mt	12.1	165	40
10 Mt	24.2	760	185
20 Mt	29.7	1040	245

Table 8.12. Gamma-ray Intensity (Roentgens), Bomb Size, and Distance from Ground Zero

Distance from bomb, miles	Bomb size					
	1 kt	**10 kt**	**20 kt**	**1 Mt**	**10 Mt**	**20 Mt**
$\frac{1}{2}$	1,000	10^4	2×10^4	2×10^6	5×10^7	3.5×10^8
1	10	10^2	200	2×10^4	5×10^5	3.5×10^6
2	0.015	0.15	0.30	30	7.5×10^2	5.2×10^3
3	3×10^{-4}	3×10^{-3}	6×10^{-3}	0.6	15	105

load size at several distances. The effective shielding of several common materials is given as the *tenth-value thickness*, the thickness needed to reduce the radiation to one-tenth its intensity (Table 8.13). A man standing in the open 1 mile from *ground zero* (the point on earth directly beneath a burst) of a 1-Mt bomb would be exposed to 20,000 R — more than the lethal dose. Standing behind 15 in. of concrete, the man would be exposed to one-tenth this, or 2,000 R — also a fatal amount.

The thermal radiation emitted in that minute or less during which the fireball is luminous represents about one-third of the total energy released. For a 1-kt bomb the radiation lasts about 0.3 s; for a 10-Mt bomb, about 30 s. Darker materials are more likely to absorb this radiation and to become hot enough for ignition. Wood will char but not burn freely. In Hiroshima trees as far as 1.7 miles from ground zero were charred; and in Nagasaki they were charred up to 2.1 miles. The danger from fire arises from substances that are more easily ignited, such as paper, rubbish, cotton fabrics. The effects of radiation on human beings can be judged from the data in Table 8.14.

The *blast wave* is a fast-moving high-pressure air column followed by a less than normal pressure region. The intensity of the

Table 8.13. Shielding Against Gamma Rays by Common Materials

Material	Tenth-value thickness, in.†
Steel	4.5
Concrete	15
Earth	22
Water	32
Wood	60

† The tenth value is the thickness of a material, used as shielding, that will reduce the gamma-ray intensity to one-tenth its initial level.

Table 8.14. Distances at Which an Exposed Person Will Suffer Thermal Radiation Burns from Nuclear Bombs

Weapon size	Distance from ground zero, miles	
	First-degree burns	Second-degree burns
1 kt	0.7	0.5
10 kt	2.0	1.5
20 kt	3	2
1 Mt	15	11
10 Mt	35	25
20 Mt	45	35

blast is expressed as the *peak overpressure*, the amount by which the high pressure exceeds the normal atmosphere of 14.7 psi. This pressure is reinforced by what is called the "Mach effect" [after Ernst Mach (1838–1916)], the union of the blast wave and its ground echo to produce twice the overpressure.

In Fig. 8.14, the way in which this overpressure drops with distance from ground zero is plotted for the six weapon loads burst at optimum heights for damage. It is possible to use the same curve for the six weapons just by adjusting the horizontal axis, or the distance scale. The time of blast-wave arrival after detonation is also indicated on the scales. With a 20-kt air burst at 1,850 ft, the blast wave drops to 1 psi at 3 miles from ground zero 12 s after detonation. All glass windows shatter with overpressures of 0.5 to 1.0 psi, or higher.

> *Example.* What is the force acting on a windowpane 20 by 20 in. if the overpressure is 0.5 psi?
>
> The area is 400 in.². The total unbalanced force is
>
> $$400 \text{ in.}^2 \times \frac{0.5 \text{ lb}}{1 \text{ in.}^2} = 200 \text{ lb}$$
>
> (Almost all the data for these effects are given in the British system.) This is the force exerted by a 200-lb man standing on the pane, which cracks.

In Hiroshima and Nagasaki no windows remained intact. At 2 miles from the blast, small glass fragments were flying around at about 70 mph. Closer to the blast, the speeds were greater, approaching velocities normally associated with bullets. The general effect on structures was that usually seen from earthquakes and hurricanes. In both Hiroshima and Nagasaki, telephone poles were snapped off at ground level.

Ignition and combustion of trash by thermal radiation, coupled with the massive blast-wave damage, presented the conditions for uncontrollable fire. The blast wave broke open gas mains and caused electric short circuits. At the same time, widespread damage in the streets made control of these fires impossible. In Hiroshima fire-fighting equipment was wrecked beyond use; rubble-cluttered streets were impassable; and almost all the firemen were casualties.

The circumstances in Hiroshima—many small conflagra-

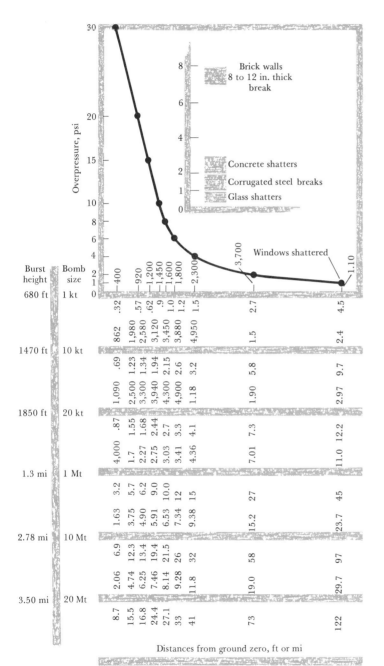

Figure 8.14. The extent of
damage by air blast from a
nuclear weapon. The
overpressure drops with distance
from ground zero. Distances (in ft
or mi) and times of arrival of the
blast effect (s) are given above and
below the shaded bars,
respectively, for six different bomb
sizes. The insert graph lists the
blast overpressures needed to
destroy four common structures.

tions and houses close together—worked to make a *fire storm*. As the small fires grew and coalesced, one immense fire covered a large part of the city. The hot-combustion-product gases rose rapidly, and air was drawn in from the periphery, creating winds up to 40 mph. The fire storm lasted for about 6 hr, consuming everything combustible within the area. Fire storms have also been observed in large forest fires and in the incendiary-bombed Axis cities during World War II (Table 8.10, Dresden). Because houses in Nagasaki were less densely spaced, a fire storm did not develop.

The amount of delayed nuclear radiation, the *fallout*, depends on the nature of the bomb and on whether or not the fireball touches or has scooped up within it some earth. A bomb can be said to be *dirty* or *clean* if its fragments have more or less radioactivity. A purely fission bomb is dirtier than a fission-and-fusion thermonuclear device. A *salted* bomb is designed especially to produce more radioactive fragments by packing it with neutron-capturing elements.

Early fallout, which settles back on the earth in the vicinity of ground zero within 24 hr, presents an immediate danger to survivors of the earlier effects. Neither alpha nor beta emissions penetrate very deeply into matter, but decaying fallout isotopes can land on food to be ingested or can be inhaled. Once taken internally these particles are quite dangerous. Because gamma rays are the most penetrating, ingestion is not necessary and shielding is required for safety (Table 8.13).

The common dose unit for radioactive damage to a living organism is the *rem*, an acronym for *roentgen equivalent man* (or *mammal*). The roentgen unit is ordinarily limited to gamma- and x-ray intensities; the rem translates any radiation dosage into the biological damage done by an equivalent gamma-ray dose. For example, energetic alpha particles can cause considerable ionization in air; unable to penetrate skin, their rem value is essentially zero.

The emission rate of early fallout will decline in time. Table 8.15 lists approximate values for this decline. For example, if the fallout has an activity of 1,000 R/hr the first hour after detonation, in 1 week the activity will have decreased to 2.3 R/hr. If fallout at first subjects a person to 1,000 rems/min, then 4 hr later he will absorb at the rate of 149 rems/min.

The damage a body sustains depends on the total accumulated dosage absorbed over an exposure time. The third column of Table 8.15 lists approximate estimations of how fast this dosage accumulates. A person living in an area contaminated

Table 8.15. Approximate Decline in Radioactivity and Accumulation of Dose over Time

Time	Relative radiation dose rate, rems	Total accumulated dose, %
1 hr	1,000	55
2 hr	440	62
3 hr	273	65
4 hr	194	68
5 hr	149	70
8 hr	85	73
10 hr	66	74
24 hr	23	80
$1\frac{1}{2}$ days	14	82
2 days	10	83
3 days	6.3	86
1 week	2.3	89
2 week	1.02	92
3 week	0.63	93
4 week	0.45	94
1 mo	0.42	94
2 mo	0.18	96
6 mo	0.049	98
1 yr	0.022	99

with fallout who eventually will absorb 2,500 rems will take up 80 percent of this, or 2,000 rems, in 24 hr. Table 8.16 lists the physiological damage from various dose levels.

The delayed fallout, or finer particles, is carried by the wind and deposited in an unpredictable pattern slowly and over wide areas. During this time most of the short half-life isotopes have decayed to a harmless level. But the more slowly decaying radioac-

Table 8.16. Radioactive Dose and Physiologic Effects

Dose, rems	Effect
0.080	None observed; this is the natural background radiation from cosmic rays and long-lived terrestrial radioactive isotopes
100	Moderate decline in white blood cells
300	Severe decline in white blood cells; risk of infection; hemorrhage; loss of hair; 100% incidence of vomiting
1,000	Lethal, death in 2 weeks; gastrointestinal tract affected; circulatory failure
5,000	Death within 2 days; central nervous system damaged; convulsions; respiratory failure

tive isotopes present still another danger. Sr^{90} is one of the more serious of these. Its decay is

$$_{38}Sr^{90} \xrightarrow{28 \text{ yr}} _{-1}e^0 + _{39}Y^{90}$$

$$_{39}Y^{90} \xrightarrow{64.2 \text{ hr}} _{-1}e^0 + _{40}Zr^{90}$$

The beta particles emitted have 0.54 and 2.26 MeV of energy. Strontium lies just below calcium in the periodic table, and the two alkaline earth metals have similar properties. In particular, their ionic radii are close in value:

$r(Sr^{2+}) = 1.12 \text{ Å}$
$r(Ca^{2+}) = 0.99 \text{ Å}$

This means strontium ions can be accommodated in the crystalline lattices of calcium compounds.

    Sr^{90} fallout eventually lands on soil; it is taken up by grass, ingested by cows, and it accumulates in milk. Milk is an important source of calcium for bones and teeth, both of which are made of calcium carbonate and phosphate; bones and teeth are about 37 percent calcium. With contaminated milk, some radioactive strontium is deposited in these structures, which presents an especially high risk for growing children. Since red blood cells are made in bone marrow, anemia, as well as damage to bones themselves, is a consequence of this isotope. In addition, cancer, possibly leukemia, and genetic damage have been attributed to the long-term presence of this beta emitter.

    The Sr^{90} dispersed throughout the world, mostly in the middle latitudes of the northern hemisphere, is the result of open-air testing of thermonuclear weapons, beginning on October 31, 1952. Table 8.17 lists estimates of the total amounts of the isotope from the start of testing to 1962. On July 25, 1963, the Soviet Union, Great Britain, and the United States agreed to stop open-air testing. Twenty-eight years are required to reduce the 1962 level by one-half.

    Another isotope also formed relatively abundantly in fission is Cs^{137}, with a 30-yr half-life. It resembles potassium and is found mostly in muscle tissue. There is a greater *turnover* rate here, compared with ions found in teeth or bones. Cesium has a *biological*

Table 8.17. Estimated Total Amounts of Sr^{90} on the Earth

Year	Mass of Sr^{90}, kg
1953	0
1955	2.74
1958	7.60
1962	15.2
(1990)†	7.6
(2018)†	3.8

† Through natural decay, with a 28-yr half-life.

half-life of about 140 days. This means that, through normal excretion, half of the total cesium in a human body is eliminated in 140 days. Therefore, Cs^{137} is not as hazardous as Sr^{90}, whose biological half-life is considerably longer (in bone, about 50 yr). A child will keep some Sr^{90} in his body for his entire life.

The amount of C^{14} in the atmosphere has increased by about one-third over its 1950 level. Through its absorption by the oceans, and slow decay, nearly a century will be required for the C^{14} level to fall to within 1 percent of the old value.

MODERN WARFARE

From observing the effects of the 20-kt bombs on the two Japanese cities and the still greater destructiveness of tested megaton devices, it is obvious that the possession of and ability to deliver thermonuclear weapons gives a nation unusual military advantages. A nation rich in nuclear weapons can destroy all or nearly all of the armed forces, industries, and (especially) civilian population of any other nation.

The defenses a nonnuclear nation can raise against such an attack are meager. Even if the civilians of Hiroshima and Nagasaki knew in advance what weapon was to be used against them and had time to build underground shelters, their casualties would still have been staggering. Nothing less than complete evacuation of these cities would have helped. Witnesses to the fire storms in German cities reported that underground shelters burst into flames at the influx of fresh air when opened several days after the fire storm had subsided. The available megaton-range bomb load makes civilian defense an even less certain endeavor.

When two adversaries have nuclear power, offensive and defensive measures are radically changed. An attack by one on a civilian target of the other will almost certainly bring a reprisal thermonuclear attack on the instigator. This can easily mean death to half the total population in both nations. A better strategy, for the purposes of an aggressor, is to employ a *counterforce capability*, i.e., launch an attack against the enemy's retaliatory potential—his missiles and bombers—not against enemy civilians and industry. The *first-strike capability* is the ability of an aggressor to accomplish just this: to prevent his victim from using force on the aggressor's civilian, military, and industrial targets.

The capacity of the first-strike victim to sustain a thermonuclear attack and still be able to retaliate is that nation's *second-strike capability.* It must make the enemy's first-strike attack as unsuccessful as possible by ensuring that some of its own thermonuclear capacity survives the first-strike attack. For this reason nations diversify their retaliatory means. Thus thermonuclear bombs can be dropped from long-range bombers. The most advanced aircraft of this type can fly halfway around the earth and carry between ten and twenty 4-Mt bombs. During times of acute international tension, it is possible always to have some of these bombers in flight, avoiding complete destruction on the ground during a surprise attack.

Missiles with thermonuclear warheads can be launched from submarines. These are about 1 Mt and have ranges up to 2,500 miles. The launching submarines are elusive targets during a surprise first-strike attack. Landbased missile sites can be protected, or *hardened,* if the missiles are launched from underground thick concrete silos; *soft sites,* in contrast, are above ground and more liable to damage. Both soft and hard sites are protected against enemy assaults by antiballistic-missile missiles, or ABM's. Designed to intercept and destroy enemy missiles and bombers, the ABM's have 1-kt warheads. Much bigger loads exploding over cities can destroy the civilian populations the ABM's are supposed to protect.

A *ballistic missile* follows a path, or trajectory, which starts from its launch site; it is then acted on only by air friction and gravity, and, provided it is not intercepted and the calculations have been correct, its trajectory ends on its target. A *guided missile* has a trajectory that can be corrected during the flight. ABM's must hit fast-moving targets and are guided to move toward these targets. Ballistic missiles are available with ranges up to 7,500 miles and loads of 1 or 2 Mt. Heavier loads are carried by shorter-ranged missiles: 5 Mt to 6,000 miles, and 10 Mt to 2,500 miles.

The precision of ballistic missiles can be impressively high and is described by its *circular error probable,* or CEP. For example, a CEP of 1 mile means that 50 percent of the launched missiles probably will fall about the intended target within a circle of radius 1 mile; the remaining 50 percent will fall outside that circle. CEP values of $\frac{1}{2}$ mile or smaller have been claimed for some ballistic missiles. The CEP describes only the trajectory probability and does not include the possibility of ABM interception.

For two reasons, an aggressor cannot be certain his first-strike attack will be 100 percent successful: his aim is not perfect, and his attack can be intercepted. To maximize his chances he must have a very large stockpile of offensive weapons and he must use all or nearly all of them.

> **Example.** Compare the chances of destroying a target with one and with two missiles if the chance of success of each missile is 50, 60, 90, and 99 percent.
>
> Suppose we let
>
> p = chance of success
> q = chance of failure
>
> and so
>
> $p + q = 1 = 100\%$
>
> A convenient device for these calculations is illustrated in Fig. 8.15. A target is destroyed by two successful hits, by a hit and then a miss, or by a miss and then a hit. The individual chances are p^2, pq, and qp, and the combined chance is $p^2 + 2pq$. For failure, two misses, the chance is
>
> $1 - (p^2 + 2pq) = q^2$
>
> The chances of destroying the target are 75, 84, 99, and 99.99 percent, respectively. In fact, no matter how many missiles are directed toward a target, there can never be a 100 percent chance of success (Prob. 8.9).

A superpower (a nation having nuclear weapons) that believes it can be the victim of a first-strike attack will also believe it must have as many retaliatory weapons as it can afford. This will make more possible the survival of some weapons for a second-strike reprisal. Since the first-strike aggressor will not have many offensive weapons left, the second strike can be made against the civilian population of the aggressor. Since thermonuclear wars can be brief (about 30 min), industrial targets are not important.

Each superpower will build up its thermonuclear capacity to maintain a reasonable second-strike capability. Unfortunately, what one superpower considers as its second-strike capability the other superpower can look upon as a first-strike threat to its own

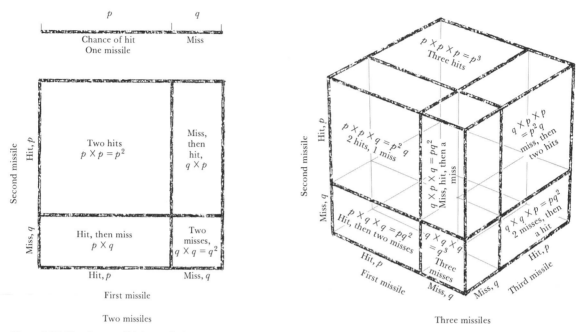

Figure 8.15. *The chances of hitting and missing a target with a missile are analyzed for the case of one, two, and three missiles. This can be done geometrically, using a line segment, square, or cube to indicate the relative probabilities of the several combinations of hits and misses. Alternatively, a purely analytical calculation involves the function $p + q$ raised to the first, second, and third power, that is, the linear function $(p + q)$, the function squared $(p + q)^2$, and the function cubed $(p + q)^3$. Even if a single missile has a 99 percent chance of hitting its target, and if three missiles are used, there is still 1 chance in 1 million that all three missiles will miss the target. Note the geometric representations of these algebraic equations*

$$p + q = 1$$
$$(p + q)^2 = p^2 + 2pq + q^2 = 1$$
$$(p + q)^3 = p^3 + 3p^2q + 3pq^2 + q^3 = 1$$

security. And the more one nation arms itself with thermonuclear bombs, the more must the other do the same. In the *arms race*, this step-by-step accumulation of first- or second-strike weapons (there is little difference) is called an "escalation." Escalation continues as long as a superpower fears its potential adversary can launch a successful first strike. Even a purely defensive measure can be interpreted as an aggressive act. An ABM system (which one superpower intends to use against a first-strike attack on itself) will be seen as a means of thwarting the second superpower's second strike after it has been attacked by the first power.

A superpower is always uncertain about both the true strength of its adversary and the reliability of performance of its own thermonuclear capability. To prevent too large an advantage to its adversary, a superpower plans by assuming the worst performance of its own and the best performance of its adversary's capability. This *worst-case analysis*—overestimating the potential enemy strength and underestimating its own—contributes even more to escalation.

The complexity of modern offensive and defensive systems necessitates several years before such a system is ready. A typical sequence is *research* (an idea evolved from basic principles), *development* (testing its practicality), next, *deployment* (the actual construction and setting up of a new system), and, last, making the system *operational* (the final modifications and adjustments always needed to make a complex system function properly). Because no system has ever been tested in a nuclear war, an uncertainty remains after all this effort. The time interval between research and operation means a superpower feels it must respond not just to its adversary's present capability but to what it anticipates for the future. Sometimes a superpower will discard a new system as impractical, but not soon enough for the other superpower to halt its countermeasures.

If a superpower can sustain a first strike and then retaliate, its second-strike capability is a *deterrent* to a first-strike attack. Because a second-strike deterrent is probably directed against civilians, each superpower civilian population is *hostage* to reprisal for its own nation's first strike. A deterrent is only effective if the other superpower knows about and believes in its existence. Such a *credible deterrent*, especially if it involves a technological innovation, contributes to escalation. If the deterrent is made known to the other superpower, it will be effective in preventing a first strike but it will stimulate the potential aggressor to a higher level of escalation. On the other hand, if the deterrent is kept confidential, the potential aggressor, ignorant of it, may be more tempted to try a first strike but may not think escalation is necessary. Also, if a superpower believes in or fears an imminent attack on itself, it may choose a *preemptive attack*, a retaliation *before* it is actually attacked.

Although exact thermonuclear-stockpile information is not available to the public, estimates have been made of the magnitude of damage each of the superpowers can inflict on the other. One set of figures cited is that the United States has enough deliverable thermonuclear bombs to kill each civilian in the Soviet Union fifteen times and the Soviet Union has the capability of killing each American eight times. Large uncertainties in these analyses are expected, but the *overkill*, the capability of killing each enemy civilian more than once, follows from escalation.

One recent innovation in weaponry is MIRV, the multiple individually-targetable reentry vehicle. This is a cluster of several warheads launched in a common missile but separable in flight. Each load is a guided missile directed toward an individual target. Estimates of the overkill ratio with MIRV are 40 and 24, respectively. Guesses of an enemy's deterrent capability can be made with reconnaissance satellites, by counting missile sites, etc. With MIRV, uncertainty exists about the exact number of warheads in each missile. With this uncertainty, a feeling of balance in deterrence between the two superpowers is harder to realize.

Contributing to the general feeling of unrest in these matters is the fear based on these uncertainties. Technology is now pitted against unpredictable human decisions, a more difficult task than technology or science controlling or investigating the natural universe, which is, in principle, predictable.

Debate within a superpower over the relative harm and good derived from escalation in defensive and offensive capabilities is an unending dialogue. What information is available is often not intelligible to a nontechnically trained citizen. Other information is classified as secret and not available to both sides of an argument even if competent. This gives an enormous advantage to arguments favoring governmental policies.

A two-superpower world, with its balance of terror, nevertheless, is probably more secure than a world with more superpowers. Each time a new country joins the superpower ranks, an added uncertainty makes decisions that much more difficult. The more nations possessing deliverable thermonuclear bombs, the greater an opportunity for an irrational decision and this kind of war.

Table 8.18 is a list of recorded catastrophes. The first is the Black Death, a bubonic plague epidemic. From church and tax records, an estimated 25 percent of all Europeans died in a 2-yr period, and it was 200 yr before Europe's population regained its preplague level. This has been described as the nearest thing to a break in the continuity of human history. The last item is the combined military-force casualties for both sides during 1937–1945. With strategic bombings, deliberate genocide, disease, malnutrition, etc., the civilian casualties were even greater. Between 35 and 40 million deaths can be attributed to World War II.

Plagues are much less likely to occur today because of mod-

Table 8.18. Recorded Catastrophes

Date	Place	Deaths	Causes
1347–1350	Europe and Asia	25,000,000	Bubonic plague epidemic, killing over one-fourth of Europeans; from 1334 to 1352 even more of the population of Europe and Asia died
1545	Cuba	250,000	Typhus
1556	Shensi, China	800,000	Earthquake
1560	Brazil	Several million	Smallpox
1672	Naples	400,000	Bubonic plague
1711	Germany, Austria	500,000	Bubonic plague
1769–1770	India	3 million	Famine
1737	Calcutta	300,000	Earthquake
1792	Egypt	750,000	Bubonic plague
1831	Europe	800,000	Cholera
1840–1862	World	Several million	Cholera
1881	China, Indochina	250,000	Typhoon
1887	Yellow River, China	950,000	Flood
1907	India	1,300,000	Bubonic plague
1911	Yangtze River, China	100,000	Flood
1909–1918	China and India	1,500,000	Bubonic plague
1914–1918	Europe	8,540,000	Military personnel killed in World War I
1921	India	500,000	Cholera
1924	India	250,000	Cholera
1926–1930	India	400,000	Smallpox
1937–1945	World	16,700,000	Military personnel killed in Sino-Japanese War and World War II; at least this many civilians, in addition, were killed or died

ern means of disease control. Except for the possibilities of germ warfare and the resistance of disease-producing organisms to drugs, for a while, at least, we have little to fear from another epidemic. We are also more prepared for floods and earthquakes. At the same time, modern warfare has become so efficient, with overkill a logical corollary, that many times more deaths can occur in a thermonuclear attack in $\frac{1}{2}$ hr than occurred in all these past catastrophes combined. *Man has the technological capacity to kill faster than he can reproduce.* The great threat to continuing human life is now not disease, but modern warfare. The natural scourges have been replaced by a potential technological catastrophe.

PEACEFUL USES OF
NUCLEAR PROCESSES

Most of the research and development cost and effort in nuclear phenomena has been directed toward making more effective weapons. However, many important scientific and technological advances have been made as a consequence of this prime military motive.

Table 8.19 lists electric energy and population data for the past several decades. Our electricity requirements are increasing, not only because our population is growing, but also because each person uses more. We cannot expect our population or our per capita consumption to continue to increase without limit. Nevertheless, there is no reason to believe the amount of needed electric energy will not continue to grow in the coming years. At present, the United States consumes about one-third the world production of electricity; as other nations develop their requirements, the total worldwide needs will grow even faster. Nuclear power seems to be the one way to supply this new demand.

Table 8.19. Electric-energy Production in the United States

Year	U.S. population	Total electric energy produced, billions of kW-hr	Per capita electric-energy consumption, MW-hr	Total energy produced % and source			
				Hydroelectric	Fuel and steam	Internal combustion	Nuclear reactor
1920	105,710,620	39.4	0.373	40.0	60.0		
1930	122,775,046	91.1	0.746	34.2	65.7	0.07	
1940	131,669,275	141.8	1.077	33.4	65.6	1.08	
1950	150,697,361	329.1	2.184	29.1	69.7	1.1	
1960	179,323,175	753.4	4.202	19.3	80.1	0.6	
1965	193,815,000	1,055.3	5.445	18.3	81.1	0.4	0.2
1970	204,765,770	1,529.6	7.470	16.2	82.0	0.4	1.4

The neutrino is the very small particle invented by Pauli and Fermi to account for mass and energy discrepancies during nuclear processes (see page 431). For example, if a neutron and a positron combine, a proton and a neutrino ν are formed:

$$_0n^1 + {_1e^0} \longrightarrow {_1p^1} + \nu$$

The neutron and positron masses, together, exceed the proton mass, and the difference can be described by the postulated neutrino. To reverse this process, an intense beam of neutrinos, as from a reactor, is needed to bombard protons:

$$_1p^1 + \nu \longrightarrow {}_0n^1 + {}_1e^0$$

This should be followed by beta decay of the neutron:

$$_0n^1 \xrightarrow{\text{12 min}} {}_{-1}e^0 + {}_1p^1$$

In 1956 Frederick Reines (b. 1918) and Clyde L. Cowan, Jr. (b. 1919) experimentally established the existence of this elusive particle. Water tanks were exposed to a presumed beam of neutrinos. They observed positrons and then electrons in the proper time sequence predicted by the above two reactions. The number of positrons and electrons varied with the pile intensity and thus with the neutrino intensity.

The isotopes of any element have almost identical chemical and physical properties. If two different isotopes of the same element are used in constructing particular molecules, we can distinguish between those molecules containing one or the other of the two isotopes. If one isotope is radioactive, its presence or absence can be detected by its alpha, beta, or gamma emission. If the isotope is stable, a mass spectrometer can be used for identification. Several examples will illustrate how isotopes have been used.

In the last chapter we saw that the oxygen in water produced during esterification comes from the carboxylic acid. This fact was established by using the stable isotope O^{18} and mass spectrometry.

Palmitic acid has the formula $CH_3(CH_2)_{14}COOH$ and contains a straight chain of 16 carbon atoms. To determine how it is formed in animals, acetic acid with the methyl carbon a radioactive isotope, $\overset{*}{C}H_3COOH$, is fed to a rat and the palmitic acid isolated. This can be broken down, or *degraded*, one carbon at a time. It was observed that alternating carbons were radioactive:

$$\overset{*}{C}H_3CH_2\overset{*}{C}H_2CH_2\overset{*}{C}H_2CH_2\overset{*}{C}H_2CH_2\overset{*}{C}H_2CH_2\overset{*}{C}H_2CH_2\overset{*}{C}H_2CH_2\overset{*}{C}H_2COOH$$

The mechanism proposed to describe palmitic acid formation is

$$\overset{*}{C}H_3CO\boxed{OH + H}\overset{*}{C}H_2COOH \longrightarrow \overset{*}{C}H_3CO\overset{*}{C}H_2COOH$$

$$\downarrow$$

$$\overset{*}{C}H_3CH_2\overset{*}{C}H_2COOH$$

$$\overset{*}{C}H_3\overset{*}{C}H_2\overset{*}{C}H_2COOH + H\overset{*}{C}H_2COOH \longrightarrow \overset{*}{C}H_3\overset{*}{C}H_2\overset{*}{C}H_2CO\overset{*}{C}H_2COOH$$

$$\downarrow$$

$$\overset{*}{C}H_3\overset{*}{C}H_2\overset{*}{C}H_2CH_2\overset{*}{C}H_2COOH$$

etc.

Acetate ion CH_3COO^-, tagged with radioactive carbon, is fed to a rat. Later cholesterol taken from the rat shows radioactivity characteristic of C^{14}. Therefore, the acetate ion is a *precursor* of cholesterol in the rat's metabolism.

Na^{24} has a half-life of 15 hr and is not found in nature. It can be prepared from the only stable isotope Na^{23} by bombardment with deuterium. The reaction is

$$_{11}Na^{23} + {}_1H^2 \longrightarrow {}_{11}Na^{24} + {}_1H^1$$

and it decays by beta and gamma emission:

$$_{11}Na^{24} \xrightarrow{15 \text{ hr}} {}_{12}Mg^{24} + {}_{-1}e^0 + \gamma$$

Isotonic saline (sodium chloride) solution which contains some Na^{24} can be intravenously injected or ingested by mouth. Then gamma and beta counters, for example, photographic film, can be placed at various body positions to monitor the rate at which the sodium ion travels in the body's circulatory system. This technique, using radioactivity to produce an image on film, is called "radioautography" or "autoradiography."

The shell of a typical hen egg is formed within the hen's body in 16 hr and contains 2 g of calcium. This is a calcium-deposition rate of 125 mg/hr. But at any time the hen will have no more than 25 mg of calcium in her blood; nor can she absorb enough calcium from her feed through her intestines to account for the amount of calcium used. Ca^{45} was used to solve this mystery. This isotope, decaying by beta emission

$$_{20}\text{Ca}^{45} \xrightarrow{\text{165 days}} {}_{-1}e^0 + {}_{21}\text{Sc}^{45} \qquad \text{(stable)}$$

was placed in the hen's feed for 1 week. Through autoradiography, Ca^{45} was detected in the bones. When a normal diet, free from Ca^{45}, was resumed, the radioactive isotope continued to be deposited in the shells. From this information, it was concluded that a hen can use one-tenth of her bone calcium to make egg shells. Her bone structure is the major repository for calcium.

Phosphorus is concentrated in our teeth and bones as phosphate PO_4^{3-} and in such organs as the brain and liver in the form of phospholipids. The total phosphorus content of the body is essentially constant. This observation can be attributed to one of two possible processes: (1) the rate of phosphorus assimilation equals the rate of phosphorus elimination from these body structures so that the phosphorus content is invariant; or (2) once phosphorus is incorporated in body tissue further ingested phosphorus is not utilized but is directly eliminated without serving any physiological purpose.

P^{32} has a half-life of 14.3 days. It is made

$$_{17}\text{Cl}^{35} + {}_0n^1 \longrightarrow {}_{15}\text{P}^{32} + {}_2\text{He}^4$$

and decays with beta emission:

$$_{15}\text{P}^{32} \xrightarrow{\text{14.3 days}} {}_{16}\text{S}^{32} + {}_{-1}e^0$$

When phosphate containing P^{32} is ingested, the radioactivity of bone, teeth, and the organs can be monitored. The radioactive emission rate first increases as P^{32} is assimilated, and then decreases as P^{32} is eliminated from these body structures. In human teeth about 1 percent of the phosphorus atoms is replaced in 250 days. This is called the "turnover rate" of phosphorus in teeth.

The total body water content can be determined by the *isotope-dilution method*. Suppose 100 cm^3 of water which is 50.0 percent heavy water D_2O is injected into a human being. Serum (blood fluid from which blood cells and clots have been removed) is collected at intervals and examined for deuterium content. After 1 hr, the D_2O content in the serum reaches a steady value (for the next 18 hr) of 0.093 percent. The injected $\text{D}_2\text{O}-\text{H}_2\text{O}$ mixture has a density of 1.05 g/cm^3. The amount of D_2O injected is

$$100 \text{ cm}^3 \text{ solution} \times \frac{1.05 \text{ g solution}}{\text{cm}^3 \text{ solution}} \frac{0.500 \text{ g D}_2\text{O}}{1.000 \text{ g solution}} = 52.5 \text{ g D}_2\text{O}$$

If the removed serum water has a density of 1.00 g/cm^3, and if the unknown total body water volume is V cm^3, then

$$V \text{ cm}^3 \text{ solution} \times \frac{1.00 \text{ g solution}}{\text{cm}^3 \text{ solution}} \frac{0.00093 \text{ g D}_2\text{O}}{1 \text{ g solution}} = 52.5 \text{ g D}_2\text{O}$$

since in only 1 hr the only water loss through skin evaporation and lung respiration is negligible. From this

$$V = 52.5 \text{ g D}_2\text{O} \times \frac{1 \text{ g solution}}{9.3 \times 10^{-4} \text{ g D}_2\text{O}} \frac{1 \text{ cm}^3 \text{ solution}}{1.00 \text{ g solution}}$$

$$= 5.65 \times 10^4 \text{ cm}^3 \text{ solution}$$
$$= 56.5 \text{ liters body water}$$

Body water can be divided into intracellular water (within the cells) and extracellular water (outside of the cells). Extracellular water consists of *plasma, or serum, water* (in the blood) and *interstitial* water (outside the cells but not in the blood). In a normal body about 5 percent of the body weight is plasma water, about 15 percent is interstitial water, and about 50 percent is intracellular water.

Solutions of molecules such as sulfate ions or sucrose, which do not ordinarily pass into the cells but which are confined to the plasma and interstitial water, can be injected. Samples of blood water are examined for dilution, and the total plasma and interstitial water content can be determined. Finally molecules which remain in the blood only can be used to assay the total plasma-water content.

ISOTOPES AS THERAPEUTIC AGENTS. Rapidly growing tissues assimilate nutrient materials more readily than slowly growing tissues. This observation suggests that radioactive isotopes can serve as *therapeutic agents* in controlling or destroying tumors.

Normally a 70-kg adult body has about 50 mg of iodine. The thyroid gland, with a mass of about 25 g, contains about 10 mg of iodine. Iodine is largely concentrated, therefore, in this organ. The radioactive isotopes I^{130} and I^{131} decay by electron and gamma emissions and have half-lives of 12.5 hr and 8.05 days, respectively. A diet with these two isotopes will result in an accumulation of the isotopes in the thyroid gland. Beta and gamma decay will damage the tissue contiguous to the radioactive isotope, but more abnormal tissue will be destroyed than normal tissue because the faster growth of the abnormal cells enhances incorporation of the isotopes.

BATTERIES FROM ISOTOPES. Some radioactive isotopes are convenient small volume-and-mass sources of energy. A 2-kg battery, using 95 g of Pu^{238} and producing 2.7 W of electric power continuously for 5 yr, has been used in satellites. A heavier battery, about 2,000 kg (4,600 lb), using 1.6 kg (3.5 lb) of Sr^{90}, can produce 60 W for 10 yr and is used in ocean weather stations.

USE OF COBALT ISOTOPES. Co^{60} is made in a reactor from natural cobalt. It decays by emitting high-energy beta and gamma particles:

$$_{27}Co^{60} \xrightarrow{5.27 \text{ yr}} _{-1}e^0 \quad + \quad \gamma \quad + _{28}Ni^{60}$$
$$\underset{\substack{0.31 \text{ and} \\ 1.48 \text{ MeV}}}{} \quad \underset{\substack{1.33 \text{ and} \\ 1.17 \text{ MeV}}}{}$$

These high-energy gamma rays can be used instead of x-rays for medical diagnosis and radiation therapy. A Co^{60} sample is considerably smaller than an x-ray machine. It may even be placed within a body at a malignant site for therapy.

Radioactive isotopes are used to initiate chemical reactions. Gamma rays can convert stable molecules to more reactive ions or radicals, making possible at room temperature and pressure a reaction which ordinarily requires high temperatures or pressures.

PROBLEMS

8.1. At the equator the mean radius of the earth is 6,378.388 km (3,963.34 miles). There are 86,400 s in one day. How fast does a point on the earth's equator move as the earth goes through one full turn each day?

Hint: If the radius is r, the circumference of a circle is $2\pi r$.

8.2. The mean earth-to-sun distance is 149,500,000 km. In 1 yr, or 3.156×10^7 s, the earth makes one complete trip around the sun. Calculate the speed of the earth going around the sun.

8.3. *a.* The calcium carbonate in a bone sample found in Arizona is converted to carbon dioxide (and a calcium salt) by treating it with acid. The C^{14} activity of the gas is found to be 4.24 counts/min-g C. How old is the bone?

b. Another bone specimen similarly treated has an activity of 5.35 counts/min-g C. How old is it?

8.4. In 1 yr 1 g of radium will emit 11.6×10^{17} alpha particles. If these are collected, the gas can be identified to be helium by its emission spectrum. The volume of helium collected in 1 yr is 4.3×10^{-5} liters (STP). Since the emitted alpha particles can be counted individually and accumulate as helium, the gas must consist of discrete particles. (This is another piece of evidence for the existence of atoms.) From the information given, calculate Avogadro's number.

8.5. The coulombic force between charged particles is described by the equation

$$f = \frac{k \, Q_1 Q_2}{r^2}$$

where $k = 8.9878 \times 10^9$ newton-m^2/coul^2, the electric charges Q_1 and Q_2 are in coulombs, and the distance between them is in meters. The gravitational force between masses is

$$f = \frac{G \, m_1 m_2}{r^2}$$

where $G = 6.668 \times 10^{-11}$ N-m^2/kg^2, the masses m_1 and m_2 are in kilograms, and the distance between them is in meters.

These data are available for nucleons:

Particle	Mass, kg	Charge, coul
Proton	1.67252×10^{-27}	1.60210×10^{-19}
Neutron	1.67482×10^{-27}	0

The distance between nucleons is $r = 10^{-15}$ m. Compare the magnitudes of these two forces acting between (*a*) two protons, (*b*) two neutrons, and (*c*) one proton and one neutron.

8.6. The data for some common elements are:

Element	Mass % in earth's crust	Most common isotope	Fraction of that isotope in element
O	46.71	$_8O^{16}$	0.99759
Si	27.69	$_{14}Si^{28}$	0.9221
Fe	5.05	$_{26}Fe^{56}$	0.9166
Ca	3.65	$_{20}Ca^{40}$	0.9697
Total	83.10		

What fraction of the earth's crust are these four common isotopes?

8.7. Of the following isotopes and their masses, which are isotopes of the same element? Which are isobars? isotones? What is the mass defect and binding energy of each? Which do you think is the most likely to be radioactive?

Isotope	Mass, amu
$_{16}S^{34}$	33.96786
$_{16}S^{36}$	34.968
$_{16}S^{38}$	35.96709
$_{18}Ar^{36}$	35.96755

8.8. In Fermi's first chain fission pile, 1 W of power was produced from 3.1×10^{10} fissions/s of U^{235}. How much of this isotope is required to produce 1 kW continuously for 1 yr?

8.9. Assume a ballistic missile has a 50 percent chance of damaging a target. How many of these missiles must be used so that the attacker is 99.99 percent confident of destroying this target?

8.10. Throughout the project to build a nuclear-chain-reaction bomb, there was a sense of emergency among Allied scientists and engineers lest the Axis powers succeed first. Estimate the basis for this fear by counting the number of Nobel Prizes won by Americans, Britons, Frenchmen, and Germans in Physics, Chemistry, and Physiology and Medicine from 1901 to 1939. Which nation seemed to be preeminent before World War II? Make a similar comparison from 1945 to the present. Has there been a significant change in the ranking according to prizes won?

8.11. The tenth-value thickness of a material is the thickness stopping 90 percent and passing 10 percent of incident gamma radiation. For example, of the gamma-ray intensity hitting a steel barrier $4\frac{1}{2}$ in. thick, only one-tenth will pass through. Through a 9-in. barrier of steel, $\frac{1}{10} \times \frac{1}{10}$, or one-hundredth, will pass through; and through

$13\frac{1}{2}$ in., one-thousandth of the radiation will go through. From this, can you suggest a way of estimating the shielding effectiveness of any thickness of steel?

8.12. Suppose a 1-Mt bomb explodes 1 mile from your shelter, the walls of which have 15 in. of concrete and 22 in. of earth. What is the gamma-radiation intensity within the shelter? Do you have a chance of surviving this prompt radiation? If it were a 10-Mt bomb at the same distance?

8.13. World War I lasted 4 yr, and World War II lasted 6 yr. Quite probably, a nuclear war could last no more than 1 hr. Obviously, time for deliberation and decision-making is severely curtailed in nuclear warfare. You can estimate the older scale of time for evaluation and action from events in World War II, which began with the invasion of Poland and the declaration of war by Britain and France. How long a time elapsed between these events? The United States entered as an active belligerent after an attack on Pearl Harbor, by declaring war on Japan and Germany, and by Germany declaring war on it. How much time elapsed between the attack and the several declarations?

8.14. What is the origin of the symbol ⊕?

Chemical Processes

9

ELEMENTARY CHEMICAL PROCESSES

As a rather good rule the systems studied by chemists are considerably more complicated and thus less amenable to precise description than systems of interest to physicists. (Living systems are still more complex, and less precisely described.) We know more about electrons, neutrons, and protons than we do about molecules. Nevertheless, scientists have been able to discover principles and laws to describe, and to help us understand, what happens in chemical processes. In this chapter we shall look at some of these, and at a few of their applications.

Chemical reactions can be classified by the way atoms or groups of atoms are combined, separated, or transferred (Table 9.1). The *molecularity* of a reaction is the least number of molecules of reactants used up. In the balanced-equation examples, the molecularities range from one (the dissociation of bromine and the isomerization of dimethyl ether) to twelve (the combustion of heptane).

During the combustion of heptane, a large number of hydrocarbon fragments and partially oxidized fragments, containing from one to seven carbons, can be detected by spectroscopy and mass spectrometry. This indicates that the combustion is not the combining of eleven oxygens and one heptane molecule, changing in one operation to seven carbon dioxide and eight water molecules; rather, carbon and hydrogen atoms are oxidized and removed from the parent heptane one or two atoms at a time.

Molecularity, then, is a measure of the overall consumption of reactants. A more detailed description of the reaction is a step-by-step account of the several individual reactions needed to convert reactants to products.

Table 9.1. A Catalog of Chemical-reaction Types

General form	General name	Examples	
$A + B \longrightarrow AB$	Addition	$H_2 + CH_2{=}CH_2 \longrightarrow CH_3{-}CH_3$	Hydrogenation
		$2Br \longrightarrow Br_2$	Recombination
$AB \longrightarrow A + B$	Decomposition	$Br_2 \longrightarrow 2Br$	Dissociation
		$CH_3{-}CH_3 \longrightarrow CH_2{=}CH_2 + H_2$	Dehydrogenation
$AB + C \longrightarrow AC + B$	Displacement	$CH_3COOH + H_2O \longrightarrow CH_3COO^- + H_3O^+$	Ionization
		$Br + H_2 \longrightarrow HBr + H$	Substitution
$AB + CD \longrightarrow AD + BC$	Double decomposition, or metathesis	$CH_4 + Cl_2 \longrightarrow CH_3Cl + HCl$	Chlorination
		$CH_3COOH + HOCH_3 \longrightarrow CH_3COOCH_3 + H_2O$	Esterification
		$C_7H_{16} + 11O_2 \longrightarrow 7CO_2 + 8H_2O$	Combustion
$A \longrightarrow B$	Rearrangement	$CH_3OCH_3 \longrightarrow CH_3CH_2OH$	Isomerization

CHAIN REACTIONS

In 1907 after studying for many years the reaction between hydrogen and bromine

$$H_2 + Br_2 \longrightarrow 2HBr$$

Max Bodenstein (1871–1942) and Samuel C. Lind (1879–1965) obtained an equation describing the speed of this reaction, based on how fast hydrogen and bromine are used up and how fast the product is formed. From the complexity of their equation it was obvious the reaction was not a simple encounter between the reactants. Thirteen years later, Jens Anton Christiansen (b. 1888), Karl F. Herzfeld (b. 1892), and Michael Polanyi (b. 1891), independent of each other, announced that the speed of the reaction can be accounted for by assuming a five-step series of reactions:

$$Br_2 \xrightarrow{\text{dissociation}} 2Br \qquad (9.1)$$

$$Br + H_2 \xrightarrow{\text{substitution}} HBr + H \qquad (9.2)$$

$$H + Br_2 \xrightarrow{\text{substitution}} HBr + Br \qquad (9.3)$$

$$HBr + H \xrightarrow{\text{substitution}} H_2 + Br \qquad (9.4)$$

$$2Br \xrightarrow{\text{recombination}} Br_2 \qquad (9.5)$$

This set of reactions is called the "mechanism" for the hydrogen-bromine reaction. Each step is called an "elementary reaction." Elementary reactions with a single reactant are said to be *unimolecular*, with two reactants *bimolecular*, and with three *termolecular*. As written, this mechanism has one unimolecular step (9.1); the others are bimolecular, and there are no termolecular steps.

Additional work on the same reaction indicated that step (9.1) is really bimolecular:

$$Br_2 + M \longrightarrow 2Br + M \qquad\qquad\qquad (9.1)$$

About 46.0 kcal/mole are required to break the bromine-bromine bond — more energy than is available at ordinary temperatures. A collision between bromine and any other molecule M and the transfer of energy into bromine is one means for that molecule to acquire sufficient energy to dissociate.

Also, if the recombination step (9.5) is only bimolecular, the 46.0 kcal/mole released remain within the formed Br_2 molecule and are sufficient to redissociate it. If a collision occurs before dissociation, some of this energy can be transferred out of the Br_2, thus stabilizing it. Consequently, step (9.5) is termolecular:

$$2Br + M \longrightarrow Br_2 + M \qquad\qquad\qquad (9.5)$$

The mechanism for this reaction is of special interest because of steps (9.2) and (9.3). In (9.2), atomic bromine is used up and atomic hydrogen produced; in step (9.3) atomic bromine is produced and atomic hydrogen used up. It is also in these two steps that the product HBr is formed. Atomic bromine and hydrogen are very reactive, and their concentrations remain quite low during the course of the overall reaction. Steps (9.2) and (9.3) are coupled: step (9.2) is followed directly by step (9.3) and then by step (9.2), and so on. Steps (9.2) and (9.3) are repeated in alternation several thousand times for each dissociation and before a recombination removes atomic bromine from active participation.

This entire mechanism has a special name, a *chain reaction*, because of steps (9.2) and (9.3). The first step (9.1) is the chain initiation step and the last the chain termination step. The fourth is a side reaction which undoes step (9.2). A chain reaction depends on a small but steady amount of highly reactive species, such as H and

Br. In both steps (9.2) and (9.3), the chain propagation steps, when one highly reactive species is used up, the other is formed, preserving the total number of these. Not all mechanisms are chain reactions. Usually chain reactions are faster.

If the propagation steps produce more highly reactive species than are consumed, there is a net increase in these, then the rate of reaction increases, and still more highly reactive species are available for the reaction. Such processes speed up continuously and are said to be *explosive chain reactions.* Chemical explosions are of this type.

A mechanism, the detailed set of elementary steps, has considerably more information than the overall balanced equation. With a mechanism a chemist can better understand how chemical reactions occur. He can see more clearly how to affect a reaction by interfering with a particular elementary step.

COLLISIONS IN GASES

As the hydrogen-bromine mechanism was described, even the first step is bimolecular. In the majority of reactions studied, there must be an encounter between two species. If the collision rate between reactants is known, we can compare this with the rate of consumption and see whether or not there is an agreement between the two rates.

We need certain information to calculate the collision rate between molecules in a gas: how many molecules there are, how fast they move, and how big a target each presents for collision. Each of these problems has been described and solved in previous chapters. We can try to calculate the number of collisions between oxygen and nitrogen molecules in air at 1 atm pressure and at 26.8°C. The result will be the number of collisions per second per cubic centimeter. However, the three bits of information must be calculated first.

> **Example.** How many molecules of O_2 and N_2 are there in 1 cm³ of air at 26.8°C and 1 atm? (Assume air is one-fifth oxygen and four-fifths nitrogen and is an ideal gas.)
>
> From the ideal gas law
>
> $$pV = nRT$$

$$\frac{n(\text{moles})}{V(\text{liters})} = \frac{p}{RT}$$

$$= 1 \text{ atm} \times \frac{\text{deg-mole}}{0.08206 \text{ liter-atm}} \frac{1}{300.0°}$$

$$= 4.062 \times 10^{-2} \text{ moles/liter}$$

$$= \frac{4.062 \times 10^{-2} \text{ moles}}{\text{liter}} \frac{6.023 \times 10^{23} \text{ molecules}}{\text{mole}} \frac{\text{liter}}{10^3 \text{ cm}^3}$$

$$= 2.447 \times 10^{19} \text{ molecules/cm}^3$$

Of these, one-fifth are oxygen, or 0.4894×10^{19} O_2 molecules/cm^3, and four-fifths are nitrogen, or 1.958×10^{19} N_2 molecules/cm^3.

Example. How fast do average oxygen and nitrogen molecules move?

The speed u depends on the temperature T and molecular weight M and can be calculated by

$$u = \sqrt{\frac{3RT}{M}}$$

In Chap. 3 we calculated the average oxygen speed at this temperature:

$$u(O_2) = 4.84 \times 10^4 \text{ cm/s}$$

At the same temperature, the speed of a nitrogen molecule will be greater and can be calculated from

$$\frac{u(N_2)}{u(O_2)} = \sqrt{\frac{M(O_2)}{M(N_2)}}$$

Thus

$$u(N_2) = u(O_2) \sqrt{\frac{M(O_2)}{M(N_2)}}$$

$$= 4.84 \times 10^4 \left(\frac{32.00}{28.02}\right)^{\frac{1}{2}} \text{ cm/s}$$

$$= 4.84 \times 10^4 \times (1.142)^{\frac{1}{2}} \text{ cm/s}$$

$$= 4.84 \times 1.069 \times 10^4$$

$$= 5.17 \times 10^4 \text{ cm/s}$$

Example. Estimate the size of an individual oxygen and nitrogen molecule. The density of solid oxygen is 1.426 g/cm³, and that of solid nitrogen is 1.026 g/cm³, both at −252.5°C.

The volume per molecule of each can be calculated:

$$\frac{32.00 \text{ g}}{\text{mole}} \frac{\text{cm}^3}{1.426 \text{ g}} \frac{\text{mole}}{6.023 \times 10^{23} \text{ molecules}}$$

$$= 3.726 \times 10^{-23} \text{ cm}^3/O_2 \text{ molecule}$$

$$\frac{28.02 \text{ g}}{\text{mole}} \frac{\text{cm}^3}{1.026 \text{ g}} \frac{\text{mole}}{6.023 \times 10^{23} \text{ molecules}}$$

$$= 4.534 \times 10^{-23} \text{ cm}^3/N_2 \text{ molecule}$$

We assume the volume per molecule does not change as the temperature is increased from −252.5 to 26.8°C. These diatomic molecules are shaped like dumbbells and spin around as they move. We cannot be sure exactly what its size is as a target, but for our purpose we can assume each molecule to be effectively spherical with an effective radius r and volume

$$v = \frac{4\pi}{3} r^3$$

From the volume we can calculate the radius:

$$r = \left(\frac{3v}{4\pi}\right)^{\frac{1}{3}}$$

For oxygen

$$r = \left(\frac{3 \times 3.726 \times 10^{-23} \text{ cm}^3}{4 \times 3.142}\right)^{\frac{1}{3}}$$

$$= \left(\frac{11.18}{12.57} \times 10^{-23}\right)^{\frac{1}{3}} \text{ cm}$$

$$= (0.8894 \times 10^{-23})^{\frac{1}{3}} \text{ cm}$$

$$= (8.894 \times 10^{-24})^{\frac{1}{3}} \text{ cm}$$

$$= 2.072 \times 10^{-8} \text{ cm}$$

For nitrogen

$$r = \left(\frac{3 \times 4.534 \times 10^{-23} \text{ cm}^3}{4 \times 3.142}\right)^{\frac{1}{3}}$$

$$= (10.82 \times 10^{-24})^{\frac{1}{3}} \text{ cm}$$

$$= 2.212 \times 10^{-8} \text{ cm}$$

Our notion of the molecular size as a target is not clear enough to warrant these four significant figures. We can say each molecule has an effective radius of 2×10^{-8} cm.

Our calculation of the frequency of nitrogen-oxygen collisions begins with following a single oxygen molecule and counting how many times in 1 s it collides with a nitrogen molecule. Figure 9.1*a* is a typical zigzag journey of this oxygen molecule: a straight-line path between collisions and change in direction at each encounter. It is evident (Fig. 9.1*b*) that a collision will occur only if a simple condition is satisfied — when the molecules pass each other and are at their closest approach, the distance from center to center can be no greater than the sum of the radii of the two molecules.

Figure 9.1*c* is a redrawn version of Fig. 9.1*a*. Here, the oxygen is represented as a larger molecule whose radius is the sum of the true oxygen and nitrogen radii. Nitrogens are represented as points having no radii. The criterion for collision can be restated: if a nitrogen point lies within the volume swept out by this enlarged oxygen, a collision will occur. There is no essential difference between Fig. 9.1*a* and *c*, but Fig. 9.1*c* makes the calculation extremely easy.

The enlarged oxygen in Fig. 9.1*c* has a radius of

$$r = r(O_2) + r(N_2)$$
$$= (2 \times 10^{-8}) + (2 \times 10^{-8})$$
$$= 4 \times 10^{-8} \text{ cm}$$

Its cross-sectional area is

$$S = \pi r^2$$
$$= 50 \times 10^{-16} \text{ cm}^2$$

In moving, the enlarged oxygen molecule sweeps out a kinked cylindrical volume. Its speed 4.84×10^4 cm/s is numerically the distance it travels in 1 s. If the kinked cylinder were straightened out, the volume swept out in 1 s would be the cross section times the length:

$$V = (50 \times 10^{-16} \text{ cm}^2)(4.84 \times 10^4 \text{ cm/s})$$
$$= 2.4 \times 10^{-10} \text{ cm}^3/\text{s}$$

The actual volume swept out in the zigzag path is equal to this. A

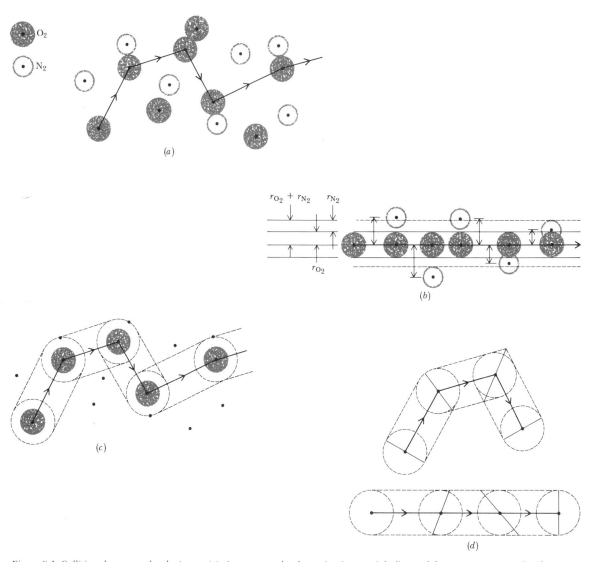

Figure 9.1. Collisions between molecules in gas. (a) An oxygen molecule moving in a straight-line path between encounters. (b) The oxygen molecule will hit only certain other molecules. At their closest approach the distance between the centers cannot be greater than the sum of the radii. Thus the centers of the colliding molecules must be within the boundary of the dashed lines. (c) Part (a) is redrawn with the moving oxygen represented as a larger sphere, whose radius is the sum of the radii of the nitrogen and oxygen molecules. The other molecules are shown as points. Collision is possible only if the points lie within the volume swept out by the larger sphere in its motion through space. (d) If the straight cylinder (bottom) is sliced and reassembled (top), the volume does not change. The volume of the straight cylinder is the cross-sectional area of the enlarged oxygen sphere times the distance or length. Thus, the swept-out volume in the kinked-path motion is also the product of the cross section and the distance. The volume swept out in 1 s by the enlarged oxygen sphere is the product of its cross section and its speed (the distance it travels in 1 s). If the cross section is S and the speed is u, the product is Su cm³/s. If there are n_2 nitrogen molecules in 1 cm³, then Sun_2 is the number of nitrogen molecules hit in 1 s by a single moving oxygen molecule. A very large number of molecules pass through this swept-out volume, but, on the average, the concentration of nitrogens at any one time is n_2.

straight-path cylinder can be sliced obliquely and reassembled as a kinked cylinder (Fig. 9.1*d*) with no change in volume.

We can calculate how many nitrogen molecules, on the average, will be contained in this swept-out volume because we know the number in 1 cm^3. This number

$$Z_1 = \frac{2.4 \times 10^{-10} \text{ cm}^3}{\text{s}} \frac{1.958 \times 10^{19} \text{ N}_2 \text{ molecules}}{\text{cm}^3}$$

$$= 4.7 \times 10^9 \text{ N}_2 \text{ molecules/s}$$

is the number of nitrogen molecules hit by a single oxygen in 1 s, about 5 billion. Many more molecules than this will pass through the swept-out volume in 1 s. But, on the average, at any time there will be 1.958×10^{19} nitrogens/cm^3; Z_1 is the number hit by the moving oxygen.

From this, we can calculate the seconds per collision—the time interval between collisions of the oxygen molecule:

$$\frac{1 \text{ s}}{4.7 \times 10^9 \text{ N}_2 \text{ molecules hit}} = 2.1 \times 10^{-10} \text{ s/collision}$$

The product of the speed and time is the distance traveled between collisions

$$(4.84 \times 10^4 \text{ cm/s})(2.1 \times 10^{-10} \text{ s/collision}) = 1.0 \times 10^{-5} \text{ cm/collision}$$

We know how many collisions a single oxygen will make with nitrogens each second. What we want to know is how many collisions occur between all oxygens and the nitrogens. Since there are 0.4894×10^{19} O$_2$ molecules/cm^3 and each one of these moves about hitting nitrogen molecules, the total oxygen-nitrogen collision frequency is

$$Z_{12} = \frac{4.7 \times 10^9 \text{ N}_2 \text{ molecules hit}}{\text{single O}_2 \text{ molecule-s}} \frac{4.9 \times 10^{18} \text{ O}_2 \text{ molecules}}{\text{cm}^3}$$

$$= 2.3 \times 10^{28} \text{ collisions/s-cm}^3$$

This lengthy and complex calculation has been enormously simplified by breaking it down into many shorter and simpler steps. Using more elaborate ideas, for example, considering molecules to have a range of possible speeds, the numbers Z_1 and Z_{12} so obtained agree quite well with our values. We can repeat the calculation using symbols rather than numbers. The effective radius of the enlarged molecule is

$$r = r_1 + r_2 \text{ cm}$$

the sum of the individual molecular radii. The cross-sectional area of this is

$$S = \pi (r_1 + r_2)^2 \text{ cm}^2$$
$$= \pi r^2 \text{ cm}^2$$

If the velocity is u cm/s, the volume swept out per second is

$$V = S \text{ cm}^2 \times u \text{ cm/s}$$
$$= Su \text{ cm}^3/\text{s}$$

Where n_2 is the number of molecules of the second kind per cubic centimeter,

$$Z_1 = Su \text{ cm}^3/\text{s} \times n_2 \text{ molecules/cm}^3$$
$$= Sun_2 \text{ molecules hit/s}$$

Then, since there are n_1 molecules of the first kind per cubic centimeter, the total number of collisions between the two kinds is

$$Z_{12} = Sun_2 \frac{\text{molecules hit}}{\text{s}} \times n_1 \frac{\text{molecules}}{\text{cm}^3}$$
$$= Sun_1n_2 \text{ collisions/s cm}^3$$

The important thing about this derived equation is that the collision frequency depends on the product of the concentrations of each kind of molecule, that is, on n_1n_2. If the concentration of one is doubled, there will be twice as many collisions; if both concentrations are doubled, there will be four times as many collisions.

The *law of mass action* follows from the equation for Z_{12}: An elementary process involving one or more molecular species will proceed at a rate that varies as the product of the concentrations of the species involved. If the process is the reaction between two like species

$$NO_2 + NO_2 \longrightarrow N_2O_4$$

the rate depends on the product

$$[NO_2][NO_2] = [NO_2]^2$$

where $[NO_2]$ is the concentration of nitrogen dioxide, and $[NO_2]^2$ the concentration squared.

THE SPEED OF CHEMICAL REACTIONS

When molecules react, we can determine the speed of the reaction by measuring how fast the products accumulate or how fast the reactants are consumed. This experimental rate can be compared with the collision rate. Then we shall know what fraction of the molecules which collide actually react.

The gasoline antiknock additive tetraethyl lead $Pb(C_2H_5)_4$ is a liquid which decomposes at 200°C:

$$Pb(C_2H_5)_4 \longrightarrow Pb + 4C_2H_5$$

Metallic lead, which boils at 1744°C, is deposited on the walls of the container. The principal reactions of the ethyl radicals are

$$C_2H_5 + C_2H_5 \longrightarrow C_4H_{10}$$
$$\text{Butane}$$

$$C_2H_5 + C_2H_5 \longrightarrow C_2H_6 + C_2H_4$$
$$\text{Ethane} \quad \text{Ethylene}$$

The rate at which the ethyl radical concentration decreases with time indicates that every collision results in a reaction.

The reaction between nitrogen and oxygen to form nitric oxide

$$N_2 + O_2 \longrightarrow 2NO$$

is not observed at room temperature, although there are about 2×10^{28} nitrogen-oxygen collisions/s in 1 cm³ of air (STP). At 1000

K (727°C), there are fewer collisions (2×10^{27}) because the gas expands and there are fewer molecules/cm³ at this higher temperature and at the same pressure; but, of these, 10,000 result in reaction. At 2000 K (1727°C), there are 8×10^{26} collisions/s-cm³ between nitrogen and oxygen, and in almost 2×10^{15} of these nitric oxide is formed.

   This reaction goes faster the higher the temperature for one important reason: a greater fraction of the collisions leads to reaction. How this comes about can be understood from the energies of the molecules involved. In Chap. 3 we saw how molecular speeds can be measured, using the rotating-wheel ensemble. At any temperature there is a characteristic average energy and average speed, but there is also a distribution of energies and speeds about the average values. Some molecules go much slower and others go much faster than the average.

   Figure 9.2 is a plot of the distribution of energies among molecules at several temperatures. The higher the temperature, the broader the distribution. If a certain minimum energy is required for reaction, then the higher the temperature, the greater the fraction of all molecules having this minimum energy. By observing how much faster a reaction goes with an elevation in temperature, a chemist can judge what this minimum energy require-

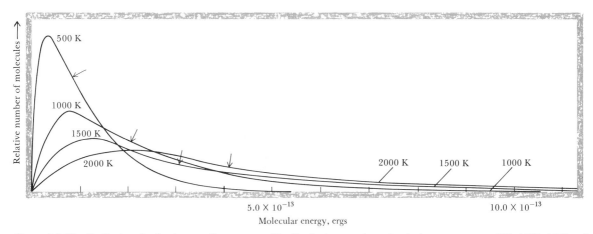

Figure 9.2. The distribution of molecules according to energy. The distributions are shown for the four temperatures: 500, 1000, 1500 and 2000 kelvin. The higher the temperature, the relatively more likely a molecule is to have a higher energy. The arrows indicate the average kinetic energies at the different temperatures. At 1000 kelvin, for example, the average energy is 2.07×10^{-13} ergs/molecule. This corresponds to a speed of 9.43×10^4 cm/s for nitrogen, and 8.82×10^4 cm/s for oxygen. If a minimum energy requirement must be satisfied for a reaction to be possible, then at higher temperatures a greater fraction of all the molecules will have at least this minimum energy. This is the principal reason why reactions go faster at higher temperatures.

ment is. The minimum amount is called the "activation energy." It can be described as an energy barrier which must be overcome for a reaction to go (Fig. 9.3).

The activation energy is related to the energy needed to break certain bonds in the reactants, and also to the energy released when products are formed. In Fig. 9.3, it can be seen that when a reaction and its reverse can both occur, one is *exothermic* (gives off heat) and the other *endothermic* (requires heat). Since the endothermic reaction has a larger activation energy, fewer collisions result in reaction. Of a pair of reversible reactions, the endothermic reaction is usually slower than the exothermic one.

The ethyl radical reactions are relatively unaffected by temperature changes, aside from higher speed and harder collisions at elevated temperatures. The activation energy is essentially zero.

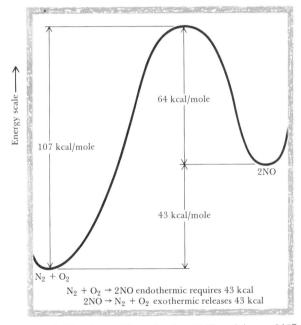

$N_2 + O_2 \rightarrow 2NO$ endothermic requires 43 kcal
$2NO \rightarrow N_2 + O_2$ exothermic releases 43 kcal

Figure 9.3. When N_2 and O_2 react to form 2NO a minimum of 107 kcal/mole is required to overcome the barrier to reaction. For the reverse process the minimum energy requirement is 64 kcal/mole. The barrier limits the fraction of molecules active or energetic enough for reaction. Both the exothermic and endothermic reactions have barriers to be overcome, and the difference in the barrier heights ($107 - 64$ kcal) is the net heat released or taken up (43 kcal) when the reaction is complete.

The formation of nitric oxide from its elements has a rather high activation energy, and its rate is significantly influenced by temperature. This reaction is important because nitric oxide is one of the set of nitrogen oxides found as air pollutants. About half of the nitrogen oxides found in air are produced in automobile engines. The higher the operating temperature, the more efficient the engine but the faster nitric oxide is produced. Thus, engine efficiency and nitrogen oxide pollution are necessarily associated.

Although the concept of activation energy was introduced here to describe the chance of reaction upon collision, it has much broader application than just in bimolecular gas reactions. For a unimolecular process, the molecule may have to assume some unusual structural distortion in order to react. The energy barrier to be overcome for distorting the molecule and then for the reaction to occur is the unimolecular activation energy.

CATALYSIS

Frequently a reaction has more than one possible mechanism. The process of converting reactants to products can go through two or more pathways. Most of the process will occur by that mechanism with the lowest activation energy.

The dehydration of tertiary butyl alcohol to isobutylene and water takes place extremely slowly if the temperature is less than 500°C. In the presence of hydrogen bromide, it can take place rapidly at 250°C. The slower mechanism has a much greater activation energy than the faster. Although both mechanisms can operate simultaneously, if hydrogen bromide is present, the faster mechanism dominates. The two mechanisms and their activation-energy diagram are given in Fig. 9.4.

In this reaction hydrogen bromide is not consumed. It merely acts as an intermediary, providing a faster alternative to the process. Such a substance is called a "catalyst," and the participation of a catalyst is called "catalysis." Since the net effect, with or without a catalyst, is the conversion of reactants to products, the energy change must be independent of the presence or absence of the catalyst. If this were not so, a violation of the first law of thermodynamics would occur.

A catalyst such as hydrogen bromide can be used to speed up many reactions. It is typical of inorganic catalysts that they can

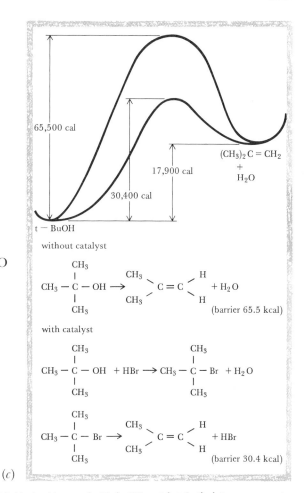

$$H_3C-\underset{\underset{CH_3}{|}}{\overset{\overset{CH_3}{|}}{C}}-OH \longrightarrow \underset{H_3C}{\overset{H_3C}{>}}C=CH_2 + H_2O$$

Tertiary butyl   Isobutylene   Water
alcohol *t*-BuOH

(a)

$$H_3C-\underset{\underset{CH_3}{|}}{\overset{\overset{CH_3}{|}}{C}}-OH + HBr \longrightarrow H_3C-\underset{\underset{CH_3}{|}}{\overset{\overset{CH_3}{|}}{C}}-Br + H_2O$$

$$H_3C-\underset{\underset{CH_3}{|}}{\overset{\overset{CH_3}{|}}{C}}-Br \longrightarrow \underset{H_3C}{\overset{H_3C}{>}}C=CH_2 + HBr$$

(b)

Figure 9.4. The dehydration of t-BuOH has a barrier of 65.5 and 30.4 kcal, without and with the HBr catalyst. In the latter case a greater fraction of the alcohol molecules are eligible for reaction; the dehydration is faster with a catalyst. But with or without a catalyst this reaction is endothermic by 17.9 kcal.

affect more than one kind of reaction. Enzymes, in contrast, ordinarily catalyze one or a very small number of reactions. They are constructed in a precise way from a definite set of amino acids linked in a unique way. The catalytic activity of each enzyme depends on the composition and geometric structure of the protein and, if present, nonprotein parts of the enzyme molecule.

Enzymes are necessary for life itself. Without them, compounds within an organism would not be broken down or built up at the proper rates to sustain life. There are specific enzymes for the many elementary chemical steps in hydrolyzing carbohydrates, fats, and proteins during digestion, joining amino acids into proteins, oxidizing molecules for energy, secretion, excretion, muscle contraction, and blood clotting, and so on.

NERVE SIGNALS
AND ENZYME ACTION

Signals, or messages, are sent through the body along fiberlike cells called *neurons*. Some neurons transmit signals from the sensory organs, such as the eyes, ears, and skin, and others transmit signals to muscles. All neurons either begin or end on other neurons and the signal may cross from one neuron to the other. A neuron at rest, not transmitting a signal, has an unusual salt composition with respect to the body fluid surrounding it. The sodium-ion concentration is much less inside the neuron than outside in the fluid, and the potassium-ion concentration is greater inside than out. The consequence of this inequality is a negative charge on the inner surface of the membrane surrounding the neuron and a positive charge on the outside of the membrane. The membrane is said to be *polarized* (Fig. 9.5*a*).

As the signal reaches the end of a particular neuron, the cationic substance acetylcholine with an accompanying anion

$$CH_3COOCH_2CH_2N(CH_3)_3 + OH^-$$

is released. This affects the membrane of the succeeding neuron or muscle, permitting sodium ions to pass into the neuron. Now, the inner membrane surface is positively charged at the point of stimulation; the membrane is said to be *depolarized* at that point. This, in turn, affects adjacent points on the same membrane, and so depolarization spreads along the length of the neuron. The signal is the motion of this depolarization.

As long as acetyl choline is present the succeeding neuron will continue to be stimulated. To prevent this, acetyl choline must be destroyed. An enzyme, acetylcholine esterase, hydrolyzes acetylcholine into acetate ion and choline:

$$CH_3COO^- + HOCH_2CH_2\overset{+}{N}(CH_3)_3$$

Then the membrane can be repolarized, ready for another signal.

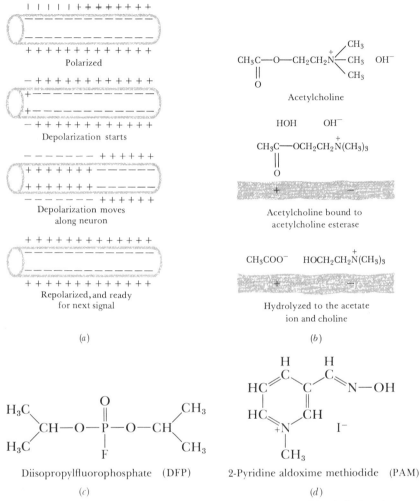

Figure 9.5. Nerve signal propagation and nerve toxin. (a) *Movement of depolarization along a neuron;* (b) *acetylcholine is necessary for depolarization but must be destroyed by the acetylcholine esterase enzyme before polarization can be restored;* (c) *DFP combines with enzyme, preventing hydrolysis of acetylcholine and causing death;* (d) *PAM, if used promptly, is an antidote for DFP, displacing it from the enzyme.*

NERVE GASES. The enzyme is believed to have two active sites by which it binds acetylcholine (Fig. 9.5b) for the hydrolysis, or splitting. One site has a negative charge for the positively charged nitrogen atom; and the other active site has a positive charge to hold the carbonyl ($\diagdown$ C=O) group. The *nerve gas* diisopropylfluorophosphate (DFP, Fig. 9.5c) is bound very tightly to the latter active site, preventing the hydrolysis of acetylcholine and thus the repolarization of the neuron. DFP, vaporized in a nerve-gas attack, can be absorbed by inhalation or through the skin or eyes, and death follows in seconds or minutes because of over-stimulation and paralysis of respiratory muscles. Substances such as DFP, which interfere with enzyme action, are often lethal in extremely minute amounts.

An effective antidote for DFP is 2-pyridine aldoxime meth-iodide (PAM, Fig. 9.5*d*). It displaces DFP from the esterase but must be administered immediately after exposure to DFP. Such *chemical warfare agents* are cheaper to make than nuclear weapons and therefore universally available to nations. However, the horror of introducing still another means of widespread killing and the uncertainties attending the delivery of a lethal vapor have led to an almost universal renunciation of this kind of weapon.

EQUILIBRIUM

If we place pure nitrogen and oxygen gas, in equal amounts, in a container at 2000 K (1727°C) and analyze the mixture, we shall find that about 1 percent of the gas is nitric oxide. Now, if we place pure nitric oxide in the container at the same temperature, analysis will show that most of the nitric oxide decomposes into its elements and, again, about 1 percent of the gas mixture will be nitric oxide. Whether we start with equal amounts of the elements or with pure nitric oxide, the composition of the resultant mixture is the same.

We can begin with various amounts of nitrogen and oxygen or nitric oxide and analyze the composition after reaction or decomposition has occurred (Table 9.2). (Since nitric oxide is quite reactive, other compounds, such as nitrogen dioxide NO_2, will be present. For simplicity we shall ignore these. Even so, nitrogen dioxide is an important air pollutant.) In Table 9.2*a*, if the initial pressure due to nitrogen alone is 1 atm and the initial pressure due to oxygen alone is also 1 atm, the total pressure in the container is 2 atm. Notice that the final pressure in *a* (the sum of the contributions from nitrogen, oxygen, and nitric oxide) is also 2 atm. There is no pressure change because there is no change in the total number of molecules during the reaction. One nitrogen and one oxygen form two nitric oxide molecules. In part *d*, beginning with pure nitric oxide, the final mixture is exactly as in *a*. Other initial and final compositions are given in *b*, *c*, and *e*.

Table 9.2. Nitrogen-Oxygen Reaction at 2000 K

		$N_2 + O_2 \longrightarrow 2NO$			
Substance	N_2	O_2	NO	$\dfrac{[NO]^2}{[N_2][O_2]}$ at equilibrium	
Initial pressure, atm	1.0000	1.0000	0		
a. Final pressure	0.9901	0.9901	0.0198	4.00×10^{-4}	
Change in pressure	-0.0099	-0.0099	$+0.0198$		
Initial pressure	10.00	1.000	0		
b. Final pressure	9.97	0.969	0.0622	4.00×10^{-4}	
Change in pressure	-0.03	-0.031	$+0.0622$		
Initial pressure, atm	1.000	10.00	0		
c. Final pressure	0.969	9.97	0.0622	4.00×10^{-4}	
Change in pressure	-0.031	-0.03	$+0.0622$		

		$2NO \longrightarrow N_2 + O_2$		
Substance	NO	N_2	O_2	
Initial pressure, atm	2.0000	0	0	
d. Final pressure	0.0198	0.9901	0.9901	4.00×10^{-4}
Change in pressure	-1.9802	$+0.9901$	$+0.9901$	
Initial pressure, atm	10.000	0	0	
e. Final pressure	0.099	4.950	4.950	4.00×10^{-4}
Change in pressure	-9.901	$+4.950$	$+4.950$	

Because some nitric oxide can decompose into its elements, and because nitrogen and oxygen can form a little nitric oxide, both reactions

$$2NO \longrightarrow N_2 + O_2$$
$$N_2 + O_2 \longrightarrow 2NO$$

must occur. When a reaction can occur in either direction — reactant-to-product and product-to-reactant — we say it is *reversible*. Both reactions may be combined and written as

$$N_2 + O_2 \rightleftharpoons 2NO$$

In the discussion of vapor pressure (Chap. 3) we saw that the constant value of the pressure of a vapor over a liquid at a fixed temperature can be described as a balance, or *equilibrium*, between two opposing processes: evaporation and condensation. We can look at the reversible nitrogen-oxygen reaction in this light, as a balance between the forward and reverse reactions.

Using the mass-action law, we can write expressions for the

rate in each direction. We can measure the amount of each of the three substances as a concentration (moles/liter or molecules/cm³) or as a pressure since one is related to the other. For generality we can designate the amount of nitrogen as $[N_2]$ and not specify how it is measured. Similarly, $[O_2]$ and $[NO]$ denote the amounts of oxygen and nitric oxide.

The rate of the forward reaction

$$N_2 + O_2 \xrightarrow{k_1} 2NO$$

depends on the number of nitrogen-oxygen encounters and on the chance of reaction at collision. By the mass-action law, this rate is

$$R_1 = k_1[N_2][O_2]$$

The factor k_1 incorporates all the necessary information, other than the product $[N_2][O_2]$, about the collision frequency and chance of reaction, such as molecular velocities and sizes and activation energies. The rate of the reverse reaction is

$$2NO \xrightarrow{k_2} N_2 + O_2$$

$$R_2 = k_2[NO]^2$$

The factor k_2 is different from k_1 because, for one thing, the energies of activation differ for the two processes. For the nitric oxide encounters, the activation energy is about 64 kcal/mole, and for nitrogen-oxygen collisions about 107 kcal/mole. This means at 2000 K one collision in 10 million between nitric oxides will result in decomposition to nitrogen and oxygen, but only one collision in 2 trillion between nitrogen and oxygen will be successful in forming nitric oxide.

Suppose we start with only nitrogen and oxygen. Initially there is no nitric oxide, so

$$[NO] = 0 \quad \text{and} \quad R_2 = k_2[NO]^2 = 0$$

There is no reverse reaction. As nitric oxide is formed, R_2 has a nonzero value. The more nitric oxide, the greater the rate R_2. At the same time, as nitrogen and oxygen are consumed, both $[N_2]$ and $[O_2]$ decrease in value; so must R_1

$$R_1 = k_1 [\text{N}_2][\text{O}_2]$$

The forward rate R_1 will progressively decrease and the reverse rate R_2 will progressively increase until they are equal:

$$R_1 = R_2$$

This means then the rate at which nitric oxide is formed is exactly equal to the rate at which it decomposes. There is no net change in the amount of nitric oxide or in the amounts of oxygen and nitrogen. The composition has reached an equilibrium.

From this last equality, and the mass-action laws,

$$k_1 [\text{N}_2][\text{O}_2] = k_2 [\text{NO}]^2$$

and

$$\frac{k_1}{k_2} = \frac{[\text{NO}]^2}{[\text{N}_2][\text{O}_2]} \quad \text{or} \quad K = \frac{[\text{NO}]^2}{[\text{N}_2][\text{O}_2]}$$

The symbol K, the ratio k_1/k_2, is the *equilibrium constant* for this reaction. Since k_1 and k_2 change with temperature, so must K.

The numerical value of the constant for this reaction at 2000 K can be determined from the equilibrium concentrations or pressures of the three substances. The last column of Table 9.2 lists these values for the five examples. As predicted, the values are the same for each example in spite of the differing initial compositions and differing final equilibrium compositions.

The equilibrium equation

$$K = \frac{[\text{NO}]^2}{[\text{N}_2][\text{O}_2]} = 4.00 \times 10^{-4}$$

can be used to determine from the composition of a mixture of these gases in which direction the reaction will go. If the concentrations or pressures are inserted into this expression, the calculated value can be compared with the equilibrium value.

Example. A mixture of nitrogen, oxygen, and nitric oxide has this composition:

$$[N_2] = 5 \times 10^{-3} \text{ atm}$$
$$[O_2] = 1 \times 10^{-3} \text{ atm}$$
$$[NO] = 4 \times 10^{-8} \text{ atm}$$

Will the reaction

$$N_2 + O_2 \rightleftharpoons 2NO$$

with this composition for each substance produce more nitric oxide or more nitrogen and oxygen?

The evaluation of the expression

$$\frac{[NO]^2}{[N_2][O_2]} = \frac{(4 \times 10^{-8})^2}{(5 \times 10^{-3})(1 \times 10^{-3})}$$
$$= 3.2 \times 10^{-10}$$

Comparing this with the equilibrium value K,

$$\frac{[NO]^2}{[N_2][O_2]} = 3.2 \times 10^{-10} < 4.00 \times 10^{-4}$$

The calculated number is too small; i.e., the numerator is too small and the denominator too large. There is too little nitric oxide and too much nitrogen and oxygen for equilibrium. Therefore some more of these elements will combine, making more nitric oxide and less nitrogen and oxygen. In this way the composition is adjusted until the calculated expression exactly equals the equilibrium value.

The equilibrium constant K for any reaction is calculated by inserting the equilibrium concentrations or pressures of all the reactants and products into an expression which we can call Q. This expression, as in the nitrogen-oxygen reaction, is a fraction in which the numerator has product concentrations raised to appropriate powers and the denominator has reactant concentrations to appropriate powers. In the last example Q was momentarily less than K but increased until $Q = K$, that is, until equilibrium was attained.

A general rule can be formulated. From the concentrations, or pressures, of the several substances, Q is evaluated and compared to K for the reaction:

1. If $Q < K$, more reactants will be used up and more products formed, increasing Q until it equals K.

2. If $Q > K$, some product is consumed to form more reactants (the reaction is reversed), decreasing Q until it equals K.
3. If $Q = K$, the reaction is at equilibrium and the concentrations of all the substances remain unchanged (see Prob. 9.1).

A catalyst or an enzyme lowers the activation energy of both the forward and reverse steps in a reversible reaction, (Fig. 9.4). It speeds up both steps equally in such a way that the equilibrium constant is not altered. Thus a catalyst will accelerate the rate at which a reaction reaches equilibrium but will not influence the amount of reactant converted at equilibrium.

IONIZATION EQUILIBRIUM IN SOLUTION

The concentration of ions in solution can be determined by several techniques. The electrical conductivity of a solution depends on the number of ions present. The voltage between specially prepared electrodes in contact with a solution varies with the hydronium-ion concentration $[H_3O^+]$. The color of indicators, including some common vegetable juices, also changes with hydronium-ion concentration.

Acetic acid (molecular weight 60.05) ionizes when mixed with water. If $\frac{1}{10}$ mole (6.005 g) is dissolved in 1 liter of water, the concentration of acetic acid is 0.1000 mole/liter but the hydronium-ion concentration is only 0.00132 mole/liter. From this, 1.32 percent of the acid molecules are ionized:

$$CH_3COOH + H_2O \longrightarrow CH_3COO^- + H_3O^+$$

Data at this and other concentrations are given in Table 9.3. As acetic acid is diluted from 0.1 to 1.00×10^{-5} moles/liter, the percent ionization increases from 1.32 to 71.1 percent. At each concentration there is a definite equilibrium for ionization. Dissociation does not continue until 100 percent ionization in the five examples of Table 9.3.

We can apply the law of mass action to this process. Collisions in liquids are more complicated than in gases because molecules are always in contact in the former. Nevertheless, the chance of an encounter between two species can be assumed to be propor-

Table 9.3. Ionization of Acetic Acid at 25°C

		Concentrations			
$[CH_3COOH]_{total}$	$[H_3O^+]$	$[CH_3COO^-]$	$[CH_3COOH]$	$\dfrac{[CH_3COO^-]}{[CH_3COOH]_{total}}$	$\dfrac{[H_3O^+][CH_3COO^-]}{[CH_3COOH]}$
1.00×10^{-1}	1.32×10^{-3}	1.32×10^{-3}	9.87×10^{-2}	0.0132	1.76×10^{-5}
1.00×10^{-2}	4.10×10^{-4}	4.10×10^{-4}	9.60×10^{-3}	0.0410	1.75×10^{-5}
1.00×10^{-3}	1.24×10^{-4}	1.24×10^{-4}	8.76×10^{-4}	0.124	1.76×10^{-5}
1.00×10^{-4}	3.40×10^{-5}	3.40×10^{-5}	6.60×10^{-5}	0.340	1.75×10^{-5}
1.00×10^{-5}	7.11×10^{-6}	7.11×10^{-6}	2.89×10^{-6}	0.711	1.75×10^{-5}

In the ionization of acetic acid in pure water $CH_3COOH + H_2O \longrightarrow CH_3COO^- + H_3O^+$, each acid molecule that dissociates forms one acetate and one hydronium ion. Thus, the hydronium and acetate ions are always present in equal concentrations: $[H_3O^+] = [CH_3COO^-]$. The undissociated acetic acid molecule concentration $[CH_3COOH]$ is simply the difference between the total amount added and the amount present as ions: $[CH_3COOH] = [CH_3COOH]_{total} - [CH_3COO^-]$.

tional to the product of their concentrations. The rate of ionization, then, is

$$R_1 = k[CH_3COOH][H_2O]$$

But because there are 55.5 moles of water in a liter, the ionization of so little acid cannot significantly affect the water concentration $[H_2O]$. It is essentially constant, as is $k[H_2O]$, which can be replaced by the constant k_1; then,

$$R_1 = k_1[CH_3COOH]$$

We can interpret k_1 as the probability that a single acetic acid will ionize; multiplied by the concentration, $k_1[CH_3COOH]$ is the actual rate of ionization.

Because ionization is not complete, the reverse process, the recombination of ions, must be considered. Its rate is

$$R_2 = k_2[CH_3COO^-][H_3O^+]$$

At equilibrium there is no net change in the concentration of any of these three species, and so

$$R_1 = R_2$$
$$k_1[CH_3COOH] = k_2[CH_3COO^-][H_3O^+]$$

and

$$\frac{k_1}{k_2} = K = \frac{[CH_3COO^-][H_3O^+]}{[CH_3COOH]}$$

To test the correctness of this equilibrium

$$CH_3COOH + H_2O \rightleftharpoons CH_3COO^- + H_3O^+$$

we must know the concentrations of the three species at several concentrations (Table 9.3). The constancy of the values of Q indicates the mass-action law is a good description of the relation between percent ionization and concentration.

From the last equilibrium equation we can see why the percent ionization increases with dilution. Suppose equilibrium is established at the first concentration:

$$[H_3O^+] = 1.32 \times 10^{-3}$$
$$[CH_3COO^-] = 1.32 \times 10^{-3}$$
$$[CH_3COOH] = 9.87 \times 10^{-2}$$

Next, if the solution is abruptly diluted tenfold, the momentary concentration of each drops to one-tenth these, or

$$[H_3O^+] = 1.32 \times 10^{-4}$$
$$[CH_3COO^-] = 1.32 \times 10^{-4}$$
$$[CH_3COOH] = 9.87 \times 10^{-3}$$

Putting these concentrations into the equilibrium equation, we find the value of Q to be less than K. (The numerator has been reduced to $\frac{1}{10} \times \frac{1}{10}$, or $\frac{1}{100}$, but the denominator only to $\frac{1}{10}$.) More undissociated molecules ionize until Q again equals K, that is, until equilibrium is again established.

Weak acids are only partially ionized; but as their concentration is decreased by dilution, the recombination process is slowed down much faster than the ionization step. With more dilution, ionization dominates. At very high dilution, even weak acids are completely ionized.

LE CHATELIER'S PRINCIPLE

For the nitrogen-oxygen reaction,

$$K = \frac{[NO]^2}{[N_2][O_2]}$$

If the volume is abruptly increased tenfold, then [NO] momentarily drops to one-tenth, and $[NO]^2$ to one-hundredth its equilibrium value. At the same time, each $[N_2]$ and $[O_2]$ drop to one-tenth, so the product $[N_2][O_2]$ drops to one-hundredth its equilibrium value. Thus, the forward and reverse reactions are equally affected, that is, reduced to one-hundredth their former values. The ratio $[NO]^2/[N_2][O_2]$ is unchanged, and so Q remains equal to the equilibrium value K. Provided the temperature is fixed, volume changes will not shift the nitrogen-oxygen reaction in either direction.

The reversible reaction in which nitrogen dioxide forms nitrogen tetroxide

$$NO_2 + NO_2 \rightleftharpoons N_2O_4$$

has the equilibrium constant

$$K = \frac{[N_2O_4]}{[NO_2]^2}$$

Its value is 1.15×10^{-6} at 1000 K. This reaction is important because both the dioxide and tetroxide are air pollutants. The larger is colorless and the smaller is reddish-brown; this is responsible for the whisky-brown-colored haze often seen in polluted air. Nitrogen dioxide has a total of 23 electrons. Like most molecules with an odd number of electrons, it is very reactive (corrosive). In water it forms nitric and other acids.

From the equilibrium constant, if

$$[NO_2] = 10^{-3} \text{ atm}$$

then

$$[N_2O_4] = K[NO_2]^2$$
$$= 1.15 \times 10^{-12} \text{ atm}$$

Now, suppose the volume of air containing these nitrogen oxides is compressed by one-half, so

$$[NO_2] = 2 \times 10^{-3} \text{ atm}$$

Then, at equilibrium,

Table 9.4. Influence of Dilution on the Extent of Reaction in Reversible Processes

Reaction	Net change in number of molecules	Shift in equilibrium when concentration or pressure is	
		Increased	Decreased
$N_2 + O_2 \rightleftharpoons NO + NO$	None	None	None
$CH_3COOH + H_2O \rightleftharpoons CH_3COO^- + H_3O^+$ or $CH_3COOH \rightleftharpoons CH_3COO^- + H^+$	+1 (water essentially constant)	To left, less ionization	To right, more ionization
$NO_2 + NO_2 \rightleftharpoons N_2O_4$	−1	To right, more association	To left, more dissociation

$[N_2O_4] = 4.6 \times 10^{-12}$ atm

When the volume is decreased, the nitrogen dioxide pressure is doubled but the nitrogen tetroxide pressure is quadrupled. There is relatively more N_2O_4 at higher pressures and relatively more NO_2 at lower pressures.

Having looked at three reversible reactions (Table 9.4) and especially at how the extent of a reaction is affected by pressure or concentration, we can try to extract a general rule. Solute concentration can be increased or decreased by solvent removal or addition, and gas pressure can be increased or decreased by volume contraction or expansion. A dilution of solution is equivalent to a decrease in gas pressure in that both are attended by fewer collisions and slower reactions.

From Table 9.4, the equilibrium is shifted with a concentration or pressure change if the reaction is accompanied by a net change in the total number of molecules. When the concentration or pressure is increased, the equilibrium is shifted in the direction resulting in fewer molecules, i.e., resulting in a lower total concentration or pressure. Conversely, if a solution is diluted or a gas pressure decreased, the number of molecules per unit volume is reduced. But the equilibrium is shifted in the direction producing more molecules or ions. In general, the equilibrium is shifted to the direction which will tend to restore the previous overall concentration or total pressure.

The influence of temperature on equilibrium can be seen from the data in Table 9.5. As the temperature increases, one equilibrium constant becomes larger and the other smaller. For the first reaction, an increase in the equilibrium constant means relatively more nitric oxide NO will be formed. In turn, this means the

Table 9.5. Equilibrium and Temperature

Reaction	$N_2 + O_2 \rightleftharpoons 2NO$	$NO_2 + NO_2 \rightleftharpoons N_2O_4$
Forward	$N_2 + O_2 \longrightarrow 2NO$ Endothermic; 43,200 cal	$NO_2 + NO_2 \longrightarrow N_2O_4$ Exothermic; 12,000 cal
Reverse	$2NO \longrightarrow N_2 + O_2$ Exothermic; 43,200 cal	$N_2O_4 \longrightarrow 2NO_2$ Endothermic; 12,000 cal

T, K	$K_{equil} = [NO]^2/[N_2][O_2]$	$K_{equil} = [N_2O_4]/[NO_2]^2$
500	2.72×10^{-18}	6.81×10^{-4}
1000	7.52×10^{-9}	1.15×10^{-6}
1500	1.06×10^{-5}	1.85×10^{-7}
2000	4.00×10^{-4}	8.73×10^{-8}

forward (endothermic) step is accelerated more than the reverse (exothermic) step. In the second reaction, the equilibrium constant decreases, corresponding to a greater acceleration of the reverse (endothermic) process.

Since an endothermic process takes in heat, it seems reasonable that it will go faster at higher temperatures. Exothermic reactions are also speeded up but relatively not as much. The general rule is that an increase in temperature will push the equilibrium in the direction of the endothermic step. A decrease in temperature will shift the equilibrium in favor of the exothermic step.

Henri Louis LeChatelier (1850–1936) combined these rules into the principle named for him: *When stress is applied to a system in equilibrium, the equilibrium shifts in the direction which will undo the result of the stress.* When the temperature is increased, the endothermic step is enhanced, absorbing heat; when the temperature is lowered, the exothermic step is promoted, liberating heat. When the total pressure is increased, the equilibrium shifts to decrease the total number of molecules and, with it, the pressure; when the total pressure is decreased, the equilibrium shifts to increase the total number of molecules and the pressure.

EQUILIBRIUM DISTRIBUTION AND SEPARATION

Equilibrium as a dynamic balance between two opposing processes has been used to describe vapor pressures and reversible chemical reactions. In this section we shall examine one more application of this concept to see how a material can be separated into its pure substances.

Water and ether are *immiscible*, that is, essentially insoluble in one another. In a separatory funnel (Fig. 9.6*a*) the water layer, being more dense, is at the bottom and the ether layer is at the top. Suppose a substance that can be dissolved in both liquids is shaken up in the separatory funnel containing some ether and water. The two liquid layers can be separately withdrawn, and the concentration of that substance determined in both water and ether. No matter how much of the solute and two solvents is put into the separatory funnel, it is common to find the ratio of the two concentrations to be a constant number.

We can designate the substance concentration in water as $[S]_w$ and the concentration in ether as $[S]_e$. The rate at which this substance leaves the water layer and goes into the ether layer depends on how many molecules in the water layer can reach the water-ether boundary. The rate, then, varies with $[S]_w$ and is

$$R_1 = k_1[S]_w$$

Similarly, the rate of molecules going in the opposite direction (from ether to water) is

$$R_2 = k_2[S]_e$$

The coefficients k_1 and k_2 depend on several things: the speed of a molecule in its solvent, the energy needed to cross the boundary, and so on. At a given temperature they are constant but, in general, not equal to each other.

When the two concentrations have reached steady values, there is no more net flow of substance. The two rates are equal:

$$R_1 = R_2$$

so

$$k_1[S]_w = k_2[S]_e \quad \text{and} \quad \frac{[S]_w}{[S]_e} = \frac{k_2}{k_1} = K$$

Here, the ratio K is called the "Nernst distribution coefficient" after Walther Nernst (1864–1941), a pioneer physical chemist.

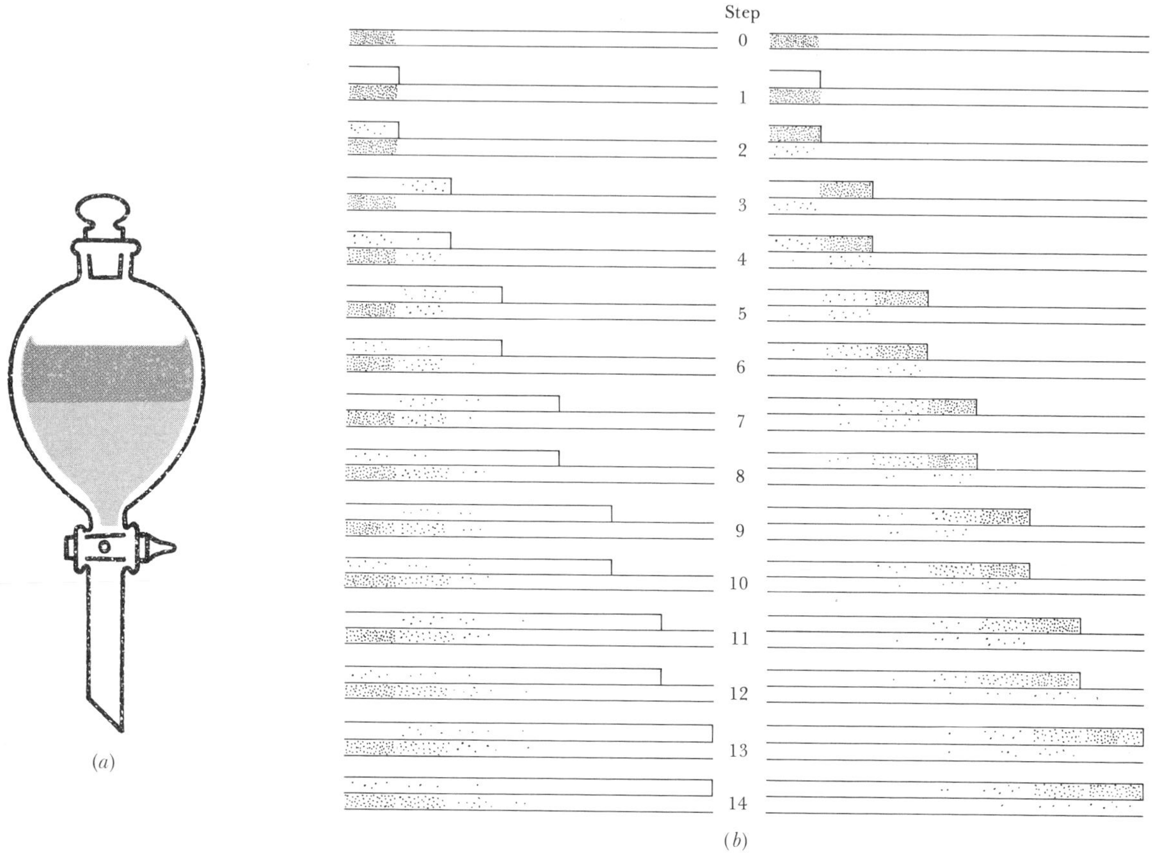

Figure 9.6. (a) *Separatory funnel containing two immiscible liquids; (b) In step 0, a pure substance is deposited on some stationary surface, e.g., paper, aluminum oxide, et al. Then (step 1) a solvent passes slowly over that surface and (step 2) a redistribution of the substance between solvent and stationary surface occurs. In each odd-numbered step the solvent flows along the surface, disturbing the equilibrium distribution; and in each even-numbered step a new equilibrium (Nernst distribution) is attained. In the example at the left the distribution at equilibrium is about 1:10, solvent:surface. At the right it is about 10:1, solvent:surface. The two substances move along the chromatographic stationary surface at different rates and are separable.*

Each substance has its own characteristic distribution coefficient for a pair of immiscible solvents. In a mixture, some substances will be more concentrated in ether and others more concentrated in water. The practically complete removal of substances from one liquid can be effected by multiple extraction. For example, 90 percent of a substance goes into ether, and 10 percent remains in water; the two liquid layers are separated. Then the water layer is shaken with fresh ether. The distribution ratio is still

9.1, of the 10 percent remaining, 9 percent goes into ether and 1 percent stays in water. Combining the two ether layers, 99 percent of the original substance is contained in ether and only 1 percent is contained in water. Extracting this again with fresh ether, 99.9 percent can be removed, leaving 0.1 percent in water.

Slight differences in Nernst distribution coefficients can be used, with multiple extractions, to separate a variety of substances. In some cases, one or more of the original substances can be made insoluble in one of the liquids. Benzaldehyde C_6H_5CHO, found in the kernel of bitter almonds, is oxidized in air to benzoic acid C_6H_5COOH. Both these aromatic substances are slightly soluble in water and very soluble in ether. To obtain pure benzaldehyde, an aqueous solution of these is made alkaline, converting the acid to a salt, which is not soluble in ether. Shaking with ether will remove only benzaldehyde, and then the ether can be evaporated, leaving the pure aldehyde.

Chromatography, an elaborate and convenient application of distribution equilibrium, is widely used for separation (Fig. 7.55). In Fig. 9.6b a simple model illustrates how two substances with different distribution coefficients travel along a chromatographic column, or layer, at different rates.

THE STEADY STATE

When hot and cold objects are placed in contact with each other, the temperature of the first drops and that of the second rises until they both are at the same temperature. It is a general empirical observation that heat will flow spontaneously from a warmer to a colder region. Heat is the random kinetic energy of the atoms and molecules in an object; the hotter the body, the greater the kinetic energy. Because the hotter molecules move more rapidly, they transfer energy faster to the cooler region than the colder molecules transfer energy to the warmer region. Heat, thus, flows (or kinetic energy is transferred) in the observed direction until the temperature (or average kinetic energy) is uniform and unchanging.

We can say that *thermal equilibrium* is established when there is no net transfer of kinetic energy in either direction. Here, too, equilibrium is a balance between two opposing processes—the transfer of energy in both directions between objects or regions within an object.

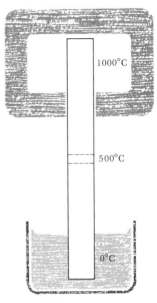

Figure 9.7. Steady-state temperature. The upper end of the metal bar is within a furnace and kept at a constant temperature of 1000°C. The lower end is within an ice-water bath and kept at 0°C. There is a continuous temperature drop along the bar from the hot upper end to the cold lower end.

Our general concept of equilibrium is a balance between opposing processes. Our indication that equilibrium has been attained is the constancy of concentrations or temperatures. However, the indication may not always be reliable.

Consider a metal bar with one end kept at the ice point and the other end kept at some high temperature (Fig. 9.7). A thin section midpoint between these extremes in temperature will have a temperature approximately the average of the high and low temperatures that will not change with time as long as the hot and cold ends are kept at constant temperatures. However, we cannot say this thin section is in thermal equilibrium in spite of its constant temperature. There are no opposing energy-transfer processes, but rather there is a net flow from the hot end to the cold end. The heat flowing into this section (the *influx* from the hot end) is exactly equal to the heat flowing out of this section (the *efflux* to the cold end). Because there is no net accumulation or loss of heat in the midpoint section, its temperature is steady.

The radioactive decay of U^{238} and its immediate daughter Th^{234} have half-lives of 4.51×10^9 yr and 24.10 days, respectively. The thorium decay rate is enormously faster than the rate at which it is produced, and therefore the amount of Th^{234} found with natural uranium can never be very high. The ratio of Th^{234} to U^{238} is the ratio of their half-lives 1.46×10^{-11}. Since uranium decays so slowly, the amount present in any sample is essentially constant during any practical period of observation. Since the ratio is fixed, the number of Th^{234} nuclei present is also essentially constant. Yet, we know this is not an equilibrium because there is no opposing thorium-to-uranium process. We also know that, eventually, all the U^{238} will be used up and, with it, all the Th^{234}. The steady number of Th^{234} nuclei in any U^{238} sample and in the entire world supply of this uranium isotope is a consequence of this formation and decay balance, with no net accumulation of Th^{234}.

The temperature of the sun is relatively steady. There is a balance between the energy produced by thermonuclear reactions and the energy lost by radiation. This, too, is not equilibrium because the reverse of the thermonuclear reactions is thought not to occur. We know that the material within the sun needed for the thermonuclear reactions will eventually be exhausted, and, with this, the solar temperature will decline. But for a very long time, the solar temperature will remain steady.

Water at 100.00°C and enclosed in a tight vessel will be in

equilibrium with a vapor at exactly 1 atm pressure. If the vessel is opened, the vapor can push back the ambient atmosphere and escape. More liquid can boil off, but the molecules evaporating are the most energetic. When they leave, the average kinetic energy (and the temperature) of the liquid drops. Boiling will cease unless additional energy is supplied to the liquid.

Since 539.55 cal are needed to evaporate 1 g of water, the addition of any amount of heat to water at its boiling point will result in the evaporation of a proportionate amount of liquid. And the vapor carries off exactly the energy added; the greater the heat put in, the more water boiled off and the more energy with it. Boiling is not an equilibrium process because the energy flux is in one direction. The temperature of the boiling liquid remains steady because there is neither an accumulation nor a decrease in the average kinetic energy of the liquid molecules.

The four cases just discussed are examples of a *steady state* rather than of equilibrium. At true equilibrium the composition and temperature of the universe should be uniform throughout. The simplest amino acid, glycine, in the presence of oxygen is completely converted to carbon dioxide, water, and nitrogen at equilibrium. Life is impossible under these conditions. Fortunately, the approach to equilibrium is often extremely slow. Glycine may be exposed to oxygen for a very long time without being completely oxidized to carbon dioxide, water, and nitrogen. Both the slowness of certain reactions and the steady state make possible those circumstances needed for life.

DIFFUSION

In discussing osmosis we saw that water spontaneously flows across a membrane from a less to a more concentrated solution. We can also describe this by saying water flows from a region of high-water concentration to a region of low-water concentration. The membrane serves as a barrier to prevent the solute from flowing in the opposite direction. If the membrane were not present, both flows (solvent and solute) would occur.

Whenever there is an unequal concentration distribution of a substance and no barrier to its free movement, there will be a spontaneous redistribution of that substance until its concentration is uniform throughout. This spontaneous process is called "diffusion." We can observe diffusion when a lump of sugar is dropped into a glass of water. After the disturbance of introducing the lump has subsided, we see patterns, or fringes, spreading up from the lump into the liquid. The refraction, or bending of light, changes with sugar concentration; differences in the refraction cause the patterns to be visible. Diffusion continues until the sugar concentration is uniform throughout—when the refraction is also uniform—and the patterns disappear. A drop of ink carefully introduced into clear water can also be seen to diffuse until the color is uniform.

Figure 9.8*a* is a simple cell designed to study diffusion. A concentrated solution is contained within the central section and, at first, separate from the outer sections of pure solvent. When the three channels are aligned, the solute can diffuse in both directions into the outer sections. The concentration at any distance along the cell and at any time can be determined by measuring some special property of the solute, e.g., absorption spectrophotometry (if colored) or electrical conductivity (if ionic). Figure 9.8*b* is a plot of the concentration profile during diffusion.

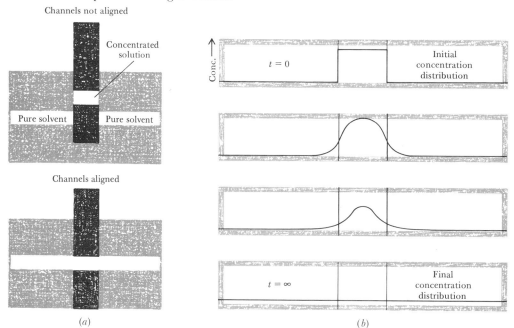

Figure 9.8. (a) Simple cell for studying diffusion. The middle section can be moved so that the three sections are aligned or not aligned. The middle section is filled with a concentrated solution and the two outer sections with pure solvent. Upon alignment diffusion begins. (b) A typical series of concentration profiles of the solute along the cell during diffusion. Initially it is confined within the middle section. At the end it is distributed uniformly along the entire cell.

An especially simple model can be used to describe this diffusion. Imagine a set of molecules allowed to move either back or forth, in one dimension only, and in a completely random way. The toss of a coin determines whether a particular molecule moves in one or the opposite direction. The entire set of molecules will move about in what is called a *random walk.*

Figure 9.9*a* is a *probability tree.* The symbols H and T represent a head and tail, respectively, in the coin tossing. All the molecules start at the origin. After one toss, they end up at either H or T; and, after two tosses, they end up at either H^2, $2HT$, or T^2. The symbol H^2 stands for HH: two consecutive heads and two consecutive advances in the positive direction. The term $2HT$ stands for HT and TH, the two ways to return to the origin, as indicated by the arrows into the $2HT$ position. The term $3HT^2$, in three coin tosses, indicates the three routes from the origin to the position $-d$, that is HTT, *THT,* and *TTH.* In general, in this model, a head advances a molecule one step in the positive direction and a tail moves a molecule one step in the negative direction. In Fig. 9.9*b,* the terms have been evaluated, assigning equal probabilities to heads and tails: $H = T = \frac{1}{2}$. The arrows indicate how the molecules are redistributed after each toss.

This *random-walk model* describes how molecules diffuse. When a nonuniform concentration exists, no special forces are needed for diffusion. It is simply a consequence of the randomness of molecular motion. Because the model is successful, this is one more corroboration of the theory that matter comprises small, rapidly moving molecules. The random-walk model can be used even when there is a bias in the coin, that is, when $H \neq T$. This corresponds to a preference for motion in one direction (Prob. 9.3) such as diffusion of ions in an electric field, or molecules and ions in a gravitational field.

THE SECOND LAW OF THERMODYNAMICS

Diffusion from regions of higher concentrations to regions of lower concentrations is a spontaneous process. The random-walk model describes diffusion in terms of chance, applied many times to each of many molecules. The conservation of energy plays no explicit part in this model. There is no restriction arising from the first law of thermodynamics preventing solute molecules, in a solution of

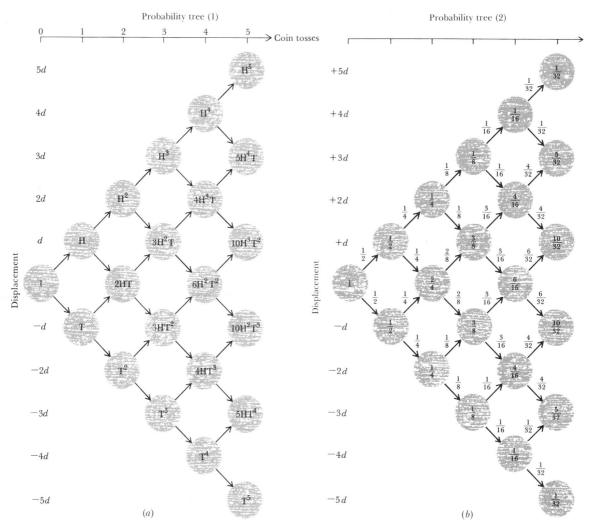

Figure 9.9. Diffusion, random-walk model, and a probability tree. (a) Starting at the origin (zero displacement and before a coin is tossed) a head (H) will move a molecule one step in the positive direction; and a tail (T) one step in the negative direction. After three tosses some molecules are displaced to 3d (by three heads, H³); some, to d (by two heads and one tail, in any order, 3H²T = HHT + HTH + THH), etc. (b) The probability tree evaluated with H = T = ½, which are the probabilities in an unbiased coin. After one toss, ½ the molecules will be at d, and the other ½ at −d. After two tosses, ¼ will be at 2d, ½ at the origin, and ¼ at −2d, etc. (c) The fractional distribution of molecules, according to the random-walk model, after 1, 2, 3, 4 and 5 coin tosses, using the numerical values of (b). Notice how these shapes resemble those in Figure 9.8b. (d) The binomial theorem for (H + T)ⁿ, the multiplication of H + T by itself n times (also see Figure 8.15). (e) Pascal's triangle for generating the binomial coefficients in (a) and (d). Any number with this triangle is the sum of the two numbers just above, one to the right and one to the left; thus 6 = 3 + 3.

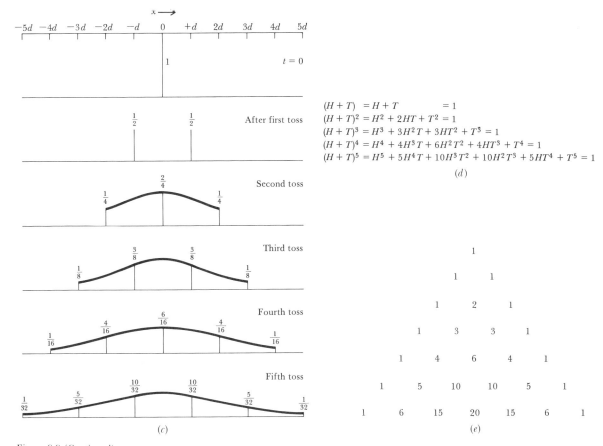

$$(H + T) = H + T = 1$$
$$(H + T)^2 = H^2 + 2HT + T^2 = 1$$
$$(H + T)^3 = H^3 + 3H^2T + 3HT^2 + T^3 = 1$$
$$(H + T)^4 = H^4 + 4H^3T + 6H^2T^2 + 4HT^3 + T^4 = 1$$
$$(H + T)^5 = H^5 + 5H^4T + 10H^3T^2 + 10H^2T^3 + 5HT^4 + T^5 = 1$$

(*d*)

Figure 9.9 (Continued).

uniform concentration, from congregating in one region of the fluid, producing high- and low-concentration regions. But it is extremely improbable that this unmixing will occur.

Similarly, if two vessels, one containing oxygen and the other nitrogen, are connected, the natural process is the spontaneous mixing, or diffusion, of the two gases. The probability that there will be any measurable unmixing in a finite time is extremely small.

We have already seen that a polymer chain above the melting point can assume any of a great number of conformations, only one of which is planar zigzag. It is extremely unlikely that a coiled polymer chain will straighten out spontaneously above its melting point.

Suppose we take a slab of metal and heat one end and cool the other. Then we can immediately place this metal in an insulated container so that heat can neither enter nor leave the metal slab. We can also arrange the container-and-slab system so that the temperature at both ends of the slab can be measured over a period of time (Fig. 9.10). We observe that the temperature of the hotter end decreases and the temperature of the cooler end increases until both temperatures are equal. Initially, the atoms at the hotter end had a greater average kinetic energy than those at the cooler end; the atoms at the hotter end, on the average, moved faster than those at the cooler end. At thermal equilibrium, when the tempera-

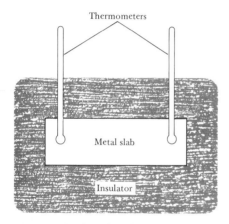

Figure 9.10. Thermal equilibrium as a natural process. The metal slab is heated at one end and cooled at the other. It is promptly placed in a well-insulated container. The temperature of the hot end drops and that of the cold end increases until the two ends are at the same temperature. The temperature throughout the entire slab is then uniform. This is a universally observed phenomenon: heat flows spontaneously from a warmer to a cooler region. Since there is no heat exchange between the slab and the surroundings during the approach to thermal equilibrium, this is an adiabatic process. No matter how long we wait, the temperature will remain uniform, providing nothing else is done to the slab. There will not be a spontaneous shift of heat, making one part of the slab hotter than another. The first law of thermodynamics is obeyed during the approach to thermal equilibrium, and it would also be obeyed during an unnatural and unseen reverse process. Therefore, the first law cannot predict that thermal equilibrium is the natural end toward which this process should go.

ture is uniform throughout the slab, the average atomic speed is the same at both ends of the slab.

Since the slab was insulated, there was no energy change in the slab during the process of heat transfer from the hotter end to the cooler end. If we make very careful measurements and perform very precise calculations, we find that the heat lost by the hotter end *exactly* equals the heat gained by the cooler end of the slab.

If, after thermal equilibrium has occurred, we wait a very long time, we *shall not* observe any difference in the temperatures of the two ends. It is extremely unlikely that the fastest-moving atoms will be found at one end of the slab and the slowest-moving atoms at the other end. Kinetic energy is transferred from one atom to the next by collision. These collisions are random processes which tend to make the local average kinetic energy the same in all regions of the slab. Although there are no energy restrictions on any one region accumulating more kinetic energy than neighboring regions in the slab, it is extremely unlikely that this will occur.

The transfer of heat from hotter regions to cooler regions can be described by the diffusion or *transport* of energy from a region of higher average energy to one of lower average energy. Our experience, consistently, is that heat spontaneously flows from a hotter region to a cooler region; heat *will not* spontaneously flow from a cooler region to a hotter region *unless* we perform some work. For example, we can pump heat from inside a relatively cool house to a hotter outside by means of an air-conditioner; we must do work to accomplish this. The natural process is for heat to flow into the house.

Since the first law of thermodynamics does not prohibit the spontaneous unmixing of a uniform mixture or the spontaneous flow of heat from a cooler region to a hotter region, another rule, the *second law of thermodynamics,* is necessary to describe which processes may occur naturally. The second law can be stated in several equivalent forms:

1. Heat does not flow naturally from a cooler region to a warmer region; work must be done to accomplish this.
2. A heat engine cannot exist which completely converts heat taken from a hot reservoir into work; some heat must be returned to another cooler reservoir.
3. In any natural process, the *entropy* (a measure of randomness) of the universe increases.

More succinctly, the two thermodynamics laws can be stated as:

The energy of the universe is constant.
The entropy of the universe increases.

The second law tells us nothing at all about the speed of a process, but it does offer criteria for determining whether or not a process can occur. We have already seen how the equilibrium constant can be used to determine in which direction a reversible reaction will go. For example, an equal mixture of hydrogen, oxygen, and water vapor will react to form more water. But the reaction at room temperature is immeasurably slow unless a catalyst or spark is present.

In an earlier section we calculated an equilibrium constant from the equilibrium concentrations of the reactants and products. Chemists have a few other methods available to them for obtaining equilibrium constants. One method uses both the heat of a reaction and the entropy change. Even though a process is endothermic (it must take up energy to occur), the increase in entropy or randomness may be sufficient to make the overall process possible. The dissociation of bromine

$$Br_2 \rightleftharpoons 2Br$$

is endothermic; that is, 48,398 cal/mole are required at 2000 K. Nevertheless, appreciable dissociation occurs at this temperature

$$K = \frac{[Br]^2}{[Br_2]} = 5$$

in part because the mixture of bromine atoms and molecules has more disorder and entropy than bromine molecules alone.

In general, there is an entropy increase for a substance as heat is added and its temperature is raised. There is also an important entropy increase during melting and during evaporation; and there is an entropy decrease during freezing and condensation. From x-ray diffraction studies, solids are known to be more regular in their structure than liquids and liquids more regular than gases. (The word *gas* was invented from the word *chaos*.) Thus, the addition of heat to water at its boiling point converts some liquid to vapor and, at the same time, increases the entropy or randomness of the liquid-and-vapor system. When heat is removed, some vapor condenses and entropy or randomness is reduced. In the rubber-band experiment (Fig. 7.50) stretching is attended by greater regularity, a decrease in entropy, and an evolution of heat; and releasing is attended by greater irregularity, an increase in entropy, and the taking up of heat.

Since diffusion is described by a random-walk model, it is accompanied by an increase in randomness and entropy. We can say that not all diffusing molecules will move into one small region because this segregation is highly regular, improbable, and has lower entropy than a well-mixed system, contradicting the second law (third statement).

Many physical and chemical processes are reversible; that is, they can occur in either direction. Suppose a motion picture of a series of collisions between billiard balls is obtained. The film can be run either forward or backward. In either direction the laws of mechanics—the conservation of energy and momentum—appear to be obeyed at each collision. In fact, it would be impossible for an observer to decide which is the correct direction. There is an ambiguity in the direction of time in these physical laws.

The observer needs other clues. A series of billiard-ball collisions begins with a cue stick hitting one ball and ends when the moving ball (or balls) comes to rest. Running the film of a billiard-ball-collision series in reverse, it is unlikely that a ball will spontaneously move from a state of rest and hit other balls until one moving ball strikes the cue stick.

Looking at a vapor trail made by a jet plane, even if the airplane is no longer visible, one can tell in which direction the plane has traveled. The vapor trail will be more spread out—its constituent droplets will have diffused more—at the older end. We know that it is improbable that diffusion will lead to a more compact trail; the longer diffusion occurs, the wider will be the range of diffusing particles. The second law, then, can be used to decide the direction of a time scale.

LOW TEMPERATURES

A heat engine run in reverse (that is, a refrigerator, see Chap. 3) extracts heat from a cold reservoir, and pumps it, with the work (energy) to run the engine, into a hot reservoir. The lowest temperature a refrigerator can attain is limited by its mechanical nature, especially the lubrication of moving parts. To go below about 225 K ($-48°C$), other means are needed.

Many chemical processes are quite sensitive to variations in temperature. Both equilibrium constants and rates of reaction are affected by temperature changes. For this reason most investigations are carried out at a convenient constant temperature. Any process which occurs at a single temperature is said to be *isothermal*. Because most processes are either exothermic or endothermic, it is necessary to either dissipate the liberated heat or absorb enough heat from the surroundings to maintain isothermal conditions. From this, in a very rapid exothermic process, it may not be possible to transfer out the liberated heat rapidly enough; heat accumulates and there is a temperature increase. Similarly, in a fast endothermic process, if too little heat flows into the system, the temperature drops. Isothermal conditions are more easily maintained for slower processes.

If the process is very rapid, or if the system is insulated from its surroundings to prevent heat flowing in or out, there will be a temperature change. A process without heat exchange between it and the surroundings is said to be *adiabatic*. An example of adiabatic cooling is the evaporation of a liquid from a poor heat-conducting surface, such as skin. In evaporation, the most energetic molecules leave and the average energy and temperature of those remaining are lower. Since skin is a poor heat conductor, heat cannot flow rapidly enough to the liquid to maintain a fixed temperature.

When a gas expands, it does work against the ambient atmospheric pressure. If the expansion is rapid, the energy for this work is obtained at the expense of the molecular kinetic energy. Gas adiabatically expanding through a nozzle or throttle will drop in temperature, provided its initial temperature is not too high. For nitrogen this limiting upper initial temperature is about 380°C; for hydrogen, about $-80°C$ (or 193 K); and for helium, $-173°C$ (or 100 K). Nitrogen can be forced through a nozzle to expand and cool repeatedly until its temperature drops to below its boiling point [$-196°C$ (or 77 K)]. In this way, liquid nitrogen can be prepared.

Liquid nitrogen at 77 K can be used to cool gaseous hydrogen sufficiently for adiabatic expansion and cooling. James Dewar (1842–1923) first liquefied hydrogen this way in 1898 and prepared solid hydrogen the next year, reaching a temperature of 14 K, or $-259°C$. Dewar also invented the vacuum bottle (in 1892) for insulating cold liquefied gases from external heat; it is now commonly used also to keep liquids hot and is called the "Dewar flask."

In 1908, Heike Kamerlingh-Onnes (1853–1926) liquefied helium, also by adiabatic expansion, and first attained the temperature of 4.2 K, or −269.0°C. In 1911 he discovered the phenomenon of *superconductivity:* at these very low temperatures, some metals have no electrical resistance. Today about three dozen metals and several hundred alloys and compounds are known to be superconductors.

When the vapor over liquid helium is rapidly pumped away, adiabatic cooling will bring the liquid down to about 0.7 K. Using the isotope He^3, the lowest temperature attainable is about $\frac{1}{3}°$ above absolute zero.

In the rubber-band experiment (Fig. 7.50) adiabatic cooling comes about when crystalline regularity is converted to amorphous randomness. A rubber band can be used to remove heat from and lower the temperature of an object (Fig. 9.11). The stretched

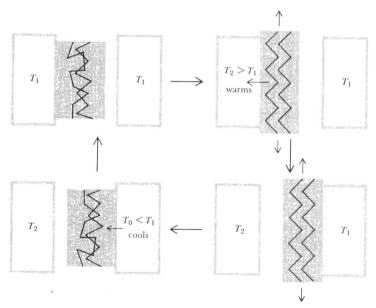

Figure 9.11. At the start of the cycle (upper left) a rubber band is in contact with one of two bodies, both at the temperature T_1. When the rubber band is stretched it releases heat which warms up one body to T_2. The stretched band is moved over to the other body and then allowed to return to its shorter length. During the contraction, the rubber-band crystallinity is destroyed and it takes heat from this second body, the temperature dropping to T_0. This cycle, in effect transferring heat from the right to the left body, can be repeated until the rubber band becomes too cold and stiff.

crystalline rubber band and object are in contact but insulated from the rest of the universe. When the rubber band is released, the heat needed to break up the crystalline regularity can come only from the object, whose temperature drops. Below a certain temperature, a rubber band becomes too stiff and brittle for this method to work. The rubber-band experiment is thus not a practical method to obtain very low temperatures, but a variation on it is.

When a substance is placed partly within a magnetic field, it may either be drawn further into or repelled from the field. Depending on its behavior, the substance is said to be either *paramagnetic* or *diamagnetic*. Paramagnetism is a consequence of unpaired electrons. Above a certain temperature (different for each substance), paramagnetism disappears and the substance is diamagnetic.

When a paramagnetic substance is within a magnetic field, there is a field inside the substance slightly greater than the external field. In a diamagnetic substance the internal field is slightly less than the external one. This suggests that atoms or ions in a paramagnetic substance act as small magnets and, under the influence of an external field, these small magnets are organized, or aligned, to produce an enhanced field. At a high enough temperature, the thermal motion of atoms, ions, and molecules is sufficient to destroy this regular alignment and, with it, paramagnetism. (In a *ferromagnetic* substance, the internal field is about a million times more intense than in a paramagnetic substance.)

As in Fig. 9.12, order can be imposed on a paramagnetic substance by an external magnetic field. Next, the substance is isolated so that no heat can flow in. When the external magnetic field is removed, alignment is destroyed and the energy required to break up this regularity comes from the thermal motion of the constituent ions and atoms. In this *adiabatic demagnetization,* the temperature of the entire paramagnetic substance drops. In 1933, William F. Giauque (b. 1895) cooled the paramagnetic salt gadolinium sulfate (initially cooled to 1 K by pumping of liquid helium) to 0.25 K. With improved techniques, this method can be used to reach a temperature of nearly 0.001 K.

The nuclei of atoms also have magnetic moments—act as small magnets—but the magnitude of these is considerably smaller than the magnetic moments of unpaired electrons. Accordingly, destruction of nuclear alignment by thermal motion is even more certain. But, for example, a fine copper wire can be cooled down to

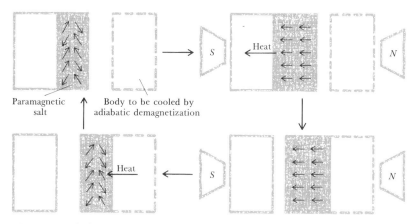

Paramagnetic
salt

Body to be cooled by
adiabatic demagnetization

Heat

Figure 9.12. Adiabatic cooling with a paramagnetic salt. An ion with an unpaired electron will act like a small magnet. In a strong magnetic field it will line up in the direction of the field; in the absence of a field, it will point in a random and varying direction. The thermal motion of other ions and molecules will promote disorder within a paramagnetic substance; and order is promoted by a magnetic field.

0.01 K by adiabatic demagnetization of a paramagnetic salt. Then the nuclear magnetic moments can be effectively aligned by an external magnetic field. If the fine copper wire is isolated from the paramagnetic salt and then adiabatically demagnetized, its temperature will drop further. In 1956, Francis Simon (1893–1956) and Nicholas Kurti (b. 1908) came within 1/100,000 K of absolute zero, or 10^{-5} K. In 1960 Kurti attained 1.2×10^{-6} K, slightly more than 1/1,000,000° above absolute zero.

It is not possible to reach absolute zero in a finite number of steps. The closer this limit is approached, the more difficult it is to remove heat and lower the temperature further. The attainment of very low temperatures is more important than merely satisfying a desire to set new records. At any temperature above absolute zero, all crystalline solids have a number of lattice imperfections or defects which affect their properties. The lower the temperature, the more nearly perfectly crystalline the substance. Then properties of a regular lattice can be studied without the complications introduced by defects. Superconductivity is one such property. If chemists know the specific heats of substances from as close to absolute zero as possible up to room temperature and higher, they can calculate equilibrium constants for reversible reactions of these substances at any temperature.

PROBLEMS

9.1. *a.* At 1000 K (727°C), the conversion of sulfur dioxide to the trioxide

$2SO_2 + O_2 \rightleftharpoons 2SO_3$ has the equilibrium constant

$$K = \frac{[SO_3]^2}{[SO_2]^2[O_2]} = 3.31$$

where the reactants and product are to be expressed as pressures and in units of atmospheres. The hot-combustion-products gas at this temperature from low-grade coal has this nonequilibrium composition:

$[O_2] = 0.10$ atm
$[SO_2] = 2.0 \times 10^{-3}$ atm
$[SO_3] = 2.0 \times 10^{-4}$ atm

If equilibrium can be established, will there be more or less sulfur trioxide?

b. The same gas also contains ozone O_3 at this pressure:

$[O_3] = 4 \times 10^{-6}$ atm

The reaction $3O_2 \rightleftharpoons 2O_3$ has the equilibrium constant

$$K = \frac{[O_3]^2}{[O_2]^3} = 7.7 \times 10^{-23}$$

Do you expect more or less ozone at equilibrium?

9.2. *a.* If the hot gases expand and the total pressure drops on leaving the smoke stack, in which direction do you expect the above two reactions to shift?

b. The first reaction is exothermic by 47 kcal in the forward direction, and the second is endothermic by 68 kcal in the forward direction. If the exhaust gas is cooled, do you expect more or less sulfur trioxide and ozone at the new equilibrium?

9.3. Use the probability tree or the binomial theorem to describe the molecular distribution after a random-walk diffusion for one, two, three, four, and five tosses of a biased coin: $H = \frac{2}{3}$ and $T = \frac{1}{3}$.

Molecular Biology and Genetics

10

Some processes occurring in living systems have been identified and described. These all function according to well-established scientific laws and principles; nothing yet has been found in living systems which contradicts those theories that are successful in describing nonliving systems.

A simplistic view of a living organism is as a collection of an enormous number of very special substances reacting with each other at just the right speeds and forming exactly those products needed to sustain life. Yet there is nothing in our present knowledge of chemistry and physics that will enable us to say that life must necessarily exist. Living systems are complicated beyond our present comprehension. Also, although chemists have characterized over three million separate pure compounds, there are numerous compounds in living organisms that have not yet been isolated, purified, or identified.

Bringing their special skills together, biologists, chemists, and physicists are learning more of what happens in these life processes. This attempt at description in terms of molecules is called "molecular biology." Deeper understanding provides us with new tools to manipulate and regulate biological functions, which, in turn, help us to control certain diseases and malfunctions. In this chapter we shall look at a few life processes, including genetics and applications, from the viewpoint of molecular biology.

LEVELS OF ORGANIZATION

The biological world can be studied at several levels of organization, namely, the molecular, cellular, organismal, population, and ecosystem levels. Each level represents a different degree of complexity. The approach to studying life may vary depending upon which level is examined. Unique models can be built and tested for each level, although some knowledge about one level may be extrapolated to others. The concept of levels of organization permits scientists to organize information and better isolate components of life for study.

All living things are made of molecules, and many life phenomena can best be understood by examination at the level of organization that deals with the properties and behavior of molecules. Methods of study at this level rely upon chemical and physical principles. Many activities observed at higher levels, such as inheritance, can be described by molecular properties. Other attributes of higher levels, e.g., the characteristic morphology of mitochondria, cannot be accounted for at present by molecular data but may be someday when sufficient information is available. However, many aspects of biology can never be understood adequately by extrapolating from the molecular level.

All higher organisms are made of smaller units called "cells"; the study of these cells and their content is called "cell biology." Cells perform most functions of the whole organism either independently or by interacting with each other. Often an activity of an organism is the sum of what each cell does. Many cells are specialized to perform particular functions; e.g., nerve cells carry messages, and muscle cells do mechanical work.

The individual plant or animal is called an "organism." The organism is made of molecules and cells, which in multicellular organisms are put together to form tissues and organs. A *tissue* is a group of cells having similar structure and function, and an *organ* consists of tissues that are grouped together to form larger functional units. Organs may interact in what is called a "system." For example, epithelial tissue, nervous tissue, and smooth muscle tissue make up a storage and digestive organ, the stomach. The stomach and other organs of the gut cooperate in the digestion and absorption of food and constitute the digestive system.

Organisms do not exist alone but as parts of larger groups called "populations." A *population* is a group of similar organisms (belonging to the same species) occupying a particular geographical area. A population is further defined as a group of organisms in which there is *exchange of genetic material* between various members of the population; in other words, the individuals within the population interbreed. Certain biological phenomena are limited almost exclusively to populations. For example, only populations can evolve.

The smallest self-sufficient unit of the biological world is the *ecosystem*. The ecosystem consists of interacting organisms living in a particular area and the physical factors of that environment, and can be considered to be the most fundamental functional unit of the biological world. An ecosystem has a biological portion called a "community" (an assemblage of plants, animals, and microorganisms) and an abiotic, or physical, portion. A community is generally described by its dominant organisms, e.g., an oak-hickory or a grassland community. The concept of the community is useful for describing and distinguishing various kinds of population assemblages, but the ecosystem concept is superior in that it includes functional relationships with the environment as well as description. In studying the biological world the largest unit to be considered is the earth. The portion of the earth which supports life, including the atmosphere, is called the "biosphere."

Dividing the biological world into levels of organization is, in part, an artificial division to help man isolate components of life for study. Many biological activities occur at every level, and these may be studied at any level. For example, respiration as a process may be studied at the molecular, cellular, organismal, or ecosystem level of organization. A biochemist might study enzymes involved; a cell biologist, the function of mitochondria; a physiologist, the exchange of gas across the lungs; and an ecologist, energy transfer in the ecosystem. Obviously the methods of study differ.

CELLS

Most plants and animals are made of millions of individual subunits of various sizes and shapes, each bounded by a membrane, called "cells." Most life processes such as metabolism, growth, and reproduction are carried out within specialized structures, or subdivisions, of the cell called "organelles." There are about twenty kinds of organelles, each performing specific functions within the cell. Some of the more important organelles are shown in the generalized cell represented in Fig. 10.1. These organelles, along with the outer boundary of the cell, consist largely of membranes.

Although the functions performed by a cell are varied and complex, it is possible to outline briefly some of its principal activi-

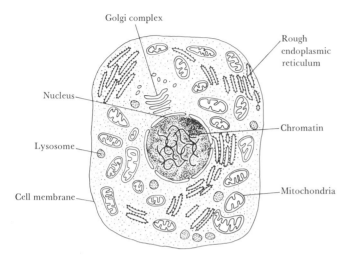

Figure 10.1. *Generalized animal cell. Cell structure consists of membranes which form various cell organelles. All cell functions are carried out by these various specialized organelles. The thread-like structures within the nucleus, when present as separated individual sections, are the chromosomes which, in turn, comprise the genetic units called genes (page 4) and contain the genetic substance DNA (Figures 7.51 to 7.54).*

ties. The *nucleus* is the largest organelle and is located centrally in the cell. It functions in the storage and retrieval of genetic information involved in the inherited characteristics of the organism. *Mitochondria* (singular, mitochondrion) break down foodstuffs and synthesize adenosine triphosphate (ATP), the principal energy compound of the cell. The *endoplasmic reticulum* is responsible for the synthesis of protein macromolecules and other important compounds. Other organelles perform other vital functions. (Functional processes of the cell will be discussed in more detail later in this chapter.)

MEMBRANE STRUCTURE

All cellular membranes contain proteins and lipids, but considerable variation exists in the kind of protein or lipid present. Studies of physical properties of cell membranes led James F. Danielli (b. 1911) and H. Davson (b. 1909) to propose a model of membrane structure consisting of three layers: an inner lipid layer sandwiched between two protein layers. They suggested that lipids were arranged so that the polar ends of the molecule were facing the protein layers (Fig. 10.2). This model was hypothetical until electron

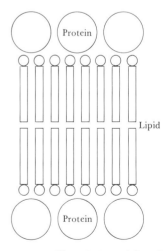

Figure 10.2. *Danielli and Davson model of cell structure. A lipid layer is found on the interior of the membrane, sandwiched between two protein layers.*

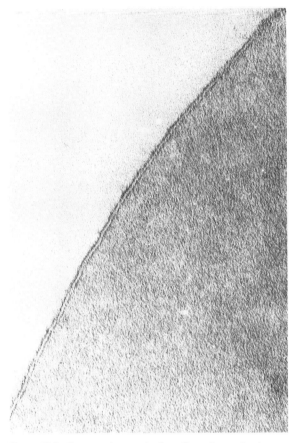

Figure 10.3. Electron micrograph of a cell membrane showing an inner layer and two electron-dense outer layers. (Courtesy J. D. Robertson.)

microscope studies revealed a trilayered structure (Fig. 10.3).

Several investigators have suggested that the cell membrane is even more complex. They propose that the membrane is made of discrete subunits situated side by side. An analogy might be made between a cell membrane and a brick wall. The brick wall is a trilayered structure, of two layers of brick with a layer of mortar between them; the bricks are definite subunits. Subunits in the membrane might not be uniform in shape or function; the membrane may contain several subunits, each performing a different function. Thus, a mitochondrial membrane performs several different functions, and these might be related to the kinds of subunits present.

ENZYME FUNCTION

Proteins are important both as structural elements and as enzymes. An enzyme is a catalyst, or an agent that speeds the rate of a reaction without itself being used up. Few chemical reactions essential to life can take place without enzymes. Theoretically, the reaction might occur but only at temperatures and pH incompatible with life. Enzymes are one of the more important functional classes of compounds in cells because they regulate the rates of chemical reactions.

Enzymes are classified and named according to the reactions catalyzed and the compounds acted upon (the *substrates*). All recently named enzymes end with a characteristic suffix *-ase*. An example of an important enzymatic reaction and nomenclature is

$$\text{Lactic acid} \xrightarrow{\text{Lactic dehydrogenase}} \text{pyruvic acid}$$

In this reaction the substrate, lactic acid, is converted to the product, pyruvic acid, catalyzed by the enzyme, lactic dehydrogenase (LDH). The name of the enzyme is derived from the substrate involved and the kind of reaction, in this case, dehydrogenation. A dehydrogenase reaction is an oxidation with transfer of hydrogens from a substrate to a coenzyme.

A *coenzyme* is an organic compound that acts as a donor or an acceptor of a substance involved in an enzymatic reaction. *Adenosine triphosphate* (ATP, Fig. 10.4) is an important coenzyme that donates phosphate groups $H_2PO_4^-$. Two coenzymes that accept and donate hydrogen atoms (or electrons) in oxidation-reduction reactions are *nicotinamide adenine dinucleotide* (NAD) and *nicotinamide adenine dinucleotide phosphate* (NADP). The oxidized form of NAD is abbreviated NAD^+, and the reduced form is NADH. The coenzyme in this reaction is nicotinamide adenine dinucleotide (NAD^+). The equation with formulas is

Lactic acid   Pyruvic acid

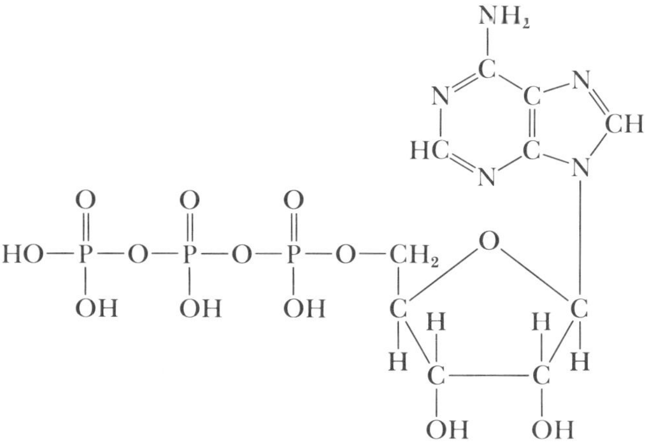

Adenosine triphosphate (ATP)

Figure 10.4. The molecule ATP is used by living organisms in a large number of reactions. Acting as an intermediary it makes possible certain reactions essential for life which, in the absence of ATP, will go either too slowly or not at all. When one of the three phosphate groups is removed, as phosphoric acid H_3PO_4, the remaining molecule is adenosine diphosphate, ADP; and when two are removed, as pyrophosphoric acid $H_4P_2O_7$, what is left is adenosine monophosphate, AMP. A general example of the role of ATP is:

$$ATP + MOH \rightleftharpoons MOP + ADP$$
$$\underline{MOP + NOH \rightleftharpoons MON + H_3PO_4}$$
$$MOH + NOH + ATP \rightleftharpoons MON + ADP + H_3PO_4$$

Another reaction will convert ADP to ATP.

The reverse reaction also occurs and is catalyzed by the same enzyme.

LDH, which is important in clinical medicine, has a number of *isozymes*, or enzymes which all catalyze the same reaction but which differ in structure and properties. If tissue is damaged by a blood clot, LDH is released into the blood. Because heart muscle has a different LDH isozyme than most other tissue, a cardiac clot can be identified by the isozyme present in blood. Also, the extent of damage may be estimated by the quantity of LDH in blood.

METABOLIC PATHWAYS

Most reactions in biological systems occur in sequence. Thus, the product of one reaction serves as the substrate in a subsequent step, and the product of the second reaction, in turn, is the substrate for a third reaction, etc. Such a series of reactions is called a "metabolic pathway."

Two types of pathways occur. *Degradative* metabolic pathways break down organic substances into simpler compounds such as carbon dioxide and water. *Synthetic* pathways elaborate more complex substances from relatively simple chemicals. Degradative pathways are generally *exergonic*, releasing potential chemical energy as heat, and synthetic pathways are *endergonic*, requiring the addition of energy. Energy is supplied to synthetic pathways by coupling to exergonic pathways. Thus, some compounds are broken down, releasing energy, and some of this energy is trapped and utilized by the synthetic pathway.

The mechanism of coupling between exergonic and endergonic pathways generally involves the coenzyme adenosine triphosphate (ATP) and two specific reactions. The first reaction, in the exergonic pathway, results in synthesis of ATP from adenosine diphosphate (ADP), and the ATP then serves as a reactant in an otherwise endergonic reaction, effecting synthesis. In the process ATP is broken down into ADP and inorganic phosphate or, in some cases, AMP and pyrophosphate. Thus, cells must have a source of energy, such as food, to sustain organic synthesis, i.e., growth. ATP also serves as the immediate source of energy for nonsynthetic processes, such as muscle contraction and active transport (the transfer of a solute from a low to a high concentration region).

An important exergonic pathway of the cell is shown in Fig. 10.5. This pathway illustrates the sequential nature of the metabolism of a compound. Glucose is converted to lactate, and ADP is converted to ATP. This pathway also metabolizes other sugars; hence the name *glycolytic* (sugar decomposition) *pathway.* The initial step in glycolysis is the transfer of a phosphate group from ATP to glucose. In subsequent steps the molecules are rearranged to create compounds which will react with ADP to form ATP. The end result is two molecules of lactate and a net gain of two ATP's for each glucose molecule.

ENZYME SPECIFICITY

A specific enzyme catalyzes each chemical reaction in the cell. This is called "enzyme specificity." For example, lactic dehydrogenase catalyzes the interconversion of lactic acid to pyruvic acid and no other reaction.

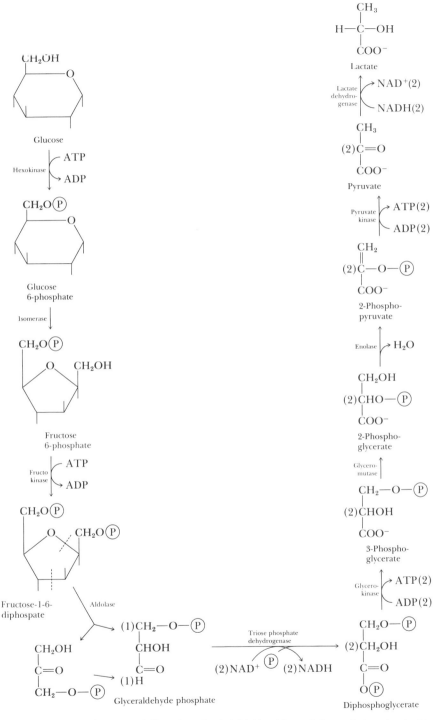

Figure 10.5. Glycolysis, a metabolic pathway for the initial breakdown of glucose. In ten separate steps glucose is phosphorylated, rearranged, split, oxidized, etc. The net reaction is

glucose + 2 ADP + 2 phosphates ⟶ 2 lactates + 2 ATP

The properties and function of each enzyme are due to its amino acid composition and sequence. The amino acid sequence within an enzyme dictates its specificity. Enzyme action is a result of properties of the subunits within the molecule and, in some cases, the coenzymes or prosthetic groups attached to the protein. Molecular shape, solubility, and other properties of proteins also depend upon their amino acid composition. A protein is programmed by the amino acid sequence to perform a particular function. Thus, a protein contains information which allows recognition of the specific substrate, and catalyzes a specific reaction.

THE GENETIC SUBSTANCE

Deoxyribonucleic acid (DNA) is located in the nucleus of the cell and, together with proteins, forms the structure of the chromosomes. (The structure of DNA is discussed in Chap. 7 and is illustrated in Figs. 7.51 to 7.54.) DNA is thought to be the source of genetic information. Wavelengths of radiation absorbed by DNA cause chemical changes in it and the consequent mutation of genes. The genetic substance must be extremely stable since each generation has the same form and function as the previous generation. Metabolically, DNA is the most stable organic compound found in a cell; it is not decomposed during the lifetime of the cell.

Genetic information must be reproduced exactly each time a cell divides. The DNA content of cells doubles just prior to cell division, and one-half the DNA goes to each daughter cell. The quantity of genetic material per cell should be constant from generation to generation. Therefore, reproductive cells—eggs and sperm—each contain one-half the genetic substance found in adult cells. The observation that a sperm cell contains half as much DNA as an adult cell is strong evidence to support the model that DNA is the genetic material.

The genetic information needed for directing the structure and function of the cell and organism is presumably coded in the chemical structure of DNA. The relation between DNA structure and genetic expression has long puzzled scientists. A rather detailed model has been proposed.

The immediate function of DNA is to serve as a template, or guide, for the synthesis of ribonucleic acid (RNA). Three types of RNA are present in cells, and all are thought to be synthesized from a DNA template. One kind of RNA, called "messenger RNA" (*m*RNA), directs the synthesis of protein by serving as a template for putting amino acids together to form a specific kind of protein. Ribosomal RNA (*r*RNA) and transfer RNA (*t*RNA) do not direct the way a protein is synthesized but are involved in protein synthesis (described next). The protein that is synthesized may be considered to be the expression of a genetic blueprint.

GENETIC TRANSCRIPTION

Genetic transcription is the process of using DNA as a template for synthesizing messenger RNA. A particular sequence of nucleotide bases in one strand of the DNA molecule determines the sequence of nucleotide bases that are to go into *m*RNA. Each nucleotide base in DNA determines one nucleotide base in *m*RNA. In this way the genetic information contained in DNA is transcribed to *m*RNA, which, in turn, will be used in directing the synthesis of protein.

In making *m*RNA, free unpolymerized nucleotides become associated with particular nucleotides in one strand of the DNA molecule by hydrogen bonding similar to that found between the two strands of DNA. The free nucleotides then condense to form *m*RNA. The hydrogen bonding between bases of different nucleotides follows a pattern. The nucleotide base guanine in DNA always pairs with cytosine in making *m*RNA, and cytosine pairs with guanine. Thymine in DNA codes for adenine in *m*RNA, but adenine codes for a nucleotide base found only in RNA: uracil. Because base pairing is always done with a complementary base, *m*RNA is not like the DNA strand copied but rather is similar to the other DNA strand, except that uracil is present in place of thymine (Fig. 10.6).

TRANSLATION

Translation is the process of protein synthesis using *m*RNA as a template. It is the second step in the reading of the genetic code by the cell. Since there are four kinds of nucleotides in *m*RNA and twenty amino acids in proteins, coding for one amino acid by one nucleotide is impossible. A sequence of more than one nucleotide must correspond to each amino acid. This nucleotide sequence is equivalent to a word in a language. In translating English into Chinese, several English letters must be used for each Chinese character. With a two-nucleotide word we can make exactly

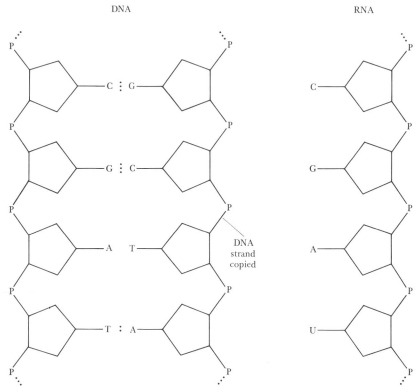

Figure 10.6. Diagram of transcription. RNA is copied from one DNA strand. The RNA structure is complementary to the DNA copied; that is, the bases are opposite. (The RNA molecule is nearly the same as the uncopied strand of DNA, differing only in the substitution of ribose for deoxyribose sugar and uracil for thymine.) A = adenine, C = cytosine, G = guanine, T = thymine, U = uracil, P = phosphate, ⬠ = ribose (RNA) or deoxyribose (DNA); :, : = hydrogen bonds.

$4 \times 4 = 16$ different words from the 4 organic bases (e.g., CC, CG, CA, CU, GC, GG, etc.). But 16 words are still too few for the 20 amino acids. With three nucleotides per word we can make

$$4^3 = 64 \text{ words}$$

which is more than enough for the 20 amino acids. Thus we have a language consisting of 64 three-letter words put together from a four-letter alphabet. For example, three adjacent nucleotides (UUC) might code for phenylalanine. Each 3 nucleotide set is called a "codon" for a special amino acid. Thus UUC is a codon for phenylalanine.

No mechanism is known whereby *m*RNA can recognize an amino acid directly, so a special mechanism must be present in the cell to read the code in *m*RNA. The following model has been proposed. Transfer RNA has been found to attach to amino acids in cells. Moreover, several *t*RNA's have been isolated, each specific for one amino acid. It is further proposed that *t*RNA has a specific nucleotide sequence in the molecule which is complementary to a base sequence in the *m*RNA. Thus, *t*RNA can recognize the code word for an amino acid, and that amino acid is attached to that *t*RNA. For example, one code word for the amino acid phenyl-alanine is UUC on the *m*RNA. Phenylalanine is attached to a *t*RNA that has a sequence AAG in its molecule. The triplet AAG is com-plementary to UUC and forms three base pairs:

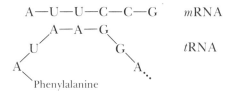

Each of the amino acids combines with its specific *t*RNA's. Each *t*RNA contains a triplet code which recognizes the triplet code word on *m*RNA. The transfer RNA's with their attached amino acids line up along the messenger RNA. An enzyme-catalyzed reaction re-sults in peptide-bond formation between the adjacent amino acids. Thus, a relatively simple translation mechanism allows the cell to use genetic information. This model, summarized in Fig. 10.7, ac-counts for the ability of DNA to determine the amino acid sequence in proteins.

CRACKING THE GENETIC CODE

An exciting breakthrough in molecular biology was solving the genetic code. This was made possible by the discovery by Severo Ochoa (b. 1905) that RNA could be synthesized in the test tube without DNA template. Synthetic *m*RNA could thus be made with a base sequence other than that found in DNA, and the base com-position could be controlled by the kind of nucleotides used in the synthesis. Synthetic *m*RNA could be added to cellfree systems con-taining ribosomes, *t*RNA's, enzymes, etc., and protein synthesis would result. One discovery by Marshall W. Nirenberg (b. 1927) was that *m*RNA containing uridine as the only base resulted in pro-tein containing phenylalanine as the only amino acid. The conclu-

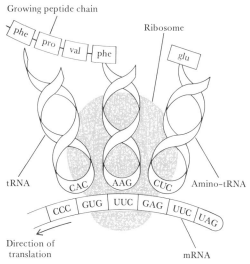

Growing peptide chain

Ribosome

tRNA

Amino-tRNA

Direction of
translation

mRNA

Figure 10.7. Translation of the genetic code. DNA in the nucleus codes for specific nucleotide sequences during messenger RNA (mRNA) synthesis. Messenger RNA moves to the ribosomes where protein synthesis is directed. The mRNA advances along the surface of the ribosome, engaging amino-transfer RNA's. Then, following addition of the attached amino acid to a growing peptide chain, the tRNA is disengaged. The code is read by attachment of specific triplets in tRNA to the code triplet in mRNA. In this way a code derived from DNA is read, and a specific protein is synthesized. Amino acids: phe = phenylalanine, pro = proline,· val = valine, glu = glutamic acid. Nucleotide bases: C = cytosine, G = guanine, U = uracil, A = adenine.

sion drawn was that UUU in *m*RNA must code for the amino acid phenylalanine. Other code words were determined using such cellfree systems, including another code word for phenylalanine UUC. The genetic code was thus elucidated.

The genetic code is universal for all organisms. Ribosomes and *t*RNA's isolated from bacteria, plants, and animals synthesize the same proteins when incubated with synthetic *m*RNA. Viral RNA added to a bacterial system results in synthesis of viral proteins. Further, genetic mutations have confirmed the universality of the code.

A *mutation* is a sudden change in the DNA in an organism which results in changes in the kind of proteins synthesized. For example, in sickle-cell anemia in man a valine is substituted for a glutamic acid in hemoglobin. Glutamic acid is coded for by either

GAA or GAG, while valine is coded for by either GUA or GUG. A base change from A to U in the middle letter accounts for the substitution of one amino acid for the other. This significant change in the properties of the molecule results in sickle-cell anemia. About ten other mutations are known for hemoglobin, each accounted for by a single base change.

ANTITUMOR AGENTS

The exact replication of cells, the integrity of an organism, and the continuity of a species all depend on the precise pairing of organic bases (Fig. 7.53). If an organic base in either DNA or RNA is replaced by another, the code for a particular amino acid sequence (e.g., in an enzyme) is altered. Through such a miscoding error, a wrong amino acid sequence and an improper, or nonfunctioning, enzyme is produced. Ordinarily, such errors are harmful to life and even fatal, but in some cases a deliberate effort is made to interfere with cell replication and cell viability. Thus sometimes the control of cancer can be managed by introducing an altered organic base.

In making RNA, the bases uracil and adenine pair (Fig. 10.6). The synthetic compound 5-fluorouracil (Fig. 10.8) is made by replacing a hydrogen with a fluorine atom at the fifth ring position. It can pair with either adenine or guanine. Normally, only cytosine pairs with guanine. If 5-fluorouracil is introduced to growing cancer cells and incorporated into messenger RNA, then

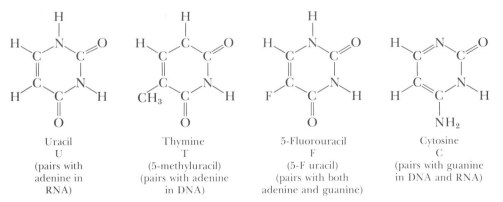

Figure 10.8. This diagram shows the keto forms of three organic bases found in nucleic acids, and one synthetic base F, which can interfere with normal growth. This compound has had some limited success in controlling malignancies.

either adenine or guanine in transfer RNA will pair irregularly with 5-fluorouracil. Studies with the bacteria *Escherichia coli* indicate the code word with 5-fluorouracil (F), that is, FAG, is mistranslated as the code word CAG, with a glutamine residue inserted erroneously in an enzyme.

Since cancer cells are among the more rapidly growing cells in the human body, 5-fluorouracil, injected intravenously, can be expected to accumulate in malignant growths. Unfortunately, red blood cells grow even more rapidly, and their production can be damaged by the same compound. The proper use of 5-fluorouracil involves careful monitoring of blood cells. This therapeutic agent is successful in controlling some but not all malignancies. Some anti-tumor agents block DNA synthesis; others, RNA synthesis; and still others interfere with protein synthesis, as is the case for 5-fluorouracil.

CONTROL OF GENETIC EXPRESSION

Each cell has thousands of different genes, each corresponding to a segment of DNA helix. Each gene specifies the kind of protein to be synthesized but not how much. Obviously not all genes are continuously expressed. Consider, for example, the cells in the human body. All cells in any human being have the same genetic composition; yet each cell has certain specific functions. A cell in the nervous system may possess the capability for secreting digestive enzymes. Why, then, does this cell not do so? The answer is that this part of the genetic reservoir is suppressed by environmental conditions. Each cell, then, has all genetic material, but only a small fraction is ever used by a single cell at one time.

The environment of the cell is the fluid surrounding both the cell and adjacent cells. This environment is determined by the cell's position within the organism. Molecules secreted by adjacent and distant cells may greatly influence the activity of a cell. Francois Jacob (b. 1920) and Jacques Monod (b. 1910) of the Pasteur Institute in Paris have proposed a molecular model to account for environmental regulation of genetic expression in bacteria. The environment of single-cell bacteria can be manipulated more easily and precisely than that of multicellular organisms. A model proposed for bacterial genetics can be applied to higher organisms.

ENZYME INDUCTION

Enzyme synthesis follows one of three patterns. If it occurs only when a suitable substrate is available for the enzyme, this process of initiating protein synthesis is called "enzyme induction." For example, bacteria grown in an environment lacking galactoside sugars such as lactose (milk sugar) do not synthesize any of several enzymes that would be required for metabolizing this kind of sugar. However, if lactose or other substrate is added to the medium, the bacteria begin making all the special enzymes for metabolizing galactoside sugars.

Various mutations affect these enzymes. Several simple mutations occur that affect one enzyme each, and these are interpreted as simple changes in the DNA which code for each protein enzyme. Another kind of mutation results in a complete failure to synthesize any enzyme regardless of whether or not substrate is present. Some mechanism must link the genes associated with glactoside metabolism. A third kind of mutation results in continuous synthesis of all the enzymes of galactoside metabolism. Apparently this gene inhibits protein synthesis, and its mutation removes this inhibition. In summary, the several genes associated with galactoside metabolism are expressed as a unit or inhibited as a unit by the action of another gene (Fig. 10.9).

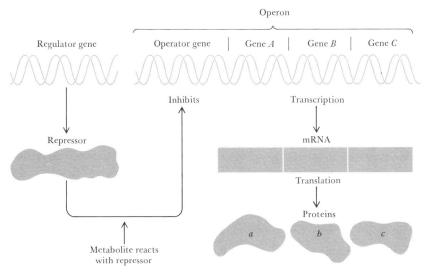

Figure 10.9. Jacob-Monod model.

A group of genes acting as a unit is called an "operon," with one gene, the *operator gene,* initiating transcription. A *regulator gene* forms a *repressor,* which inhibits the operator gene and thus the operon. In the presence of a substrate, the repressor is inactivated and the inhibition of the operator gene is removed; thus transcription proceeds. *m*RNA will be produced from the operon, with a segment corresponding to each gene. The *m*RNA directs the synthesis of a protein corresponding to each gene.

ENZYME REPRESSION

A second pattern of enzyme synthesis is called "enzyme repression." Here enzyme synthesis is stopped, or repressed, by the presence of a particular substance. Enzymes involved in synthetic pathways producing the particular compound responsible for repression may be repressed. Enzyme repression prevents synthesis of an enzyme or group of enzymes that would catalyze production of a substance already present in the cell. For example, histidine is an amino acid required by cells for normal growth and metabolism. If histidine is present in the cell, enzyme synthesis is repressed so that additional histidine is not produced. This avoids wasteful synthesis of enzymes and checks formation of unneeded histidine.

A simple description of the mechanism for repression involves the Jacob and Monod model. Histidine combines with an inactive repressor substance, resulting in its activation. The active repressor then inhibits the operator gene, blocking synthesis of proteins that serve as enzymes in the biosynthesis of histidine. Enzyme repression involves activation of an inactive repressor, whereas enzyme induction follows from inactivation of an active repressor.

CONTINUOUS SYNTHESIS

A third pattern of enzyme synthesis is called "continuous synthesis," which takes place under all circumstances. The enzymes involved in continuous synthesis are called "constitutive enzymes." All of the enzymes of glycolysis (Fig. 10.5) are constitutive enzymes; they are required by organisms at all times.

In general, degradative pathways breaking down special substrates are controlled by enzyme induction; synthetic pathways for making required products, e.g., histidine, are controlled by repression; and enzyme systems basic to the life of the cell are usually constitutive enzymes.

INHIBITION OF ENZYMES

Regulation of enzyme synthesis is not the only mechanism affecting enzyme activity in the cell. Control of enzyme activators and inhibitors and the availability of coenzymes and substrates, among other processes, influence the activity of enzymes.

Enzyme activators are ions or other substances that are required for converting an inactive enzyme to an active state; for example, in the stomach the hydrogen ion activates the digestive enzyme pepsin. *Inhibitors* are substances which interfere with normal enzymatic activity. In one kind of inhibition, the end product of a synthetic metabolic pathway inhibits an enzyme at the beginning of the pathway. The activity of the enzyme is thus regulated by the final product, which is advantageous to the organism. This product is not synthesized when in adequate supply; the enzyme is inhibited. If the substance is deficient, there is no inhibition and the metabolic pathway proceeds to synthesize the product. A generalized example is

$$\text{Inhibits} \longrightarrow$$
$$A \xrightarrow{\text{Enzyme 1}} B \xrightarrow{\text{Enzyme 2}} C \xrightarrow{\text{Enzyme 3}} D \xrightarrow{\text{Enzyme 4}} E$$

Coenzymes such as ADP and NAD affect enzyme activity. For example, the function of lactic dehydrogenase is controlled by the availability of NADH, and the activity of the whole glycolytic pathway depends upon ADP concentrations. If ADP is in high concentration, i.e., if a lot of ATP is being split to ADP, the rate of glycolysis increases. Glycolysis converts ADP to ATP (Fig. 10.5).

The concentration of reactants directly affects enzyme activity. Obviously, if no reactant or substrate is present, the enzyme will exhibit no activity. As substrate concentration increases, the activity of an enzyme increases. However, at high substrate concentrations the enzyme is saturated; it is acting upon a maximum amount of substrate. Further increase in substrate concentration has no effect upon enzyme activity.

PRE–MENDELIAN INHERITANCE

For centuries man has recognized that the physical appearance of offspring is inherited from parents. Man has used the concept of inheritance for developing most of the common domesticated cattle, horses, pets, fruit, vegetables, and grains we know today. Breeding for variety in plants and animals has been accomplished using such techniques as selection, inbreeding (i.e., mating siblings), and crossing. Breeding was a highly developed art long before the science of genetics became established. Obviously many aspects of heredity must have been understood even though the mechanisms of inheritance were not known.

For 2,000 yr the accepted mechanism for inheritance was the model proposed by Aristotle. The embryo coagulated from a mixture of semen and menstrual fluid, and the resulting traits were a blend of contributions from the two parents. In 1677 one of the first great microscopists, Anton van Leeuwenhoek (1632–1723), observed sperm cells in semen and proposed the *homunculus theory*. The homunculus, or miniature man, was supposedly curled up in the sperm, and development followed placement in a suitable environment, the uterus. The characteristics of the resulting individual were contributed largely by the father, but the mother influenced the appearance of the child because of strong influences of the uterus on the embryo. In a second theory, *ovism*, the homunculus was believed to originate in the egg.

After scientists established that embryo development occurred only following union of two gametes, a blending theory similar to Aristotle's returned to prominence. *Gametes* are sex cells: an *egg* and a *sperm* in animals and an *egg* and a *pollen grain* in plants. This model stated that the characteristics of the offspring result from an equal mixing of fluids from the sperm and egg. With this model one would predict that crossing a white horse with a black horse would produce a gray horse. Although there was some evidence supporting this model, it was eventually disproved.

MENDEL'S EXPERIMENTS

In the mid-nineteenth century, Gregor Mendel performed what are now considered to be classical experiments in inheritance. Mendel selected the garden pea for his experiments and examined seven different traits in detail. Although we do not know the model

Mendel intended to test, we might suppose it was blending inheritance.

Mendel's experimental design was to cross varieties of peas that had distinct traits and then examine the offspring for the presence of these distinct traits. The traits were observed in the first and second filial generations. The *first filial generation F_1* are the immediate offspring of a cross; the *second filial generation F_2* are the offspring resulting from self-breeding of the F_1 generation. The data from Mendel's first crosses are given in Table 10.1. A total of seven different crosses were made. Examine these data on the traits of parents and F_1 offspring and decide whether you would accept the blending theory of inheritance.

Table 10.1. Data from Mendel's Pea Experiment

Cross†	F_1
Smooth × wrinkled seeds	100% smooth seeds
Yellow × green seeds	100% yellow seeds
Inflated × constricted pod	100% inflated pods
Green × yellow pod	100% green pods
Axial × terminal flower	100% axial flowers
Red × white flower	100% red flowers
Tall × dwarf	100% tall

† In each cross one variety of pea, e.g. smooth seeded, was bred, or crossed, with another, wrinkled seeded. This breeding is represented by smooth × wrinkled. The result of each cross is reported as percent of the offspring F_1 having one or the other trait of the parents.

In every case the appearance of the individuals of the F_1 generation was identical to one of the parents. The trait that is expressed in the F_1 generation is called a "dominant" trait, while the trait that disappears is called a "recessive" trait. The disappearance of a trait in the F_1 generation was an unexpected result in view of scientific knowledge of the time. Although Mendel sent reprints of his work to many other scientists, 44 yr elapsed before such a radical finding was accepted by the scientific community.

Why did the recessive trait disappear in the F_1 generation? Was this trait destroyed? In Mendel's second series of experiments, he allowed the pea plants to self-pollinate. Peas and other plants are bisexual, with both male and female organs in one individual. In self-pollination the male organ produces pollen which fertilizes the egg in the female organ on that same plant. The traits in the second filial generation F_2 were recorded and are given in Table 10.2.

Table 10.2. Data from Mendel's Second Pea Experiment

$F_1 \times F_1$			F_2		Ratio Experimental	Ratio Whole-number
Smooth seed	5,474	smooth	1,850	wrinkled	2.959	3:1
Yellow seed	6,022	yellow	2,001	green	3.009	3:1
Inflated pod	822	inflated	299	constricted	2.75	3:1
Green pod	428	green	152	yellow	2.82	3:1
Axial flower	651	axial	207	terminal	3.14	3:1
Red flower	705	red	224	white	3.15	3:1
Tall	787	tall	277	dwarf	2.84	3:1

The recessive trait, which disappears in the F_1 generation, reappears in some offspring of the F_2 generation. We must conclude that it was present in the F_1 generation but was not expressed. Mendel concluded that inheritance must be based upon transfer of discrete units, now called "genes," from one generation to the next. Further, these genes behave in a predictable way. The recessive trait invariably disappears in the F_1 generation and always reappears in the F_2 generation in about one-fourth of the individuals. Mendel accounted for this behavior by postulating that genes are inherited in pairs, one from each parent; these might be two dominants, two recessives, or a dominant and a recessive.

The term *allele* is used to designate either member of a gene pair. Thus, Mendel studied seven genes, with two alleles for each gene. A dominant allele is designated by a capital letter, and a recessive allele is indicated by the lower case of the same letter. For example, the alleles of the gene for seed color in peas are abbreviated Y for yellow and y for green.

Genotype is the description of an organism by its allele content. A genotype might be *homozygous* dominant if both alleles are dominant, homozygous recessive if both alleles are recessive, or *heterozygous* if the alleles are different. *Phenotype* is the description of the appearance of the organism. For example, the phenotype for seed color is either yellow or green. Green seeds are always homozygous recessive in genotype, but yellow phenotypes may have a genotype of either YY or Yy.

Most organisms, including peas, have two alleles for each gene. Each organism is an offspring of its parents and is a parent of its own offspring. To maintain the constancy of two alleles of each gene, each parent contributes a gamete, or sex cell, such as an egg or pollen sperm, containing only one allele of each gene. A gamete

Parent *YY*

Gametes   *Y*   *Y*

*F*₁ offspring

All *Yy* genotypes

All yellow-seeded phenotypes

Parent *yy*

y   *Yy*   *Yy*

Offspring

y   *Yy*   *Yy*

Figure 10.10. Analysis of Mendel's first experiment crossing yellow-seeded pea plants (YY) with green-seeded pea plants (yy). The homozygous parents each contribute either Y or y gametes. All resulting F₁ offspring are heterozygous Yy with yellow seeds.

receives from the parent one or the other of a pair of alleles for each gene for each trait. The offspring resulting from union of two gametes will have two alleles, one from each parent, for each gene.

In Mendel's cross of unlike parents, the F_1 offspring received one dominant allele and one recessive allele (Fig. 10.10). Therefore, the phenotypes of F_1 offspring were identical to the parent contributing the dominant allele. Mendel's 3 : 1 ratio in the F_2 generation can be accounted for by simple combination probability (Fig. 10.11). Any gamete has a 50 percent chance of receiving a recessive allele from its F_1 parent. The probability for combining two gametes, each with a recessive allele, is one in four. Thus there is a one-fourth chance for a recessive phenotype in the F_2 generation. The probabilities for other genotypes and phenotypes are given in Fig. 10.11.

These probabilities are identical to the head-and-tail probabilities in coin tossing. With one toss of a coin, either a head *H* or a

F_1 Parent *Yy*, self-pollinated

Gametes   *Y*   *y*

Y   *YY*   *Yy*

Offspring

y   *Yy*   *yy*

F_2 offspring

$\frac{1}{4}$ *YY* genotype = yellow-seed phenotype

$\frac{2}{4}$ *Yy* genotype = yellow-seed phenotype

$\frac{1}{4}$ *yy* genotype = green-seed phenotype

Figure 10.11. Analysis of Mendel's second experiment. The F_1 offspring (from Fig. 10.10) are each self-pollinated and become the parents of the F_2 generation. In one-fourth of the cases a gamete carrying a Y allele combines with a gamete carrying another Y allele. Another one-fourth result from combination of two gametes, each carrying a recessive allele. One-half the offspring result from combination of unlike gametes. The phenotypes of the offspring are three-fourths yellow-seeded and one-fourth green-seeded, a 3 : 1 ratio. (See also Figures 8.15 and 9.9 for other applications of probability.)

tail T can appear with equal chance. In a second toss, a head or tail can appear, still in equal probability and unaffected by the first toss. Considering two tosses, four combinations—$HH, HT, TH,$ and TT—are all equally probable, where HH represents two heads, HT represents a head followed by a tail, etc. The respective probabilities of the combinations HH, HT or $TH,$ and TT are $\frac{1}{4}, \frac{2}{4}, \frac{1}{4}$; these are the respective probabilities in the F_2 generation of $YY, Yy,$ and yy. Mendel's ratios can be described as the consequence of independent random pairing of alleles.

MENDELIAN INHERITANCE IN MAN. The principles of genetics first proposed by Mendel have been shown to apply to all organisms, including man. The individuality of each human being is due to the combined action of hundreds of genes. Many of the alleles of these genes are transferred from generation to generation in simple Mendelian fashion. Some simple examples of inheritance are given here. In following sections, more complex examples of inheritance are given, again using man as a representative organism.

Hair on the middle segment of a finger is inherited as a dominant trait. Absence of hair on this middle segment indicates a homozygous recessive individual, and presence of hair indicates a homozygous dominant or a heterozygous individual.

Example. What is the genotype of your two parents if they have mid-digital hair and you do not?

Your genotype must be hh, with one h allele coming from your father and the other from your mother. They each have an h allele and also an H allele since they both express the dominant trait. Therefore their genotypes are both Hh.

The disease Huntington's chorea results in severe nervous disorders and death. This trait is dominant, but it is not expressed until late in life. The disease is extremely rare and is known only in the heterozygous state.

Example. The famous folk singer and songwriter Woody Guthrie died of Huntington's chorea. What is the probability that Woody's son Arlo Guthrie inherited this trait?

The probability of receiving the H allele from his father is $\frac{1}{2}$, and the probability of receiving it from his healthy mother is 0. Therefore Arlo has a 50 percent chance of having the dominant allele.

MULTIPLE ALLELES

A number of genes have three, four, or more alleles, each expressing a different trait for the particular gene; these are called "multiple alleles." The ABO blood types in man are inherited as multiple alleles.

Blood groups are important in blood transfusions. The blood type must be determined before transfusion to avoid possible agglutination of the blood cells in the recipient's blood system. *Agglutination* is a clumping of cells into a large cluster (Fig. 10.12). It is an immune response caused by reaction of antibodies in the *plasma,* or fluid phase, of the blood with antigens in red blood cells. An *antibody* is an organic substance secreted by the body which inactivates a foreign protein. An *antigen* is a foreign protein that elicits an antibody reaction.

Three alleles, *A*, *B*, and *O*, exist for determining blood types. Each allele determines a specific antigen on an erythrocyte (red blood cell). Since only two alleles can be present in one organism, six combinations are possible (Table 10.3). An *O* allele determines a very weak, ineffective antigen. If the allele pair is *AB*, both antigens are present. There are four phenotypes: blood type *A* from *AA* or *AO*; blood type *B* from *BB* or *BO*; blood type *AB* from genotype *AB*; and blood type *O* from *OO*.

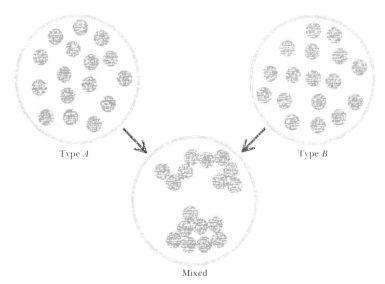

Type *A*

Type *B*

Mixed

Figure 10.12. Agglutination observed after mixing bloods of different types. Cells are clumped because of a reaction between antibodies in the plasma and antigens on blood cells.

Table 10.3. Genotypes and Phenotypes of the Human ABO Blood Types

Genotype	Antigens	Phenotype blood type	Antibodies
AA	*A*	*A*	Anti-*B*
AO	*A* and *O*	*A*	Anti-*B*
BB	*B*	*B*	Anti-*A*
BO	*B* and *O*	*B*	Anti-*A*
AB	*A* and *B*	*AB*	None
OO	*O*	*O*	Anti-*A* and anti-*B*

Each individual has red blood cells with *A*, *B*, or *O* antigens, and antibodies in plasma against whatever antigens are absent from the red blood cells. For example, type-*A* blood has *A* antigens on its erythrocytes and anti-*B* antibodies in its plasma.

With the introduction of a donor's blood into a recipient's blood, the antibodies and antigens of the donor and recipient are mixed. The donor's antibodies are absorbed in the recipient's body cells. However, the significant reaction is between the donor's antigens and the recipient's antibodies. A type-*A* recipient, with anti-*B* antibodies, cannot receive blood from a type-*B* donor with *B* antigens, because anti-*B* antibodies agglutinate the erythrocytes carrying the *B* antigens. Type-*A* blood can receive blood only from type-*A* (genotype *AA* or *AO*) and -*O* donors.

Type-*O* blood has anti-*A* and anti-*B* antibodies and thus cannot receive blood from type-*A*, -*B*, or -*AB* donors. A type-*O* donor has no effective antigens, and so his blood can be received by all types of blood. A type-*AB* recipient has no antibodies and can receive any type blood. Type-*AB* blood is called a "universal recipient," and type-*O* blood is called a "universal donor."

Blood groups have been an important factor in legal decisions involving paternity cases. Since the blood type is inherited from the parents, knowledge of the blood types of the two parents and the offspring might help determine whether or not a certain man is the father of a particular child. For example, two parents with type-*O* blood could not have a child with type *B* blood. However, although blood typing can prove that a man is *not* the father of a child, it can never prove that he is the father. For example, parents with genotypes *AO* (type-*A* blood) and *BO* (type-*B* blood) may have children with genotypes *AB*, *AO*, *BO*, or *OO* (all four phenotypes).

In a famous case of the 1940s, Charlie Chaplin was accused of being the father of a particular child. The mother's blood was type *A* (genotype *AA* or *AO*), Chaplin's blood was type *O* (genotype

OO), and the child's blood was type *B* (genotype *BB* or *BO*). The judge ruled that Chaplin was liable. Was he the child's father?

MULTIPLE GENES

Frequently, traits are not inherited in a simple manner. Characteristics such as stature are not due to a single gene but rather are the result of numerous genes acting in concert. Inheritance involving more than one gene is called "quantitative inheritance," and the several genes controlling a phenotype are called "multiple genes."

Blackness of the human skin is due to the presence of melanin pigment and is the result of genetic expression of multiple genes. These genes have incomplete dominance. In *incomplete dominance* the heterozygous state is intermediate between the two homozygous conditions. We arbitrarily assign a capital letter to designate alleles for blackness and a lowercase letter to designate alleles for whiteness.

The number of genes involved in determining color types is not known but can be estimated by the number of phenotypes. If there were only one pair of alleles, three phenotypes corresponding to the genotypes *AA*, *Aa*, and *aa* would result. However, there are more than three phenotypes. With two genes, each with two alleles, five types are possible, assuming equal expression of both genes. These would range from four doses of the gene *AABB* to none *aabb*. The observed number of color types exceeds five.

The model best fitting the observed color types and their distribution in mixed races is five genes, with two alleles for each gene. These have genotypes ranging from *AABBCCDDEE* (blackest) to *aabbccddee* (whitest), with 241 intermediate genotypes. If the five genes are equal in their expression, which appears to be the case, eleven phenotypes result. A genotype homozygous for black in all five genes would have ten doses of the gene; a genotype homozygous in four genes and heterozygous in one gene would have nine doses of the gene; etc. The effects of the gene are accumulative. Thus, melanin deposition in human skin is inherited as a quantitative trait controlled by action of multiple genes.

ENVIRONMENTAL INFLUENCES ON INHERITANCE

Genetic expression may depend upon the environment. The concept of environment includes not only the environment external to the organism but also the cellular environment.

Green in a pea pod results from chlorophyll synthesis controlled by a dominant allele (Table 10.1). In darkness, however, chlorophyll synthesis does not take place and the pod is yellow regardless of which alleles are inherited.

It has been suggested that the tone of the singing voice in man is controlled by a single pair of alleles. During childhood, males and females do not differ in their singing voices, indicating similar genotypes. At sexual maturity changes in hormone secretion result in a deepening of the voice in the male, caused by a change in the environment of the cells. Soprano in the female and bass in the male are both homozygous for one allele. The heterozygote is either a mezzosoprano or baritone. The homozygote for the other allele is an alto or tenor. The genotypes and phenotypes are given in Table 10.4. This gene also has incomplete dominance.

Table 10.4. Genotype and Phenotype of the Singing Voice in Man as Influenced by Sex Hormones

Genotypes	SS	Ss	ss
Phenotype			
Female	Soprano	Mezzosoprano	Contralto
Male	Bass	Baritone	Tenor

MUTATIONS

Occasionally a trait appears in an individual that was not present in either parent or any ancestor. This new trait may then be passed on to offspring. The sudden appearance, disappearance, or modification of a genetic trait is called a "mutation." Hugo De Vries, a Dutch geneticist (1848–1935), described numerous cases of new traits appearing in the evening primrose and called these "mutations." All mutations are a result of changes in either the composition or the quantity of DNA. A *gene mutation* involves changes in a single base (or only a few bases) in DNA, which, in turn, results in a change in a single amino acid in the protein coded by the DNA. The stabilities of genes vary. One gene may be extremely stable while another may mutate more frequently.

Mutations may also involve chromosomes, the cellular structures in the nucleus which contain DNA. Deletion or addition of a portion of a DNA molecule or loss or addition of whole chromosomes are called "chromosomal mutations." Mutations of chromosomes include:

1. Loss of a fragment of a chromosome
2. Addition of an extra segment to a chromosome
3. Loss of a chromosome
4. Gain of a chromosome
5. Gain of extra sets of chromosomes

The occurrence of a mutation of a gene from one allele to the opposite allele is a rare event. In most instances, appearance of a trait unseen in the previous generation is not a mutation but rather a recombination of alleles inherited but undetected for several generations. There are, however, thousands of well-documented cases of genetic mutations in bacteria, plants, and animals.

All cases of Huntington's chorea in the United States have been traced to ancestors immigrating to this country; mutations are very rare. A gene associated with chondrodystrophic dwarfism mutates more frequently. Chondrodystrophic dwarfism is an abnormality characterized by a normal trunk and shortened limbs. Approximately 1 birth in 10,000 has this trait. Only one-fifth of these dwarfism cases can be attributed to inheritance; the remainder are assumed to result from mutation.

There is no single mechanism or cause of mutation. Mutations may occur spontaneously or may be induced by a variety of chemical agents or by radiation. An agent causing mutation is called a "mutagen." Mutagens can interfere with the DNA organic bases, resulting in code changes, and they can cause chromosome breakage.

CHEMICAL MUTAGENS. Any chemical that reacts with DNA may be a mutagen. The chemical warfare agent mustard gas $S(CH_2CH_2Cl)_2$, first used against British troops at Ypres, July 1917, is a powerful vesicant, or blistering agent. It is also a mutagen.

Many commonly used chemicals may react with DNA, but these normally do not enter the nucleus in significant quantities. However, as technology supplies more kinds of chemicals in food, insecticides, cosmetics, etc., the exposure to potential mutagens will increase (Fig. 10.13). The genetic safety to human beings of new additives may be difficult to assess. A tenfold increase in mutation

Figure 10.13. DDT has been a major influence in the control of typhus; however, the indiscriminate spraying of crops has made DDT a serious environmental hazard. Here a Guatemalan child is sprayed to help control a typhus epidemic (courtesy of the United Nations).

frequency might go unnoticed for a number of years. A suitable experiment using animals might require one million animals to detect such a change in the mutation rate. Safety is presently checked by administering doses to test animals that are several thousand times the expected dose in humans.

RADIATION. Irradiation with ultraviolet, alpha, beta, gamma, and x-rays is potent in producing mutations. Ultraviolet radiation causes mutations by a photochemical reaction involving the bases of DNA. The light of these wavelengths has enough energy to break chemical bonds, and the resultant damaged molecule has altered genetic properties. In multicellular organisms, however, this radiation is relatively harmless because of its low penetration. On the superficial layers of the body, ultraviolet radiation causes other photochemical reactions such as suntan, sunburn, and, in susceptible persons, skin cancer, but its genetic effects are minimal.

Radiation from alpha, beta, gamma, and x-rays passing through matter strikes atoms, displacing electrons and ionizing molecules. The ionized molecule becomes very reactive, and if the molecule is DNA, a mutation may result. A hit on a base in DNA may result in a gene mutation; a hit on the backbone of DNA may cause a chromosomal mutation. Also, these ionizing radiations may activate a molecule near DNA, and this molecule then reacts with DNA, producing a mutation. Sources of radiation include cosmic rays, radiation from decay of isotopes, x-rays, and nuclear fission.

An important question concerning exposure to radiation is "What is a safe dose?" There are considerable data suggesting a linear relationship between dose of radiation and mutation rate (Fig. 10.14). The curve does not extrapolate to zero; i.e., there is a natural nonzero mutation rate even in the absence of exposure to radiation and chemical mutagens. This is the residual spontaneous mutation rate. Note that the curve suggests that all doses of radiation above zero cause some increase in mutation rate. If true, there is no safe dose of radiation; all exposures affect mutation rates. However, some work indicates that a minimum, or *threshold*, of radiation is needed to induce certain kinds of mutations. Below this threshold, the damage is repairable.

Recommendations concerning tolerable or safe levels of radiation are based upon a compromise between benefit from use

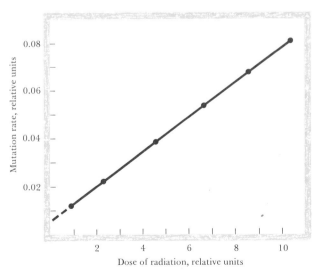

Figure 10.14. Effects of radiation dose upon mutation rates of genes (all arbitary units).

of radiation and harm from potential mutations. A slight increase in mutation rate is a risk worth taking if the benefit is to detect tuberculosis. Sunbathing increases exposure to cosmic radiation, but this genetic risk is usually taken for the supposed value of suntan. Different building materials have different quantities of natural radioactivity. One has a greater exposure to radiation in a brick house than in a wooden house and least exposure in a straw house. However, the relative safety of these building materials as shelter is in the reverse sequence, and the small effect on mutation rate is a risk worth taking.

DOWN'S SYNDROME. A group of symptoms called mongolism or Down's syndrome, after the English physician (John L. H. Down, 1828–1896) who described it, has been attributed to a change in chromosome number. This disease of humans is characterized by slanted eyes, small heads, large tongues, gaping mouths and peculiar handprints. Individuals have a friendly disposition and a capacity for memorization and mimicry of music, but other mental capacity is low. These individuals are very susceptible to infection and have a high mortality, but due to modern antibiotics many now survive into adulthood. Human beings normally have 46 chromosomes consisting of 23 pairs. The 23 different kinds of chromosomes can be individually identified and numbered 1 to 23. Down's syndrome may be caused by an extra chromosome 21 making 47 chromosomes or by attachment of a segment of chromosome 21 to another chromosome. In both cases the symptoms are similar, but inheritance differs. The extra chromosome may result from a failure to shuffle the chromosomes properly during the final cell division in the formation of the egg, so that the mother contributes two chromosome 21s and the father one chromosome 21 to the child. This is a random event and is unlikely to be repeated in a particular family. The occurrence is more frequent in older mothers, happening in 1 to 2 percent of the births. Childbearing is therefore not recommended in women past 40. Such a Down's syndrome idiot would be expected to pass the extra chromosome to approximately one-half the offspring. Sufficient reproduction has been observed to confirm this prediction.

Down's syndrome caused by an attached chromosome is also inherited. A mutation in the pedigree results in attachment of all or part of chromosome 21 to another chromosome, and this person becomes a carrier. Live progeny of a carrier have one chance in

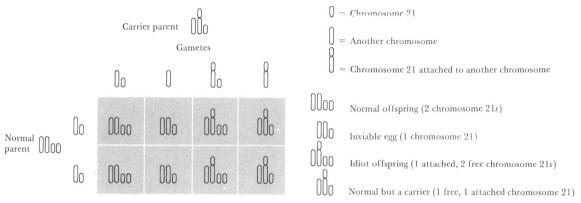

Figure 10.15. *Inheritance of Down's syndrome caused by a chromosome 21 attached to another chromosome. Mating a normal individual to a carrier having one attached and one free chromosome 21 could result in four possible types, depending upon whether one, two, or three doses of chromosome 21 are received. Only three of the possibilities have been observed. It is assumed that a fertilized egg with only one dose of chromosome 21 is inviable and never develops.*

three of receiving the attached chromosome and two other chromosome 21s (Figure 10.15). This individual is a Down's syndrome idiot. There is a one in three chance that a live offspring is a carrier, and a one in three chance of the offspring being normal. Thus two different chromosomal mutations may result in a similar phenotype.

MUTATION AND EVOLUTION

Mutation is a significant force in evolution. Chromosomal mutations may result in the addition or loss of DNA to the organism and may directly affect the expressed traits of the organism. Down's syndrome is an example of a chromosomal mutation which is detrimental to the survival of the organism. Chromosomal mutations may also increase the quantity of DNA available for subsequent gene mutations. Gene mutations introduce new traits into the population.

Mutations are usually of no benefit to an organism. Any random change in the genetic composition is likely to be detrimental. A random change in the wiring of a television set would probably decrease the quality of the picture. However, a very small proportion of mutations may be of value to the organism in that they enhance the ability of that organism to survive or reproduce. A beneficial mutation, then, is one that gives a competitive advantage to the individual having the mutation. The new (mutated) gene would be passed on to at least some of the offspring, giving the off- spring a competitive advantage over those individuals that do not possess this gene. Because more of these individuals survive, or reproduce, than individuals without the beneficial gene, the gene becomes more frequent each generation and indeed may com- pletely eliminate the alternate gene. The process by which the gene becomes more frequent is called "natural selection" if it is occur- ring in nature and "selective breeding" if the selection is being practiced by man. Several mutations over many generations, together with the process of selection, result in a change, or *evolu- tion*, of the organism.

POPULATION GENETICS

A *population* is defined as a set of organisms living in a common location and naturally mating with each other. The subject of the genetic composition of populations and the behavior of genes in them is called "population genetics." The collection of all the genes for all traits in a population is the *gene pool* for that population. A census of the relative abundance of alleles in the gene pool for each gene is called the "gene frequency." Specifically, popu- lation genetics deals with measurement of gene frequency, mu- tation rates, and natural selection and is thus akin to the study of evolution.

THE Rh FACTOR

A few percent of all newborn infants suffer from a disease called "erythroblastosis fetalis," a disease characterized by anemia caused by destruction of erythrocytes (red blood cells). Infants suffering from this disease have been found to carry an antigen on their erythrocytes that is similar to an antigen found in rhesus-monkey blood. This antigen is called the "rhesus factor," or "Rh factor." An individual carrying this antigen is designated *Rh positive*. However, not everyone carries the Rh antigen, and those that do not are termed *Rh negative*.

The mother of an erythroblastosis fetalis infant is always Rh negative. Erythroblastosis fetalis is caused by incompatibility between an Rh positive fetus and an Rh negative mother. Specifically, the mother elaborates antibodies against the Rh antigens of the fetus. Antibody synthesis, however, follows a different pattern than that found for the *ABO* blood groups. Instead of conducting continuous synthesis, the mother makes antibodies only after exposure to the antigen, which is the most common pattern of antibody formation in immunity.

Immunity is the resistance of the body to infection by other organisms. Bacteria and viruses which cause disease have specific organic substances on their cell surfaces. These specific substances serve as antigens and elicit antibody synthesis. Antibodies attack antigens and inactivate the invading foreign organism. This immunity against a particular organism may be retained for a considerable length of time after infection. Artificial immunization is generally accomplished by injecting isolated antigens, or dead cells, of a potentially harmful foreign organism. In erythroblastosis fetalis, antigens on the erythrocytes elicit antibody formation, and these antibodies destroy the erythrocytes as if they were foreign organisms.

The placenta is a structure that exchanges nutrients, respiratory gases, and wastes between the fetus and the mother; the fetus depends upon the placenta for life. Exchange of substances across the placenta occurs through membranes; the fetus and mother have separate blood systems that do not normally mix. The Rh antigens on the red blood cells do not normally cross the placental membrane, and the Rh negative mother should not be exposed to Rh antigens. However, temporary breaks in the membrane allow some mixing of the bloods especially during birth of the baby. If there is Rh incompatibility, maternal antibody production results. Generally several pregnancies are required before these antibodies build up to dangerous levels.

An obvious measure to avoid erythroblastosis fetalis in later children is simply contraception. It is possible, however, to prevent development of antibodies in the mother by injecting Rh antibodies into the mother shortly after birth. These antibodies block further production of antibodies in the maternal system by interfering with the action of the Rh antigens which are present in high concentrations immediately after birth. Such treatment might eliminate this disorder in the near future. If measures are not taken to prevent formation of Rh antibodies, subsequent pregnancies might produce infants that are severely affected by erythroblastosis fetalis. An infant so affected usually can be treated by complete replacement of its blood with blood that does not contain Rh antibodies.

Rh factor may also be important in blood transfusions. An Rh negative individual receiving a transfusion of Rh positive blood reacts by making antibodies against the Rh antigen. Subsequent transfusions with Rh positive blood may result in death.

Genetic studies of rhesus antigens indicate that Rh positive is inherited as a simple dominant trait. The alleles are designated Rh^+ for the dominant allele and Rh^0 for the recessive allele. The three genotypes are Rh^+Rh^+, Rh^+Rh^0, and Rh^0Rh^0. Rh blood types are inherited independently of other blood groups.

THE HARDY–WEINBERG LAW. Persons with Rh positive blood types are more common in most human populations than persons with Rh negative blood. The Rh^+ allele has a higher frequency than the Rh^0 allele. Of the phenotypes in nonnative Americans, 2.25 percent are Rh negative and 97.75 percent are Rh positive. Other populations have characteristic distributions of Rh phenotypes. Within any population the Rh distribution remains rather constant from generation to generation. In 1908 Godfrey H. Hardy (1877–1947) and Wilhelm Weinberg (1862–1937) independently formulated a theoretical basis for this constancy.

Suppose p and q represent the fraction of alleles in the gene pool that are Rh^+ and Rh^0, respectively; thus,

$$p + q = 1$$

Then, in a population of prospective fathers, p is the fraction of Rh genes containing the Rh^+ allele; also, in the prospective mothers, p is the fraction of Rh genes with the Rh^+ allele. In each case q represents the fraction of Rh genes containing the Rh^0 allele.

A certain fraction of the gametes formed by the parent generation carry Rh^+ alleles; this fraction equals p. Gametes combine to form zygotes, which become the individuals of the offspring generation. The fraction of Rh genes carrying Rh^+ alleles in the offspring generation is again p. Similar constancy in the frequency of the Rh^0 allele is observed from one generation to the next. Here the fraction q represents the fraction of Rh genes with the Rh^0 allele in the parents, gametes, and offspring. The ratio of one allele to the other remains constant from generation to generation because the parent population makes gametes with allele frequencies identical to their own.

The offspring, taken collectively, have a genotypic distribu-

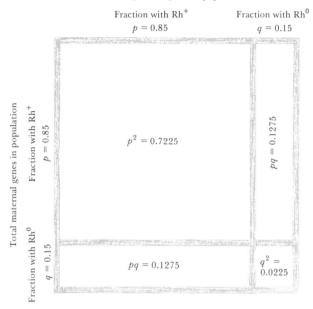

Figure 10.16. Graphic presentation of the Hardy-Weinberg equation. The fraction of Rh genes with the Rh⁺ allele is p, and the fraction with the Rh⁰ allele is q. The fraction of offspring with two Rh⁺ alleles is represented by p²; the fraction having two recessive alleles is q². The fraction 2pq represents the heterozygous individuals. The area of each rectangle is proportionate to the size of the fraction.

tion that can be predicted from the frequencies of alleles in the parent population (Fig. 10.16). In a certain fraction of offspring two Rh⁺ alleles are present; this fraction equals p^2. Similarly, the fraction of individuals with two Rh⁰ alleles equals q^2. The fraction having both alleles is $2pq$.

$$p^2 + 2pq + q^2 = 1$$

This equation is a formula for the frequencies of genotypes in a population if two alleles exist for the gene. This is the Hardy-Weinberg equation. The combination probabilities it describes can be applied to coin tossing. If several hundred coins are tossed randomly on the floor, the fraction landing heads will be $p = \frac{1}{2}$. If coins are withdrawn randomly in pairs, the fraction of pairs with two heads will be $p^2 = \frac{1}{4}$. With tails, the fractions are the same. The fraction of pairs with one head and one tail equals $2pq$ or $\frac{1}{2}$.

Example. What are the genotype frequencies for Rh blood types when the fraction p (for Rh$^+$) equals 0.85 and the fraction q (for Rh^0) equals 0.15 (the values for nonnative Americans)?

The fraction of individuals homozygous for the Rh$^+$ allele is

$$p^2 = 0.85 \times 0.85 = 0.7225$$

The fraction of homozygous recessive individuals is

$$q^2 = 0.15 \times 0.15 = 0.0225$$

The fraction of heterozygous individuals is

$$2pq = 2 \times 0.85 \times 0.15 = 0.255$$

The sum of the three fractions should equal 1:

$$p^2 + 2pq + q^2 = 0.7225 + 0.255 + 0.0225 = 1.000$$

The Hardy-Weinberg equation assumes random mating. An example of nonrandom mating is in height. Tall people tend to select tall mates, and short people tend to select short mates. Therefore alleles for tallness are not randomly paired with alleles for shortness. The frequency of heterozygous individuals is reduced; that is, $2pq$ is less than predicted. The frequency for tallness is greater than predicted by p^2, and the frequency for shortness is greater than predicted by q^2.

Random mating generally occurs between individuals in regard to blood groups. The frequencies of Rh$^+$ and Rh^0 alleles appearing in gametes and offspring are shown in Fig. 10.16. These frequencies are the same as in the parental generation. *The Hardy-Weinberg law* states that the frequency of alleles within a gene pool does not change from generation to generation. Gene frequencies in offspring populations are similar to gene frequencies in parent populations.

The Hardy-Weinberg law holds even in cases where the Hardy-Weinberg equation may not fit the data. In the case of nonrandom mating, e.g., the selection of tall mates by tall people, the Hardy-Weinberg equation does not hold but the Hardy-Weinberg law does. The frequency of tall alleles in one generation will equal the frequency in the next.

EXCEPTIONS TO THE HARDY–WEINBERG LAW. There are important exceptions to the Hardy-Weinberg law. Mutations may add alleles to a population. Mutation rates, however, are low, and changes in gene frequency from one generation to the next are slight. If many generations are examined, or if the mutation involves a rare phenotype, a change in allele frequency might be detected.

Genes may mutate in either direction. An allele F mutates to f with a particular frequency, but f may mutate back to F. The mutation rate from F to f is the product of the number of F alleles times the chance that any single F allele will mutate. Similarly the rate from f to F is the product of the number of f alleles times the individual chance that one f will mutate to F. If the rate $F \rightarrow f$ exceeds the rate $f \rightarrow F$, the number of f alleles must increase and the number of F alleles will decrease. Then the rate $F \rightarrow f$ will decline and the rate $f \rightarrow F$ will increase until the two rates become equal. At equilibrium the number of F alleles changing to f alleles equals the number of f alleles mutating to F. There is no net change in the number of either F or f alleles.

In small populations, marked differences in gene frequency between generations sometimes occur. Assume a population size of two, with genotypes of GG for one individual and Gg for the other; that is, $p = 0.75$ and $q = 0.25$. Three offspring might have genotypes of GG, GG, and Gg. Now $p = 0.83$ and $q = 0.17$. If the individual with the g allele happens to die, the frequency of G becomes 100 percent. Such shifts in allele frequencies in small populations are appropriately called "genetic drift." Genetic drift commonly occurs in very tiny populations or in populations that have wide fluctuations in numbers. The best example is in insects that have few overwinter survivors. The genetic composition of thousands of individuals in the summer may be a result of an accidental survival of one or two individuals with unusual traits.

LAW OF NATURAL SELECTION. The most important factor affecting changes in gene frequencies is *natural selection,* which is the differential transfer of alleles from one generation to another caused by differences in survival of individuals or in reproductive success. Natural selection is the most important exception to the Hardy-Weinberg law. Natural selection may be negative if the presence of an allele reduces the ability to survive and reproduce or positive if the allele enhances survival and reproduction.

A model for natural selection was developed by Charles Darwin (1808–1882), who proposed that all species of organisms on earth descended from common ancestors by the process of natural selection. Darwin did not originate the idea of evolution, but his model of natural selection provided a mechanism to account for evolutionary changes. Although Charles Darwin was ignorant of the mechanism of inheritance, his theory of natural selection is compatible with genetic principles. Today we know that natural selection cannot account completely for evolution; mutation must provide variability in traits. Although Darwin observed variations in traits in nature, he was unable to account for their origin.

SICKLE-CELL ANEMIA

The genetic disease sickle-cell anemia is common in Africans or persons of African descent. This disease is characterized by abnormally shaped erythrocytes (red blood cells). These abnormal cells are fragile and have a short lifespan. This condition can result in a shortage of red blood cells, a disease called "anemia."

The abnormal blood cells are caused by a hemoglobin which differs from the normal hemoglobin by substitution of a single amino acid. Individuals homozygous in this abnormal hemoglobin have severe anemia and greatly reduced survival and reproduction. (It would be expected that the allele would be eliminated from the population in a few generations, but this is not the case.) Heterozygous individuals have both kinds of hemoglobin but no anemia. Individuals with this abnormal hemoglobin are resistant to malarial attack caused by a parasite which infects erythrocytes, and survival is enhanced in tropical climates.

SEXUAL REPRODUCTION

Sexual reproduction allows combinations and recombinations of genes by placing two or more genes together in one organism to effect a new phenotype. We may speculate that very light or very dark skin pigmentation in man arose because of recombination of genes. Assume that each of the five alleles for light skin originated independently by mutation. Two individuals might have been found that had genotypes of *AaBB* and *AABb* for two of the five sets of genes. Both would have lighter than normal (dark) skin because these genes have incomplete dominance. Mating might result in genotypes of *AaBb*, *AaBB*, *AABb*, and *AABB*. The *AaBb* individual would be lighter than either parent and may have survival advantages in cold climates. A mating of one *AaBb* with another in a future generation can result in some individuals with a genotype of *aabb* and still lighter phenotype. These individuals would have even greater survival advantages, and these alleles would tend to become more frequent in northern populations. Thus, a sequence of mutation, recombination, and selection can account for evolution of white skin from black skin.

Sexual reproduction increases the range of phenotypes within a population and thus gives evolutionary advantages to the organisms practicing it. As an example of how recombination can affect new phenotypes, more than 99 percent of the "sudden appearances of new traits" which Hugo De Vries first described as mutations were caused by recombinations of genes already present. Although his concept of gene mutation was correct, his data are now accounted for by new combinations, not new genes.

The physiological, anatomical, and, especially, behavioral features of sexual reproduction, which are often those most readily noticed, only serve to ensure that the process takes place. Many organisms have come to depend exclusively upon sexual reproduction; therefore, the urge to reproduce must be great in order to ensure that it occurs. Many plants and a number of animals can reproduce either sexually or asexually. For example, potatoes normally reproduce asexually by sprouts from the tuber. [The evolutionary significance of having both capabilities is not clear, but perhaps it allows propagation of a desirable combination once it has occurred, without being lost by obligatory reproduction.]

MITOSIS

In multicellular organisms the fertilized egg is the precursor of the thousands of cells within the organism. Each cell must receive an exact copy of all DNA contained in the fertilized egg cell. Each time a cell divides into two cells, DNA is replicated and each of the two daughter cells receives exactly one-half. The cellular process distributing the DNA correctly into the daughter cells is called "mitosis." Mitosis also results in a distribution of approximately equal quantities of cytoplasm between the two daughter cells (see p. 589).

Mitosis acts upon chromosomes which carry genes. Replication of DNA is associated with formation of a new chromosome called a "daughter chromosome." During cell division, mitosis en-

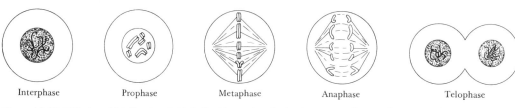

Interphase   Prophase   Metaphase   Anaphase   Telophase

Figure 10.17. Mitosis, artificially separated into five phases. In mitosis, chromosomes line up along a midplane of the cell and then migrate to opposite ends of the cell. This is followed by division of the cell into two, each with identical genetic material.

sures that the daughter chromosome and the chromosome copied go to different cells. Most organisms have more than one chromosome per cell, and the behavior of these chromosomes must be coordinated during mitosis.

Mitosis is arbitrarily separated into phases (Fig. 10.17). In *interphase,* cells are not actively dividing and chromosomes are found as chromatin. *Chromatin* material is an elongated, diffuse form of the chromosomes. During *prophase,* chromatin material shortens into chromosomes, revealing newly formed chromosomes (and DNA). Each daughter chromosome is paired with the chromosome copied. In *metaphase* these chromosomes arrange along an equatorial plane in the middle of the cell. In *anaphase* the chromosome pairs separate and one complete set moves toward each pole of the cell. In *telophase* the cell divides into two and the two cells then enter another interphase. Thus a cell divides into two cells, with the mitotic process providing for proper distribution of genetic material to the two daughter cells.

Mitosis is almost identical in plants and animals, differing only in the method of cell division in telophase. In animals the cell is pinched into two cells (Fig. 10.17), whereas in plants a plate forms between the two daughter cells (Fig. 10.18).

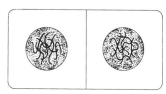

Figure 10.18. In plants, cells are separated following mitosis by the formation of a plate between the cells.

MEIOSIS

In sexual reproduction the egg, or female sex cell, fuses with a male gamete to form the *zygote.* The zygote, or fertilized egg, and the adult organism have twice as much DNA as either gamete. Meiosis is a process accompanying gamete formation whereby both chromosome number and DNA are reduced by one-half. Meiosis is a necessity for sexual reproduction because it maintains constancy in the quantity of DNA from generation to generation.

The behavior of chromosomes in meiosis (Fig. 10.19) superficially resembles mitosis. In meiosis one replication of DNA and

First division
I

Second division
II

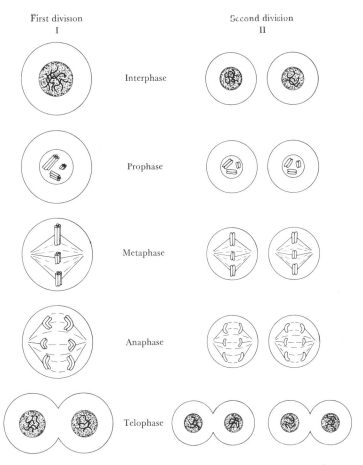

Interphase

Prophase

Metaphase

Anaphase

Telophase

Figure 10.19. Meiosis, artificially separated into phases. Chromosomes replicate once, and the cell divides twice, resulting in a 50 percent reduction in chromosome number.

the chromosomes takes place during two cell divisions, thus achieving chromosome reduction. Meiosis is separated into phases, which are given the same names as in mitosis except that the phases are identified as I for the first division and II for the second division. The phases of meiosis are: prophase I, metaphase I, anaphase I, telophase I, interphase, prophase II, etc.

Chromosomes and DNA replicate before meiosis begins. In prophase I the chromosomes shorten and maternal chromosomes pair with their equivalent, or homologous, paternal chromosome. In man, for example, chromosome 21 from the mother pairs with chromosome 21 from the father. At this stage there are two maternal and two paternal chromosomes for each kind of chromosome. This group of four chromosomes is called a "tetrad" (Fig. 10.20). The chromosome groups line up along the equatorial plane of *metaphase I* and separate in *anaphase I*. The maternal chromosomes of any group go to one resulting cell, and the paternal go to the other. For example, the maternal chromosome 11 and its daughter go to one cell, and the paternal chromosome 11 and its daughter go to the other cell. Each group, however, behaves independently of other groups, and so one cell does not receive all maternal chromosomes; for example, one cell may receive maternal chromosomes 11, 13, and 16 and paternal chromosomes 12, 14, and 15.

DNA is not replicated preceding the second meiotic division, and so the reduction in chromosome number occurs here. The second meiotic division proceeds through metaphase II, anaphase II, etc. The final result of this division and the biological goal of meiosis is the reduction in chromosome number.

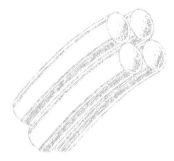

Figure 10.20. Tetrad observed in prophase I. A tetrad consists of four chromosomes: two maternal chromosomes and two paternal chromosomes. These separate in meiosis, ultimately, one going to each of four cells.

DEVELOPMENT

The goal of sexual reproduction is fertilization of one gamete by another. A new individual develops from this union. A general pattern of development in plants and animals is for initial stages of embryonic growth to occur within the body of the female, which provides protection for the developing embryo and, in some cases, nourishment. In mammals, including man, and in some other animals, the placenta provides for exchange of nutrients and waste products between the mother and embryo. In higher plants, nutrition is obtained from endosperm. *Endosperm* is a highly nutritious food stored within a seed. Some fertilized eggs develop outside the body of the parent; generally these are large eggs with considerable stores of nutrients, i.e., yolk, within the egg.

Embryonic and postembryonic growth of the organism takes place in an orderly, sequential manner. In a tracheophyte plant (higher plant) fertilization of the egg by a nucleus from a pollen grain initiates embryonic development (Fig. 10.21). Cells increase in number by a series of mitotic divisions and grow by assimilating endosperm in the seed. As the embryo develops, cells differentiate into specialized structures. *Differentiation* is change in cell structure or function during development. Some cells differentiate into cotyledons, others into epicotyls, and still others into the hypocotyls.

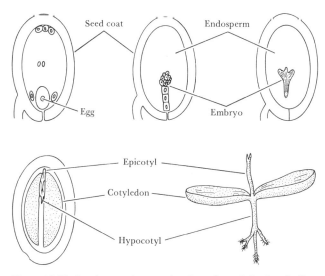

Figure 10.21. Development in a tracheophyte plant. Following fertiliza-tion by a nucleus from a pollen grain, the egg divides by mitosis to form an embryo. In embryonic development differentiation of cotyledons, epicotyls, and hypocotyls occurs. Following seed germination, epicotyl differentiates into stems and leaves, and hypocotyl becomes roots.

Development continues following seed germination. Epicotyl be-comes the stem and leaves, and hypocotyl becomes the root.

In animals, development is initiated by fertilization of the egg by the sperm. Early development in *Amphioxus* (Fig. 10.22) is entirely by mitosis, with no growth in size. Cells become smaller as their numbers increase. Nutrition depends entirely upon the yolk in the egg. *Amphioxus* is an inconspicuous primitive chordate and is chosen as an example because it clearly shows the major develop-mental events of vertebrate embryology. Developmental patterns differ among different animals, but the processes are similar in all multicellular animals although they might look different. An early stage in *Amphioxus* development is a hollow ball of cells called a "blastula." The cavity in the blastula is called a "blastocoel." *Gastrulation* is an invagination into the blastocoel and results in another cavity, the archenteron (the precursor of the gut). Gastru-lation is analogous to forcing a rubber ball to fold inward by press-ing in with one's finger. A double layer embryo called a "gastrula" results.

The outer layer of the gastrula is called the "ectoderm"; the inner layer is called the "endoderm." Next, migration of cells from

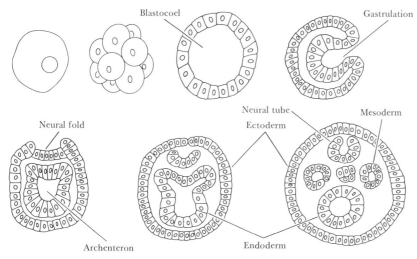

Blastocoel Gastrulation

Neural tube
Neural fold   Ectoderm   Mesoderm

Archenteron   Endoderm

Figure 10.22. Early development in a chordate animal Amphioxus. A series of mitotic divisions results in formation of the hollow blastula. Gastrulation, or invagination into the blastocoel, produces a gastrula with two layers, endoderm and ectoderm. Mesoderm is a third basic layer to differentiate. These three layers further differentiate into specific adult organs. Neural tube (to become the nervous system) forms by folding of ectoderm. Similar processes occur in the development of all multicellular animals, although details differ considerably.

the endoderm forms a middle layer of cells called "mesoderm." The ectoderm, endoderm, and mesoderm are the three primary germ layers of most animals. These germ layers are the precursors of all the tissues and organs of adults. Differentiation of ectoderm results in skin and the nervous system. Endoderm gives rise to the gut and all organs attached to the gut, such as the liver, pancreas, and lungs. Mesoderm becomes muscle, bone, and the circulatory and excretory systems.

Differentiation is affected by the cellular environment, which is a function of the position of a cell in the developing organism. A cell located in the surface layer obviously has a different cellular environment than a cell in the interior. Subtle differences in environment may have profound effects upon the kind of cell that differentiates. A large part of environmental influences consists of interaction with other cells. The action of cells in one tissue upon differentiation of cells in another tissue was called "embryonic induction" by Hans Spemann (1869–1941). As a tissue grows and makes contact with another tissue, a chemical substance, possibly RNA, may be released, which diffuses to the tissue to be induced. The inducing agent acts upon cells and affects dif-

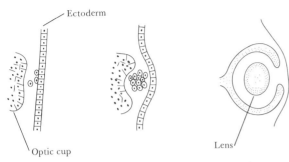

Figure 10.23. Embryonic induction of the lens in the vertebrate eye. The optic cup underlying ectoderm induces ectodermal cells to migrate and differentiate into lens.

ferentiation. For example, growth of the optic cup induces ectoderm to differentiate into a lens for an eye (Fig. 10.23). A chemical (RNA) diffuses from the optic cup and induces the ectodermal cells. It can be shown that any ectodermal cells can become lens. Transplantation of the optic cup to another site on the body results in lens formation at this site.

GENES AND DEVELOPMENT

In the fertilized egg, DNA has all the genetic information needed to direct cellular activities in the embryo and adult. Little genetic information is expressed in the egg, and only a small part is expressed in any one cell. During development, the *genome*, or genetic content, of the cell must unfold as a result of differentiation. This unfolding of the genome is a sequential expression of various portions of the genome. Further, the genome unfolds differently in different cells since only certain genetic traits are expressed in any one cell. Development, then, can be viewed as differentiation of genetic expression.

The molecular mechanisms controlling genetic expression during development have long puzzled biologists. A mechanism must be present which turns genes on and off at appropriate times. For expression of endoderm, a particular part of the genome is turned on; when this tissue differentiates, these genes, having done their particular job, are turned off and another part of the genome is turned on.

The regulatory system proposed for control of protein synthesis in bacteria by Jacob and Monod may be applied to developing systems. This model can account for embryonic induction if the embryonic-inducing substance acts upon the repressor substance, which, in turn, affects the operator gene and operon. Sequential secretion of embryonic inducers effects sequential production of different *m*RNA's. Genes are therefore sequentially expressed. Thus, differentiation is an alteration in codon expression.

Development can be viewed as a series of embryonic inductions: tissue *A* affects differentiation of *B* into *C*; *C* induces *D* to differentiate to *E*; *E* affects *F*; etc.

PHENYLKETONURIA—A GENETIC DISEASE. The genetic disorder phenylketonuria (PKU) is one cause of mental retardation in man. Several aspects of this disease will serve to summarize genetic principles. In normal metabolism the amino acid phenylalanine is converted to tyrosine by addition of a hydroxyl group to the benzene ring (Fig. 10.24). The reaction is catalyzed by the enzyme phenylalanine hydroxylase. If DNA coding is faulty, i.e., if the gene is defective, the protein synthesized on the ribosomes of the endoplasmic reticulum is not an active enzyme. This protein will not convert phenylalanine to tyrosine.

One result is a deficiency in tyrosine, and because tyrosine is a precursor of melanin, there is reduced pigmentation. Some tyrosine is supplied in the diet. Further, phenylalanine is converted by an alternate pathway to phenylpyruvic acid (a phenylketone). This substance is toxic to developing cells in the brain and results in mental retardation. Phenylpyruvic acid appears in the urine; hence the name phenylketonuria. PKU may be detected clinically by reduction of ferric chloride by the urine.

At present there is no cure for the mental retardation caused by PKU; abnormal development has occurred irreversibly. However, the disease can be prevented if diagnosed early. A special diet, low in phenylalanine, i.e., no meat, eggs, fish, or natural milk, must be followed strictly for the three or four years during which the nervous system develops. Phenylpyruvic acid has also been reported to cross the placenta and kill or retard a normal embryo in a PKU mother.

PKU IN THE GENETIC POOL

PKU is inherited as a simple recessive trait. In the homozygous recessive (*kk*) no active phenylalanine hydroxylase is pro-

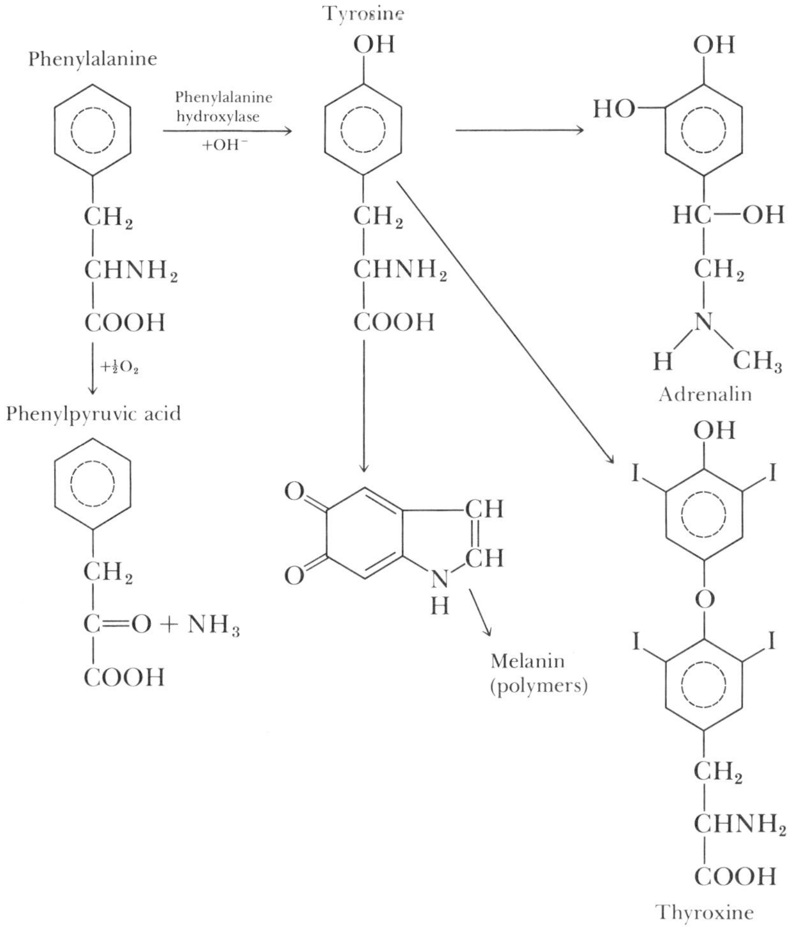

Figure 10.24. Phenylalanine is normally converted to tyrosine, catalyzed by the enzyme phenylalanine hydroxylase. In PKU the gene is defective, and this enzyme is inactive. Phenylalanine is converted to phenylpyruvic acid (a phenylketone) by an alternate pathway, and this compound accumulates. Two hormones, adrenalin from the adrenal gland and thyroxine from the thyroid gland, are also derived from tyrosine.

duced. In the heterozygous state (Kk) a subnormal amount of active enzyme is present, which can be detected using special tests. The heterozygous individual has no PKU symptoms.

Using the Hardy-Weinberg formulas we can calculate the distribution of genotypes and frequency of alleles in the population. PKU appears in approximately 1 birth in 10,000. Also, there are about 20,000 PKU mentally retarded persons in the United States in a population of 200 million (20,000 per 200 million = 1 per 10,000). From either figure the frequency of the recessive genotype can be estimated as 1 per 10,000, or 0.0001. Solving the Hardy-Weinberg equations:

$$p^2 + 2pq + q^2 = 1 \qquad \text{Genotypes}$$
$$p + q = 1 \qquad \text{Alleles}$$

If q^2 equals 0.0001, then q is the square root of this number

$$q = \sqrt{q^2} = \sqrt{0.0001} = 0.01$$

and p is the difference between q and 1

$$p = 1 - q = 1 - 0.01 = 0.99$$

Thus, 0.01 (or 1 percent) of the alleles is for the recessive allele and 0.99 (or 99 percent) are for the normal trait.

The other genotype frequencies can be calculated from the allele frequencies. The frequencies of the three genotypes are

$$kk = q^2 = (0.01)^2 = 0.0001 \qquad \text{(PKU)}$$
$$Kk = 2pq = 2(0.99)(0.01) = 0.0198$$
$$KK = p^2 = (0.99)^2 = 0.9801$$

Thus, 1 person in 10,000 has PKU (0.01 percent); another 198 persons in 10,000 (1.98 percent) has one allele for PKU; and 98.01 percent are homozygous dominant.

Before prevention of PKU by diet, the homozygous recessive individuals had almost no reproductive success. Therefore, these alleles were not passed to the next generation and were eliminated from the population. The rate at which the allele was eliminated can be estimated. For 4 million births/yr, 400 infants (1 in 10,000) would be expected to have PKU. Since a PKU individual then could not reproduce, this represents two PKU alleles lost from the gene pool for each individual, or 800 PKU alleles lost from a total of 400 individuals. The Hardy-Weinberg frequency would be expected to be altered by this amount per year.

However, the fraction of recessive genes in the population has remained fairly constant. For this to be possible, there must be a mutation of K to k at a rate that exactly balances the decrease caused by poor reproduction. Therefore 800 mutations must occur each year. In 4 million births, the parents donated a total ot 8 million alleles $(K + k)$. Of these, 99 percent are the dominant allele K:

$$8 \times 10^6 \times 0.99 = 7.92 \times 10^6$$

Of the 7.92×10^6 K alleles, 800 mutate from K to k. The fraction of K mutating to k each year is

$$\frac{800}{7.92 \times 10^6} = 1.01 \times 10^{-4}$$

At present, most states require routine testing for PKU in newborn infants. By detecting PKU, mental retardation can be prevented and these individuals can live normal lives. The loss of the alleles from the population is reduced from 800/yr to almost zero. Therefore, we can expect an accumulation of recessive alleles for PKU, that is, an increase in the fraction q. This is because the mutation rate remains at 0.000101 and each year these alleles are added to those already present. In 100 yr the frequency of k alleles will be increased by

$$100 \times 0.000101 = 0.0101$$

and the frequency of k will be

$$0.01 + 0.0101 = 0.0201$$

The frequency of K will decrease to 0.9799, and the Hardy-Weinberg distribution of the genotypes will be:

Genotype	Fraction
KK	$p^2 = (0.9799)^2 = 0.9602$
Kk	$2pq = 2(0.9799)(0.0201) = 0.0394$
kk	$q^2 = (0.0201)^2 = 0.0004$

Thus, in 100 yr the fraction of k recessive alleles will double and the number of homozygous kk recessives will be expected to quadruple.

Man can and is altering evolutionary relationships. Although saving homozygous recessive individuals from mental retardation is certainly a worthy program, we should be aware of the possible evolutionary consequences. Several thousand years in the future (if man is lucky and is still around) the frequency of k allele may be very large. There is no guarantee that sufficient resources will always be available for preventing the genetic disease, in which case a large number of mentally retarded individuals would result. Until then, however, *PKU* remains largely an inconvenience to the parents of affected children.

There is small danger that a fraction q of gene k will become 100 percent. The mutation rate of 800/yr is a net rate of K to k. The reverse mutation k to K also occurs with a certain frequency. As the population having k increases, the number mutating to K will increase until the number mutating from k to K exactly equals the number mutating from K to k. At this equilibrium, the frequency of each gene will have a steady value.

GENETIC COUNSELING

Another aspect of PKU deserves attention. The kk homozygous individual can result from three sets of parents:

Parent's genotype	Offspring's genotype	Fraction with PKU
$KK \times KK$	KK	0
$KK \times kk$	Kk	0
$Kk \times Kk$	KK, Kk, kk	$\frac{1}{4}$
$Kk \times kk$	Kk, kk	$\frac{1}{2}$
$kk \times kk$	kk	1

Since heterozygous individuals can be detected, an opportunity to advise against certain marriages is presented; this is called *genetic counseling*. As long as counseling is against certain matches in marriage rather than against marriage completely for certain persons, the allele frequency will not be affected. Rather, by appropriate mate selection, the distribution of genes can be controlled so as to avoid homozygous recessive combinations. Genetic counseling against reproduction by couples who both have a recessive allele, if effective, would reduce current recessive allele frequencies and eliminate PKU phenotypes. Each year in the United States about eight hundred heterozygous PKU children are born of these parents. If these parents did not reproduce, a total of 1,600 recessive alleles would be eliminated from the gene pool each year: 800 for the heterozygous children and 800 for the PKU children that would not be born.

Where the suspicion of a genetic disease exists, as from a family history of genetic defects, a genetic counselor can examine the chromosomes of the developing embryo. He removes amniotic fluid (the liquid surrounding the embryo) which contains some cells given off by the embryo. For several dozen genetic defects he can tell the prospective parents definitely whether or not their developing embryo will be born with the defect. Although the idea of abortion is not now universally accepted, it is an option available to parents of a defective embryo. They can choose to carry to birth only those conceptions, governed by chance, which are free of the defect.

THE ORIGIN OF LIFE

Natural scientists now believe that the earth has a finite age. Analysis of the decay of radioactive elements gives an estimate for the age of the earth and moon of about 4 billion years. The earth was first molten, or melted rock, and then cooled sufficiently for formation of a crust of solid rock. At later stages, as further cooling occurred, water condensed on the surface to become the sea. The atmosphere above this primordial ocean differed from the present-day atmosphere. Scientists believe that our present atmosphere of nitrogen and oxygen is the result of millions of years of biological activity. The primitive atmosphere is thought to have consisted of water vapor, ammonia, methane, and hydrogen. Life arose in this environment.

In 1925 the Russian biochemist Aleksandr Ivanovitch Oparin (b. 1894) proposed an elegant model to account for this spontaneous generation of life. He proposed that superheated water vapor reacted with carbon compounds to give rise to simple organic hydrocarbons. Hydrocarbons were converted to alcohols, aldehydes, ketones, and organic acids. These hydrocarbon derivatives reacted with ammonia to form amines and other nitrogenous compounds. The early seas became rich in these organic materials, and some of these molecules in solution entered chemical condensation reactions to become polymers and macromolecules. These macromolecules came together as colloidal particles and gels. These more complex systems were thought to be able to adsorb and absorb organic material from their environment and thus increase in size and complexity. These complexes were continually dispersed and reassembled as temperatures and other conditions fluctuated. More stable molecular arrangements persisted in favor of less stable assemblages. Thus, selective processes resulted in development of stable physicochemical organization.

A similar model was developed independently in 1928 by John Haldane (1892–1964), a British biologist. Haldane suggested that synthesis of organic molecules resulted from activation of morganic compounds by ultraviolet light. These organic molecules, synthesized in the atmosphere, became dissolved in the oceans, giving them the consistency of soup. At some point in time (according to both Oparin and Haldane) certain molecular groupings exhibited properties of primeval organisms, i.e., the ability to metabolize other organic compounds and assemble macromolecules for self-replication. These primitive organisms might have later acquired the capability for synthesis of organic compounds from inorganic materials.

Although such models describing the origin of life on earth are highly speculative, nevertheless several observations support these models. In 1953, Stanley Lloyd Miller (b. 1930) published the results of an experiment performed under primitive earth conditions. A gaseous mixture of methane, ammonia, water vapor, and hydrogen was recirculated past an electric discharge (duplicating lightning) for a period of 1 week. This apparatus generated the amino acids, glycine, alanine, aspartic acid, and amino-butyric acid, as well as hydrogen cyanide and several unidentified compounds.

Similar synthesis has been experimentally verified using ultraviolet radiation. Other experiments in laboratories using various modifications have yielded formaldehyde, formic acid, benzene, and a variety of other organic compounds, including almost all of the amino acids. Ribose, deoxyribose, guanine, and adenine—compounds which are important in nucleic acid synthesis—have been identified in flasks duplicating primitive earth conditions. Adenine is noteworthy in that it is obtained in good quantity and can be readily converted to adenosine triphosphate (ATP).

Furthermore, the condensation of these compounds to macromolecules has been reported. Under primitive earth conditions proteinoid (proteinlike) compounds are readily formed. Some of these artificially synthesized proteinoids have enzymatic activities, such as catalyzing the breakdown of glucose.

Finally, Oparin and others have succeeded in forming *microspheres* from proteinoids by simply boiling, decanting the aqueous supernatant liquid layer, and cooling. A microsphere is a hollow spherical structure, bounded by a proteinoid membrane, and resembles certain bacteria. The membrane has permeability and other properties similar to biological membranes. One mg of proteinoid may yield up to 10 million microspheres. These microspheres have been observed to divide, but they do not grow. Upon warming, microspheres redissolve. Microspheres lack many biological characteristics and cannot be considered as living.

These observations suggest that several of the events required in the spontaneous generation of life occur under presumptive primordial conditions. Furthermore, the probability of each of the several steps occurring is high; i.e., they are likely to occur under the proper conditions. We can speculate, however, that other events leading toward generation of life are less likely to occur. Assemblage of a complete metabolic pathway, for example, is a highly improbable event. Assemblage of all the metabolic pathways, together with systems to synthesize these pathways and reproduced the "organism," may seem so unlikely as to never occur. Nevertheless, life did originate.

We shall attempt to account for the generation of life using a statistical model. Primitive conditions suitable for generation of life existed on the earth for a period of approximately 1 billion years. During this period, organic chemicals were accumulated and aggregated to form microspheres. One mg might have generated 10 million (10^7) microspheres in 1 liter of seawater. The oceans contain about 1.5×10^{21} liters, and therefore more than 10^{28} ($1.5 \times 10^{21} \times 10^7$) microspheres would have been possible. Daily temperature fluctuations might have dispersed 1 percent of these, which would have reformed the following day, that is, 10^{26} microspheres/day, or $3.65 \times 10^2 \times 10^{26} = 3.65 \times 10^{28}$/yr. In 1 billion yr a total of more than 10^{37} microspheres would have been possible ($10^{28} \times 10^9$).

It is highly improbable that any one microsphere would have had the capability of synthesizing protein enzymes; a version of the model presented in Fig. 10.7 would have been required. Let us assume that, by chance alone, 1 microsphere in 1 billion possessed this capability. Also highly unlikely is the probability that a microsphere capable of synthesizing proteins would have been able to synthesize specific enzymes of the glycolytic pathway (Fig. 10.5). Let us assume that there was one chance in one billion for such a microsphere to have the specific metabolic system for breaking down glucose. (The probability of any random microsphere having both specific capabilities is the product of the two individual probabilities, that is, $10^{-9} \times 10^{-9} = 10^{-18}$, or 1 chance in 1 billion billion.)

Before this microsphere can be considered to be a biological organism it must be capable of self-replication or reproduction. Let us assume that the probability of the microsphere having the capacity for DNA replication, the machinery for separating DNA (as in mitosis), and the ability to divide was 1 in 10 billion (10^{-10}). The chance of all these occurring in the *same* microsphere as the complete protein-specific synthesizing system is vanishingly small, i.e., 10^{-28}, or 1 chance in 10^{28}. However, only such a microsphere could be considered to be alive. We might further speculate that only about 1 microsphere in 1,000 would escape thermal conditions and other environmental hazards and become successful. Thus we conclude that the chance for any one microsphere to become alive and successfully reproduce was only one in 10^{31}.

By comparing this hypothetical probability (10^{-31}) with the estimated number of microspheres (10^{37}), we can get an idea of the possibility for spontaneous generation. We would predict that one case of spontaneous generation would occur approximately every 1,000 yr—frequent enough to account for life on earth. The only proof is that we are here.

Although life probably arose by spontaneous generation in some primordial time, this process is not as likely to occur again. First of all, the physical and chemical conditions on earth now differ significantly from primitive conditions; the atmosphere no longer has significant quantities of hydrogen, ammonia, and methane, and temperatures and radiation levels are reduced. Secondly, the earth is now populated with bacteria, plants, and animals which would immediately consume or decay the very first chemicals leading toward spontaneous generation. Even experiments conducted in laboratories under primitive conditions are unlikely to show spontaneous generation unless our guesses of the probabilities involved are wrong by a factor of around 10^{20}.

Ironically, most natural scientists rejected the idea of spontaneous generation in the eighteenth and nineteenth centuries. Prior to this time, many observations—maggots arising from decaying meat, mice from grain, worms from mud, bacteria from broth—all supported spontaneous generation. The *coup de grâce* was delivered to the older spontaneous-generation theories by the great French scientist Louis Pasteur. Refining earlier methods, Pasteur prepared flasks with long curving swan's necks. Each flask was half filled with broth and boiled. The neck was left open to the air, so that as the broth cooled, air filled the other half of the flask. Because dust, bacteria, and spores settled out in the long neck, the broth remained sterile for months, proving that spontaneous generation did not occur. By the time Pasteur completed these studies in 1864, nothing remained of the earlier theories of spontaneous generation. Natural scientists were left without a rational account of the origin of life until the publication of Oparin's book in 1925.

PROBLEMS

10.1. In which layer of a cellular membrane would you expect to find enzymes?

10.2. A patient is admitted to the hospital with severe pains in his chest. A laboratory test shows high levels of LDH (lactic acid dehydrogenase) in his blood. What disease would you suspect?

10.3. Explain how a genetic defect might result in an unusual form of myoglobin, a hemoglobin found in red muscle.

10.4. DNA is now accepted as the chemical responsible for storage of genetic information. What evidence supports this theory?

10.5. What part of the DNA molecule is used for reading the genetic code?

10.6. How does ionizing radiation cause genetic mutations?

10.7. What were the two versions of the blending theory of inheritance that were accepted by scientists in the past?

10.8. Account for the disappearance of a recessive trait in a test cross. Discuss the disappearance in terms of DNA control of protein synthesis.

10.9. If a homozygous dominant parent (involving a simple gene) were crossed with its heterozygous offspring, what proportion of offspring would be expected to express the recessive trait?

10.10. In coin tossing what is the probability of tossing three successive heads? Two heads and one tail?

10.11. Sex determination in human beings is based on the inheritance of X and Y chromosomes. An individual with two X's is a female, and an individual with an X and a Y is a male. What is the probability of any particular sperm having an X chromosome? A Y chromosome? What is the probability of an egg, or ovum, receiving a Y chromosome? An X chromosome?

10.12. If you have an AB blood type and your father has blood type B and your mother blood type A, what possible genotypes are your parents? What other phenotypes might be possible in your brothers and sisters?

10.13. Why is ultraviolet radiation so much more effective in causing death in bacteria than in man?

10.14. How many different possible genotypes might be produced from a marriage between a black person with genotype $AaBBCCddEE$ and a white person with genotype $aabbCcddee$? How many phenotypes could be distinguished?

10.15. A family has three children, two of which have Down's syndrome. What is the probability that the normal child is a carrier, i.e., has an attached chromosome 21? What would be the probability of a fourth child being completely normal, i.e., neither an idiot nor a carrier?

10.16. What are three situations in which the Hardy-Weinberg law is not true?

10.17. What role is played by mutations in the evolution of new kinds of organisms?

10.18. Plant breeders, attempting to create new kinds of plants, often use mutagenic agents. Why?

10.19. Why has the allele for sickle-cell anemia not been eliminated from the human population? With mosquito control in the tropics, what would you predict concerning the future of this trait in human populations?

10.20. Man's ancestors probably all had abundant body hair. Why do you suppose man became nearly naked during the course of his evolution?

10.21. Why is sexual reproduction important for the genetic well-being of a species?

10.22. How may genetic expression vary during embryonic development?

10.23. How does the universality of the genetic code support the theory that all life on earth descended from a common ancestral species?

10.24. What evidence supports the hypothesis that life arose spontaneously on earth? What additional information is needed to "prove" that this hypothesis is true?

Environmental Biology   **11**

In this final chapter we shall look at a system which is both enormously complicated and quite important: the earth with all the interactions among living organisms and their environment. To cope with so great a complexity we shall do just what natural scientists do. We shall break up the total system into smaller and more manageable units; that is, we shall analyze it. The importance of this analysis will become evident as we study the several problems associated with man and his environment.

Environmental biology is the study of interactions between living systems and their environment. This includes the study of evolution, ecosystems, and populations. In this chapter we shall study some of the concepts of classification based upon the evolutionary origin of different organisms, including man. Although the question of whether evolution has taken place is no longer controversial, we shall discuss some of the evidence supporting this theory.

Individual organisms have interactions with their environment. The absorption of energy by plants involves sunlight striking plants, which trap energy by a molecular mechanism. The process of manufacturing or obtaining food can be considered an aspect of environmental biology.

The study of the environment in which organisms live and the ways that organisms interact with this environment and with each other is called "ecology." Ecology is sometimes defined as the study of interrelationships between populations of organisms and their biological and physical surroundings. Ecology deals largely with the population and ecosystem levels of biological organization. Air and water pollution are considered to be in the area of ecology because of the disturbances to the environment of man and other organisms.

Population biology deals with the growth and control of populations of organisms, including man. Continued uncontrolled population growth in man compounds the effects of pollution, threatens the depletion of many natural resources, and heightens famine and poverty.

CLASSIFICATION

Most kinds of organisms can be categorized into groups, and the natural science dealing with the placement of organisms into natural groups is called "classification," or "taxonomy." Classification is based upon evolutionary relationships between different kinds of organisms. The principal problem of classification is determining evolutionary relationships. We shall now discuss the way in which a group is assigned a name and some of the methods used in deciding evolutionary kinships.

All members of one kind of organism collectively constitute a *species.* This is the basic unit of classification. The modern method of naming species using a generic and a specific name was proposed by Carolus Linnaeus (1707–1778). This binomial nomenclature is now used universally. Each species of organism is given a unique two-part name in Latin, for example, *Homo sapiens* for man. Species closely related to each other evolutionarily are given the same generic name but a different specific name. For example, an extinct primate species which is considered the immediate progenitor of man is named *Homo erectus. Homo sapiens* and *Homo erectus* are placed in the same genus because of their close evolutionary relationship.

Organisms with similar evolutionary origins are said to be related, and degrees of relatedness are expressed in terms of distance from a common ancestor. The more direct the descent, the closer the relationship. Categories higher than genus and species are used to designate more distant relationships in a hierarchy analogous to kinships in human society. In human society, brother signifies a closer relationship than cousin, and nationality a closer relationship than race. In the biological world all species descended, or evolved, from ancestral species. In biological classification, genus, family, order, class, phylum, and kingdom express decreasing degree of relatedness and increasing size of category. Members of the same genus evolved from a common ancestor relatively recently; members of the same family (but different genera) had a more distant common ancestor; and members of the same order had even more distant common ancestry; etc.

Classification would be simple if accurate pedigrees, or records of ancestry, of present-day forms were known. Most ancestral species are now extinct. However, there is a relatively small number of imperfectly preserved fossils which are of value in establishing ancestral lines. *Fossils* are remains or impressions of organisms in rocks. Organisms die and are covered with mud, silt, or sand. The body decomposes, but an impression may remain in the substratum, which subsequently becomes stone. The hard tissues, i.e., bone, teeth, or shells, may persist, although the material making up these hard tissues is usually replaced by other minerals. Seldom is the body preserved, but by using our knowledge about anatomy and function of modern organisms it is possible to reconstruct rather realistic models of prehistoric organisms from impressions and fossilized bones. The science dealing with the study of fossils is called "paleontology."

In practice biological classification depends upon fossils to only a limited degree. Few extinct animals or plants and almost no microorganisms have fossil remains. The taxonomist must depend upon present-day material to determine evolutionary relationships. Classification using modern forms is based largely upon *morphological* (structural) characteristics, but various aspects of biochemistry, physiology, behavior, and ecology may also be used. Organisms with similar structure are considered to be related evolutionarily. A fish, frog, bird, and dog may all appear to differ markedly, but upon closer examination it is found that all have hinged jaws, two pairs of appendages, vertebrae, a similar central nervous system, and many other morphological characteristics in common. In addition, all have similar hemoglobins, similar hormones, and similar reproductive behavior.

We decide that these animals have more traits in common than can be accounted for by random chance, and thus they must have evolved from a single ancestor. We therefore classify them into a single inclusive group, the *vertebrates*. All classification is based upon such deductive logic. However, there is a pitfall to avoid. Structures, behavioral traits, and physiological characteristics may be similar but may *not* be indicative of a similar ancestry. Birds, bees, and bats all have wings and, in fact, other features in common; however, this does not prove a recent common ancestry.

A better interpretation is that wings evolved more than one time in different groups of animals.

Classification is useful because it organizes information about the biological world. Information about one species may be extrapolated to a closely related form but not to a more distant relative. For example, chickens have been found to excrete uric acid in order to eliminate nitrogenous wastes. (Uric acid is the white material seen mixed with their feces.) Taxonomists classify chickens as *Aves,* or birds, a class of vertebrates. Excreta from approximately 1 percent of the 8,500 species of birds have been analyzed, and all have been found to contain uric acid. We conclude that all birds excrete nitrogenous wastes as uric acid. Thus, we examine only a few individuals, sometimes only one, and determine the trait for the thousands of species in that group. Knowledge of taxonomy also allows us to set limits upon such generalizations.

KINGDOMS

Most large conspicuous organisms can be classified into one of the two best known kingdoms: plants or animals. Some biologists fit all organisms into these two kingdoms. However, there are a number of organisms which differ significantly from what are commonly called plants or animals. These share other traits in common and can conveniently be classified into a third kingdom, *Monera*.

Monera are relatively simple in organization and structure. They are considered primitive and may have been the first organisms on earth, ancestors to both plants and animals. *Monera* are characterized by a lack of cell organelles and the presence of mucopeptides in the cell wall. (A mucopeptide is a molecule with both amino acids and amino sugars.) *Monera* are generally single-celled organisms and are classified into two subgroups: bacteria and blue-green algae.

The *plant kingdom* includes all organisms with typical cell organelles, chlorophyll in chloroplasts, and cellulose cell walls. (A cell wall is a rigid layer outside the cell membrane.) Most plants lack the power of locomotion. Since chlorophyll is green and cellulose cell walls impart rigidity, plants are readily recognized as green, rigid, nonmotile organisms. Some plants have secondarily lost their chlorophyll and are therefore colorless; and others are microscopic, and rigidity is not noticeable.

The *animal kingdom* includes all organisms having typical cell organelles, lacking chlorophyll and cell walls, and possessing the power of locomotion. Animals range from microscopic single-celled organisms to complex multicellular organisms such as man.

There are some problems with this simple three-kingdom classification. Some organisms have traits of both plants and animals. For example, some flagellated organisms (Fig. 11.1) have chlorophyll and the power of locomotion. Some biologists consider these to be plants while others consider them to be animals. Some green flagellates lose their chlorophyll permanently if kept in the dark. It is unthinkable to call them plants before they lose their chlorophyll and animals afterward. The problem of classification of these organisms is difficult to resolve.

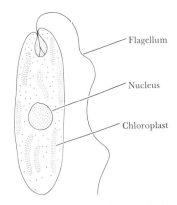

Figure 11.11. Euglena, a single celled organism which has chlorophyll but also has a flagellum and is capable of movement. It might be classified as either a plant or an animal.

BACTERIA

Bacteria are microscopic single-celled *Monera* that lack cellulose cell walls. Although nuclei, endoplasmic reticulum, mitochondria, and chloroplasts are absent, the functions performed by these organelles in higher organisms are carried out by homologous systems in the cytoplasm (the fluid of the cell surrounding the nucleus and other organelles) of bacteria.

Bacteria are best known for their activity as pathogens to man and his domesticated animals. However, as a group bacteria are beneficial to man's welfare because they are responsible for decay of dead organisms and thus the return of essential chemical nutrients to the ecosystem. Bacteria also break down the organic material of feces.

The earth today is in an approximate *steady state* in regard to rates of synthesis and breakdown of organic compounds. Bacteria and other microorganisms decompose organic material as rapidly as plants synthesize them. During long periods of the earth's history, a steady state did not exist; bacterial decomposition did not keep pace with synthesis, and large quantities of organic material accumulated. These materials are now coal, oil, and natural gas deposits and are accurately termed *fossil fuels.*

Several species of bacteria are capable of reducing atmospheric nitrogen to nitrate and nitrite, a process called "nitrogen fixation." The best known nitrogen-fixing bacteria live in symbiotic relationships with legumes (clover, peas, beans, etc.). Nitrogen-fixing bacteria are valuable in agriculture since nitrates and nitrites serve as plant nutrients.

ARTHROPODS

The *Arthropoda* are considered to be the most successful phylum of animals, having many marine and terrestrial representatives. The body of an arthropod is segmented, although most of the vital organ systems are not repeated in each segment. Arthropods have jointed legs, generally a pair for each segment. Frequently these appendages are modified into specialized structures such as mouth parts, or they are entirely absent from a particular segment.

Arthropods have a calcified shell covering most of the body parts. This shell, or *exoskeleton*, provides support for the body. The exoskeleton, providing support, and the jointed appendages, providing motility, allowed arthropods to invade the terrestrial environment and become extremely successful. They are the most plentiful phylum of animals, both in terms of numbers of species and numbers of individual organisms. Typical arthropod structure is shown in Fig. 11.2. Arthropods are represented by such common animals as insects, spiders, ticks, scorpions, crabs, crayfish, lobsters, shrimp, barnacles, centipedes, millipedes, sow bugs, and 700,000 other species (most of these are insects).

Because insects are the most abundant kind of terrestrial

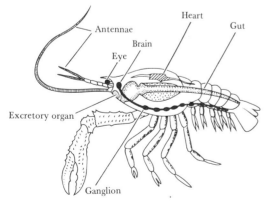

Figure 11.2. Diagrammatic presentation of some structural features of an arthropod, the crayfish. The body is segmented, with muscles, ganglion, and jointed appendages repeated on most segments. Most organs, however, are not segmental.

animal, they are man's principal competitor for food. Man has developed potent chemical agents in order to eliminate some of these insect competitors. These chemicals are of several kinds and are called "pesticides" because they kill pests. Unfortunately, they are not specific for pests; all insects are sensitive to pesticides. There are a large number of insect species which are predacious upon other insects. Application of a pesticide often may kill all of the predators but only most of the prey. The prey species may then multiply rapidly without control by a predatory species. Consequently the pest situation might be worse after pesticide treatment than before.

Repeated use of a pesticide is required to control the pest; thus biological control of a pest is replaced entirely by chemical control. This practice, however, creates another problem. Repeated use of a pesticide results in buildup of pesticide levels in the environment and in crops. Pesticides also affect other animals, including man, and such agricultural practices result in toxic accumulations in man (see p. 632).

MAN

Man is a vertebrate animal and has evolved from lower forms in much the same manner that all organisms have evolved. The characteristics of modern man are the results of natural selection for features that have had survival or reproductive advantages.

Fossils from all parts of the world indicate a succession of manlike creatures, each becoming more human with the passage of time. The so-called "missing link" between man and the apes does not and did not exist. Man did not evolve from any ape we know today. Rather, man and the modern apes evolved from a common ancestor, each passing through several intermediate species, all now extinct. Thus there are several intermediate species, only some of which are known. The known progenitors of man are discussed below.

The evolution of man probably began 30 million yr ago. The earliest known manlike creatures were *Kenyapithecus* (found as fossils in Kenya) and *Ramapithecus* (found as fossils in India). These are about 14 million yr old. A few fossils of each are known, consisting only of jaw fragments, but these indicate a creature that is more manlike than apelike. The shape of these jaw fragments is arched rather than horseshoe-shaped as in apes (see

Fig. 11.3). The cranial cavity of these early hominids was probably around 400 cm³.

About 2½ million yr ago *Australopithecus* appeared in Africa. Evidence suggests that several species of this genus existed. Different areas of the world may have been inhabited by different species, and each species may have evolved through a series of distinctly different forms. The fossil record is not complete, but nevertheless it appears that one form of *Australopithecus* evolved into *Homo* and the other species became extinct. Numerous fossils of skulls, jaws, and limb bones and a large number of tools have been discovered, many by L. S. B. Leakey (b. 1903) at Olduvai Gorge in Tanzania. East Africa has many sites containing fossils of nearly every hominid, the richest being the Olduvai Gorge site discovered by Leakey and his wife in 1959.

The cranial cavity of *Australopithecus* was approximately 500 cm³. The articulation of the skull with the vertebral column and the position of attachment points for muscles suggest that the head was held erect and that *Australopithecus* walked erect. He was only about 4 ft tall. Muscle attachments to the jaws are similar to Homo sapiens, and this suggests the possibility of speech. Bones of small birds and mammals are common at campsites, and we conclude that this manlike creature was a meat-eater. Although he used tools, his skill at fashioning tools was probably very limited.

Approximately 700,000 yr ago the first true man evolved. This man has been named *Homo erectus* and is best known from fossils discovered in Java (Java man) and China (Peking man). Fossils have also been discovered in Africa, including Olduvai Gorge. The skull capacity of this man was about 1000 cm³, or about twice that of *Australopithecus*, but was not as great as that of modern man.

Homo erectus was similar to modern man in size and bone structure, but his jaws were massive, resulting in a projecting face. He had a receding forehead and no chin. *Homo erectus* made and used tools and hunted large animals as well as small animals, including other men. He used fire, although it is not clear whether he used it for cooking or for warmth. *Homo erectus* was no doubt the immediate progenitor of man.

Fossils indicate that about 100,000 yr ago Homo sapiens was abundant over all of Europe, the Near East, and Central Asia. In Europe this human evolved into the Neanderthal type, named after the first fossil found in the Neander Valley in Germany in 1856. Neanderthal man was characterized by a large brain

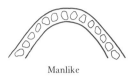
Manlike

Apelike

Figure 11.3. Jaw fragments.

size, 1450 cm³, slightly larger than the average brain size for modern man. A receding forehead and chin, a heavy brow ridge, and a massive jaw all were his primitive traits. He was about 5 ft tall, with characteristic bowed legs. Neanderthal man lived largely in caves, made his own tools, and hunted large animals.

Because he buried his dead with gifts of food and tools, we might suppose that Neanderthal man believed in the hereafter. This would suggest that he had a highly developed language and culture. Neanderthal man suddenly disappeared and became extinct about 25,000 yr ago. It is highly unlikely that modern man arose from this race of the species. Rather, Neanderthal man must be a divergent offshoot from the main evolutionary pathway leading to modern man.

The direct pathway to modern man was probably through the subspecies known as Cro-Magnon man. Cro-Magnon man appeared about 50,000 yr ago and probably existed contemporaneously with Neanderthal man during the latter part of the Neanderthal man's existence. The appearance of Cro-Magnon man in Europe coincided with the disappearance of Neanderthal man and may have caused it. Cro-Magnon man is almost identical to modern man in physical appearance. A modern man's brain size, the facial features of a high forehead and full chin, and a height of 6 ft were his physical characteristics. Cro-Magnon man was skillful at making tools for cutting and sewing. He is most famous for his exquisite paintings of animals, which are found in caves throughout Europe.

Modern man does not differ significantly from Cro-Magnon man in physical appearance or mental capacity. Rather, modern man is distinguished by his cultural development. The agricultural revolution, which began only about 10,000 yr ago, the industrial revolution, and the communications revolution must all be considered part of man's evolution. Man is what he is today because of the physical and cultural evolution that has occurred over the past 30 million yr and is continuing today. The probable evolutionary history of man is summarized in Fig. 11.4.

VIRUSES

Viruses differ markedly from other forms of life, and at present it is not possible to classify viruses into any conventional group. Viruses are not cells but consist of nucleic acids with a protein coat

Years ago

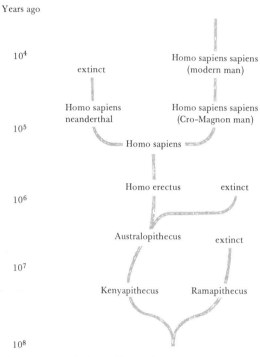

Figure 11.4. Evolutionary history of man.

(Fig. 11.5). Outside of living cells, which they parasitize, viruses are lifeless chemicals. Upon contact with a cell, a virus sticks to the cell and its nucleic acid is injected into the cell. The viral nucleic acid takes over the metabolic machinery of the host cell and directs it to manufacture new viral particles. These new viral particles may be released and may infect other cells.

Viruses infest all major groups of organisms: bacteria, plants, and animals. The protein coat, which sticks to the cell of the host, is specific, and therefore specific kinds of viruses infect only

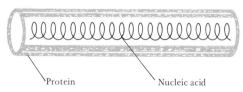

Protein   Nucleic acid

Figure 11.5. Viruses consist of nucleic acids with a protein coat. They are invisible to all but the most powerful of microscopes.

specific hosts. Influenza, chicken pox, the common cold, and yellow fever are viral diseases in man. Some forms of cancer may be a result of infestation of cells by viruses, which causes breakdown of the normal control systems within the cell.

Viruses might be considered to be primitive because of their simple organization. However, their parasitic life history suggests that they probably evolved from cellular organisms relatively recently. If so, they are not primitive.

BIOCHEMICAL EVOLUTION

Because evolutionary changes have as their basis changes in DNA molecules, it should be possible to find specific examples of molecules that have evolved. Although the isolation and analysis of DNA is a formidable task, variations in DNA can be monitored by variations in RNA or proteins whose composition is determined by the DNA. In particular, the amino acid sequence in a hormone will vary from species to species. This variation is a consequence of the DNA coding, which varies with species. A *molecular* study of this kind can be used to corroborate the *macroscopic* patterns of species' development and relation, e.g., skeletal form.

The posterior pituitary gland, the neurohypophysis, releases two hormones, the neurohypophyseal (NHP) hormones. In man and other mammals (except the pig) these hormones are oxytocin and vasopressin. The former is involved in the contraction of smooth muscles; the latter functions in the constriction of capillaries and in body water and salt balance.

Oxytocin has the structure illustrated in Fig. 11.6a. A box is drawn about each amino acid residue. For convenience the structure can be written as shown in either Fig. 11.6b or c. There is a disulfide bridge between the sulfur atoms in cysteine (CySH)

$$CH_2 - CH - COOH$$
$$\quad| \qquad |$$
$$SH \quad NH_2$$

The other amino acids are tyrosine (Tyr), leucine (Leu), isoleucine (Ile), glutamine (Gln), asparagine (Asn), proline (Pro), and glycinamide [Gly(NH_2)]. Corresponding hormones in other vertebrates (animals with backbones) contain, in addition to these, the amino acids serine (Ser), lysine (Lys), phenylalanine (Phe), and arginine (Arg).

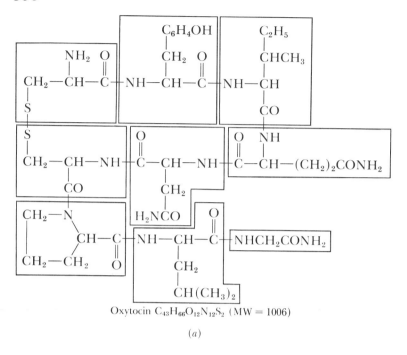

Oxytocin $C_{43}H_{66}O_{12}N_{12}S_2$ (MW = 1006)

(a)

CyS—Tyr—Ile
CyS—Asn—Gln
Pro—Leu—Gly(NH$_2$)

(b)

CyS—Tyr—Ile—Gln—Asn—CyS—Pro—Leu—Gly(NH$_1$)

(c)

Figure 11.6. (a) The structure of oxytocin; (b) a simpler way to represent the structure of oxytocin; (c) another simple representation. structure of oxytocin; another simple representation.

The NHP hormones isolated and analyzed from a variety of vertebrates all have nine amino acids. The structures are listed in the Table 11.1*a* and *b*. From the table, the following facts are evident: each hormone has a disulfide bridge from the first to the sixth amino acid; the only changes in amino acid residue are at the fourth and eighth or at the third and eighth positions; and the parent, or ancestral, hormone (the hormone of the common ancestor to these vertebrates) probably had, at positions 1, 2, 5, 6, 7, and 9, the amino acids present at those positions common to the hormones listed.

Other substances studied include hemoglobin, which contains over 500 amino acid residues. There appears to be not much,

Table 11.1

a. Smooth-muscle-contraction Hormone

Vertebrate	Hormone	Amino acid sequence								
		1	2	3	4	5	6	7	8	9
Cartaliginous fish	Glumitocin	CyS-Tyr-Ile-Ser-Asn-CyS-Pro-Gln-Gly(NH_2)								
Bony fish	Isotocin	CyS-Tyr-Ile-Ser-Asn-CyS-Pro-Ile-Gly(NH_2)								
Amphibians	Mesotocin	CyS-Tyr-Ile-Gln-Asn-CyS-Pro-Ile-Gly(NH_2)								
Mammals (except pigs)	Oxytocin	CyS-Tyr-Ile-Gln-Asn-CyS-Pro-Leu-Gly(NH_2)								
Pigs	Oxytocin	CyS-Tyr-Ile-Gln-Asn-CyS-Pro-Leu-Gly(NH_2)								

b. Capillary-contraction Hormone

Vertebrate	Hormone	Amino acid sequence								
		1	2	3	4	5	6	7	8	9
Cartaliginous fish	Vasotocin	CyS-Tyr-Ile-Gln-Asn-CyS-Pro-Arg-Gly(NH_2)								
Bony fish	Vasotocin	CyS-Tyr-Ile-Gln-Asn-CyS-Pro-Arg-Gly(NH_2)								
Amphibians	Vasotocin	CyS-Tyr-Ile-Gln-Asn-CyS-Pro-Arg-Gly(NH_2)								
Mammals (except pigs)	Arginine vasopressin	CyS-Tyr-Phe-Gln-Asn-CyS-Pro-Arg-Gly(NH_2)								
Pigs	Lysine vasopressin	CyS-Tyr-Phe-Gln-Asn-CyS-Pro-Lys-Gly(NH_2)								

if any, variation in the *number* of residues present in the analogous proteins of widely dissimilar species. In general, the more closely related two species are, the fewer the differences in amino acid residues. In the evolutionary development measured by amino acid sequences, the replacement of a single amino acid residue by another can be considered a *biochemical evolutionary change unit.*

Since some of these proteins are present in very small amounts, extracting them from an organism can be a major task. It is not uncommon to have to use about ten thousand glands from smaller animals to obtain about 5 mg of a particular protein. (A single drop of water has a mass of about 50 mg.)

PHOTOSYNTHESIS

Although there are numerous examples of evolutionary changes in biochemical molecules, many biochemical systems are remarkably similar throughout the living world. The molecules involved in photosynthesis and glucose metabolism are similar in all organisms, and the enzymes associated with these systems have relatively minor variations in their amino acid composition. The genetic code (studied in Chap. 10) is the same for all organisms. The biochemical uniformity suggests that all life evolved from a common ancestor.

The study of photosynthesis and metabolism of food serves as a background for the study of ecology in later sections. The relation of sunlight energy to plants is one of the key components in any ecosystem. The organic molecules manufactured by plants provide all of the food necessary for maintaining an ecosystem. To better understand some of the ecosystem aspects of food energy, we shall discuss why food is important, how it is manufactured, and some of the ways it is used.

Considerable energy is required to manufacture these organic compounds. Furthermore, energy is required to maintain the highly organized structure of the cell. Many organisms also expend energy in the transport of substances within them or in physical work such as movement. For life to continue to exist, there must be regular addition of energy to biological systems. To be useful to biological systems energy must be supplied in a particular form. Two forms of energy may be utilized by organisms: sunlight and potential chemical energy.

Radiation energy from sunlight of a particular wavelength can be absorbed by green plants and converted to potential chemical energy of organic molecules. A maximum of about one-third of absorbed radiant energy may be converted to chemical energy. However, most wavelengths of light are not absorbed, and the efficiency of conversion of those that are is generally less than 20 percent in nature. The overall efficiency in nature is on the order of 1 percent. The process of light absorption coupled with synthesis of organic molecules is called "photosynthesis."

Organisms that are capable of photosynthesizing contain a number of molecular species, including the light-absorbing pigment chlorophyll. Only *Monera* and plants contain these necessary molecules. Because chlorophyll absorbs specific wavelengths of light, the plant or *Monera* takes on the color of the reflected light, the complementary color of the wavelengths absorbed, i.e., green. Thus, an organism carrying out photosynthesis can be recognized at a glance. Green color indicates that light is being absorbed for photosynthesis.

Light striking chlorophyll excites the molecule, forcing electrons out of the molecule. These electrons are passed to an iron-containing compound called "ferredoxin" (Fig. 11.7). Then the electrons are passed to oxidized NADP (nicotinamide adenine dinucleotide phosphate). Electrons are restored to chlorophyll from water, which then decomposes to hydrogen ions and oxygen. This oxidation of water occurs via intermediates, i.e., iron-containing cytochromes, and is coupled with the formation of ATP from ADP and inorganic phosphate. The above reactions all occur only in the presence of light and have been called the "light reaction."

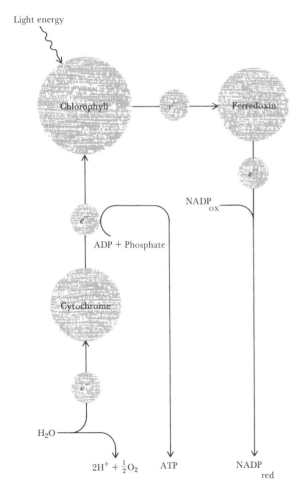

Figure 11.7. The light reaction in photosynthesis.

Reduced NADP and ATP are used to synthesize glucose from hydrogen ions and carbon dioxide. This synthesis involves about fifteen steps, some of which are common to the glycolytic pathway (Fig. 10.5). These normally occur only in light, but because they can occur immediately after being placed in the dark or in the dark in the laboratory by addition of ATP and reduced NADP, they are called the "dark reaction."

The overall reaction of photosynthesis is

$$\text{Sunlight} + 6CO_2 + 6H_2O \longrightarrow C_6H_{12}O_6 + 6O_2$$

The significance of photosynthesis is that sunlight energy is captured and used to synthesize organic compounds containing potential chemical energy. These chemical compounds are used as food by the plants.

OTHER FOOD AND ENERGY SOURCES

A few organisms, namely some bacteria, obtain energy from inorganic chemical reactions. Conversion of an inorganic chemical is coupled with synthesis of organic molecules. These bacteria are called "chemosynthetic" bacteria.

Chemosynthetic bacteria are not abundant today, but in the geological past some species were numerous and very active. All important iron-ore deposits on earth are the result of the activity of bacteria that converted dissolved ferrous iron Fe^{2+} to insoluble ferric iron Fe^{3+}. The ferric ion precipitated as various salts and accumulated in deposits hundreds of feet thick on the bottom of the sea. Subsequent geological uplifts made these accessible to mining by man. The bacteria coupled this chemical reaction to synthesis of organic molecules in much the same way that light energy is utilized in photosynthetic organisms. Thus they were able to utilize an inorganic ion to obtain energy to manufacture food.

Most organisms that utilize chemical energy as an energy source obtain organic molecules that were synthesized by other organisms. This is done by consuming the bodies, either dead or alive, of other organisms.

Organisms which manufacture their own food, whether by photosynthesis or chemosynthesis, are called "autotrophs." Organisms that rely upon other organisms for food are called "heterotrophs." All green organisms are autotrophs, including most plants. Most bacteria, fungi, and animals are heterotrophs.

METABOLISM

Food of both heterotrophs and autotrophs is broken down in orderly metabolic pathways. Part of one metabolic pathway (glycolysis) was given in Fig. 10.5. In the breakdown of glucose by glycolysis the potential chemical energy stored in glucose is transferred first to ATP and then to biological processes requiring energy. This potential energy eventually appears in the form of heat. Of the energy in glucose, about 40 percent appears in ATP; the remainder appears as heat. The efficiency of conversion for energy from glucose to ATP is thus approximately 40 percent.

Breakdown of ATP to ADP and inorganic phosphate is a strongly exergonic (energy-producing) reaction. Most work done by a biological system is coupled directly or indirectly with the breakdown of ATP. This work includes such activities as organic synthesis, transport across membranes, and muscle contraction. In each process some of the potential energy in ATP is transferred to provide energy to do work. The remainder of the energy is lost as heat. Thus, additional loss of heat occurs as ATP is utilized as an energy source. The overall efficiency for converting potential chemical energy in glucose to some other form of energy is approximately 10 percent. Finally, any energy used to do internal work in a biological system, e.g., pumping blood, also appears as heat. Thus, most of the potential chemical energy in glucose and other food appears as heat.

The *metabolic rate* is the measure of the rate at which the potential chemical energy of foods is converted to heat by the body. The metabolic rate is generally expressed in Calories (1000 cal) of heat. An average 70-kg man in the United States uses about 2800 Calories of energy per day. The energy content of food can be estimated by burning in a calorimeter.

The potential chemical energy of food can also be expressed in calories. In an adult in a steady state, the potential chemical energy of food ingested and assimilated equals the metabolic rate. Thus the caloric intake of an average 70-kg man in the United States should be 2800 Cal. If food intake exceeds the metabolic rate, the individual is no longer in a steady state and gains weight. If the metabolic rate exceeds the rate of food ingestion, the individual loses weight. Weight control in human beings and all organisms involves manipulating, either consciously or unconsciously, the food intake (e.g., by dieting) or the metabolic rate (e.g., by exercise). The objective is to balance food intake with the metabolic rate to maintain a desired weight. People who become obese do so because this balance is not maintained.

The metabolic rate under resting conditions and with an empty stomach is called the "basal metabolic rate" (BMR). Thyroid hormones increase the basal metabolic rate, and therefore measurement of BMR allows the clinician to assess thyroid-gland functioning.

Measurement of metabolic rate using calorimetry is not always the most convenient method. Indirect methods are much easier than measuring the heat output directly. Oxygen is required in precise quantities for the breakdown of food material, and precise quantities of carbon dioxide are produced. The quantity of oxygen taken up is measured, and since this is directly proportional to the heat liberated from complete breakdown of food, the metabolic rate can be determined. Likewise, carbon dioxide excretion is proportional to heat liberated in metabolism, and measurement of carbon dioxide excretion rate is a good measure of the metabolic rate. Thus, the rate of oxygen uptake and the rate of carbon dioxide excretion give good estimates of the metabolic rate.

AMINO ACIDS AND PROTEINS

In addition to organic materials used for energy, organisms require nucleotides, amino acids, fatty acids, and a variety of other organic compounds either as building blocks for growth or as part of enzyme systems. Every required organic compound is synthesized by some organism, although all organisms cannot make all required organic nutrients. Many heterotrophs, including man, are not able to synthesize a number of these required organic nutrients, and therefore these must be obtained in the diet. In man, these include a number of amino acids, a few fatty acids, and a group of miscellaneous compounds which are collectively called "vitamins."

The 20 amino acids in proteins must all be available if protein synthesis is to occur. In general, plants are able to manufacture all 20 amino acids from inorganic materials. Animals are unable to make any of these amino acids from inorganic materials. Therefore, animals have a dietary requirement for amino acids. These are obtained by hydrolyzing ingested plant or animal proteins and absorbing the amino acids. Consumption of 60 to 70 g of protein per day satisfies this requirement. Grains have approximately 10 percent protein; meats and fish are about 25 percent protein.

Most other foods have 1 to 2 percent protein. The 70 g of protein can be obtained by consuming about 225 g (8 oz) of meat per day plus grain products, fruits, and vegetables.

In addition to a quantitative requirement for protein (70 g/day), there is also a qualitative requirement. Some amino acids can be made in the body using organic molecules as starting materials. In general, amino acids are synthesized by transferring the amine group from one organic acid to another. Some specific amino acids cannot be formed in sufficient quantities in the body and must be obtained in the diet. These are the so-called "essential amino acids." All twenty amino acids are required for protein synthesis, and of these eight are considered essential in that they must be obtained in the diet. A high-quality protein is one rich in the essential amino acids. A low-quality protein, such as in corn, lacks one or more of these essential amino acids.

VITAMINS

Vitamins are miscellaneous compounds which are required by organisms in minute quantities. Some are lipids, others are carbohydrates, and still others are complex nitrogen-containing compounds. A vitamin may be defined as an organic structure which is required by the body but which cannot be synthesized by that organism. A vitamin for one organism, then, may not be a vitamin for another either because it is not required or because this organism may be capable of synthesizing the compound. A vitamin may not be a single chemical entity. Several related compounds may all have the activity of a single vitamin; they are all converted into one active agent in the body.

Human vitamins are classified into two groups: the fat- and water-soluble vitamins (Table 11.2). Most vitamins perform specialized functions in the body. Vitamin A is important for maintaining normal integrity of epithelial tissues, such as skin and the linings of the digestive and respiratory systems. Vitamin A is also one of the visual pigments found in the eye and is important for night vision. Thus, a vitamin may have different unrelated functions within the same organism. Vitamins may also have different functions in different organisms. Vitamin D, another fat-soluble vitamin, functions in calcium metabolism and thus maintenance of bone.

Two vitamins are associated directly with blood: Vitamin K is important in the blood-clotting mechanism of the body; vitamin

Table 11.2. Important Vitamins in Man

Vitamin	Function	Sources
Fat-soluble		
A	Maintenance of membranes, visual pigment	Vegetables, sea food, liver
D	Calcium metabolism of teeth and bones	Seafood, eggs, milk
K	Blood clotting	Intestinal bacteria, vegetables
Water-soluble		
Thiamine (B_1)	Thiamine pyrophosphate, a coenzyme in carbohydrate and lipid metabolism	Cereals, beans
Riboflavin (B_2)	FAD, a coenzyme in respiration	Eggs, meat, milk, green leaves
Niacin (nicotinic acid)	NAD and NADP, coenzymes in photosynthesis and respiration	Poultry, fish, liver
Pyridoxine (B_6)	Coenzyme for decarboxylation of amino acids	Liver, cereal, milk
Cobalamin (B_{12})	Erythrocyte formation	Meat, fish
Ascorbic acid (C)	Antioxidant, maintains intercellular cement of teeth and bones	Citrus fruit

B_{12} is responsible for maturation of erythrocytes. Other vitamins are indirectly involved.

The several B vitamins (except B_{12}) are all either coenzymes or are converted to coenzymes for enzymes of carbohydrate, amino acid, and lipid metabolism. These are essential to the life of all organisms. They are not considered vitamins in all organisms only because many organisms are capable of synthesizing them. A list of some important vitamins and their functions and chief dietary sources is given in Table 11.2.

MINERALS

Every organism contains various chemical elements either as inorganic molecules or as part of organic compounds. These elements must be obtained from the environment or diet. *Mineral nutrition* deals with the inorganic matter required by an organism.

Of all the atoms in organic matter, the majority are carbon, hydrogen, and oxygen. Proteins and nucleic acids also contain large amounts of nitrogen and phosphorus. Autotrophs manufacture organic matter and thus require these elements in the inorganic form. In photosynthesis carbon and oxygen are obtained from carbon dioxide in air and hydrogen from water in soil. Nitrogen is assimilated as nitrate, nitrite, or ammonium ion and phosphorus as phosphate ion, all from the soil. Heterotrophs obtain carbon, hydrogen, and nitrogen only as preformed organic material. Heterotrophs may obtain phosphorus either as inorganic phosphate ion or as a phosphate group attached to an organic molecule. Oxygen is a component of most organic material in the diet and is also obtained as a gas from the air for respiration.

In addition to the elements that make up organic matter, some elements are required in major amounts to provide a proper ionic environment within the organism. The fluid within the cell in almost all organisms has a high concentration of potassium ion, which is generally accompanied by phosphate ion. Autotrophs must obtain potassium from the soil. Heterotrophs obtain potassium ion by eating either autotrophs or other heterotrophs. Multicellular animals have a relatively large amount of fluid bathing the cells. This fluid is a solution of sodium and chloride ions. Animals, then, require sodium and chloride in addition to potassium ions.

In agriculture, growth of plants may be limited by one or all of the three major inorganic nutrients (nitrogen, phosphorus, and potassium) required from soil. *Fertilizer* is a mixture of inorganic compounds (or organic compounds that break down) containing one or more of these major nutrients. Application of fertilizer to plant crops increases the availability of nutrients that might be limiting and thus increases growth and yield.

HABITATS AND NICHES

We have examined some of the food requirements of individual organisms, but organisms also have requirements for particular kinds of environments. Organisms need proper conditions in terms of physical factors and food quality. Every organism or population exists within an environment consisting of both physical and biological components. The immediate environment surrounding any organism is termed its *habitat*, which is equivalent to the organism's home. A description of a habitat usually includes both the physical and biological components. The habitat of a marine fish is the ocean, together with algae, other fishes, whales, etc. The habitat of a woodpecker is a forest, together with other organisms making up the forest and the physical conditions found in the forest.

Every species of organisms is adapted evolutionarily to live in a particular habitat. For example, woodpeckers evolved special mechanisms for capturing insects that live in wood. Woodpeckers now require the forest habitat for their existence; they do not survive on the prairie habitat. The same kind of generality applies to other species in regard to their habitats. However, some species of organisms may occupy a wide variety of habitats. For example, salmon are fish with a complex life cycle involving several different kinds of habitat. Life begins in a stream; they then migrate down rivers to the ocean; and, following several years, they return again to their home stream to reproduce. However, salmon can live in habitats differing somewhat from the native habitat. Several species of salmon have been introduced into the Great Lakes, and they have been very successful. The salmon has the capability to survive in a variety of habitats, including habitats in which they do not naturally occur.

It might appear that man, too, can tolerate a wide range of conditions; he is present in almost every habitat on earth. However, man is not truly adapted to these environments. Rather, man modifies his habitat to make it suitable for his survival; for example, he introduces vegetation and animals and builds houses.

Many organisms utilize natural isolated areas to successfully survive in regions that would be otherwise unsuitable. These isolated areas differ from the prevailing environment in their physical conditions in terms of temperature, moisture, sunlight, etc., and they are called "microhabitats"; the unusual environmental conditions within microhabitats are called "microclimates." A house can be considered a microhabitat. The classical example of a microhabitat is a brook. A brook fed by a natural spring has a relatively constant temperature summer and winter, and organisms that are normally found much farther north (or south) may be found within the brook. The microclimate within the brook allows survival within a region that would otherwise be unsuitable. The underside of a decaying log is a microhabitat for a salamander. The burrow of a kangaroo rat in the desert is a microhabitat for the kangaroo rat and several uninvited guests.

Every organism performs a particular function in its habitat or, more correctly, in its ecosystem. Some plants photosynthesize organic material using full sunlight; others use diffuse light, as in the shade of a large tree. Some organisms break down organic matter in the food they have consumed; others break down material in dead bodies in which they live. In short, every organism does his own thing. The term *niche* is used for describing what an organism does in its environment. Thus, the two concepts, habitat and niche, tell where an organism lives and what it does. A bullfrog and a bullrush occupy the same habitat at the edge of a pond, but they occupy entirely different niches. The bullrush is a green plant and synthesizes food. The bullfrog is a carnivorous vertebrate that feeds upon flying insects which it catches with a flick of its tongue.

Frequently, organisms occupy the same habitat and an *overlapping niche*, which is a niche in which the two organisms do some things in common but not others. For example, the bullfrog and the dragonfly both live in the habitat around a pond, and both catch flying insects. However, the frog catches those insects that fly close to the ground, while the dragonfly catches high flying insects. There is considerable overlap in the kinds of insects captured, and thus the bullfrog and dragonfly occupy overlapping niches. In addition, the larval, or immature, forms of both the bullfrog and dragonfly occupy different habitats and niches than the adults. The tadpole and larval dragonfly both live within the water (habitat) and both feed upon aquatic organisms. These larval stages, however, do not occupy even overlapping niches. The tadpole is a herbivore, feeding upon live and dead vegetation, whereas the dragonfly larvae is a predator, feeding upon small aquatic animals.

It is often stated that two organisms never occupy the same habitat and the same niche at the same time. This is called the "exclusion principle." Simultaneous occupancy in the same habitat and niche results in direct competition. Exclusion is based upon the premise that in a niche and habitat occupied by two species, one species always outcompetes the other and eliminates it from the habitat. The exclusion principle has been tested in laboratory situations and found to hold true. For example, introduction of two species of fruit flies (genus *Drosophila*) into a jar containing one kind of food invariably results in elimination of one of the species. One kind of fly outcompetes the other.

Sometimes one species prevails, and sometimes the other prevails. It appears that physical conditions, such as light and humidity, determine the winner. A consistent winner species is obtained in every case where conditions are held constant. However, if the experiment is done on the sill outside the laboratory window, many times neither species will be eliminated. An equilibrium is reached with both species present. On the windowsill conditions vary such that first one species has the advantage and then the other. (One fruit fly may have a slight advantage at night, and the other may have an advantage during the day.) Neither species can eliminate the other through competition. Moreover, with the variety of microhabitats and foods such as found in nature, the exclusion principle model becomes extremely difficult to test. This example shows that a controlled laboratory experiment, no matter how well conceived or executed, does not always reveal the correct model for what is occurring in nature.

A larger unit of a habitat is called a "biome," which is a large geographical area having a uniform general climate and occupied by a particular type of vegetation. For example, the *prairie biome* in North America occupies a region in the central United States and Canada that receives lesser amounts of rainfall and is covered by grasses and small herbs. Under natural conditions, the range of the American bison corresponded to the prairie biome. Today much of this biome is used for wheat and cattle culture by man. The *deciduous forest biome* occupies a region receiving greater rainfall, with the predominant vegetation being trees that lose their leaves in winter. As is true for all biomes, the deciduous forest biome is not uniform in species composition. Its trees may be elms, oaks, hickories, ashes, walnuts, or maples, depending upon local soil and climatic conditions.

The *tundra* is a biome in the arctic, containing mostly low shrubs and various inconspicuous plants, many of which are present only during the short summer. The *taiga* is a biome between the tundra and deciduous forest. It is covered with forests of conifers (evergreens) such as pine, spruce, and fir. The geographical distribution of the principal biomes in North America is shown in Fig. 11.8. The *rain forest* and savanna are two tropical biomes. (A *savanna* consists of widely spaced trees with grass in between.) Temperature and rainfall are the two most important factors determining the kind of biome that occupies a geographical area.

ECOSYSTEMS

An *ecosystem* consists of various populations of organisms interacting with each other and with their physical environment. An ecosystem is the functional unit of ecology. All interacting organisms living in a particular habitat, together with the nonliving

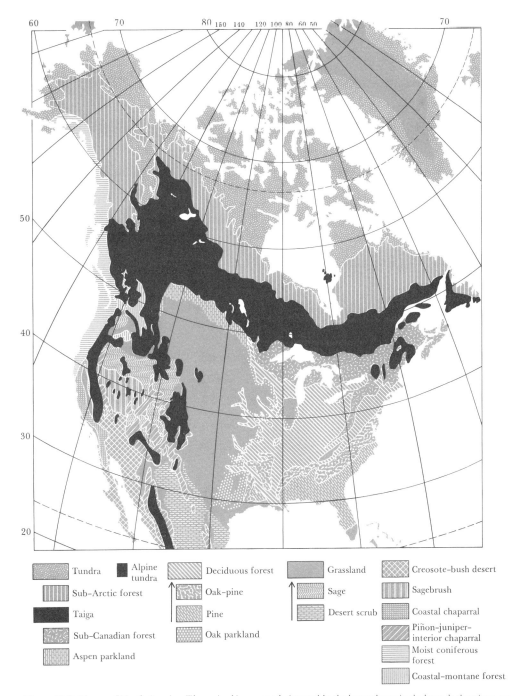

Figure 11.8. Biomes of North America. The major biomes are designated by the larger boxes in the legend; the minor, or intermediate, biomes by small boxes. (Pitelka, Amer. Midland Naturalist, Vol. 25, pp. 113–137, 1941.)

Legend:

Tundra · Alpine tundra · Deciduous forest · Grassland · Creosote-bush desert

Sub-Arctic forest · Oak-pine · Sage · Sagebrush

Taiga · Pine · Desert scrub · Coastal chaparral

Sub-Canadian forest · Oak parkland · Piñon-juniper-interior chaparral

Aspen parkland · Moist coniferous forest · Coastal-montane forest materials of that environment, make up the structure of an eco-system. The ecosystem concept emphasizes relationships between different organisms and between organisms and their environ-ment.

An ecosystem consists of four components: producers, con-sumers, decomposers, and abiotic substances. Autotrophic orga-nisms are *producers;* they produce the organic matter required by all organisms in the ecosystem. Most producers are green plants. *Consumers* are large heterotrophs, namely animals, that ingest the bodies of other organisms. *Decomposers* are microscopic hetero-trophs, i.e., bacteria and fungi, that decay (break down) organic material external to their bodies. Consumers and decomposers both return inorganic nutrients to the environment. *Abiotic sub-stances* consist of the minerals, water, gases, and sunlight required by various organisms within the ecosystem. They are the nonliving part of the ecosystem.

Most ecosystems contain all four components. Abiotic mate-rial is essential as raw material for food production by producers. Producers are required to manufacture food. Decomposers are es-sential for regenerating abiotic materials. Without these three com-ponents an ecosystem would not be able to function. Consumers are not required but are usually present.

The physical limits of an ecosystem are often indistinct. It is difficult to decide where interactions cease. The boundaries of an ecosystem must sometimes be set arbitrarily. Usually, however, an ecosystem is considered in general terms, without reference to its boundaries. For example, a pasture ecosystem consists of soil, air, grasses, insects, rabbits, and cows and is not clearly separated from adjacent pastures, woodlots, etc. Obviously interactions occur across fences.

The ecologist, however, is not as interested in defining the limits of the ecosystem as in describing it. He might choose an area of study within the pasture and measure the rate of photosynthesis, the quantity of grass eaten by rabbits and cows, and the rela-tionships between grass seeds and birds and mice. The results are reported as processes and interactions occurring per square meter of the pasture. It is of secondary importance how far into the dis-tance these functional relationships extend. Ecosystems range in size from the minute to an area the size of an ocean.

Man belongs to various ecosystems; he is a consumer. Under primitive conditions man's ecosystem was well defined; it coincided with his immediate habitat. Modern man's ecosystem includes the whole area from which food and resources are gathered. This may include a large geographical area. For example, urban man's habitat is a city; his ecosystem is the city plus the region supplying food and other resources.

Man has modified and altered his habitat and ecosystem. In agriculture, man attempts to simplify the ecosystem for his own welfare. He changes the abiotic environment by tilling the soil and adding fertilizer. He selects the species of producer by planting seeds and killing weeds. He eliminates undesirable consumers such as bison, wolves, insects, and rodents. He speeds decomposition by plowing under the residue from the previous year's crop.

Agriculture has created a number of ecosytem types, all of which include man as a consumer. A cornfield ecosystem consists of corn, various insects, decomposers, and man, as well as the abiotic substances. A pasture ecosystem consists of grasses, beef cattle, insects, birds, rodents, decomposers, man, and abiotic substances. These simple, man-directed ecosystems are less stable than natural, more complex ecosystems. The instability is due to the lack of interrelationships that provide checks and balances upon each population in the ecosystem.

A pond is a classical example of an ecosystem. It is a discrete ecosystem with clearly defined boundaries. Because of its relatively small size, the pond is amenable to easy study. Populations of organisms in this ecosystem can be measured quantitatively using simple methods and inexpensive equipment.

Ecosystem structure in a pond consists of producers, consumers, decomposers, and abiotic material. Producers are of two types: plants rooted to the bottom in shallow water and algae suspended in the water throughout the pond. Plants growing in shallow water include cattails, water lilies, and bullrushes. In large ponds, lakes, and oceans the most important producers are the myriad of tiny algae suspended in the water below the surface. These plants serve as the food base for the ecosystem. Algae are fed upon by minute arthropods such as water fleas and copepods (Fig. 11.9).

Organisms suspended in midwater are called "plankton." Algae are called "phytoplankton"; the animals are called "zooplankton." Insect larvae (such as dragonfly larvae), minnows, and sunfishes feed upon this minute zooplankton. These predators are, in turn, fed upon by large predators such as bass and walleye pike. The unconsumed bodies of all dead organisms, both plants and animals, fall to the bottom of the pond. Here bacteria and fungi

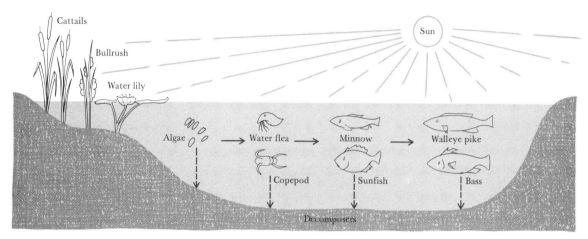

Figure 11.9. Pond ecosystem. Producers are the cattails, bullrushes, and water lilies in the shallow water and algae in the remainder of the lake. Consumers include a series of organisms: (1) water fleas and copepods, (2) minnows and sunfishes, and (3) bass and walleye pike. Decomposers are found on the bottom, and abiotic materials are dissolved in the water. The organisms are not drawn to scale.

decompose the organic matter, releasing abiotic materials to the environment. These abiotic materials, or nutrients, largely go into solution in the pond water and are again available to organisms. A small part of the dead material remains in the bottom sediments and is lost to biological utilization. Nutrients lost to the bottom muds are restored by addition of nutrients from the surrounding terrestrial environment, which enter the pond in the runoff of water from the watershed of the pond.

Most abiotic materials (minerals and gases such as oxygen and carbon dioxide) are dissolved in the water. There is some exchange of gases between the pond and the atmosphere. Sufficient sunlight penetrates into the water to provide the light essential for photosynthesis by the submerged aquatic plants. The pond ecosystem, summarized in Fig. 11.9, is self-sufficient, requiring only the addition of sunlight energy and some nutrients and water from the surrounding watershed.

FOOD WEBS

Food manufactured by autotrophs is used by the autotrophs themselves and by all heterotrophs in the ecosystem. Plants are eaten by *herbivores* (plant eaters); herbivores, in turn, are consumed by *carnivores* (meat eaters). The bodies of all organisms after death are used by decomposers. Thus, food manufactured by producers is incorporated first into the body of the plant, then into herbivores, and then into the body of a carnivore; finally it is broken down by decomposers to abiotic nutrients, with some of the organic material incorporated into the bacterial cell or fungi. The series of organisms in which each becomes food for the succeeding organism is called a "food chain."

A simple food chain consists of one species each of producers, herbivores, carnivores, and decomposers. For example, the simple pasture ecosystem might consist of grass as the producer, cattle as the herbivore, and man as the carnivore:

Grass $\longrightarrow$ cattle $\longrightarrow$ man $\longrightarrow$ decomposers

An example not involving man might be

Herbs $\longrightarrow$ rabbits $\longrightarrow$ foxes $\longrightarrow$ decomposers

The pond ecosystem (Fig. 11.9) has a food chain containing five steps:

Algae $\longrightarrow$ water fleas $\longrightarrow$ minnows $\longrightarrow$ bass $\longrightarrow$ decomposers

Carnivores in the food chain are farther from the food base than herbivores. Carnivores are considered to be on a different *trophic level,* or food level, than herbivores, and herbivores are on a different level than green plants. All populations in an ecosystem can be placed into various trophic levels, depending upon how many steps they are from the food base. The trophic level of a population refers to the number of steps taken in the transfer of food from the producer to consumer. Producers are in trophic level 1; they are the food base. Herbivores are in trophic level 2; primary carnivores are in trophic level 3; etc. Food chains and trophic levels for several ecosystems are analyzed in Fig. 11.10.

Describing the ecosystem in terms of food chains and trophic structure is useful because it allows us to make predictions concerning growth of particular populations. For example, a fisheries biologist might predict the yield of fish in a lake if he knows the trophic level occupied by the fish and the rate of photosynthesis.

Decomposers do not fit at the end of the food chain, as implied above. Decomposers act upon all dead bodies regardless of

Tropic level	Food chain			
	Meadow	Forest	Prairie	Pond
4		Falcons ↑		Bass ↑
3	Foxes ↑	Songbirds ↑	Wolves ↑	Minnows ↑
2	Rabbits ↑	Insects ↑	Bisons ↑	Water fleas ↑
1	Herbs	Trees	Grass	Algae

Figure 11.10. Food chains and trophic levels of four natural ecosystems. In each case the decomposers are omitted.

the trophic levels; therefore they belong on all trophic levels, beginning with trophic level 2. The most abundant source of nutrition for decomposers are the bodies of the most abundant organisms, the producers. There are many more tons of dead plants than there are of dead men for decomposers to utilize.

Most consumers feed upon various kinds of food, often from more than one trophic level. For example, man might be placed on trophic level 2 when eating potatoes, on trophic level 3 when consuming beefsteak, and on trophic level 5 when eating salmon. Our model of the ecosystem as a simple food chain is not tenable with observations of ecosystems. Simple food chains are not observed in nature, although man sometimes tries to create them, with man at the end. We replace the food-chain model of the ecosystem with a *food-web model*. In the food-web model all the trophic relationships of all the organisms are considered (Fig. 11.11). Food webs are generally complex, with each organism

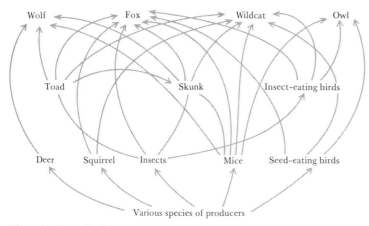

Figure 11.11. Food web in a deciduous forest. In nature, most organisms consume many kinds of food. (The species of producers and decomposers are omitted for simplicity.)

sharing many trophic interrelationships, both as food for several other organisms and as a consumer of several other organisms.

BIOLOGICAL PRODUCTION

The sum of the growth rates of all individuals in a population is referred to as *production*. Production is sometimes defined as the rate of biomass, or quantity of living matter, formation. This is analogous to man's concept of production in agriculture, e.g., beef and wheat production.

The concept of biological production can be applied to trophic levels as well as to populations. We speak of plant production or of herbivore production in an African savanna where several herbivores are present. *Trophic level production* is the sum of the productions of the individual populations. *Herbivore production* in a pasture is the sum of the growth of rabbits, cows, rodents, seed-eating birds, insects, etc. *Primary production* is the rate of production in autotrophic organisms, chiefly green plants. Production in all subsequent trophic levels depends upon primary production. Primary production provides the food for the entire ecosystem. There are two kinds of primary production: gross and net primary production.

Gross primary production (GPP) is the rate of synthesis of food from inorganic substances; it is equivalent to the photosynthetic rate. Part of the food synthesized by the producer is used in the producer's own metabolism. The food broken down in metabolism results in release of abiotic nutrients back into the environment. The rate of this food use equals the rate of respiration (R). Thus, the producers synthesize and break down organic matter at particular rates. The difference between gross primary production and the rate of respiration equals producer growth, and this rate of producer growth is called "net primary production" (NPP). The relationship between the three rates is

$$NPP = GPP - R$$

or

$$GPP = NPP + R$$

Net primary production is the food base for all consumers and decomposers, i.e., the heterotrophs. This food supports energy requirements for metabolism and growth in heterotrophs. Any particular population of heterotroph does not consume all the food that is available or absorb all the food that is consumed. The same applies to any trophic level; there is a certain amount of food that is not utilized. Furthermore, of the food absorbed, only a part is used in the heterotroph's own growth; the rest is lost in respiration. The rate at which food is converted into biomass in a heterotroph population is called "heterotroph production," or secondary production.

The rate of biomass formation at any trophic level is called "trophic level production," e.g., production at trophic level 3. Trophic level production is the sum of the production of all populations at that trophic level. A simple model describes the relationship between the rate of production at a particular trophic level and the rate of production at the previous trophic level:

$$P_t = P_{t-1} - U - R$$

where P_t = production at a particular trophic level
P_{t-1} = production at the previous trophic level
U = unutilized and unabsorbed food
R = respiration

This model states that the production at one trophic level equals the production at the previous trophic level minus the food that is not utilized and minus the food that is metabolized in respiration. A more complete model for all trophic levels, including quantitative relationships, is developed and presented in Fig. 11.12.

The efficiency for the conversion of energy in food into work or growth is about 10 percent. This efficiency applies to the ecosystem: Total growth (production) at any trophic level is about 10 percent of the production at the previous trophic level. In the grassland ecosystem, about 7,000 kg of grass are required to grow 700 kg of beef, which is enough beef to grow one 70-kg man. If man consumes grains as his primary food, the overall efficiency will be somewhat greater. Here 7,000 kg of vegetation has about 3,500 kg of usable food (man does not eat straw), which is sufficient to produce five 70-kg men. Man may also use aquatic ecosystems for food. Here the terminal predator is consumed. In an aquatic ecosystem 7,000 kg of algae might support 700 kg of trophic level 2; 70 kg of minnows; 7 kg of terminal carnivores; and

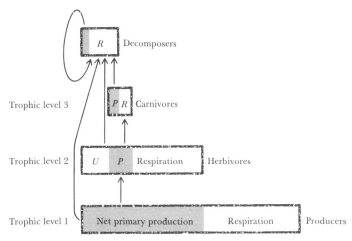

Figure 11.12. Food-pyramid model for the ecosystem. Production at one trophic level serves as the food for the next higher trophic level. Of the food available (represented by the length of the bar), part is unutilized, part is dissipated in respiration, and the remainder becomes the production at that level. Decomposers utilize food that is not used at other levels. (U = unutilized; P = production; R = respiration rate.)

only 0.7 kg of man. Based upon trophic level theory and efficiencies, it can be predicted that the oceans of the world will not yield significant quantities of food despite their large size. The maximum yield from all the oceans could supply only about 3 percent of the world's food requirements. However, oceans do yield harvests of high-quality protein (Problems 11.22 to 11.24).

Because of these limited efficiencies, trophic level production is always less at high trophic levels than at lower trophic levels. Because primary production is the food base and the terminal predator can be thought of as the apex of the trophic structure, the trophic level model of the ecosystem has been called the "pyramid model." This model of the ecosystem is shown in Fig. 11.12.

Insertion of decomposers into the pyramid model of the ecosystem distorts the pyramid shape because decomposers metabolize food that is not utilized by other trophic levels, including food from all trophic levels. Note also that decomposers ultimately decay their own production and that of the terminal carnivores.

Many organisms cease growing after reaching the adult size, and no net growth means zero production. The entire food intake in an adult appears as respiration. A population containing mostly adults has low production, while a population of immature orga-

nisms has a relatively larger production. In animal and plant hus-
bandry, advantage is taken of this ecological principle to maximize
yield. Beef cattle, hogs, chickens, alfalfa, and timber are harvested
toward the end of their growing periods. By maintaining a population of immature individuals, production is increased relative to
the quantity of food available, which increases profit. Adults are
maintained only as a source of replacement stock.

Production is reported in terms of either energy or mass.
The units of energy used are Calories, and the units of mass are kilograms (or pounds); thus, production is expressed in the same
units as food. (Recall that the quantity of food eaten by a man can
be given in terms of Calories or in terms of mass of food eaten.)
Production is always reported on the basis of a unit area and a unit
of time, for example, pounds per acre per year or calories per
square meter per day.

BIOLOGICAL–GEOLOGICAL CYCLES

Many elements are taken up from the environment in the form of
salts, gases, etc., and incorporated into biomass. These elements
are returned to the environment when organic matter is broken
down during respiration and decomposition. Thus they cycle
through the biological and abiotic parts of the ecosystem. Each essential element has its own cycle.

Elements also cycle geologically. They are transported from
land into rivers and into oceans. In the oceans, sediments form on
the sea floor. These sediments become thousands of feet thick and
turn into *sedimentary rock*. Periodically, every few million years,
forces within the earth result in an uplift of the sea floor. The sedimentary rock is lifted above sea level, and this material again
becomes subject to erosion. Then another cycle of erosion–
sedimentation begins.

Most biological and geological cycles intercommunicate.
During erosion of a mineral from rock to the ocean, the mineral
may be incorporated first into trees, later into fish in a river, and
finally into marine organisms. A part of geological cycles may in-
volve gases which pass to and from the atmosphere via organisms.
Biological activity may enhance or impede a geological cycle.
Chemosynthetic bacteria speed the sedimentation of ferric iron.
Carbonates are deposited in limestone largely as a result of biologi-

cal activity. Vegetation retards erosion of the land by providing a protective covering. Finally, elements may cycle repeatedly in an ecosystem and thus not be available for transport to the sea.

Some elements may bypass the geological cycle. Sea birds and migratory fish may return elements to the nonmarine environment. In Chile sea birds consume fish at sea and deposit excrement, *guano*, on land. Some deposits of excrement are hundreds of feet deep, and these are mined and used for fertilizer and gunpowder. Most wars fought since the eighteenth century have relied upon this *Chile saltpeter* (sodium nitrate, $NaNO_3$). However, in most cases the return of elements to the land is not as spectacular as in the case of Chile saltpeter.

THE CARBON CYCLE

The carbon cycle involves the atmosphere, land, fresh water, and the oceans and is dominated by biological activities. Carbon dioxide in the atmosphere or in water is used by plants in photosynthesis and fixed as organic carbon. Carbon dioxide is returned to the environment by the process of respiration.

A relatively small portion of the organic carbon becomes trapped in sediments, and this carbon is transformed into coal, oil, and natural gas by heat and pressure as the deposits become deeply buried. Many rocks, such as shale, also have significant quantities of deposited carbon. This carbon does not become available again until geological uplift and erosion expose these deposits. Man, however, accelerates this process by actively seeking and extracting these deposits to burn as fuel, thus quickly returning carbon dioxide to the atmosphere.

Rain falling through the atmosphere reacts with carbon dioxide to form carbonic acid:

$$H_2O + CO_2 \rightleftharpoons H_2CO_3$$

Rainwater, containing carbonic acid, may fall upon exposed sedimentary rock, such as limestone or marble, and the carbonic acid dissolves the calcium carbonate in the rock forming the bicarbonate:

$$H_2CO_3 + CaCO_3 \rightleftharpoons Ca(HCO_3)_2$$

Calcium bicarbonate is soluble in water and is carried into streams and then into the sea. Photosynthetic activity of green plants in the ocean uses up carbon dioxide. This removes carbon dioxide from solution, which shifts the reaction toward the insoluble calcium carbonate:

$$Ca(HCO_3)_2 \rightleftharpoons H_2CO_3 + CaCO_3$$

$\downarrow$ Precipitates

Removed by photosynthesis

The insoluble carbonate settles to the bottom and becomes limestone deposits. (In addition to the activity of green plants, many animals deposit calcium carbonate in their shells, and these shells are an important component of limestone.) Ultimately geological uplift brings carbonate deposits above sea level, and then another cycle of erosion and sedimentation begins. The carbon cycle is summarized in Fig. 11.13.

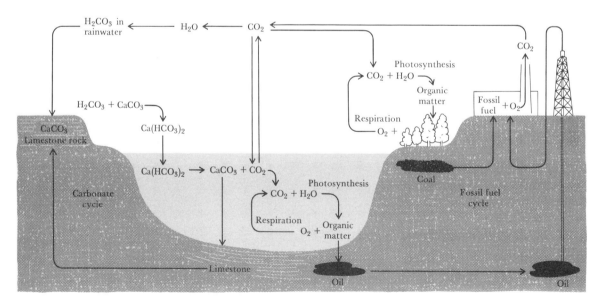

Figure 11.13. Carbon cycle. Carbon cycles biologically from carbon dioxide to organic carbon via photosynthesis and from organic carbon to carbon dioxide via respiration. This biological cycle ties in with the carbonate and the fossil-fuel cycles (two geological cycles).

OTHER ELEMENTAL CYCLES

Nitrogen, phosphorus, oxygen, silicon, iron, and most other elements essential to biological systems have biogeological cycles, which differ more or less from the carbon cycle. Many of these cycles involve specific kinds of organisms. Diatoms, because of their silica shell, are central in the silicon cycle. After the carbon cycle, the nitrogen and phosphorus cycles are perhaps the most important cycles to biological organisms.

The time scale of most geological cycles is on the order of several million years. We should be aware of the magnitude of time involved as we alter these cycles in our technology. Man is greatly speeding up the return of carbon to the atmosphere by burning fossil fuels. Phosphorus deposition in the oceans is increased by mining and agriculture. One specific case is the addition of phosphorus (as phosphates) to household detergents. The phosphate detergent is flushed directly into rivers and lakes, and from here it enters the oceans. Elements deposited in the sea bottom may be lost from terrestrial ecosystems for millions of years.

ECOLOGICAL SUCCESSION

Every organism lives and reproduces only in a particular kind of environment; each organism is unique in regard to its environmental requirements. An organism can survive only within a certain range of physical conditions; for any one condition this range is called the "range of tolerance." If an organism is exposed to conditions beyond its range of tolerance, it dies.

The species of organisms present in a particular geographical location may change with time. This change in community, or ecosystem, is caused by alterations in the physical and biological environment due to biological activities. Change in community over time is called "ecological succession." In ecological succession there is a progressive change in the kind of organisms living in an area. One community replaces another, and, in turn, is replaced by another. Ultimately, a stable condition is reached in which there is no further succession. The stable community is called the "climax community," or "climax ecosystem." The kind of climax community depends upon the prevailing climate, the soil type, and the organisms available for populating the area.

Succession is a result of modification of the environment by a community. One kind of modification results in a change such that the organism can no longer survive in that environment. For example, cedars, sycamores, and cottonwoods are relatively early stages in a succession. As these trees grow in size, they modify the soil and shade the ground. However, seedlings of these species cannot survive in the shade; i.e., the environmental conditions exceed the range of tolerance for shade for these species. Other species can tolerate the shade and eventually replace those that cannot.

Another kind of modification occurs when a community alters an environment so that the environment becomes suitable for invasion by another community. For example, grass growing on a sand dune alters the stability of the soil and allows growth of shrubs and trees. Thus, the modification caused by grass makes the environment suitable for trees, and the trees make the environment unsuitable for grass.

Succession may be rapid, occurring over a few months, or extremely slow, taking place over thousands of years. Succession can be repeated each season, for example, in a pond, where algae begin rapid growth in the spring when phosphorus concentrations are high. As phosphorus is depleted from the water, the environment becomes unfavorable for algae species which require high levels of phosphorus. Species with efficient phosphate uptake mechanisms replace those without them. Algae also excrete materials which affect the growth of other organisms, including other members of the same species. These inhibitory substances render the environment unsuitable for species that are susceptible to them, and species that can tolerate these substances become dominant. Thus, succession of algae species occurs over a season.

Old-field succession is an example of ecological succession occurring over hundreds of years. Old-field succession begins when a plot of ground is abandoned from cultivation. During the first season following abandonment, annual weeds and grasses are predominant. These are replaced by plants and shrubs that have longer life cycles, e.g., blackberries, broom sedge, cedar, and various grasses. These are replaced by the pioneer trees: elms, sycamores, and cedars. Shade produced by these trees favors growth of young oak and hickory seedlings. An oak-hickory forest becomes the climax community in many areas in the eastern half of the United States. However, in areas with sufficient moisture, oak and hickory are replaced by maple and beech. For example, a beech-maple climax occurs in the northeastern United States.

Man modifies the environment in numerous ways. One can speculate as to whether man is modifying his environment in such a way that it becomes unsuitable for his own survival. Pollution is one obvious example of this possibility (pollution is discussed in the following section). Man also simplifies ecosystems in attempts to maximize the food yield. These ecosystems often consist of a single species of crop. Single-species crops encourage outbreaks of animals called "pests," which consume these crops. Man temporarily controls these pests using chemicals; however, evolutionary processes create pesticide-resistant strains of the pest.

Man's activities are directly and indirectly producing changes in the communities found on the earth. We are beginning to worry whether any of these changes will alter man's position in the biological community.

POLLUTION ECOLOGY

Pollution might be defined as the dumping of man's wastes into the environment. However, because man is a part of nature, any definition of pollution cannot consider man's activities as totally unnatural events. We shall further define pollution as the loading of the ecosystem with materials at rates faster than they can be recycled. Pollution may be caused by man or by natural events, such as volcanos or floods. Any of man's activities may be considered as polluting if that activity overloads the ecosystem. For example, dumping excrement from one million persons into the Mississippi River is considered to be pollution. Any material that causes pollution is called a "pollutant." Before considering details of water and air pollution, it might be well to discuss some general consequences of pollution.

In ecological succession, one community of organisms replaces another in time. The mechanisms causing ecological succession in natural communities are analogous to man-made changes in his ecosystem. Is pollution changing the physical environment so that it may no longer be suitable for habitation by man? This appears to be the case. Is the environment created by man more favorable for survival of species other than himself? Will pesticide-resistant insects replace man? It is conceivable that they could, ironically, through the resistance they build to materials designed to kill them.

Pollution also affects the biogeological cycles. Materials may be removed from one ecosystem and added to another, and biological production may be reduced in both, decreasing man's food supply. Specifically this might be caused by erosion of minerals from the soil and their accumulation in aquatic ecosystems.

Several substances are being rapidly depleted from the land and becoming pollutants in the water. For example, mercury is a metal that is used in many industries as a catalyst in chemical reactions. However, mercury is a mineral resource in very short supply. In addition, mercury is a serious pollutant. For example, Wisconsin, Michigan, and Canada have recently banned the sale of fish caught in several bodies of water because of their high levels of mercury content. The very industries that face a shortage of mercury have been dumping it into the aquatic ecosystem. There are many such cases of pollution coupled to depletion.

WATER POLLUTION

Water pollution includes a variety of man-created problems adversely affecting the aquatic environment. Pollutants of the aquatic ecosystem include some materials that are toxic to life and others that subtly affect aquatic ecosystems by changing physical conditions or communities.

Toxic pollutants are more noticeable than others because they cause massive numbers of fish to die. The toxicity of a substance is expressed as the dose which will kill a stated percentage of a group of test organisms. The dose producing a 50 percent mortality is called the "median lethal dose" (LD50) and is used most frequently in determining the toxicity of drugs, food additives, insecticides, etc., in laboratory mammals.

The dose of a pollutant received by an aquatic organism living in a solution of the pollutant is difficult to determine. Instead, the concentration in the environment which produces 50 percent mortality is determined. This is called the "median lethal concentration" (LC50). The LC50 is determined by adding the pollutant (or drug or pesticide) to a series of aquaria, each at a different concentration. After a suitable time period (usually 96 hr) the mortality is recorded for each concentration. The concentration causing 50 percent mortality is observed or estimated (Fig. 11.14).

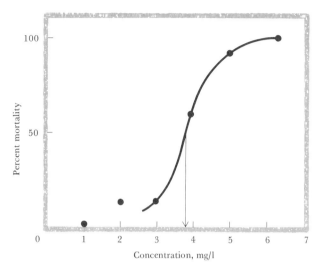

Figure 11.14. Measurement of toxicities. The LC50 toxicity is determined for a test substance. Six aquaria are prepared with concentrations ranging from 1 to 6 mg/l. The percent mortality is plotted to estimate the concentration at which 50 percent of the animals would be expected to die. This concentration is 3.8 mg/l, and this is taken as the LC50.

Toxic pollutants include salts of cadmium, iron, zinc, copper, mercury, lead, and silver, as well as sulfide, nitrate, nitrite, arsenic, cyanide, and sulfate; and organic compounds such as phenol, alcohol, insecticides, and herbicides. (The toxicities of various pesticides are listed in Table 11.3. Pesticide pollution is discussed in detail in the next section.) These pollutants are often discharged into aquatic ecosystems because recovery processes are inadequate.

Table 11.3. Toxicities of Various Pesticides for Fish

Pesticide	LC50 in μg/liter (rainbow trout)†	LC50 in μg/liter (bluegill)†
Aldrin	14	10
Chlordane	22	54
DDT	8	7
Dieldrin	6	14
Lindane	30	61
Malathion	100	120
Parathion	2,000	56
Zectran	6,500	16,000

† Twenty-four-hr period; long-term exposure increases the toxicity about tenfold.
Source: Charles G. Wilber, "The Biological Aspects of Water Pollution," p. 119, Charles C Thomas, Springfield, Ill., 1969.

Mercury contamination is one such example. Also, toxic materials may be used intentionally to kill aquatic organisms. For example, copper sulfate is used to remove organisms growing on cooling coils in power plants. Generally, little effort is made to recover the materials used.

In addition, the corrosion of structures built in the water may contaminate the water, especially with copper and zinc. For example, a nuclear power plant on the eastern coast of the United States caused copper pollution of an estuary (an arm of the sea at the lower end of a river) because of an error in salinity determination. Saline or brackish waters are more corrosive than fresh water. In an estuary the less dense fresh water is usually found on the surface above a brackish layer of water. In this case, salinity was determined for this surface water, but in power plant operation the deeper saline water was used to cool and corrode copper heat-exchange structures. As a result, the oysters in the estuary turned green from accumulation of copper salts.

Pollution may also cause changes in physical or chemical conditions in the environment. *Thermal pollution* is the disposal of excess heat into the environment. (See also Figures 5.2 and 8.10.) Power-generating plants are the most common source of heat pollution. Too-high temperatures in the environment will kill organisms that are especially sensitive to high temperatures. For example, salmon and trout have a relatively low tolerance for high water temperatures. Moreover, the young fish are often much more sensitive to temperature than the adults, but their death may not be noticed until it is discovered that there are no adults remaining. Thus, because high temperatures may differentially kill certain sensitive species, thermal pollution can alter the community of the aquatic ecosystem. Elimination of certain species can upset the balance between producers and herbivores and between herbivores and carnivores. Some species may be eliminated, and others may become very abundant.

In addition to lethal effects of thermal pollution, temperature increases also affect the metabolic rates of most aquatic organisms. Almost all aquatic organisms have a body temperature identical to that of the environment. Metabolic rates increase at higher body temperatures because higher temperatures increase the rates of all chemical reactions in the cells. A 10°C rise in temperature increases the metabolic rate by about two to three times (Fig. 11.15). Because metabolic rate is a measure of oxidation of food, increased metabolic rates mean increased food requirements.

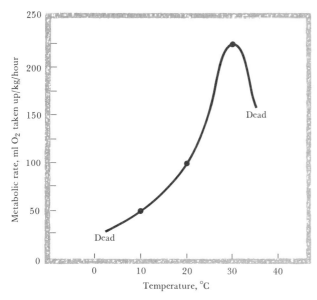

Figure 11.15. Metabolic rate of aquatic organisms as affected by temperature. In this example of a marine fish, the metabolic rate doubles for every 10°C rise in temperature. The animal dies at about 35°C, the upper lethal temperature, and also at 2°C, the lower lethal temperature.

Thus, fish trapped in the warm water discharge from a power plant might starve to death because of insufficient food to maintain their higher rates of metabolism.

Many pollutants affect aquatic organisms by changing the quantity of oxygen dissolved in water. There is relatively little oxygen in water compared to air. A liter of air has 210 ml of oxygen; 1 liter of water at 20°C has only 6.5 ml of dissolved oxygen. The small quantity of oxygen in water can be depleted relatively easily. (A further aspect of thermal pollution is the decrease in gas solubility with increase in temperature; for example, 1 liter of water contains 6.5 ml of oxygen at 20°C, 5.3 ml at 30°C, and 4.6 ml at 40°C.)

Reducing chemicals, such as sulfites from paper mills, use up oxygen from water. Microorganisms also use oxygen from the water and release carbon dioxide. The intensity of bacterial activity is directly related to the quantity of organic matter being decomposed. The most important source of organic matter in aquatic ecosystems is untreated or partially treated domestic sewage. The greater the quantity of sewage, the greater the oxygen depletion in the water.

Oxygen in water comes from two sources: photosynthesis by green plants which exist in the water and diffusion from the air. Photosynthesis occurs only in light and may provide significant quantities of oxygen to replenish the supply during the day. Diffusion of oxygen through the interface between water and air is relatively rapid. However, diffusion of oxygen through a large water mass is very slow. Oxygenation (addition of oxygen) at any distance from the surface depends upon mixing and turnover of the water from the surface layers.

Aquatic ecosystems differ in their mixing characteristics. Lakes and ponds are mixed rather poorly by wind-generated currents (Fig. 11.16). Because water is mixed slowly in lakes, organic pollution is very harmful to lake ecosystems. Streams and rivers are well mixed by water flowing downstream. Streams are well oxygenated and make ideal sites for man to dispose of his organic pollution. A river, because of its bacteria, has a tremendous capacity for metabolizing an organic load. However, even rivers are often loaded with sewage beyond their capacity to remain oxygenated.

Organisms differ in their tolerances for low oxygen levels. Unfortunately, the aquatic species that man finds most desirable are unable to survive in low-oxygen water; these include trout, salmon, pike, bass, sunfish, and crappie. Species that are less valuable to man are more tolerant of low-oxygen water; these include carp, gar, bullheads, and others. A few organisms, namely, some bacteria, can withstand complete absence of oxygen; these are

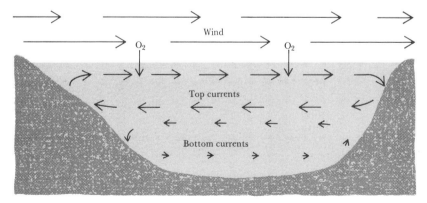

Figure 11.16. Currents in a lake are generated by winds blowing over the surface. These currents are responsible for mixing water and oxygenating deeper waters. The upper layers of the lake are better mixed than the lower layers.

called "anaerobes." Organic wastes can thus cause changes in communities which are not considered desirable.

Zonation is usually found in the vicinity of pollutant discharge. The effect of the pollutant is less at greater distances from the point of discharge, owing to its dispersal or metabolism. Different communities are found at various distances from the source of the pollution. Zonation in a river polluted with domestic sewage is shown in Fig. 11.17.

Addition of nutrients to water is another form of pollution. Nutrient pollution may be caused by inorganic substances, such as phosphates (found in many detergents), or by organic compounds. Decomposition of organic material releases abiotic substances into the environment. Depending upon the organic matter, carbon dioxide and compounds of potassium, nitrogen or phosphorus are liberated and become raw materials for plant growth. In the soil, the elements potassium, phosphorus, and nitrogen all are in short supply, and therefore plant growth is limited. In aquatic ecosystems, phosphate appears in the lowest concentration, and addition of phosphates to the aquatic ecosystem has a great impact on growth. (Sewage and many detergents are high in phosphates; phosphorus also comes from agricultural use of phosphate fertilizers.)

Nutrient enrichment of water usually increases primary production and thus increases the food base. In fact, fertilization of ponds is practiced intentionally in order to increase fish production. However, fertilization—intentional or otherwise—often does

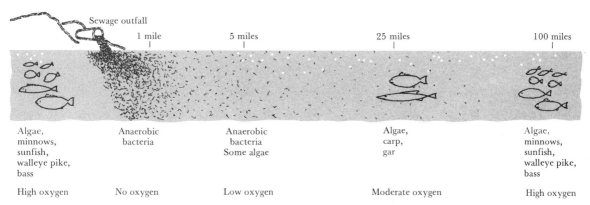

Figure 11.17. Community zonation and oxygen gradient in a river downstream from a sewage outfall. Oxygen is depleted below the outfall by the metabolic activity of bacteria decaying the organic matter. The biological community found at any point downstream is determined largely by the tolerance of individual species to low oxygen.

not produce the predicted results. Plants differ in their abilities to utilize nutrients in low concentrations. Many species that are the base of desirable food chains are able to use nutrients from low concentrations, but at higher concentrations other species become dominant. In some cases, these do not serve as a food for desirable edible fish. For example one genus of algae, *Cladophora*, is abundant in fertile water and forms thick slimy mats on the surface of the water (Fig. 11.18*a*). These algae are not only unsightly but are unsuitable as food for zooplankton and most fish. Thus, fertilization may actually decrease fish production.

Lakes become more fertile naturally over thousands of years because nutrients gradually enter the lakes from watersheds. Lakes pass through a number of successional stages (Fig. 11.18*b*). The process of enrichment, or aging, of a lake is called "eutrophication." The earliest stage is an *oligotrophic* lake (a lake with scant

(*a*)

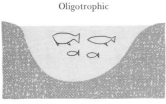
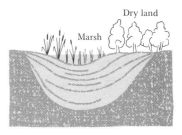

Oligotrophic

Eutrophic

Marsh

Dry land

Bass, walleye pike, sunfishes

(*b*)

Bullhead, carp, sunfishes

Figure 11.18. (*a*) *Slimy mats of algae covering water surface.* (*b*) *Successional stages in a lake. Addition of nutrients and silt to a lake gradually fills the lake. The biological communities come and go depending upon the conditions found in the lake.*

food); the next stage is *a eutrophic lake* (a lake that is rich in food). The following stage is a *marsh*. Finally, the lake becomes dry land. Each stage in this eutrophication process is characterized by particular communities of plants and animals. (The most desirable species of fish are found in the oligotrophic stage, e.g., bass, walleye pike, and pike.)

Man's activities have greatly accelerated the rate of eutrophication, or aging, of lakes. Addition of nutrients from sewage and other sources directly enriches the water, and lakes may gradually become filled with sedimentation from this enrichment. Erosion accelerated by forest fires, agriculture, highway construction, etc., fills lakes rapidly. Especially vulnerable are man-made lakes located on rivers. Artificial impondments built for flood control or hydroelectric power have life expectancies of only 100 to 200 years owing to sedimentation.

PESTICIDE POLLUTION: DDT

Good insecticides are needed because insects do considerable damage to agriculture and are carriers of disease. In 1873 the compound 1,1-bis(parachlorophenyl)-2,2,2-trichloroethane was first prepared. Its long name describes its structure exactly but with 34 letters, this name is written far more often than spoken. A shorter but less descriptive name is dichlorodiphenyltrichloroethane. From this comes the acronymic and better known DDT.

DDT

In 1938 the Swiss chemist Paul Müller (1899–1965) observed that DDT was a very effective insecticide, yet apparently nontoxic to human beings. In late 1943 the Allied forces, fighting around Naples, discovered cases of typhus among the native Italian population. Typhus is caused by a bacterialike microorganism, a rickettsia; it is spread from person to person by body lice and fleas.

The symptoms of typhus include chills, fever, headaches, severe body pains, and delerium and may last from a few weeks to a few months. About 70 percent of untreated typhus cases die. This disease is very contagious provided the proper *vectors* (carriers) are present. Crowding of people and inadequate sanitary facilities (conditions commonly found in armies and among the very poor) make a typhus epidemic a serious threat to a fighting force.

In January 1944 Müller's compound was used on the civilian population of Naples. The infected lice were destroyed, and the epidemic stopped. In 1948 Müller was awarded the Nobel Prize. The compound DDT is a major weapon against both disease and crop destruction. (In Chap. 10, Fig. 10.13 shows a child being sprayed with DDT.)

DDT has an important characteristic: it is relatively stable. It is also fat-soluble. Predator animals take up DDT in their diet, accumulating the insecticide in their body fat. As we go up the food chain, from algae and insects to fish, birds, and man, the concentration of DDT increases in fat deposits.

> **Example.** Algae can contain 1 μg DDT/100 g. This 100 g of algae is consumed by zooplankton to produce 10 g of zooplankton. Now the 1 μg of DDT is in 10 g of living organisms. This 10 g of zooplankton is enough food for 1 g of minnow; the DDT concentration is 1 μg/g of minnow. In turn, bass eat minnows, and the DDT concentration is 10 μg/g of bass. When man eats the pesticide-loaded bass, the DDT is further concentrated, now in the fatty deposits of the last consumer. Because of the several steps in most aquatic ecosystems, some substances can be concentrated ten-thousandfold in the terminal consumer. (Terrestrial ecosystems usually have shorter food chains, and the accumulated concentration is commonly less.)

DDT has certain obvious disadvantages. It is too good an insecticide in that it kills the natural enemies of harmful insects, both other insects and birds. For example, DDT interferes with calcium deposition in eggshells in some birds. These bird species face eventual extinction from DDT because their young cannot survive in the delicate shells. Dutch elm disease, a fungus infection spread by bark beetles, threatens to destroy elms but can be controlled by DDT. Thus, with DDT we can have either elm trees or robins but probably not both.

Carried by streams, rivers, wind, and clouds, DDT is distributed worldwide. It has even been found in Antarctic snow.

Although we are not certain what high levels of DDT will do in a human body, it is unlikely to be beneficial. Because of its persistence and distribution, heavy use of DDT in one area affects people all over the world. For this reason the free use of DDT has been proscribed by several states, and fish from certain of the Great Lakes have been condemned for human consumption. Gradually, as better insecticides are developed, the use of DDT will be confined to controlling disease. In India, for example, it is estimated that four or five million persons annually are saved from death by malaria because the carrier mosquitoes are killed by DDT. In India the short-range management of malaria is more important than whatever long-range effects DDT may have.

The very toxic pre-World War II insecticides, such as arsenic and copper compounds, nicotine, and kerosene, are called "first-generation insecticides." The less toxic DDT and related compounds are called "second-generation insecticides." Because most insects are not pests, and because many are beneficial, there is growing interest in *third-generation insecticides,* which are designed to attack only particular harmful species of insects. One example is the *juvenile hormone* (or a synthetic substitute), which prevents sexual maturity and reproduction in the specific target pest. Another example is a sex attractant, which lures specific pests into traps. Because third-generation insecticides are unnatural uses of natural substances, it is unlikely that pests will develop a resistance, or immunity, against them (as has been the case with DDT). A still more elegant pest-control method is also becoming increasingly important: using a pest's natural enemies to keep its population at a low and safe level.

AIR POLLUTION

Each day we take into and exhale from our lungs about 14,000 liters of air. Air is a mixture of gases (about four-fifths nitrogen, one-fifth oxygen, and some rare gases), varying amounts of water vapor and carbon dioxide, and, depending upon where we live, pollen, waste gases, and fine solids from industry and automobiles. This section deals with some problems associated with the undesired components, or pollutants, found in the air. The principal pollutants are carbon monoxide, oxides of sulfur, oxides of nitrogen, hydrocarbons, photochemical smog, and particulate matter.

The level, or concentration, of air pollution can be determined in the case of solid pollutants by pumping a measured volume of air through filters which trap the solids. The amount of solids trapped can be determined by weighing or by chemical analysis and expressed as micrograms of solids per cubic meter of air. Gaseous pollutants are pumped through reactive solutions. Sulfur dioxide, for example, reacts with and bleaches the color of a number of dyes. (Sulfur dioxide is also used as a bleach for many canned and prepared fruits; maraschino cherries are bleached by sulfur dioxide from red-brown to pale yellow and then redyed a typical brilliant red.) The change in the absorption of light of a dye solution can be related to the amount of sulfur dioxide contained in a measured amount of pumped air. The gaseous level of a pollutant is expressed in parts per million (ppm), that is, in grams of pollutant/1 million g of air.

Monitoring air pollution levels is desirable to catalog what is in polluted air, relate pollutant levels to adverse effects, and follow trends in pollution levels to predict future emergency situations. Not all pollutants are monitored; only those pollutants that pose a threat to human health and that can be analyzed by automatic rather than manual methods are monitored. Levels of pollutants are now regularly reported along with weather reports by many radio and TV stations.

Pollutant levels are not constant; there are seasonal cycles. In northern United States more fuel is burned in winter, increasing air pollution; in southern United States, fuel consumption (for electricity generation and air conditioning) is greater in summer. Also, there are week-day and weekend cycles from factory and automobile-traffic activity. Finally, there are diurnal, or 24-hr, cycles caused by rush-hour traffic and home heating and lighting requirements.

In addition, there is a very important diurnal cycle due to a weather condition called "nocturnal inversion." *Inversions* occur when layers of cool dense air are trapped below layers of warmer lighter air. Gas volumes increase with an increase in temperature. A given gas sample will occupy a larger volume when warmed than when cool; accordingly, the mass per unit volume, or density, is less for warm gas than for cool gas. When two air masses are contiguous, the warmer less dense mass will rise and the cooler more dense mass will descend. (See also Fig. 3.15.)

During the day the sun's rays are absorbed by the earth, which becomes warm. The earth, in turn, warms the air layer adjacent to it. This layer, being warmer than air at higher altitudes, will rise up, to be replaced by descending cooler air, which, in turn, warms and rises. In this way pollutants are dispersed into a large volume of air, and concentrations do not build up (Fig. 11.19). At night the earth loses heat by radiation, and the earth's surface cools. A layer of cool air forms next to the earth; this layer is more dense than layers at higher altitudes, and so the daytime convection stops. This *nocturnal inversion* (Fig. 11.20) of a cool layer under a warmer layer permits an accumulation of air pollutants. However, this diurnal cycle is observed only when wind disturbances are minimal.

Occasionally, temperature inversions remain during the daylight hours, too. Under these conditions, pollutants accumulate during both night and day. Continuous inversions are common in climates where cool air from a body of water moves inland to continuously regenerate an inversion layer.

AIR POLLUTION DISASTERS. Any discussion of the effect of pollution on human health must include three well-documented disasters. These occurred in Belgium in 1930, in Pennsylvania in 1948, and in London in 1952; and they all resulted from prolonged temperature inversions.

The first episode occurred along the 15-mile-long, heavily industrialized Meuse River Valley in Belgium during the first week in December, 1930. A day-and-night temperature inversion trapped pollutants from homes and factories. After a few days' accumulation, symptoms of coughing and painful breathing were recorded. Thousands of residents became ill, and 60 persons died. Most of the victims were elderly persons or persons with histories of respiratory and heart disease.

During the last week of October, 1948, the entire northeastern United States was under a temperature inversion. In Donora, Pennsylvania, there was fog, a high sulfur dioxide level, and no wind. Symptoms were coughing and breathlessness. In a population of 14,000, 6,000 became ill. The death rate in Donora was ordinarily about 16 deaths/mo at this time of year; but at the height of the pollution, in one day alone 15 persons died. The victims were all older persons, and most had preexisting cardiorespiratory diseases. Autopsies revealed bronchial and lung damage attributed to the pollutants.

The third episode took place in London. From December 5

Figure 11.19. Daylight unstable atmospheric conditions. Warm, less dense air rises by convection; cool, more dense air descends and replaces the warm air. Pollutants are mixed and dispersed into the upper atmosphere.

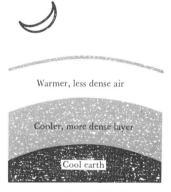

Figure 11.20. Nocturnal inversion. Under these conditions, a layer of cool dense air forms next to the cool earth. There is no mixing of ground air, and pollutants may be trapped in the inversion layer. (See also Fig. 3.15).

to 9, 1952, inversion and a heavy fog prevailed. In 1 week there were 4,000 deaths in excess of what was normal at that time of year. Again, fatalities were predominantly among the aged with cardiorespiratory disorders. A decade later a similar situation developed, but this time the problem was recognized. The elderly and persons with cardiorespiratory disorders were advised to stay indoors; those who ventured out wore mouth-and-nose gauze masks to filter out particulate matter. Consequently, the toll of human lives was considerably less in 1962 than in 1952.

POLLUTION EFFECTS

The effects of pollutants on human health ordinarily are not studied under laboratory conditions. Relations are established by examining pollutant levels and reported mortality and disease rates. A *correlation* is said to exist between a pollutant and a specific health problem when an increase in the incidence of one is attended by an increase in the other. However, a correlation does not necessarily establish a cause-and-effect relation. For example, if pollution is correlated with lung cancer, then pollution may or may not cause lung cancer. [It is necessary, therefore, to identify specific carcinogens (cancer-producing agents) in polluted air.]

Many studies have correlated air pollution with specific diseases of the cardiorespiratory systems. Air pollution and cigarette smoking correlate with chronic bronchitis, an inflammation of the bronchi and bronchioles. In England and Wales there is a correlation between sulfur dioxide levels and mortality attributed to chronic bronchitis. The quantity of air-borne particulate pollution is also correlated with chronic bronchitis. Emphysema, a disease in which alveoli of the lungs are destroyed, is correlated with both air pollution and cigarette smoking. Several studies report correlations between air pollution and lung cancer and between cigarette smoking and lung cancer. For example, the incidence of lung cancer among urban heavy smokers is about twice that among rural heavy smokers. In nonsmokers, the incidence of lung cancer among urban dwellers is also about twice as great as among rural dwellers. The larger the city, the greater the incidence of lung cancer. These correlations strongly suggest that air pollution (and cigarette smoking) causes disease and mortality.

Carbon monoxide CO, rather than carbon dioxide CO_2, is produced by the partial oxidation of carbon-containing fuels due to an insufficiency of oxygen. Carbon monoxide is a completely colorless and odorless gas, and it is moderately toxic. In man and animals carbon monoxide combines with hemoglobin and prevents adequate oxygen-hemoglobin combination. Interfering with the oxygen-carrying capacity of the blood in this way may result in death. At lower than lethal levels, carbon monoxide produces headaches, reduced alertness, decreased muscular coordination, and reduced visual acuity.

Exposure to carbon monoxide concentrations of 120 ppm for 1 hr or 30 ppm for 8 hr constitutes a serious health hazard to persons with cardiovascular disorders, anemia, and respiratory diseases. The U.S. Public Health Service sets a value of 10 ppm carbon monoxide as being unsafe. Carbon monoxide levels in the passenger compartment of automobiles in heavy city traffic has been measured as 20 to 40 ppm. In some cities (for example, Chicago) the average CO level along expressways is 10 ppm. In addition to direct danger of carbon monoxide, an increased number of accidents may result from decreased driver alertness. How alert can a driver be after spending hours in a traffic jam like the one shown in Fig. 11.21?

The oxides of sulfur (sulfur dioxide SO_2 and sulfur trioxide SO_3) enter the atmosphere from paper and pulp mills and from combustion of sulfur-containing fuels. These oxides are responsible for metal and stone corrosion. Steel corrodes about four times faster in urban areas where sulfur-containing fuel is used. The obelisk popularly called "Cleopatra's Needle" was brought to Central Park in New York in 1881. It has sustained more corrosion during its nine decades in New York than during 3,000 yr in Egypt.

Figure 11.21. A typical New York City traffic jam. (Chemical and Engineering News.)

In addition, sulfur oxides are injurious to vegetation, animals, and man. Concentrations of 0.04 ppm are believed to damage health. The average concentration in many cities exceeds these levels. When breathed in, sulfur oxides produce coughing and difficulty in breathing. (These oxides were largely responsible for deaths in the day-night temperature inversions discussed in the previous section.) A specific disorder caused by prolonged breathing of sulfur-dioxide-polluted air is chronic bronchitis (see pp. 127–129).

Nitrogen oxides (NO and NO_2) are formed from oxygen and nitrogen at high temperatures and pressures. Motor vehicles, especially high-compression internal-combustion engines, generate about one-half of the nitrogen oxide pollution in the United States. Nitrogen oxides are one of the causes of smog; nitrogen dioxide NO_2 gives a whiskey-brown color to smog. Nitrogen oxides react with water to form nitrous and nitric acid, which are highly corrosive (see Table 9.2).

Hydrocarbons are added to the air by petroleum refineries, automobile gas tanks, and automobile exhaust pipes. Some of these hydrocarbons, such as benzopyrene (Fig. 7.10), are identical to those found in cigarette smoke and are thought to be factors in lung cancer. Breathing the air in cities with moderate carcinogenic (cancer-producing) hydrocarbon levels is equivalent to smoking six cigarettes per day; in more polluted cities breathing the air is equivalent to smoking a pack of cigarettes daily.

Photochemical smog is formed in sunlight by reaction between hydrocarbons and oxides of nitrogen. Smog is a mixture of several gases, including nitrogen dioxide, ozone, and a variety of olefins and aromatic organic compounds. Sunny climates are particularly vulnerable to this variety of smog. Smog is worst in large cities that rely upon the automobile for transportation. Southern California has ideal climatic conditions for generating and trapping smog. Smog causes serious damage to vegetation; crops are affected many miles from metropolitan areas. Forests in the mountains 100 miles from Los Angeles have been attacked by air pollution. Smog is irritating to eyes, nose, and throat, and it may contribute to the onset of cancer and other respiratory diseases.

Particulate matter is the collective name for an assortment of finely divided solid particles, such as iron dust from steel mills, carbon soot from incomplete combustion of fuels and refuse, asbestos fibers from a variety of sources, and many other man-made particles. Pollen is a natural air-borne particulate. Particulate matter constitutes the most noticeable filth in the air in cities. It settles out on buildings, clothing, home furnishings, parked cars, and vegetation.

Particulate pollutants may be a significant health hazard. Particles are inhaled and deposited in the respiratory passageways and in the lungs. Large particles are trapped in the moist surfaces in the upper respiratory tract. They are less dangerous than the intermediate-size particles, which are carried into the lungs and deposited there. Small particles (less than $0.5\ \mu$ in diameter) are respirable; i.e., they can be exhaled from the respiratory system. Some kinds of particulate matter, for example, asbestos, have been linked to lung cancer (Problem 11.30).

In addition, particulate matter absorbs sulfur dioxide. Ordinarily sulfur dioxide, which is water-soluble, is absorbed by moisture in the mouth, nose, and upper throat, where it merely causes breathing difficulties. When absorbed in particulate matter, some of it is carried deeper into the bronchial and alveolar regions of the lungs. Here sulfur dioxide causes more serious damage, including bronchitis. The simultaneous combination of high levels of sulfur oxides and particulate matter in the air is believed to be an especially dangerous condition.

THE SOURCES OF POLLUTANTS

The tonnage of pollution produced in the United States in 1966 is shown in Table 11.4. Carbon monoxide is the chief pollutant, and its principal source is the automobile. These figures may be misleading as far as public health is concerned because of differences in toxicity of different pollutants. Sulfur oxides, for example, are about five hundred times more toxic than carbon monoxide. Thus, in terms of impact upon public health, 25 million tons of sulfur oxides are potentially 100 times more harmful than 72 million tons of carbon monoxide.

Table 11.4. Principal Atmospheric Pollutants in the United States in 1966

Type	Million tons/yr	Percent
Carbon monoxide	72	52
Sulfur oxides	25	18
Hydrocarbons	18	13
Particulate matter	12	8.5
Nitrogen oxides	12	8.5
Total	139	100.0

SOURCE: Sources of Air Pollution and Their Control, *U.S. Public Health Service Publ. No. 1548*, U.S. Government Printing Office, Washington, D. C., 1966.

Table 11.5. Sources of Air Pollution

Pollution	Annual amounts in million tons from:					
	Motor vehicles	Industry	Power plants	Space heating	Refuse disposal	Total
Carbon Monoxide	66	2	1	2	1	72
Sulfur Oxides	1	9	12	3	less than one	25
Nitrogen Oxides	6	2	3	1	less than one	12
Hydrocarbons	12	4	less than one	1	1	18
Particulate Matter	1	6	3	1	1	12
Total	86	23	19	8	3	139

SOURCE: The Sources of Air Pollution and Their Control, *Public Health Service Publication No. 1548.* U.S. Government Printing Office, Washington, D.C., 20402, $0.35, 1966.

The sources of pollution in the United States in 1966 are reported in Table 11.5. Transportation, principally by motor vehicles, is the chief source of pollutants, contributing about 60 percent of the total tonnage of pollutants. Industry, which is the most conspicuous polluter, contributes only about 20 percent of the total amount. Electric-power plants produce 12 percent of the total tonnage. However, there is a difference in the kind of pollutant produced by each source. The gasoline engine is chiefly responsible for carbon monoxide, hydrocarbons, and nitrogen oxides, whereas the main source of sulfur dioxide is the electric-power plant.

TECHNOLOGY AND POLLUTION

As population and economy grow, the level of pollution in the United States is expected to increase proportionally. If the quantity of pollution is related to population, the amount of pollution should double in about 63 yr (the estimated time for doubling the United States population). Most evidence suggests that pollution is more related to the level of economic development than to variations in population. Assuming an annual economic growth of 4 percent (considered by economists to be a healthy growth rate), the quantity of pollution would be expected to double in 18 yr.

Levels of pollution in several cities already exceed what is considered to be safe. Therefore, some urgency exists for pollution abatement, not only to improve the present situation, but to prevent further deterioration. If we do nothing, we can expect more serious and more frequent disasters of the Meuse Valley, Donora, and London variety.

AIR–POLLUTION CONTROL

Pollution abatement involves social, economic, legal, and ethical considerations in addition to technological problems. Technological solutions cost money, and they must ultimately be paid for by the citizen. Money in antipollution devices buys a certain amount of benefit—corrosion is reduced, lives are saved, and sickness is prevented. Initially, the more money spent, the greater the benefit. Beyond a certain point, however, further expenditure of money does not return as much benefit: the more money spent, the less benefit per dollar.

 Unhappily, we cannot assign a monetary value to benefits from pollution control; illness and death cannot be measured by dollars and cents. The determination of the optimal expenditure for pollution control is not amenable to cost-benefit analysis. There is no simple answer to the question, "How much pollution control is needed?"

 We shall now discuss some of the technological aspects of pollution abatement. Gaseous pollutants can be prevented from entering the atmosphere by absorption in liquids, absorption in solids, condensation to liquids by cooling, or complete combustion to harmless products. Particulate matter can be removed by filtration or electrostatic precipitation. These methods all remove pollutants by altering their chemical or physical states. Each method has its advantage; no one method can remove all pollutants.

 Gaseous pollutants can be removed by several methods. A gas scrubber (Fig. 11.22) removes gas by bubbling the contaminated mixture through a solution containing chemicals which will react with or dissolve the pollutant. Also, the contaminated air may be passed through a chamber containing a solid which will absorb the pollutant (Fig. 11.23). In many cases the gaseous pollutant is an incompletely oxidized product of combustion, e.g., carbon monoxide and hydrocarbons. The effluent containing the pollutant may be passed through a combustion chamber or afterburner to completely oxidize the pollutant (Fig. 11.24). Some effluents contain valuable products that escape to become pollutants. These can be recovered by cooling the gaseous mixture (Fig. 11.25) or by compressing it. The product is recovered as a liquid or solid.

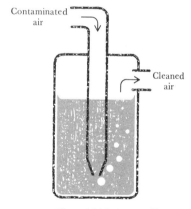

Figure 11.22. Gas scrubber. Pollutant dissolves in the liquid as the contaminated air is bubbled through the solution.

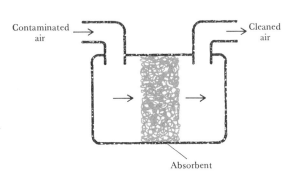

Figure 11.23. Pollutant removal by a solid absorbent. Polluted air is passed through a solid material which reacts with the gaseous pollutant and removes it from the contaminated air.

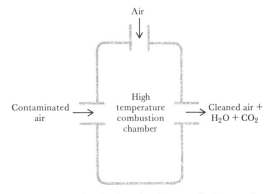

Figure 11.24. Combustion chamber, or afterburner, for burning pollutants. Incompletely oxidized materials, that is, carbon monoxide and hydrocarbons, are completely burned to form carbon dioxide and water.

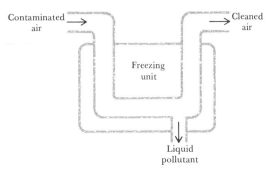

Figure 11.25. Antipollution cooling unit. The pollutant is liquefied or solidified by cooling (or compressing), allowing recovery of possibly valuable by-products.

Liquid and solid pollutants can be removed by filtration. Coarse particles are easily filtered, or settled out. Fine dust, however, must first be directed at high speed into a mist of liquid drops in what is called a "Venturi scrubber" (Fig. 11.26). The dust-and-droplet particles are larger than the original dust particles and can be more easily filtered. Electrostatic precipitators charge fine dust particles, which are then attracted to charged plates from which they can be removed (Fig. 11.27).

Automobiles are responsible for a very large proportion of the air pollution in the United States, and therefore reduction in emission of pollutants by automobiles would significantly reduce levels of pollution. All new automobiles sold in the United States must now be equipped with antipollution devices. The principal

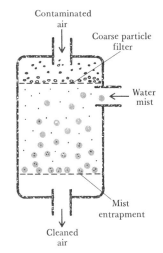

Figure 11.26. Venturi scrubber. Coarse particulates are removed by a filter. Water mist is sprayed into the partially cleaned air. The water particles absorb fine dust, and the water-dust particles are trapped by a second filter.

type is an afterburner which converts hydrocarbons and carbon monoxide in the exhaust to harmless carbon dioxide and water. (Nitrogen oxide emission is not reduced by an afterburner.)

Considerable progress has been made toward reducing pollutants from manufacturing and electric-power plants. Sulfur dioxide from sulfur-containing fuels (e.g., coal) has been reduced by using low-sulfur or sulfurfree fuels in high pollution areas and by removing some of the sulfur oxides from the combustion products before exhausting them into the atmosphere. About 99 percent of the dust and ash can be removed from most smoke stacks at a very reasonable cost. Many industries have installed electrostatic precipitators or Venturi scrubbers and have achieved these results.

A number of industrial pollution-control programs have yielded valuable marketable by-products from the contaminants removed. The red smoke from steel mills contains high concentrations of iron. This dust can be sintered (heated to conglomerate the finer particles into coarser particles) and then used as a raw material for steel in open-hearth furnaces. Fine ash may contain significant unburned carbon, which can be used as a fuel. Ash can also be recovered and sold to cement and abrasive manufacturers. Many valuable materials escape during the manufacturing process. Thus the cost of recovering what would otherwise be pollutants may be offset by their commercial value. For example, chlorine users can recover liquid chlorine from waste gases at a cost of about one-third of its market value. For many marketable pollutants, however, the more recovered, the greater the supply and the lower the market price.

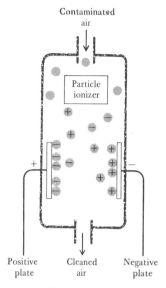

Figure 11.27. Electrostatic precipitator. Fine particles are ionized and then trapped on charged plates from which they can be removed.

POPULATION PROBLEMS

The problem of population growth, along with economic growth, is related to pollution problems. Although higher densities of people do not necessarily mean more intense levels of pollution, nevertheless there is some correlation between dense populations and severe pollution. This is most evident in comparing industrialized regions having different population densities.

The exhaustion of natural resources, such as natural gas and petroleum, is related to the number of people utilizing these resources. If populations are dependent upon supplies of these resources for heating homes or transporting food, then eventually these resources will be depleted and alternative resources will have to be developed. However, more and more people will cause more rapid exhaustion of resources, and the time for development of alternative resources will be shortened. Man's present urban system is very dependent upon fossil fuels for transporting food, and as yet there is no practical substitute. Continued population growth will continue to place strain on our natural resources.

A third area where continued population growth may be critical is in food supply. The production of food continues to increase throughout the world, but the increase is barely enough to feed the growing populations. There is little promise that undernourishment and malnutrition can be eliminated if present population growth continues. With more people there is greater danger of a worldwide famine triggered by crop failures. (Problems 11.19–11.24)

Some of the problems caused by continued population growth might be less acute if we could predict exactly how the population was going to change in the future. Proper planning of pollution-abatement programs, such as sewage-treatment facilities, require some information concerning future populations. Political or economic decisions concerning conservation vs. development of a resource (as in the North Slope oil deposits in Alaska) are based, at least in part, on the knowledge that future populations will require these resources. There is considerable value in being able to predict future population trends in man.

Presently, population predictions are based upon information about birth and death rates and their effects on population growth. These have been very unreliable in the past and probably will be even more so in the future. Organisms other than man have population growths that follow particular patterns. These patterns might be useful in predicting the future growth of man's population.

POPULATION–GROWTH MODELS

Every individual organism within a population ultimately dies and must be replaced if the species is to survive. Replacement is achieved by reproduction. To always insure adequate replacement, all species possess a capacity for reproduction in excess of the minimum required to merely replace dying individuals. This reproductive potential insures against species extinction during periods when mortality rates may temporarily increase. If the reproductive potential is realized, the number of individuals in the population increases; i.e., there is population growth. In nature population growth is a phenomenon that occurs for limited periods of time.

There are two basic models for population growth: the sigmoid, or S-shaped, curve and the J-shaped curve. The sigmoid growth model (Fig. 11.28) fits growth patterns for species that have well-developed mechanisms for controlling population size. This pattern of growth is observed after such a species is introduced into a suitable but previously unoccupied habitat or after a catastrophic reduction in population. Initially, growth is slow because relatively few reproductive individuals exist. As the population size increases with time, growth becomes explosive. The steep part of the curve in Fig. 11.28 corresponds to this population explosion. At high population densities, i.e., many individuals per unit of available habitat and niche, self-regulating mechanisms reduce population growth. (These regulating mechanisms are discussed in the final section of this chapter.) A steady state is reached where there is no further increase in numbers; i.e., there is zero population growth. At this level, the environment can support the population indefinitely.

The second model for population growth, the J-shaped curve, fits growth patterns of species that have poorly developed mechanisms for regulating population size. Initially the pattern of

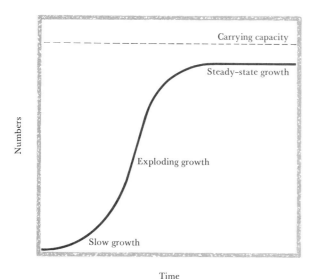

Figure 11.28. Sigmoid, or S-shaped, growth curve for a population.

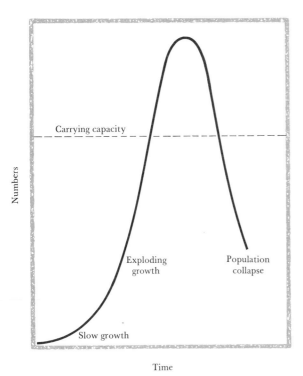

Figure 11.29. *J-shaped curve for population growth.*

population growth is identical to the sigmoid-growth model. There is a slow growth period, followed by a period of population explosion (Fig. 11.29) At some point, the population collapses; i.e., numbers drastically decrease.

There are various causes for population collapse, depending upon the species and the conditions. One cause for population collapse is that the population exceeds the *carrying capacity* of the habitat. The carrying capacity is the maximum number of individuals of that species that can be supported indefinitely by a particular ecosystem. Collapses from this cause are characterized by over-exploitation of resources; the population uses up materials or food that have accumulated in the ecosystem. A classical example is the deer herd of the Kaibab National Forest in Arizona. Between 1907 and 1923 deer predators, i.e., mountain lions and wolves, which normally regulate deer numbers, were systematically exterminated. The deer population exploded, increasing by twenty-fivefold by 1925. The deer ate every living plant within their reach, and the ecosystem was effectively depleted of its producers; i.e., there were no plants left to manufacture food for deer. The deer population starved to death, and the population declined 90 percent. The observed pattern of growth fits the J-shaped curve.

Population growth in insects that are pests to many of man's crops often follows the J-shaped model, but the population collapse is not usually due to depletion of the resources. Under ideal food and weather conditions, insect populations explode, but then they collapse because of cold or rain (or spraying of pesticide).

The population-growth model fitting human population growth is of great concern to mankind today. Man's population growth is presently in the exploding phase. It is not yet possible to determine whether the observed growth fits the S- or J-shaped-curve model. Speculations are possible. The J-shaped model generally fits population growth for species that overexploit their environmental resources. Currently man appears to be depleting the earth of its mineral, fossil fuel, and soil resources (Problem 11.36).

Man has experienced several population crashes in the past. The Yucatan peninsula of Mexico was once densely populated with highly cultured Maya Indians. For causes that are unclear, the population and civilization collapsed, leaving behind many deserted cities. Beginning in 1348, bubonic plague, or Black Death, spread over Europe. This disease is caused by a bacterium transmitted from rodents to man by fleas. Large numbers of people died (25 million in Europe alone). The population in some areas declined by 75 percent; England's population decreased by one-half. In 1846 the potato crop, upon which the Irish peasant depended, was destroyed by potato blight. A potato famine resulted, in which the population declined from 8 to 4 million, with 1 million people starving to death.

Other evidence suggests that human growth curves follow the sigmoid model. Until modern medicine became available, many countries remained in approximate steady state in regard to population. Population was controlled by infant mortality, disease, tribal warfare, human sacrifice, and various sexual taboos. Europeans conquered and colonized most of the earth, and with this conquest they brought many supposed reforms in culture, morality, and public health. These changes effectively removed control of population growth, and consequently populations exploded.

Part of the debate concerning the growth model for human population hinges upon the maximum carrying capacity of the earth for man. There is no agreement among natural scientists concerning the carrying capacity of the earth. One view is that the world population has already surpassed the earth's capacity to support the present number of people for any length of time. If this is true, we can expect a population crash within the next two or three generations. Others believe that the world population will soon exceed the carrying capacity and will be followed by a population collapse.

A slightly more optimistic view is that man's population will level off near the maximum carrying capacity (following the sigmoid model), presumably regulated at this level by increased death rates from starvation and disease. A fourth possibility is that population will level off at some point below the carrying capacity, regulated by control of birth rates. At present, this last possibility seems remote. Finally, there are two possibilities that a population crash might occur independently of the carrying capacity. Thermonuclear warfare would effectively reduce world population, as would a mass epidemic such as plague (or a virulent influenza). Indeed, it is an increasingly more common view: the future for the world's human population looks bleak.

BIRTH AND DEATH RATES

Reproductive success in populations is measured as reproductive rates, or birth rates. The *birth rate* is defined as the number of new individuals born to a population during a unit of time. The most widely used time units are the year and the generation. One *generation* is the elapsed time between birth and reproduction. A generation time is often taken as the difference between the average age of parents and offspring. The generation time varies between species, being about 30 yr in man.

Birth rates may be stated as the absolute number of births per unit of time or as the average number of births per parent. In human populations, the birth rate is generally expressed as the average number of births per year for every 1,000 members of the population. In the United States this number is about 19 births per 1,000 members of the population per year. Other modern industrialized nations have similar birth rates, the lowest being Hungary, with 13 per 1,000. On the other hand, the so-called underdeveloped nations of the world have birth rates around 40 per 1,000 per year. The world average birth rate has been estimated to be 35 per 1,000 per year.

Not all newborn individuals become adult members of the population. Many young die, some immediately after birth and others before reaching maturity. In many species the number of deaths of young individuals is a large percentage of the total births. In these cases the birth rate has little significance in assessing population dynamics. A more important parameter is the number of young reaching maturity, or recruited into the adult population. *Recruitment* is the rate at which reproductively mature individuals are added to the population. The term is seldom used in developed human society because of relatively high survival rates of the young.

Every individual organism ultimately dies. The rate at which individuals die is called the "death rate." Death rates are commonly reported in the same terms as birth rates, i.e., the annual number of deaths per 1,000 population. The death rate in the United States is currently about 9 per 1,000 per year. Other industrialized nations have similar death rates. Surprisingly, many underdeveloped countries have death rates similar to developed nations (Ceylon has about 9 per 1,000; Taiwan, about 6 per 1,000). This is because developed countries have exported their medicine and health-care technology to these underdeveloped countries. As a result mortality has greatly decreased in every country of the world over the last 100 yr.

The growth rate for any population is the difference between the birth rate and the death rate:

Growth rate = birth rate − death rate

In all nonhuman populations examined over any significant time interval, the population growth is zero. Stated another way, birth rates always balance death rates, providing the observations are made over enough generations. In these populations the phenomenon of positive population growth is a temporary situation, resulting from unusual conditions. The same principle probably applies to human populations.

Population growth is analogous to growth of money in a bank account upon which compound interest is paid. In the United States the difference between a birth rate of 19 per 1,000 and a death rate of 9 per 1,000 is an annual population growth rate of 10 per 1,000, which is equal to a 1 percent increase per year. In 1970 the United States population was 205 million. The growth rate of 1

Table 11.6. Annual Birth, Death, and Growth Rates for Selected Populations of the World †

Country	Birth rate	Death rate	Growth rate	Doubling time, yr
Mexico	45	9	36	19
Brazil	42	12	30	23
Egypt	44	15	29	24
Ceylon	33	8	25	28
Argentina	23	8	15	47
U.S.S.R.	18	7	11	63
U.S.A.	18	7	11	63
France	17	7	10	70
W. Germany	18	12	6	116
Japan	18	7	11	63
Sweden	16	8	8	87
Hungary	13	10	3	231
World	34	14	20	35

SOURCE: "World Population Data Sheet," Population Reference Bureau, Washington, D.C. 1968.

† All rates are in numbers per 1,000 population per year. Doubling time is the number of years required for the population to increase by twofold.

percent means a 2.05 million increase in 1970 to 207 million. In 1971, the increase will be 2.07 million; in 1972, 2.09 million; in 1973, 2.11 million; etc. By 1980 the population will reach 226 million, with an annual increase of 2.26 million. In 63 yr the United States population will double, providing the present trends continue.

Countries with high birth rates and relatively lower death rates have a high rate of population growth because the difference between the birth and death rates is greater. For example, despite a death rate of 22 per 1,000, India has an annual population growth of 20 per 1,000: [42 per 1,000 (birth rate) − 22 per 1,000 (death rate) = 20 per 1,000 (growth rate).] The population of India is expected to double in about 30 yr.

Other countries have significantly reduced death rates without concurrent reductions in birth rates. For example, Mexico has a birth rate of about 45 per 1,000 and a death rate of 9 per 1,000 — an annual population growth of 3.6 percent. Mexico's population will double in about 19 yr. Birth and death rates of some representative countries are given in Table 11.6.

REGULATION OF POPULATIONS

In 1798, Thomas Robert Malthus (1766–1834) published a pioneer paper entitled "An Essay on the Principle of Population." Malthus indicated the potential for population growth and suggested that war, pestilence, and famine prevented this potential from being realized. Today this model for potential growth is accepted; it has been observed in many natural and laboratory populations of bacteria, plants, and animals, including man.

However, limitations of population growth by war, disease, and starvation are only three of several controlling factors of population growth. In this section we shall discuss several additional mechanisms for regulating population numbers.

Regulation means control, or homeostasis, of numbers. Not all populations are regulated; those that are not generally follow the J-shaped model for population growth (Fig. 11.29); these populations will not be considered further. Two models for population regulation will be discussed in detail.

The predator-prey model is one mechanism which accounts for regulation of some populations. A *predator* is a consumer which captures, kills, and devours another species, called "prey." In a simple model the rate at which a prey species is consumed exactly equals the rate of population increase of the prey if the predator were absent. Thus, population growth does not occur because the predator harvests the excess numbers. However, observations in nature fail to support this simple model. Reproduction and survival of the prey species are seldom constant; sometimes survival is good, sometimes it is poor. For a given population of predator, food requirements remain approximately constant under all conditions; i.e., predators each eat about the same number of prey each day. This means that predators remove a small fraction of the total prey if the prey population is large and a large fraction if the prey population is small.

For example, assume populations of 10 predators and 10,000 prey and a predation rate of 10 prey per day per predator. The prey are reduced by $\frac{10}{10,000}$, or 0.1 percent/day. At low prey numbers of 100 the predator consumes $\frac{10}{100}$, or 10 percent of the population per day. Thus, at low prey densities predators significantly reduce prey numbers (causing extinction after 10 days in this hypothetical example), whereas at high prey densities the predation has no significant effect. For regulation to be effective, there must be high predation rates at high prey population densities and low predation rates at low prey population densities.

In some cases the predator population increases along with the prey. Generally, however, predators have longer generation times and slower growth rates than the prey. Therefore there is a lag between an increase in prey density and in predator density, and imprecise regulation. For example, consider the relationship between snowshoe-hare populations and lynx populations in Canada. Increase in hare populations results in an increase in lynx population, but there is usually a lag of about 1 yr before the change occurs (Fig. 11.30). Presumably the lynx population exerts some regulation of the hare population, but the control is not very precise and marked oscillations occur in the population densities of the two species. This second predator-prey model predicts instability, and instability is observed.

A more complex model for regulating populations containing multiple species predicts better regulation than in single-prey–single-predator populations. In this complex model, a predator consumes several different kinds of prey. As the population of one prey species increases, the predator changes his diet to include a larger proportion of this species. Thus, the rate of predation increases as the prey species multiplies. The immediate effect is an inhibition of population growth. Because simultaneous outbreaks are unlikely to occur in several species, the greater the number of prey species, the less likely that all will experience population explosions at the same time. Regulation of insect populations by birds appears to fit this model of regulation.

One of the reasons that man has problems with insect pests in his crops is because of the simplified ecosystems he creates. Man

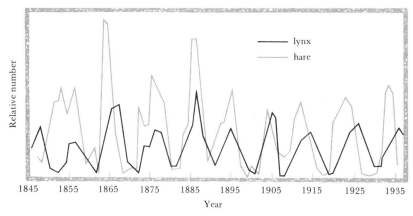

Figure 11.30. Population cycles of a prey, the snowshoe hare, and a predator of the hare, the lynx. (Data from the number of pelts collected by the Hudson Bay Company. Figure redrawn from: D. A. MacLulich, Fluctuations in the Numbers of the Varying Hare, Lepus americanus, University of Toronto Studies, Biol. Ser., No. 43, 1937.)

plants a single crop (for example, corn); only a few species of insects feed upon this crop. Predators may feed upon these insect pests, but the control is inadequate because of the high reproductive potential of the pest insects compared with that of predaceous birds and insects. Man resorts to chemical control, which further simplifies the ecosystem.

Another model of population regulation involves control of the number of individuals participating in reproduction. Many vertebrates stake out territories during the breeding season. A *territory* is an area which is defended rigorously against incursions by other members of the species. This defense of territory seldom involves physical battle or fighting; various behavioral mechanisms warn intruders of property rights. A bird's song may sound beautiful to our ears, but for the bird it serves an aggressive function—the song drives away trespassers. The bright colors of birds and fishes have long puzzled students of evolution. At one time bright color was thought to function in attracting a female to the more brightly colored male. It is now believed that the basic function of the bright colors is so that the male will be conspicuous to intruders into his territory. Generally, the male defends the territory—he has brighter color and does all the singing. Territories are sometimes defended against incursions by other species.

Territories serve to limit the number of breeding pairs. The available habitat is divided into territories, each occupied by one pair. As long as the area of suitable habitat for a species remains constant, the number of breeding pairs will be the same. This set number of reproducing adults produces a certain number of young each year, usually more than enough to replace mortality losses. In the following year, the extra individuals cannot find a territory, and thus they do not reproduce. These excess individuals are usually forced out of the desirable habitat into unsuitable environments, where they may die of exposure, starvation, or predation. This model of population regulation is based upon control of the birth rate; only so many individuals are added to the population each year.

Some plants also exhibit a phenomenon similar to territoriality. Desert plants often secrete a substance into the soil which inhibits the growth of seeds. Thus, once an individual plant is established, no other plants will grow in the surrounding area. This allows the established plant to obtain a maximum amount of the water that falls on the area. The population density of the plants is related to rainfall: the greater the rainfall, the greater the leaching, or washing out, of the inhibiting substances from the soil. Thus, areas which receive more rainfall can support a larger population.

Other mechanisms also function in regulating populations. There are cases where the incidence of disease and parasites increases as the population becomes larger. Changes in the physiology of some species occur at high population densities, resulting in spontaneous abortions or infertility. Some species resort to cannibalism, infant neglect, or fighting to achieve population control. There are few cases of general starvation in nonhuman populations; usually some other mechanism operates to prevent the population from reaching densities near the carrying capacity. However, in civilized man, there appears to be no specific population-regulating mechanism. Man remains behaviorally and physiologically reproductive until mass starvation occurs to regulate his populations.

Changes in human fertility can be achieved using several modern methods, but as yet these methods have not effectively regulated populations in any country of the world. Cultural changes in breeding patterns appear to be more effective than contraceptive methods. For example, Ireland has achieved approximate population stability because many persons elect to forego reproduction entirely or postpone it until later in life. Of all species, one would expect that human beings would be most likely to achieve population stability by methods involving birth control rather than death control.

Figure 11.31. The irony of this photograph is evident. The Cuyahoga became so polluted that a stray spark ignited the oil and debris, causing a large fire.

PROBLEMS

11.1 Suppose a lawn produces an average of 400 liters O_2 per acre each day; and that a man consumes 600 liters O_2 daily; how much lawn is required so that this man does not diminish the total world supply of oxygen.

11.2. Why are fossils important in taxonomy?

11.3. Why do pharmaceutical companies use rats and mice instead of insects, which are cheaper, more abundant, and easier to care for, to test new drugs?

11.4. Why do natural scientists not agree on kingdom classifications?

11.5. What niche do bacteria occupy in an ecosystem?

11.6. Why do farmers plant legumes as a crop to maintain soil fertility?

11.7. A large brain and head causes problems during birth. What evolutionary changes in behavior might have allowed a newborn infant with an immature brain and small head, as occurs in modern man?

11.8. Why are viruses not readily classified with other life forms?

11.9. In terms of light utilized in photosynthesis, why do chlorophyll-containing plants appear green?

11.10. Why is the name "dark reaction" applied to a reaction that occurs primarily in the light?

11.11. The stored chemical energy of food appears mostly as what form of energy after metabolism?

11.12. Why is oxygen consumption used to estimate the utilization of stored chemical energy of foods?

11.13. One species of worm has evolved which lives only in the warm, moist felt mats covering bars in certain German beer halls. The worms receive nourishment from the sugar and alcohol in the beer spilled on the mat. Using this example, distinguish between the concepts of habitat and niche.

11.14. High elevations have much cooler temperatures than lower elevations. What kind of biome would you expect to find on top of a very high mountain in the temperate zone of North America?

11.15. Although man has a very limited tolerable microhabitat, he has a wide variety of niches. What are the more important niches occupied by man?

11.16. What are some of the ways in which man alters his ecosystem to maximize food production for himself?

11.17. If a man consumes food equivalent to 2800 kcal, loses 300 kcal in his excrement, and has a respiration rate of 2400 kcal/day, will he gain, lose, or maintain his body weight? If his weight changes, calculate the daily change, assuming that 1 g of fat has a food value of 9 kcal.

11.18. The nutritional energy value of fat is 9.0 Cal/g; for both proteins and carbohydrates, it is 4.0 Cal/g. (The nutritional Calorie is 1,000 cal, or 1 kcal.) For comparison purposes, the energy value of a food is stated for 100-g portions (about $3\frac{1}{2}$ oz avoir) of the food. Calculate the energy values of the following foods, given (in this order) the percent protein, fat, and carbohydrate for each:

a. Cod liver oil (0, 100, 0)   *f.* White sugar (0, 0, 100)
b. Roast beef (20, 19, 0)   *g.* Broiled chicken (22, 2, 0)
c. Gelatin (84, 0.1, 0)   *h.* Butter and oleomargarine (1, 84, 0)
d. Caviar (30, 20, 0)   *i.* Ice cream (5, 12, 20)
e. Bacon (10, 65, 0)   *j.* Peanuts (25, 40, 22)

11.19. The total land area on earth is about 32 billion acres (32×10^9 acres). Of this, 8 billion acres are potentially arable and 3.5 billion acres are actually cultivated and support 3.5 billion people (the present world population). What is the average number of acres per person. Estimate from this the upper limit to the earth's population, assuming present nutritional levels are maintained. Assuming the earth's population doubles in 35 to 40 yr, approximately when will we reach this limit?

11.20. With an average around-the-clock sun radiation of 0.5 cal/min-cm², given 1 acre $= 4.05 \times 10^3$ m², how much energy reaches an average acre of land each year? An unusually intensively cultivated acre of land will yield an annual crop of about 1,800 kg of dried organic matter (mostly carbohydrate), which has stored energy of 4000 cal (4 Cal)/g. What is the total annual energy stored as food? What fraction of the sunlight is stored as food? Under these optimal agricultural conditions, how many persons can be fed from the annual harvest of 1 acre, assuming a purely vegetable diet? (A person requires about 2500 Cal of food energy each day.) Can you suggest why all the incident sunlight is not used to make food?

11.21. If the energy stored as food is only one-third the optimal amount in Prob. 11.20, and if one-half the vegetation is used for grazing (assuming cattle store 10 percent of what they ingest), then how many persons can be fed from the produce of 1 acre? With 8 billion potentially arable acres and an equal number of acres potentially suited for grazing, using the data from this problem, calculate the upper limit that can be placed on the earth's population.

11.22. The annual fishing harvest from 0.55 million km² in the North Sea is about 1.4 million metric tons of herring, haddock, small cod, and other small fish. (A metric ton is 1,000 kg.) The primary production rate from photosynthesis is 1,500 g of sea-plant matter/m²-yr. What percent of the primary production ends up as these small fish? Estimate their trophic level. Estimate the North Sea catch if only larger fish were caught.

11.23. The combined ocean area is 360 million km². However, the average primary production rate in the open sea is about one-third the North Sea rate. Calculate the total annual primary photosynthetic rate for the combined oceans. Using the North Sea trophic efficiency for the small fish of the open sea, estimate the total small-fish tonnage.

11.24. Aquaculture, or aquatic farming, is to fishing what land farming and grazing are to hunting. Seafood, particularly carp and oysters, can be intensively raised in confined regions and, incidentally, can ecologically make use of sewage. The annual aquaculture harvest of the People's Republic of China (mostly carp) is estimated at nearly 2 million metric tons. Assume fish is 20 percent protein and 5 percent fat; and that a man needs 60 g protein and 2500 Cal daily (see the nutritional energy values listed in Problem 11.18). For how many persons is the harvested protein sufficient? If 1 million km² of temperate waters can be devoted to aquatic farming, raising 50 tons of fish/km², how many persons can live on the harvested protein? How many can live on the harvested energy? Is the sea a better source of protein or of energy?

11.25. Trace possible pathways that might be taken by an atom of carbon as it travels from the land to the sea.

11.26. Where do the nutrients ultimately come from that accumulate in a lake during eutrophication?

11.27. Compare the ideal maximum efficiencies of a fossil-fuel power generator, with an upper operating temperature of 600°C, and a nuclear-power reactor, with an upper temperature of 320°C, where the lower temperature of each is 45°C. Next compare the relative amounts of heat returned to the colder reservoir by each power generator for equal amounts of electricity generated. (In actual practice, the overall efficiency is not more than 80 percent of the ideal maximum.) Which type of generator presents the greater problem of thermal pollution to water?

11.28. Why does DDT become more concentrated in higher trophic levels, such as in man?

11.29. Why is sulfur dioxide a more serious air pollutant than carbon monoxide even though only about one-third as much is produced?

11.30. In a statistical investigation of the relationship between mortality and smoking, Hammond and Horn used almost 188,000 men, 95,000 of whom were regular cigarette smokers and 93,000 of whom were not. The men in the two groups were matched with respect to age, occupation, urban or rural dwelling, etc. During the 44-mo study period, 488 of the regular cigarette smokers and 54 of the noncigarette smokers died of cancer of the lung or upper respiratory tract. Express the mortalities in deaths per 1,000 men. What is the relative chance of this type of cancer occurring in cigarette smokers as opposed to nonsmokers? In another study of 370 workers exposed to asbestos dust at high occupational levels, there were 24 deaths (all were regular cigarette smokers) from lung and upper-respiratory-tract cancer. Matching the 370 workers carefully with regular cigarette smokers from the first (and larger) group, Hammond and coworkers expected only 2.98 deaths. What is the relative chance of finding lung and respiratory tract cancer among regular cigarette-smoking asbestos workers as opposed to cigarette smokers who are not exposed to high levels of asbestos fiber? Compare the chance of finding this cancer among regular cigarette smokers exposed to asbestos to the chance of finding it among nonsmokers who are not so exposed. Are persons other than asbestos workers exposed to this fiber? How?

11.31. Prepared foods, in contrast with dishes made in the home entirely from natural produce, contain many additives. Among these are the following: flavoring agents, acidulants for tartness, preservatives against bacterial and mold growth, antioxidants to retard fat oxidation and rancidity, artificial sweeteners, emulsifiers to keep aqueous and fatty liquids well mixed, thickeners, colors, anticaking agents to prevent moisture absorption and caking, moisture-retaining agents to prevent loss of water, and nutritional supplements. Examine the labels of prepared foods, such as cold cereals, bread, candy, frozen dinners, and ice cream, and identify the substances other than the main natural food. Why do we use more food additives now than in former years?

11.32. Songbirds in nature establish territories which are treated as private property. One function of territoriality is to regulate population. Predict the population-growth curve for a songbird such as a robin if it were introduced into temperate Australia.

11.33. With a population of 205 million, a birth rate of 19.4 per 1,000, and a death rate of 9.4 per 1,000, what is the annual rate of increase in the population of the United States? How many births and how many deaths are there each year? What is the net annual increase in population?

11.34. Can you suggest a way of calculating the years needed for doubling a population, given the annual growth rate.
Hint: In the equally tempered music-scale problem (Chap. 1) we knew the frequency doubled in 12 steps and calculated the fractional increase from one note to the next. Here we know the fractional increase and wish to calculate the number of years.

11.35. In discussing overpopulation, the statement has been made more than once that if the earth's population continues to grow at its present rate (doubling every 35 to 40 yr), then in 1,500 yr the combined mass of mankind will exceed the mass of the earth. Comment on this statement, using scientific laws established no later than the time of Lavoisier.

11.36. For each person in the U.S., annually about $\frac{1}{2}$ ton of steel is produced. The total known world reserve of iron is about 3×10^{10} tons. If all the earth's 3.5 billion people used steel at the level of the United States, how long would the iron reserve last? Suggest some alternatives to doing without steel and iron.

11.37. In recent years Hungary has had a population of 10.3 million, with birth and death rates of 13.1 and 10.7 per 1,000 population. How many births and deaths are there each year? What is the annual increase in population? In 1967 in Hungary there were 18.4 legal abortions per 1,000 population. What is the ratio of legal abortions to live births? If the abortions had not been performed, and if one-half of these resulted in live births, what would have been the birth rate, the growth rate, and the annual increase in population?

11.38. Suggest why it is that technologically more advanced countries have lower birth rates than countries that are still largely agricultural.

A Brief Afterword

Throughout this text we used the word "describe" rather than "explain." The difference between description and explanation is the difference between questions beginning with "how" and those beginning with "why." Anyone who has known a small boy has learned that there is no end to a why-because question-and-answer sequence. Describing is the more modest accomplishment, and most scientists prefer to think of their work in this way.

Through science, our ideas about physical reality are enormously expanded. What we know through our senses has been extended by measuring and by creating models. Our reality continually is redescribed in terms of models more remote from our immediate sensory experiences. There seems to be no end in sight to the advance of science as it pushes back the frontier between the unknown and what we call knowledge. Yet, if the uncertainty principle remains valid, we are limited in what we can know about the very simplest systems.

The science which describes very simple and very isolated parts of the universe is more exact. Physics is more exact than chemistry, and chemistry more than biology. This is because subject complexity increases in that order. Presently, the more complicated something is, the less completely can we describe it. This is the power and the limitation of the scientific method.

In both science and the arts creativity is an intuitive leap from relative chaos to relative order. But there is an important difference between a work of art and a work of science. If Shakespeare had never lived, his particular account of "Hamlet" would never have been written; without Rembrandt, not one of his self-portraits would exist; no Stravinsky, no "The Rite of Spring." There is a unique relationship between an artist and his art.

In contrast, coincidental discovery is common in science: electromagnetic induction was discovered independently by three men; the periodicity of the elements by at least two; the calculus was invented by both Isaac Newton and Gottfried Leibniz. If Einstein had not existed, it is highly unlikely that a single other person could have made his contributions to Brownian motion, blackbody radiation, atomic heats, the photoelectric effect, and special and general relativity. But it is also unlikely that other scientists, individually or in teams, would not have made all these discoveries before very long. The entire structure of natural science, in this sense, is independent of the scientists who built it. This independence gives us confidence in the essential correctness of our description of the universe.

As we more powerfully and demandingly manipulate the universe, ours is an increasingly more interdependent world. Two millenia ago Rabbi Simeon ben Yohai wrote of a ship's passenger who took a drill and began boring beneath his own place.

> "What are you doing?" his fellow passengers said to him.
> "What business is it of yours," he replied: "am I not boring under my own place?"
>
> —*Midrash, Leviticus Rabba, chap. 4*

What we do affects not only ourselves but persons all over the globe and not just our own but future generations.

Science has contributed to the creation of problems so complex that they are not amenable to simple scientific description alone. At the same time, science and technology have given us greater choices in acting or not acting than were available even a few years ago. We can do more harm as well as more good. With this, more responsibility and greater wisdom in making decisions are needed.

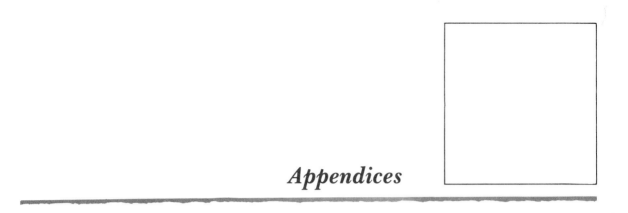

Appendices

Logarithms

TABLE I   COMMON LOGARITHMS

n	0	1	2	3	4	5	6	7	8	9
10	0000	0043	0086	0128	0170	0212	0253	0294	0334	0374
11	0414	0453	0492	0531	0569	0607	0645	0682	0719	0755
12	0792	0828	0864	0899	0934	0969	1004	1038	1072	1106
13	1139	1173	1206	1239	1271	1303	1335	1367	1399	1430
14	1461	1492	1523	1553	1584	1614	1644	1673	1703	1732
15	1761	1790	1818	1847	1875	1903	1931	1959	1987	2014
16	2041	2068	2095	2122	2148	2175	2201	2227	2253	2279
17	2304	2330	2355	2380	2405	2430	2455	2480	2504	2529
18	2553	2577	2601	2625	2648	2672	2695	2718	2742	2765
19	2788	2810	2833	2856	2878	2900	2923	2945	2967	2989
20	3010	3032	3054	3075	3096	3118	3139	3160	3181	3201
21	3222	3243	3263	3284	3304	3324	3345	3365	3385	3404
22	3424	3444	3464	3483	3502	3522	3541	3560	3579	3598
23	3617	3636	3655	3674	3692	3711	3729	3747	3766	3784
24	3802	3820	3838	3856	3874	3892	3909	3927	3945	3962
25	3979	3997	4014	4031	4048	4065	4082	4099	4116	4133
26	4150	4166	4183	4200	4216	4232	4249	4265	4281	4298
27	4314	4330	4346	4362	4378	4393	4409	4425	4440	4456
28	4472	4487	4502	4518	4533	4548	4564	4579	4594	4609
29	4624	4639	4654	4669	4683	4698	4713	4728	4742	4757
30	4771	4786	4800	4814	4829	4843	4857	4871	4886	4900
31	4914	4928	4942	4955	4969	4983	4997	5011	5024	5038
32	5051	5065	5079	5092	5105	5119	5132	5145	5159	5172
33	5185	5198	5211	5224	5237	5250	5263	5276	5289	5302
34	5315	5328	5340	5353	5366	5378	5391	5403	5416	5428
35	5441	5453	5465	5478	5490	5502	5514	5527	5539	5551
36	5563	5575	5587	5599	5611	5623	5635	5647	5658	5670
37	5682	5694	5705	5717	5729	5740	5752	5763	5775	5786
38	5798	5809	5821	5832	5843	5855	5866	5877	5888	5899
39	5911	5922	5933	5944	5955	5966	5977	5988	5999	6010
40	6021	6031	6042	6053	6064	6075	6085	6096	6107	6117
41	6128	6138	6149	6160	6170	6180	6191	6201	6212	6222
42	6232	6243	6253	6263	6274	6284	6294	6304	6314	6325
43	6335	6345	6355	6365	6375	6385	6395	6405	6415	6425
44	6435	6444	6454	6464	6474	6484	6493	6503	6513	6522
45	6532	6542	6551	6561	6571	6580	6590	6599	6609	6618
46	6628	6637	6646	6656	6665	6675	6684	6693	6702	6712
47	6721	6730	6739	6749	6758	6767	6776	6785	6794	6803
48	6812	6821	6830	6839	6848	6857	6866	6875	6884	6893
49	6902	6911	6920	6928	6937	6946	6955	6964	6972	6981

TABLE I **COMMON LOGARITHMS** — Cont.

Appendix A
665

n	0	1	2	3	4	5	6	7	8	9
50	6990	6998	7007	7016	7024	7033	7042	7050	7059	7067
51	7076	7084	7093	7101	7110	7118	7126	7135	7143	7152
52	7160	7168	7177	7185	7193	7202	7210	7218	7226	7235
53	7243	7251	7259	7267	7275	7284	7292	7300	7308	7316
54	7324	7332	7340	7348	7356	7364	7372	7380	7388	7396
55	7404	7412	7419	7427	7435	7443	7451	7459	7466	7474
56	7482	7490	7497	7505	7513	7520	7528	7536	7543	7551
57	7559	7566	7574	7582	7589	7597	7604	7612	7619	7627
58	7634	7642	7649	7657	7664	7672	7679	7686	7694	7701
59	7709	7716	7723	7731	7738	7745	7752	7760	7767	7774
60	7782	7789	7796	7803	7810	7818	7825	7832	7839	7846
61	7853	7860	7868	7875	7882	7889	7896	7903	7910	7917
62	7924	7931	7938	7945	7952	7959	7966	7973	7980	7987
63	7993	8000	8007	8014	8021	8028	8035	8041	8048	8055
64	8062	8069	8075	8082	8089	8096	8102	8109	8116	8122
65	8129	8136	8142	8149	8156	8162	8169	8176	8182	8189
66	8195	8202	8209	8215	8222	8228	8235	8241	8248	8254
67	8261	8267	8274	8280	8287	8293	8299	8306	8312	8319
68	8325	8331	8338	8344	8351	8357	8363	8370	8376	8382
69	8388	8395	8401	8407	8414	8420	8426	8432	8439	8445
70	8451	8457	8463	8470	8476	8482	8488	8494	8500	8506
71	8513	8519	8525	8531	8537	8543	8549	8555	8561	8567
72	8673	8579	8585	8591	8597	8603	8609	8615	8621	8627
73	8633	8639	8645	8651	8657	8663	8669	8675	8681	8686
74	8692	8698	8704	8710	8716	8722	8727	8733	8739	8745
75	8751	8756	8762	8768	8774	8779	8785	8791	8797	8802
76	8808	8814	8820	8825	8831	8837	8842	8848	8854	8859
77	8865	8871	8876	8882	8887	8893	8899	8904	8910	8915
78	8921	8927	8932	8938	8943	8949	8954	8960	8965	8971
79	8976	8982	8987	8993	8998	9004	9009	9015	9020	9025
80	9031	9036	9042	9047	9053	9058	9063	9069	9074	9079
81	9085	9090	9096	9101	9106	9112	9117	9122	9128	9133
82	9138	9143	9149	9154	9159	9165	9170	9175	9180	9186
83	9191	9196	9201	9206	9212	9217	9222	9227	9232	9238
84	9243	9248	9253	9258	9263	9269	9274	9279	9284	9289
85	9294	9299	9304	9309	9315	9320	9325	9330	9335	9340
86	9345	9350	9355	9360	9365	9370	9375	9380	9385	9390
87	9395	9400	9405	9410	9415	9420	9425	9430	9435	9440
88	9445	9450	9455	9460	9465	9469	9474	9479	9484	9489
89	9494	9499	9504	9509	9513	9518	9523	9528	9533	9538
90	9542	9547	9552	9557	9562	9566	9571	9576	9581	9586
91	9590	9595	9600	9605	9609	9614	9619	9624	9628	9633
92	9638	9643	9647	9652	9657	9661	9666	9671	9675	9680
93	9685	9689	9694	9699	9703	9708	9713	9717	9722	9727
94	9731	9736	9741	9745	9750	9754	9759	9763	9768	9773
95	9777	9782	9786	9791	9795	9800	9805	9809	9814	9818
96	9823	9827	9832	9836	9841	9845	9850	9854	9859	9863
97	9868	9872	9877	9881	9886	9890	9894	9899	9903	9908
98	9912	9917	9921	9926	9930	9934	9939	9943	9948	9952
99	9956	9961	9965	9969	9974	9978	9983	9987	9991	9996

A Census of the 268 Stable Isotopes

appendix **B**

A Census of the 268 Stable Isotopes

(Listed according to atomic number Z, mass number A, and neutron number N)

Z	A	N = A − Z	$\dfrac{N}{Z}$ for most common isotopes
1	1, 2	0, 1	$\frac{0}{1} = 0.00$
2	3, 4	1, 2	$\frac{2}{2} = 1.00$
3	6, 7	3, 4	$\frac{4}{3} = 1.33$
4	9	5	$\frac{5}{4} = 1.25$
5	10, 11	5, 6	$\frac{6}{5} = 1.20$
6	12, 13	6, 7	$\frac{6}{6} = 1.00$
7	14, 15	7, 8	$\frac{7}{7} = 1.00$
8	16, 17, 18	8, 9, 10	$\frac{8}{8} = 1.00$
9	19	10	$\frac{10}{9} = 1.11$
10	20, 21, 22	10, 11, 12	$\frac{10}{10} = 1.00$
11	23	12	$\frac{12}{11} = 1.09$
12	24, 25, 26	12, 13, 14	$\frac{12}{12} = 1.00$
13	27	14	$\frac{14}{13} = 1.08$
14	28, 29, 30	14, 15, 16	$\frac{14}{14} = 1.00$
15	31	16	$\frac{16}{15} = 1.07$
16	32, 33, 34, 36	16, 17, 18, 20	$\frac{16}{16} = 1.00$
17	35, 37	18, 20	$\frac{18}{17} = 1.06$
18	36, 38, 40	18, 20, 22	$\frac{22}{18} = 1.22$
19	39, 41	20, 22	$\frac{20}{19} = 1.05$
20	40, 42, 43, 44, 46	20, 22, 23, 24, 26	$\frac{20}{20} = 1.00$
21	45	24	$\frac{24}{21} = 1.14$
22	46, 47, 48, 49, 50	24, 25, 26, 27, 28	$\frac{26}{22} = 1.18$
23	51	28	$\frac{28}{23} = 1.22$
24	50, 52, 53, 54	26, 28, 29, 30	$\frac{28}{24} = 1.17$
25	55	30	$\frac{30}{25} = 1.20$
26	54, 56, 57, 58	28, 30, 31, 32	$\frac{30}{26} = 1.15$
27	59	32	$\frac{32}{27} = 1.19$
28	58, 60, 61, 62, 64	30, 32, 33, 34, 36	$\frac{30}{28} = 1.07$
29	63, 65	34, 36	$\frac{34}{29} = 1.17$
30	64, 66, 67, 68, 70	34, 36, 37, 38, 40	$\frac{34}{30} = 1.13$
31	69, 71	38, 40	$\frac{38}{31} = 1.23$
32	70, 72, 73, 74, 76	38, 40, 41, 42, 44	$\frac{42}{32} = 1.31$

A Census of the 268 Stable Isotopes (Continued)

(*Listed according to atomic number Z, mass number A, and neutron number N*)

Z	A	$N = A - Z$	$\dfrac{N}{Z}$ for most common isotopes
33	75	44	$\frac{44}{33} = 1.33$
34	74, 76, 77, 78, 80, 82	40, 42, 43, 44, 46, 48	$\frac{46}{34} = 1.35$
35	79, 81	45, 46	$\frac{45}{35} = 1.29$
36	80, 82, 83, 84, 86	44, 46, 47, 48, 50	$\frac{48}{36} = 1.33$
37	85, 87	48, 50	$\frac{48}{37} = 1.30$
38	84, 86, 87, 88	46, 48, 49, 50	$\frac{50}{38} = 1.32$
39	89	50	$\frac{50}{39} = 1.28$
40	90, 91, 92, 94, 96	50, 51, 52, 54, 56	$\frac{50}{40} = 1.25$
41	93	52	$\frac{52}{41} = 1.27$
42	94, 95, 96, 97, 98, 100	52, 53, 54, 55, 56, 58	$\frac{56}{42} = 1.33$
43			
44	96, 98, 99, 100, 101, 102, 104	52, 54, 55, 56, 57, 58, 60	$\frac{58}{44} = 1.32$
45	103	58	$\frac{58}{45} = 1.29$
46	102, 104, 105, 106, 108, 110	56, 58, 59, 60, 62, 64	$\frac{60}{46} = 1.30$
47	107, 109	60, 62	$\frac{60}{47} = 1.28$
48	106, 108, 110, 111, 112, 113, 114, 116	58, 60, 62, 63, 64, 65, 66, 68	$\frac{66}{48} = 1.38$
49	113	64	$\frac{64}{49} = 1.31$
50	112, 114, 115, 116, 117, 118, 119, 120, 122, 124	62, 64, 65, 66, 67, 68, 69, 70, 72, 74	$\frac{70}{50} = 1.40$
51	121, 123	70, 72	$\frac{70}{51} = 1.37$
52	120, 122, 123, 124, 125, 126, 128, 130	68, 70, 71, 72, 73, 74, 76, 78	$\frac{78}{52} = 1.50$
53	127	74	$\frac{74}{53} = 1.40$
54	124, 126, 128, 129, 130, 131, 132, 134, 136	70, 72, 74, 75, 76, 77, 78, 80, 82	$\frac{78}{54} = 1.44$
55	133	78	$\frac{78}{55} = 1.42$
56	130, 132, 134, 135, 136, 137, 138	74, 76, 78, 79, 80, 81, 82	$\frac{82}{56} = 1.46$
57	139	82	$\frac{82}{57} = 1.44$
58	136, 138, 140	78, 80, 82	$\frac{82}{58} = 1.41$
59	141	82	$\frac{82}{59} = 1.39$
60	142, 143, 145, 146, 148, 150	82, 83, 85, 86, 88, 90	$\frac{82}{60} = 1.37$
61			
62	144, 147, 148, 149, 150, 152, 154	82, 85, 86, 87, 88, 90, 92	$\frac{90}{62} = 1.45$
63	151, 153	88, 90	$\frac{90}{63} = 1.43$
64	152, 154, 155, 156, 157, 158, 160	88, 90, 91, 92, 93, 94, 96	$\frac{94}{64} = 1.47$

A Census of the 268 Stable Isotopes (Continued)

(Listed according to atomic number Z, mass number A, and neutron number N)

Z	A	$N = A - Z$	$\dfrac{N}{Z}$ for most common isotopes
65	159	94	$\frac{94}{65} = 1.45$
66	156, 158, 160, 161, 162, 163, 164	90, 92, 94, 95, 96, 97, 98	$\frac{98}{66} = 1.48$
67	165	98	$\frac{98}{67} = 1.46$
68	162, 164, 166, 167, 168, 170	94, 96, 98, 99, 100, 102	$\frac{98}{68} = 1.44$
69	169	100	$\frac{100}{69} = 1.45$
70	168, 170, 171, 172, 173, 174, 176	98, 100, 101, 102, 103, 104, 106	$\frac{104}{70} = 1.49$
71	175	104	$\frac{104}{71} = 1.46$
72	176, 177, 178, 179, 180	104, 105, 106, 107, 108	$\frac{108}{72} = 1.50$
73	180, 181	107, 108	$\frac{108}{73} = 1.48$
74	180, 182, 183, 184, 186	106, 108, 109, 110, 112	$\frac{110}{74} = 1.49$
75	185	110	$\frac{110}{75} = 1.47$
76	184, 186, 187, 188, 189, 190, 192	108, 110, 111, 112, 113, 114, 116	$\frac{116}{76} = 1.53$
77	191, 193	114, 116	$\frac{116}{77} = 1.51$
78	194, 195, 196, 198	116, 117, 118, 120	$\frac{117}{78} = 1.50$
79	197	120	$\frac{120}{79} = 1.52$
80	196, 198, 199, 200, 201, 202, 204	116, 118, 119, 120, 121, 122, 124	$\frac{122}{80} = 1.53$
81	203, 205	122, 124	$\frac{124}{81} = 1.53$
82	206, 207, 208	124, 125, 126	$\frac{126}{82} = 1.54$
83	209	126	$\frac{126}{83} = 1.52$

Answers to Selected Problems

appendix **C**

Chapter 1

1.1. 18.8, 9.00, 0.0036, 15,330, 75

1.3. 1.875×10, 9.004, 3.65×10^{-3}, 1.5327×10^4, 7.49×10

1.5. 1.2730, 0.9544, $0.5623 - 3$ or -2.4377, 4.1855 (This table of logarithms is limited to numbers with four significant numbers at most.) 1.8745

1.7. The former, 152.2 ± 0.3, is more precise than the latter, 152.4 ± 0.6 pounds.

1.9. 4,608 tablets

1.11. 4.64 m

1.13. After 17.71 or 18 full years.

1.15. 1.26×10^{17} s

1.17. 2×10^{14}

Chapter 2

2.1. 291; 310.4; 173; 297.13 Kelvin

2.3. 1.25 atm

2.5. 20.43 cm³

2.7. 126.1°C

2.9. Per gram of phosphorus there are 0.7746, 1.033, and 1.291 grams of oxygen. These numbers are in the ratio 1.000, 1.334, and 1.667; or 1, 4/3 and 5/3; or 3, 4 and 5.

2.11. 35.9%

2.13. $W = 9$ cal; $Q_2 = 10$ cal

Chapter 3

3.1. 719 m/s

3.3. 57.9 g H_2O

3.5. *b*

Chapter 4

4.1. 1.3 s
4.3. 4×10^{13} km
4.5. 26.4 pounds

Chapter 5

5.1. 3300 s or 55 min
5.3. 1.95×10^{18} Hz
5.5. 4.49×10^{-23} g; about 2.5 Å
5.13. 6100 K
5.15. 5800 K

Chapter 6

6.1. HI
6.3. 1.275 Å
6.5. Exothermic; 41,700 cal

Chapter 7

7.3. $CH_3C(CH_3)_2C(CH_3)_2C(CH_3)_2CH_3$
7.5. Chlorobenzene; 1,2-(or ortho-), 1,3-(or meta) and 1,4-(or para) dichlorobenzene; 1,2,3-, 1,2,4-, and 1,3,5-trichlorobenzene; 1,2,3,4-, 1,2,3,5- and 1,2,4,5-tetrachlorobenzene; 1,2,3,4,5-pentachloro-ben-zene or, more simply, pentachlorobenzene; and 1,2,3,4,5,6-hex-achlorobenzene or just hexachlorobenzene.
7.7. CH_3—CH—CH_3

   $\quad\quad\quad\;|$

   $\quad\quad\;\;OH$

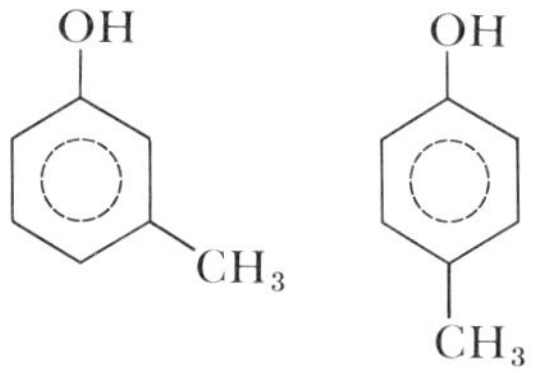

7.11. $CHBrClCH_3$
7.13. ser-cySH-ser-gly-cySH-gly-ala-leu-leu-thr
7.15. $4^{10,000} = 10^{6021}$
7.17. At the IpH, a compact coil because of mutual attraction between op-positely charged functional groups; above and below the IpH, an ex-tended conformation because of mutual repulsion between like-charged groups. Negatively charged and extended conformation.
7.19. A simple test for biodegradability: bury the object in soil and exam-ine it for decay several months later.

Chapter 8

8.1. 464 m/s

8.3. 10,700 yr; 8740 yr

8.5. Electrostatic repulsion for p,p pair; 1.24×10^{37} stronger than gravitational attraction; no classical electrostatic forces with neutrons.

8.7. Isotopes S-34, S-35 and S-36; isobars S-36 and Ar-36; isotones S-34 and Ar-36; mass defects -0.03214, -0.032, -0.03291, -0.03245; binding energies 291.84, 299.78, 308.70, 306.70 Mev; binding energies per nucleon 8.584, 8.565, 8.575, 8.519; S-35 is an even-odd isotope, has a half-life of 87.2 days; the others are even-even and stable.

8.9. 14

8.11. Steel wall thickness plotted against the logarithm of the fraction transmitted.

Chapter 9

9.1. More SO_3 at equilibrium; less ozone.

9.3. For five tosses, 32/243 at $x = +5d$; 80/243 at $x = +3d$; 80/243 at $x = +d$; 40/243 at $x = -d$; 10/243 at $x = -3d$; and 1/243 at $x = -5d$.

Chapter 10

10.1. The protein layers

10.3. A genetic defect can result in alterations in the amino acid composition of myoglobin, changing the structure and function of the molecule.

10.5. The bases

10.7. A coagulation of menstrual fluid with semen; blending of fluids from the sperm and egg

10.9. 0

10.11. 1/2 or 0.5; 0.5; 0.5; 0.5

10.13. In man only his surface is exposed and the bulk of his body shielded; an entire bacterium is exposed.

10.15. 1/2 or 0.5; 1/3 or 0.33

10.17. Mutations provide new traits, or additional chromosomal material.

10.19. The allele for sickle-cell anemia gives the carrier protection against malaria. The allele will be eliminated (unless medical science counteracts its sickle-cell effects).

10.21. Sexual reproduction results in new combinations of traits that when acting together enhance survival.

10.23. All organisms possess the same genetic code because they inherited it from a single common ancestor.

Chapter 11

11.1. 1.5 acres

11.3. Rodents are more closely related to man than are insects, and would be expected to react to drugs more like man than would insects.

11.5. Decomposers

11.7. An infant with a smaller head has a better chance to survive the birth process but requires the care and attention of a stable family.

11.9. Blue and red light is absorbed; what is not absorbed is transmitted through, or reflected by the leaves, and appears green.

11.11. Heat

11.13. The habitat is the moist environment of the felt; its niche is its role as a beer consumer.

11.15. Herbivore of roots and grains; fruit eater; carnivore of grass-eating mammals; combinations of these; fish consumers.

11.17. At the base, 2×10^4 kg of grass; on this, 4,000 kg of steers; on top, 225 kg of carnivores.

11.19. 900; 251; 337; 300; 625; 400; 106; 760; 208; 548

11.21. 1.06×10^{13} cal/yr; 7.2×10^9 cal/yr; 6.8×10^{-4} or 0.068 percent; 8 men; in photosynthesis only blue light (at 430 nm) and red light (at 680 nm) is absorbed to be used; the rest is lost by reflection, as heat, for evaporation, etc. (a good forest may utilize 0.6 percent of the incident light).

11.23. 0.17 percent; between 3 and 4; 0.14 million metric tons.

11.25. 1.53×10^8 metric tons; 3.06×10^{13} g protein and 7.65×10^{12} g fat; 1.4×10^9 or 1.4 billion men; 76 million men; protein.

11.27. Limestone, soil, grass, cow, bacteria, atmosphere, raindrop, alga, zooplankton, minnow, bass, river water, Gulf of Mexico (one possible example).

11.29. 64 percent; 46 percent; the nuclear power plant returns twice as much heat to the cold reservoir; the nuclear power plant.

11.31. 3.3×10^5 gal by car, 5.4×10^4 gal by diesel bus; 960,000 and 3,200 lb CO from car and bus; 170,000 and 9,700 lb hydrocarbon; 36,000 and 12,000 lb NO_x; CO reduced by 92 percent; hydrocarbons, 63 percent; NO_x, 34 percent; neither hydrocarbons nor CO, but some NO_x formed when electric sparks heat up air; more SO_2 from fossil fuel power plant.

11.33. 5.14 and 0.58 deaths per thousand men, or 1.40 and 0.16 deaths per thousand per year; 8.9; 8.0; 71; yes, the general population is exposed to air-borne asbestos fibers from wearing of automobile brakes and from fire-proofing of construction, and especially from the demolition of older buildings so treated.

11.35. An S-shaped curve

11.37. 10.0/1000; 3.98 million; 1.93 million; 2.05 million.

11.39. About 108 to 124 yr

11.41. 17 yr; other metals (e.g., Al), synthetics (e.g., plastics), and the recycling of iron

11.43. 135,000 births, 110,000 deaths, 25,000 net annual gain; 1.40; 22.3/1000, 11.6/1000, 119,000 net annual gain.

Selected Readings

appendix **D**

Chapter 1

Astin, A. V.: Standards of Measurement, *Scientific American*, **218:**50 (June, 1968).

Lord Ritchie-Calder: Conversion to the Metric System, *Scientific American*, **223:**17 (July, 1970).

Chapter 2

Boyle, Robert: Pressure–volume Relations in a Gas, in M. Shamos (ed.), "Great Experiments in Physics," Holt-Dryden, New York, 1959.

Hall, M. B.: Robert Boyle, *Scientific American*, **217:**96 (August, 1967).

Toulmin, Stephen, and June Goodfield: "The Architecture of Matter," chap. 8, Harper and Row, New York, 1962.

van Melsen, Andrew G.: "From Atoms to Atom," chap. 4, Harper and Row, New York, 1960.

Chapter 3

Jenkins, D. S.: Fresh Water from Salt, *Scientific American*, **196:**37 (March, 1957).

Partington, J. R.: "A Short History of Chemistry," chaps. 7–8, Harper and Row, New York, 1960.

Snyder, A. E.: Desalting Water by Freezing, *Scientific American*, **207:**41 (December, 1962).

Chapter 4

Cooper, Leon N. "An Introduction to the Meaning and Structure of Physics," chaps. 16–18, Harper and Row, New York, 1968.

Dart, Francis E.: "Electricity and Electromagnetic Fields," Merrill, Columbus, 1966. A well-written introductory survey in less than 100 pages.

Gamow, G.: Gravity, *Scientific American*, **204:**94, (March, 1961).

Morrison, P. and E. Morrison: Heinrich Hertz, *Scientific American*, **197:**98 (December, 1957).

Pollard, Ernest C. and Douglas C. Huston: "Physics, An Introduction," chaps. 5–6, Oxford, New York, 1969.

Rush, J. H.: The Speed of Light, *Scientific American*, **193:**62 (August, 1955).

Shamos, M. (ed.): "Great Experiments in Physics," Holt-Dryden, New York, 1959. (Chapters on Thomas Young, Charles Coulomb, Henry Cavendish, Michael Faraday, James Clerk Maxwell, and Heinrich Hertz.)

Sharlin, H. I.: From Faraday to the Dynamo, *Scientific American*, **204:**107, (May, 1961).

Chapter 5

Baranger, M. and R. A. Sorenson: The Size and Shape of Atomic Nuclei, *Scientific American*, **221:**58 (August, 1969).

Gamow, G.: The Principle of Uncertainty, *Scientific American*, **198:**51 (January, 1958).

⎯⎯⎯: The Exclusion Principle, *Scientific American*, **201:**74 (July, 1959).

Huffer, Charles M., Frederick E. Trinklein, and Mark E. Bunge: "An Introduction to Astronomy," chap. 12, Holt, Rinehart, and Winston, New York, 1967.

Nier, A. O. C.: The Mass Spectrometer, *Scientific American*, **188:**68 (March, 1953).

Schrodinger, E.: What is Matter? *Scientific American*, **189:**52 (September, 1953).

Shamos, M. (ed.): Holt-Dryden, New York, 1959. (Chapters on Henri Becquerel, J. J. Thomson, Robert A. Millikan, James Chadwick, Michael Faraday, Albert Einstein, Max Planck, and Neils Bohr.)

Weisskopf, V. F.: How Light Interacts with Matter, *Scientific American*, **219:**60 (September, 1968).

Chapter 6

Holliday, L.: Early Views on Forces between Atoms, *Scientific American*, **222:**116 (May, 1970).

Partington, J. R.: "A Short History of Chemistry," chaps. 10, 12, Harper and Row, New York, 1960.

Horrigan, Phillip A.: "The Challenge of Chemistry," chaps. 4, 5, McGraw-Hill, New York, 1970.

Chapter 7

Adams, E.: Barbiturates, *Scientific American*, **198:**60 (January, 1958).

Amerine, M. A.: Wine, *Scientific American*, **211:**46 (August, 1964).

Barron, F., M. E. Jarvik, and S. Bunnell, Jr.: The Hallucinogenic Drugs, *Scientific American*, **210:**29 (April, 1964).

Bragg, Sir Lawrence: "X-Ray Crystallography, *Scientific American*, **219:**58 (July, 1968).

Bridgman, P. W.: Synthetic Diamonds, *Scientific American*, **193:**42 (November, 1955).

Charles, R. J.: The Nature of Glasses, *Scientific American*, **217:**126 (September, 1967).

Crawford, B., Jr.: Chemical Analysis by Infrared, *Scientific American*, **189:**42 (October, 1953).

Grinspoon, L.: Marihuana, *Scientific American*, **221:**17 (December, 1969).

Kornberg, A.: The Synthesis of DNA, *Scientific American*, **219:**64 (October, 1968).

Lambert, J. B.: The Shape of Organic Molecules, *Scientific American*, **222:**58 (January, 1970).

Le Corbeiller, Philippe: Crystals and the Future of Physics, in James R. Newman (ed.), "The World of Mathematics," Simon and Schuster, New York, 1956.

Mason, B. J.: The Growth of Snow Crystals, *Scientific American*, **204:**120 (January, 1961).

Merrifield, R. B.: The Automatic Synthesis of Proteins, *Scientific American*, **218:**56 (March, 1968).

Mirsky, A.: The Discovery of DNA, *Scientific American*, **218:**78 (June, 1968).

Mott, Sir Nevill: The Solid State, *Scientific American*, **217:**80 (September, 1967).

Pauling, Linus, and Roger Hayward: "The Architecture of Molecules," Freeman, San Francisco, 1964. (A handsomely illustrated picture book, with 57 color plates.)

Chapter 8

Alfven, H.: Antimatter and Cosmology, *Scientific American*, **216:**106 (April, 1967).

Bahcall, J. H.: Neutrinos from the Sun, *Scientific American*, **221:**28 (July, 1969).

Deevey, E. S., Jr.: "Radioactive Dating, *Scientific American*, **186:**94 (February, 1952).

Glasstone, Samuel: "The Effects of Nuclear Weapons," United States Atomic Energy Commission, Washington, 1964.

Gough, William C., and Bernard J. Eastlund: The Prospects of Fusion Power, *Scientific American*, **224:**50 (February, 1971).

Hamilton, Andrew: The Arms Race: Too Much of a Bad Thing, *The New York Times Magazine*, p. 34, October 6, 1968.

Hardin, Garrett (ed.): "Science, Conflict and Society," chaps. 4–7; 38–47, Freeman, San Francisco, 1969. (Possibly the best collection of writings on modern warfare, reprinted from *Scientific American* articles.)

Hogerton, J. F.: The Arrival of Nuclear Power, *Scientific American*, **218:**21 (February, 1968).

Johnson, G., and H. Brown: Nonmilitary Nuclear Explosions, *Scientific American*, **199:**29 (December, 1958).

Leachman, R. B.: Nuclear Fission, Scientific American, **213:**49 (August, 1965).

Peierls, R. E.: The Atomic Nucleus, *Scientific American*, **200:**75 (January, 1959).

Pollard, Ernest C., and Douglas C. Huston; "Physics, An Introduction," chaps. 9, 10, Oxford, New York, 1969. (Some excellent and interesting accounts by the actual investigators.)

Reynolds, J. H.: The Age of the Elements in the Solar System, *Scientific American*, **203:**171 (November, 1960).

J. Schubert, "Radioactive Poisons," *Scientific American*, **193:**34

Schurr, S. H.: Energy, *Scientific American*, **209:**110 (September, 1963).

Scoville, Herbert, Jr.: The Limitation of Offensive Weapons, *Scientific American*, **224:**15 (January, 1971).

Shankland, R. S.: The Michelson-Morley Experiment, *Scientific American*, **211:**107 (November, 1964).

Smyth, Henry DeW.: "Atomic Energy for Military Purposes," Princeton, Princeton, 1945.

Taylor, T. G.: How an Eggshell Is Made," *Scientific American*, **222:**88 (March, 1970).

Telegdi, V. L.: Hypernuclei, *Scientific American*, **206:**50 (January, 1962).

Energy and Power, *Scientific American*, **225** (September, 1971). (An entire issue on this subject.)

Chapter 9

Dawkins, M. J. R.: The Production of Heat by Fat, *Scientific American*, **213:**62 (August, 1965).

Eccles, Sir John: The Synapse, *Scientific American*, **212:**56 (January, 1965).

Holum, John R.: "Introduction to Principles of Chemistry," chaps. 6–8, Wiley, New York, 1969.

Meselson, M. S.: Chemical and Biological Weapons, *Scientific American*, **222:**15 (May, 1970).

Oster, G.: The Chemical Effects of Light, *Scientific American*, **219:**158 (September, 1968).

Phillips, D. C.: The Three-Dimensional Structure of an Enzyme Molecule, *Scientific American*, **215:**78 (November, 1966).

Chapter 10

Bearn, A. G., and J. L. German III: Chromosomes and Disease, *Scientific American*, **205:**66 (November, 1961).

Braun, A. C.: The Reversal of Tumor Growth, *Scientific American*, **213:**75 (November, 1965).

Crick, F. H. C.: The Genetic Code, *Scientific American*, **207:**66 (October, 1962).

_____: The Genetic Code III, *Scientific American*, **215:**55 (October, 1966). *See also Nirenberg, below.*

Dobzhansky, T.: The Genetic Basis of Evolution, *Scientific American*, **182:**32 (January, 1950).

Gorini, L.: Antibiotics and the Genetic Code, *Scientific American*, **214:**102 (April, 1966).

Hanawalt, P. C., and R. H. Haynes: The Repair of DNA, *Scientific American*, **216:**36 (February, 1967).

Henahan, John F.: "Men and Molecules," Crown, New York, 1966.

Lerner, Michael I.: "Heredity, Evolution and Society," Freeman, San Francisco, 1968.

Macalpine, I., and R. Hunter: Porphyria and King George III, *Scientific American*, **221:**38 (July, 1969).

Nirenberg, M. W.: The Genetic Code II, *Scientific American*, **208:**80 (March, 1963).

Watanabe, T.: Infectious Drug Resistance, *Scientific American*, **217:**19 (December, 1967).

Yanofsky, C.: Gene Structure and Protein Structure," *Scientific American*, **216:**80 (May, 1967).

The Living Cell, **205,** entire issue (September, 1961).

Chapter 11

Boerma, A. H.: A World Agricultural Plan, *Scientific American*, **223:**54 (August, 1970).

Bormann, F. H.: The Nutrient Cycles of an Ecosystem, *Scientific American*, **223:**92 (October, 1970).

Brower, L. P.: Ecological Chemistry, *Scientific American*, **220:**22 (February, 1969).

Ehrlich, Paul, and Anne Ehrlich: "Population, Resources, Environment," Freeman, San Francisco, 1970.

Esposito, John C.: "The Vanishing Air," Grossman, New York, 1970.

Hardin, Garrett (ed.): "Science, Conflict and Society," chaps. 16, 17, 22, 33, 34, 36, Freeman, San Francisco, 1969. (A good set of *Scientific American* articles about pollution, pesticides, population, abortion, etc.)

Helfrich, Harold, Jr.: "The Environmental Crisis," Yale, New Yaven, 1970.

Hinch, Nylds: Air Pollution, *J. Chem. Educ.*, **46:**93 (1969).

Keifer, David M.: Population, *Chemical and Engineering News*, p. 118, October 7, 1968; p. 90, October 14, 1968.

McDermott, W.: Air Pollution and Public Health, *Scientific American*, **205:**49 (October, 1961).

Morriman, D.: The Califaction of a River, *Scientific American*, **222:**42 (April, 1970).

Peakall, D. B.: Pesticides and the Reproduction of Birds, *Scientific American*, **222:**72 (April, 1970).

Pirie, N. W.: Orthodox and Unorthodox Methods of Meeting World Food Needs, *Scientific American*, **216:**27 (February, 1967).

Powers, C. F., and A. Robertson: The Aging Great Lakes, *Scientific American*, **215:**94 (November, 1966).

Pratt, C. J.: Chemical Fertilizers, *Scientific American*, **212:**62 (June, 1965).

Sanders, Howard J. Food Additives, *Chemical and Engineering News*, p. 100, October 10, 1966; p. 108, October 17, 1966.

Scrimshaw, N. S.: Food, *Scientific American*, **209:**72 (September, 1963).

Simpson, D.: The Dimensions of World Poverty, *Scientific American*, **219:**27 (November, 1968).

Turner, James S.: "The Chemical Feast," Grossman, New York, 1970.

Wecker, S. C.: Habitat Selection, *Scientific American*, **211:**109 (October, 1964).

Woodwell G. M.: Toxic Substances and Ecological Cycles, *Scientific American*, **216:**24 (March, 1967).

"*Cleaning Our Environment: The Chemical Basis for Action*," American Chemical Society, Washington, 1969.

The Ocean, *Scientific American*, **221** (September, 1969). (An entire issue devoted to resources, etc.)

The Pill, *Consumer Reports*, p. 314 May, 1970.

"Resources and Man," National Academy of Sciences–National Research Council, Freeman, San Francisco, 1969.

Index